CW00828838

UNCERTAINTY IN INTELLIGENT SYSTEMS

UNCERTAINTY IN INTELLIGENT SYSTEMS

edited by

Bernadette Bouchon-Meunier
LAFORIA
University of Paris VI
Paris, France

Llorenc Valverde
Department of Information Sciences
University of the Balears
Palma(Balears), Spain

Ronald R. Yager
Machine Intelligence Institute
Iona College
New Rochelle, NY, U.S.A.

1993

NORTH-HOLLAND
AMSTERDAM • LONDON • NEW YORK • TOKYO

ELSEVIER SCIENCE PUBLISHERS B.V.
Sara Burgerhartstraat 25
P.O. Box 211, 1000 AE Amsterdam, The Netherlands

ISBN: 0 444 81508 2

This book is printed on acid-free paper.

Printed in The Netherlands

Preface

The material in this volume was selected from presentations made at the Fourth, IPMU, International Conference on Information Processing and Management of Uncertainty in Knowledge-Based Systems held on Mallorca in the Balearic Islands in Spain. This collection of papers, in addition to being chosen based upon the quality of the work, focuses on the issue of uncertainty in intelligent systems. The management of uncertainty is becoming more and more appreciated as a central problem that must be addressed in the construction of intelligent systems. The contents of this book reflect the main approaches being used, fuzzy set methods, probability theory, evidence theory and network models.

We would like to thank the contributors to this volume as well as the other participants at the IPMU conference for making this book possible. We also acknowledge our debt to the hidden legion of reviewers who helped select the valued from the valueless.

Last but not at all least we would like to express special thanks to Lotfi A. Zadeh, recipient of the First Kampe De Feriet Award for Contributions to the Theory and Management of Uncertainty, who in both intellectual matters and interpersonal behavior has inspired.

B. Bouchon-Meunier

L. Lalverde

R.R. Yager

TABLE OF CONTENTS

CHAPTER 6: UNCERTAINTY IN SOCIAL AND BEHAVIOURAL SCIENCES

CHAPTER 1:

NETWORK BASED REASONING

Uncertainty in Intelligent Systems
B. Bouchon-Meunier et al. (Editors)

Valuation Networks, Decision Trees, and Influence Diagrams: A Comparison

Prakash P. Shenoy

School of Business, University of Kansas, Summerfield Hall, Lawrence, KS 66045-2003, USA
pshenoy@ukanvm.cc.ukans.edu

1. INTRODUCTION

Recently, we proposed a new method for representing and solving Bayesian decision problems based on the framework of valuation-based systems [Shenoy 1992b, 1993a]. The new representation is called a valuation network, and the new solution method is called the fusion algorithm. In this paper, we briefly compare valuation networks to decision trees and influence diagrams. For symmetric decision problems, valuation networks are more expressive than both decision trees and influence diagrams. We also compare the fusion algorithm to the backward recursion method of decision trees and to the arc-reversal method of influence diagrams. For symmetric decision problems, the fusion algorithm is more efficient than the backward recursion method of decision trees, and more efficient and simpler than the arc-reversal method of influence diagrams.

An outline of this paper is as follows. In section 2, we give a statement of the *Medical Diagnosis* problem. In section 3, we describe a decision tree representation and solution of the *Medical Diagnosis* problem highlighting the strengths and weaknesses of the decision tree method. In section 4, we describe an influence diagram representation and solution of the *Medical Diagnosis* problem highlighting the strengths and weaknesses of the influence diagram technique. While the strengths of the influence diagram technique are well known, its weaknesses are not so well known. For example, it is not well known that in data-rich domains where we have a non-causal graphical probabilistic model of the uncertainties, influence diagrams representation technique is not very convenient to use. Also, the inefficiency of the arc-reversal method has never before been shown in the context of decision problems. In section 5, we describe a valuation network representation and solution of the *Medical Diagnosis* problem highlighting its strengths and weaknesses vis-á-vis decision trees and influence diagrams. For more details of the comparison, see [Shenoy 1993b].

2. A MEDICAL DIAGNOSIS PROBLEM

A physician is trying to determine a policy for treating patients suspected of suffering from a disease D. D causes a pathological state P that in turn causes symptom S to be exhibited. The physician first observes whether or not a patient is exhibiting symptom S. Based on this observation, she either treats the patient (for D and P) or does not. The physician's utility function depends on her decision to treat or not, the presence or absence of disease D, and the presence or absence of pathological state P. The prior probability of disease D is 10%. For patients known to suffer from D, 80% suffer from pathological state P. On the other hand, for patients known not to suffer from D, 15% suffer from P. For patients known to suffer from P, 70% exhibit symptom S. And for patients known not to suffer from P, 20% exhibit symptom S. We assume D and S are conditionally independent given P. Table 1 shows the physician's utility function.

Table 1. The physician's utility function for all act-state pairs.

Physician's Utilities (υ)		States			
		Has pathological state (p)		No pathological state (~p)	
		Has disease (d)	No disease (~d)	Has disease (d)	No disease (~d)
Acts	Treat (t)	10	6	8	4
	Not treat (~t)	0	2	1	10

3. DECISION TREE REPRESENTATION AND SOLUTION

A popular method for representing and solving Bayesian decision problems is decision trees. Decision trees have their genesis in the pioneering work of von Neumann and Morgenstern [1944] on extensive form games. Decision trees graphically depict all possible scenarios. The decision tree representation allows computation of an optimal strategy by the backward recursion method of dynamic programming [Zermelo 1913, Bellman 1957, Raiffa and Schlaifer 1961]. Raiffa and Schlaifer [1961] call the dynamic programming method for solving decision trees "averaging out and folding back." Figure 1 shows a decision tree representation of the *Medical Diagnosis* problem. Figure 2 shows its solution.

The strengths of decision trees are its simplicity and its flexibility. Decision trees are based on the semantics of scenarios. Each path in a decision tree from the root to a leaf represents a scenario. These semantics are very intuitive and easy to understand. Decision trees are also very flexible. In asymmetric decision problems, the choices at any time and the relevant uncertainty at any time depend on past decisions and revealed events. Since decision trees depict scenarios explicitly, representing an asymmetric decision problem is easy.

The weaknesses of decision trees are its modeling of uncertainty, its modeling of information constraints, and its combinatorial explosiveness in problems in which there are many variables. Since decision trees are based on the semantics of scenarios, the placement of a random variable in the tree depends on the information constraints. Also, decision trees demand a probability distribution for each random variable conditioned on the past decisions and events leading to the random variable in the tree. This is a problem in diagnostic decision problems where we have a causal model of the uncertainties. For example, in medical decision problems, symptoms are revealed before diseases. For such problems, decision trees require conditional probabilities for diseases given symptoms. But, assuming a causal model, it is easier to assess the conditional probabilities of symptoms given the diseases [Shachter and Heckerman 1987]. Thus a traditional approach is to first assess the probabilities in the causal direction and then compute the probabilities required in the decision tree using Bayes theorem. This is a major drawback of decision trees. There should be a cleaner way of separating a representation of a problem from its solution. The former is hard to automate while the latter is easy. Decision trees interleave these two tasks making automation difficult. This drawback of decision trees can be alleviated by using information sets [see Shenoy 1993d].

Figure 1. A decision tree representation of the *Medical Diagnosis* problem.

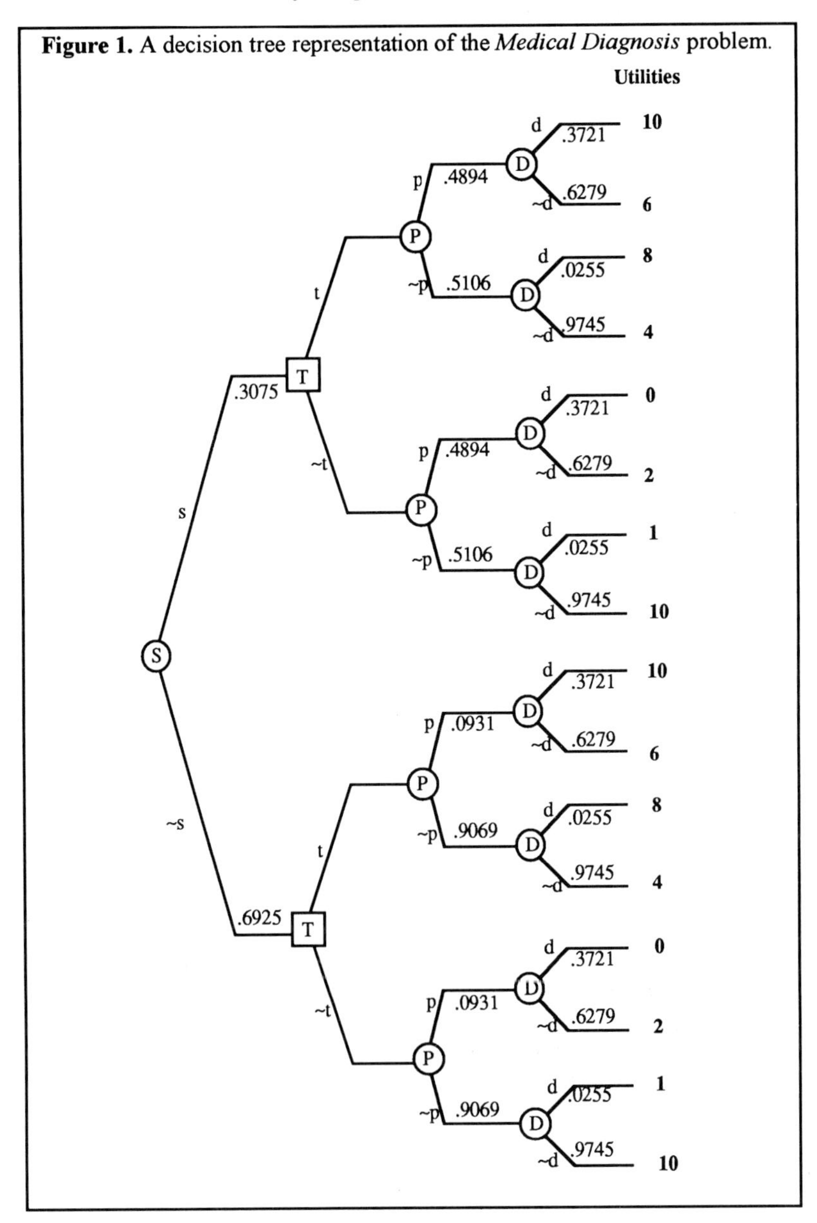

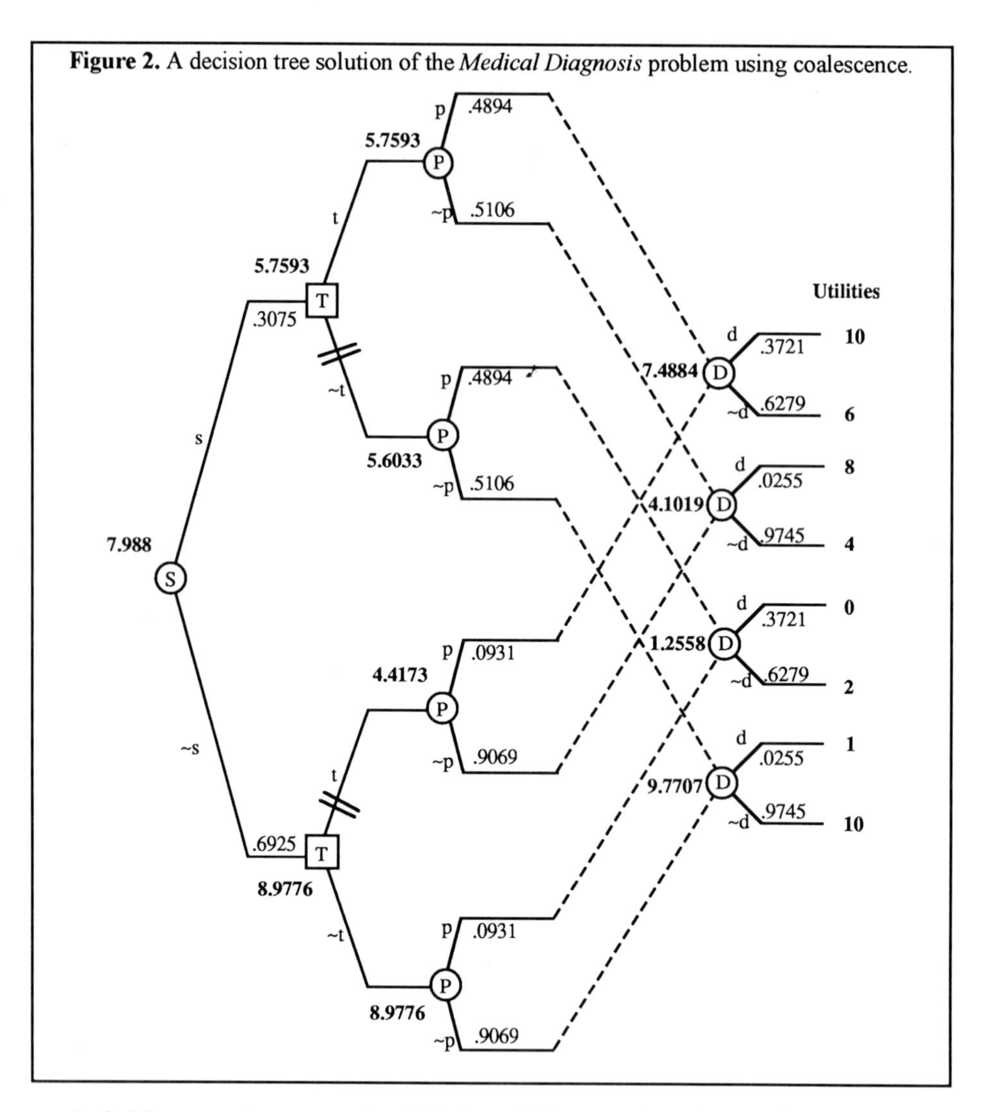

Figure 2. A decision tree solution of the *Medical Diagnosis* problem using coalescence.

In decision trees, the sequence in which the variables occur in each scenario represents information constraints. In some problems, the information constraints may only be specified up to a partial order. But the decision tree representation demands a complete order. This overspecification of information constraints in decision trees makes no difference in the final solution. However, it may make a difference in the computational effort required to compute a solution.

The combinatorial explosiveness of decision trees stems from the fact that the number of scenarios is an exponential function of the number of variables in the problem. In a symmetric decision problem with n variables, where each variable has 2 possible values, there are 2^n

scenarios. Since decision trees depict all scenarios explicitly, it is computationally infeasible to represent a decision problem with, say, 50 variables.

The strength of the decision tree solution procedure is its simplicity. Also, if a decision tree has several identical subtrees, then we can make the solution process more efficient by coalescing the subtrees [Matheson and Roths 1967].

The weakness of the decision tree solution procedure is the preprocessing of probabilities that may be required (before the decision tree representation). A brute-force computation of the desired conditionals from the joint distribution for all variables is intractable if there are many random variables. Also, although preprocessing is required for representing the problem as a decision tree, some of the resulting computations are unnecessary for solving the problem [Shenoy 1993b].

4. INFLUENCE DIAGRAM REPRESENTATION AND SOLUTION

Influence diagram is another method for representing and solving decision problems. Influence diagrams were initially proposed as a method only for representing Bayesian decision problems [Miller et al. 1976, Howard and Matheson 1981]. A motivation behind the formulation of influence diagrams was to find a method for representing decision problems without any preprocessing. Subsequently, Olmsted [1983] and Shachter [1986] devised methods for solving influence diagrams directly, without first having to convert influence diagrams to decision trees. In the last decade, influence diagrams have become popular for representing and solving decision problems [Oliver and Smith 1990]. Figure 3 shows an influence diagram representation of the *Medical Diagnosis* problem. Figure 4 shows its solution.

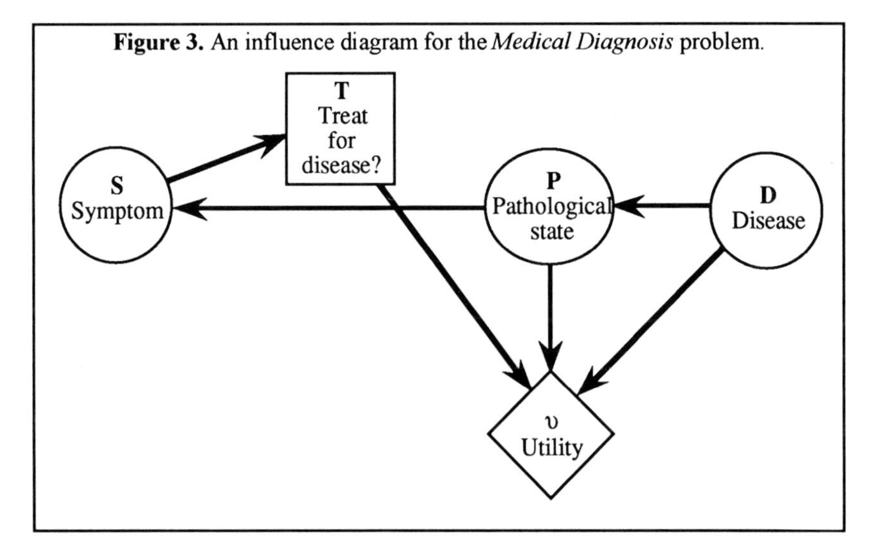

Figure 3. An influence diagram for the *Medical Diagnosis* problem.

Figure 4. The solution of the influence diagram representation of the *Medical Diagnosis* problem.

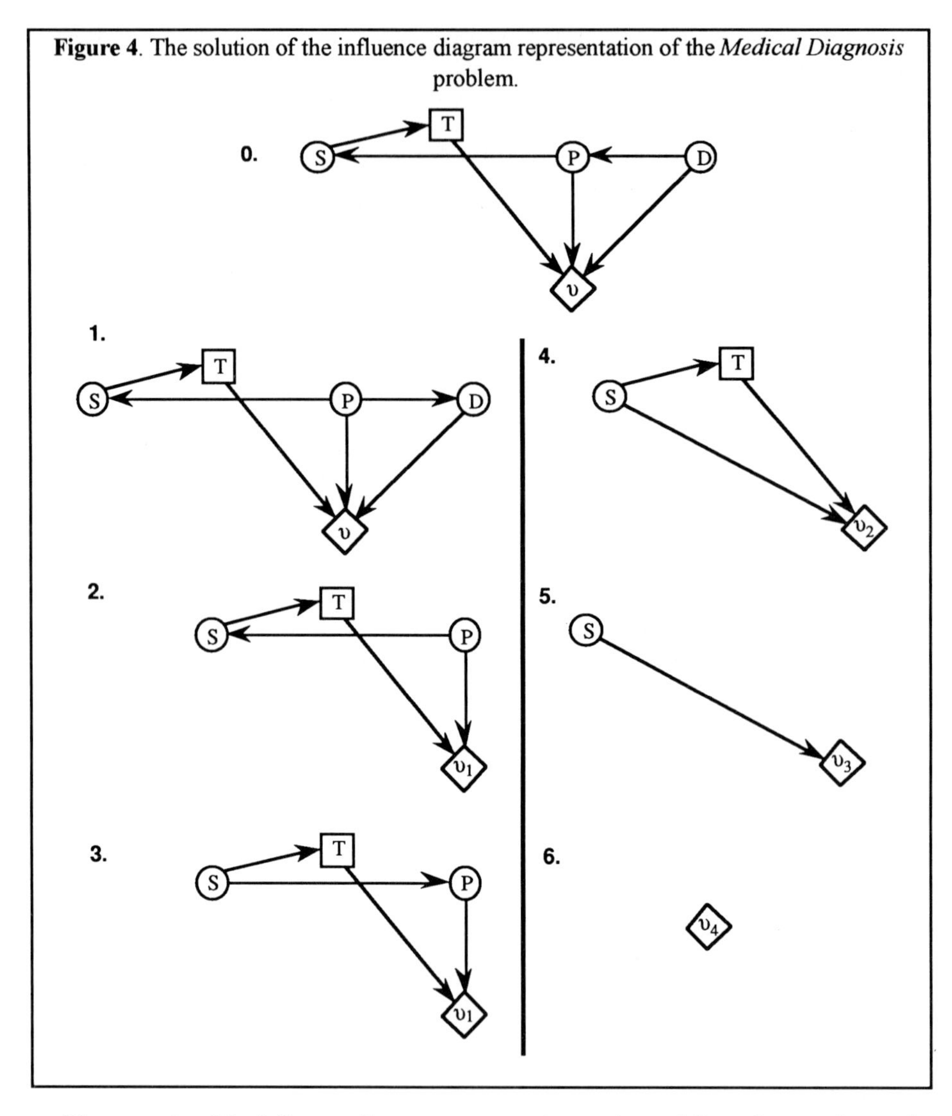

The strengths of the influence diagram representation are its modeling of uncertainty and its compactness. Influence diagrams are based on the semantics of conditional independence. Conditional independence is represented in influence diagrams by d-separation of variables [Pearl et al. 1990]. Practitioners who have used influence diagrams in their practice claim that it is a powerful tool for communication, elicitation, and detailed representation of human knowledge [Owen 1984, Howard 1988, 1989, 1990].

Influence diagrams do not depict scenarios explicitly. They assume symmetry (i.e., every scenario consists of the same sequence of variables) and depict only the variables and the sequence up to a partial order. Therefore, influence diagrams are compact and computationally more tractable than decision trees.

The weaknesses of the influence diagram representation are its modeling of uncertainty and requirement of symmetry. Influence diagrams demand a conditional probability distribution for each random variable. In causal models, these conditionals are readily available. However, in other graphical models, we don't always have the joint distribution expressed in this way [see, e.g., Darroch et al. 1980, Wermuth and Lauritzen 1983, Edwards and Kreiner 1983, and Kiiveri et al. 1984]. For such models, before we can represent the problem as an influence diagram, we have to preprocess the probabilities, and often, this preprocessing is unnecessary for the solution of the problem [Shenoy 1993b].

Influence diagrams are suitable only for decision problems that are symmetric or almost symmetric [Watson and Buede 1987]. For decision problems that are highly asymmetric, influence diagram representation is awkward. For such problems, Call and Miller [1990], Fung and Shachter [1990], Covaliu and Oliver [1992], and Kirkwood [1993] investigate representations that are hybrids of decision trees and influence diagrams, and Olmsted [1983], and Smith et al. [1989] suggest methods for making the influence diagram representation more efficient.

The strength of the influence diagram solution procedure is that, unlike decision trees, it uses local computation to compute the desired conditionals in problems requiring Bayesian revision of probabilities [Shachter 1988, Rege and Agogino 1988, Agogino and Ramamurthi 1990]. This makes possible the solution of large problems in which the joint probability function decomposes into small functions.

The weakness of the arc-reversal method for solving influence diagrams is that it does unnecessary divisions. The solution process of influence diagrams has the property that after deletion of each variable, the resulting diagram is an influence diagram. As we have already mentioned, the representation method of influence diagrams demands a conditional probability distribution for each random variable in the diagram. It is this demand for conditional probability distributions that requires divisions, not any inherent requirement in the solution of a decision problem.

5. VALUATION NETWORK REPRESENTATION AND SOLUTION

Valuation network is yet another method for representing and solving Bayesian decision problems [Shenoy 1992b, 1993a]. Like influence diagrams, valuation networks depict decision variables, random variables, utility functions, and information constraints. Unlike influence diagrams, valuation networks explicitly depict probability functions. Valuation networks are based on the semantics of factorization. Each probability function is a factor of the joint probability distribution function, and each utility function is a factor of the joint utility function. The solution method for valuation networks is called the fusion algorithm. Figure 5 shows a valuation network representation of the *Medical Diagnosis* problem. Figure 6 shows its solution.

Figure 5. A valuation network for the *Medical Diagnosis* problem.

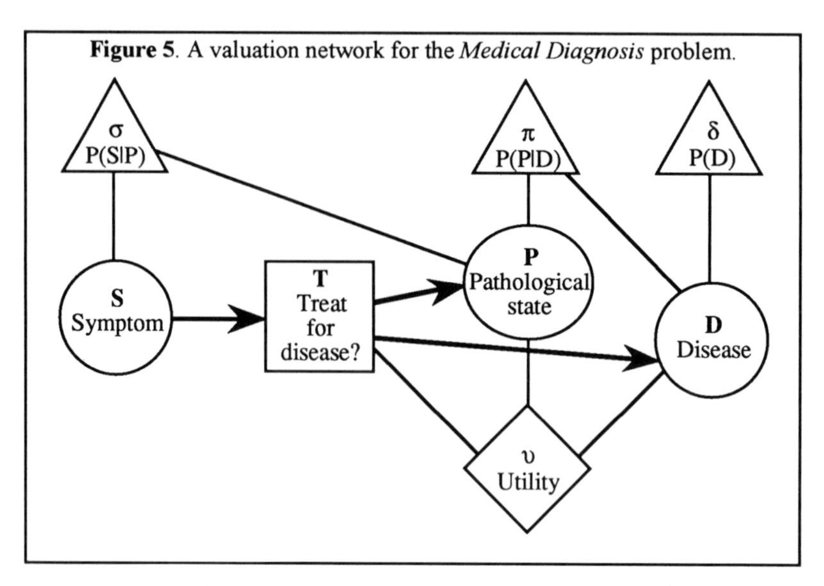

The strengths of valuation networks are its expressiveness for modeling uncertainty, its compactness, and its simplicity. Unlike decision trees and influence diagrams, valuation networks do not demand specification of the joint probability distribution function in a certain form. All probability models can be represented directly without any preprocessing.

Like influence diagrams, valuation networks are compact representations of decision problems. They do not depict scenarios explicitly. Only variables, functions, and information constraints are depicted explicitly. Of course, like influence diagrams, valuation networks assume symmetry of scenarios.

Valuation networks are very simple to interpret. Each probability function is a factor of the joint probability distribution function, and each utility function is a factor of the joint utility function. Thus it treats probability functions and utility functions alike. Given the factors of the joint probability distribution, we can easily recover the independence conditions underlying the joint probability distribution using separation of variables [Shenoy 1993c]. Also, the information constraints are represented explicitly.

A weakness of the valuation network representation is that, like influence diagrams, it is appropriate only for symmetric and almost symmetric decision problems. For highly asymmetric decision problems, the use of valuation networks is awkward. Another weakness of valuation networks is that the semantics of factorization are not as well developed as the semantics of conditional independence. Hopefully, current research on the semantics of factorization will address this shortcoming of valuation networks [Shenoy 1989, 1991, 1992a, 1992c].

The strengths of the fusion algorithm for solving valuation networks are its computational efficiency and its simplicity. The fusion algorithm uses local computations, and it avoids unnecessary divisions. This makes it always more efficient than the arc-reversal method of influence diagrams. It is also more efficient than the backward recursion method of decision trees for symmetric decision problems. The fusion algorithm is also extremely simple to understand and execute.

Figure 6. The fusion algorithm for the *Medical Diagnosis* problem.

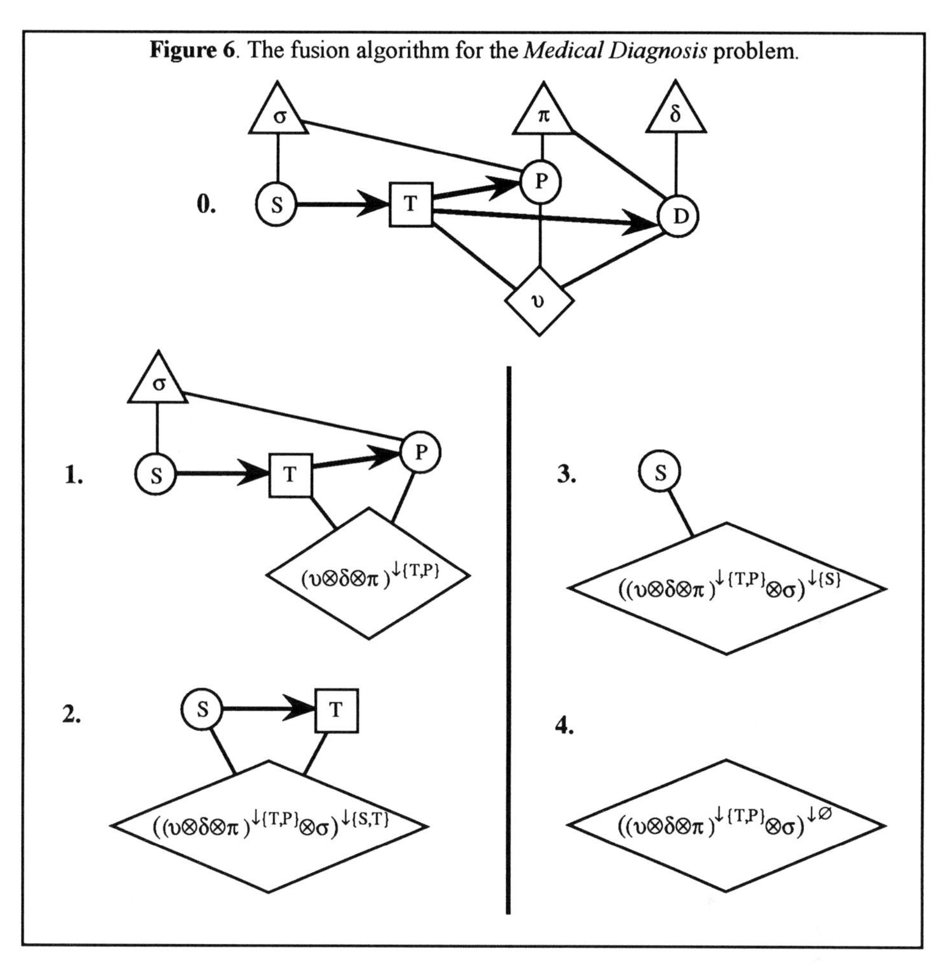

The weakness of the fusion algorithm is that, in the worst case, its complexity is exponential. This is not surprising because solving a decision problem is NP-hard [Cooper 1990]. In problems with many variables, the fusion algorithm is tractable only if the sizes of the frames on which combinations are done stay small. The sizes of the frames on which combinations are done depend on the sizes of the domains of the functions and on the information constraints. We need strong independence conditions to keep the sizes of the potentials small. And we need strong assumptions on the joint utility function to decompose it into small functions. Of course, this weakness is also shared by decision trees and influence diagrams.

ACKNOWLEDGMENTS

This work was supported in part by the National Science Foundation under grant SES-9213558. I am grateful for discussions with and comments from Ali Jenzarli, Pierre Ndilikilikesha, Anthony Neugebauer, Geoff Schemmel, and Leen-Kiat Soh.

REFERENCES

Agogino, A.M. and K. Ramamurthi. 1990. Real time influence diagrams for monitoring and controlling mechanical systems. In Oliver, R. M. and J. Q. Smith (eds). *Influence Diagrams, Belief Nets and Decision Analysis*. 199–228. John Wiley & Sons, Chichester.

Bellman, R. E. 1957. *Dynamic Programming*. Princeton University Press, Princeton, NJ.

Call, H. J. and W. A. Miller. 1990. A comparison of approaches and implementations for automating decision analysis. *Reliability Engineering and System Safety*. **30**, 115–162.

Cooper, G. F. 1990. The computational complexity of probabilistic inference using Bayesian belief networks. *Artificial Intelligence*. **42**, 393–405.

Covaliu, Z. and R. M. Oliver. 1992. Formulation and solution of decision problems using decision diagrams. Unpublished manuscript. University of California at Berkeley, CA.

Darroch, J. N., S. L. Lauritzen, and T. P. Speed. 1980. Markov fields and log-linear interaction models for contingency tables. *The Annals of Statistics*. **8**, 522–539.

Edwards, D. and S. Kreiner. 1983. The analysis of contingency tables by graphical models. *Biometrika*. **70**, 553–565.

Ezawa, K. J. 1986. Efficient evaluation of influence diagrams. Ph.D. dissertation. Department of Engineering-Economic Systems, Stanford University.

Fung, R. M. and R. D. Shachter. 1990. Contingent influence diagrams. Unpublished manuscript. Advanced Decision Systems, Mountain View, CA.

Howard, R. A. 1988. Decision analysis: Practice and promise. *Management Science*. **34**(6), 679–695.

Howard, R. A. 1989. Knowledge maps. *Management Science*. **35**(8), 903–922.

Howard, R. A. 1990. From influence to relevance to knowledge. In Oliver, R. M. and J. Q. Smith (eds.). *Influence Diagrams, Belief Nets and Decision Analysis*. 3–23. John Wiley & Sons, Chichester.

Howard, R. A. and J. E. Matheson. 1981. Influence diagrams. Reprinted in Howard, R. A. and J. E. Matheson (eds.). 1984. *The Principles and Applications of Decision Analysis*. **2**, 719–762. Strategic Decisions Group, Menlo Park, CA.

Kiiveri, H., T. P. Speed, and J. B. Carlin. 1984. Recursive causal models. *Journal of the Australian Mathematics Society, Series A*. **36**, 30–52.

Kirkwood, C. W. 1993. An algebraic approach to formulating and solving large models for sequential decisions under uncertainty. *Management Science*. To appear.

Matheson, J. E. and W. J. Roths. 1967. Decision analysis of space projects: Voyager Mars. Reprinted in R. A. Howard and J. E. Matheson (eds.). 1984. *Readings on The Principles and Applications of Decision Analysis*. **1**, 446–475. Strategic Decisions Group, Menlo Park, CA.

Mellouli, K. 1987. On the propagation of beliefs in networks using the Dempster-Shafer theory of evidence. Ph.D. dissertation. School of Business, University of Kansas, Lawrence, KS.

Miller III, A. C., M. W. Merkhofer, R. A. Howard, J. E. Matheson, and T. R. Rice. 1976. Development of Decision Aids for Decision Analysis. Final Technical Report DO #27742. Stanford Research Institute, Menlo Park, CA.

Ndilikilikesha, P. 1991. Potential influence diagrams. Working Paper No. 235. School of Business, University of Kansas.

Oliver, R. M. and J. Q. Smith (eds.). 1990. *Influence Diagrams, Belief Networks, and Decision Analysis*. John Wiley & Sons, Chichester.

Olmsted, S. M. 1983. On representing and solving decision problems. Ph.D. dissertation. Department of Engineering-Economic Systems, Stanford University.

Owen, D. L. 1978. The use of influence diagrams in structuring complex decision problems. Reprinted in Howard, R. A. and J. E. Matheson (eds.). 1984. *Readings on The Principles and Applications of Decision Analysis*. **2**, 765–772. Strategic Decisions Group, Menlo Park, CA.

Pearl, J., D. Geiger, and T. Verma. 1990. The logic of influence diagrams. In Oliver, R. M. and J. Q. Smith (eds.). *Influence Diagrams, Belief Nets and Decision Analysis*. 67–88. John Wiley & Sons, Chichester.

Raiffa, H. 1968. *Decision Analysis: Introductory Lectures on Choices Under Uncertainty*. Addison-Wesley, Reading, MA.

Raiffa, H. and R. Schlaifer. 1961. *Applied Statistical Decision Theory*. MIT Press, Cambridge, MA.

Rege, A. and A. M. Agogino. 1988. Topological framework for representing and solving probabilistic inference problems in expert systems. *IEEE Transactions on Systems, Man, and Cybernetics*. **18**(3), 402–414.

Shachter, R. D. 1986. Evaluating influence diagrams. *Operations Research*. **34**, 871–882.

Shachter, R. D. 1988. Probabilistic influence diagrams. *Operations Research*. **36**, 589–604.

Shachter, R. D. and D. E. Heckerman. 1987. A backwards view for assessment. *AI Magazine*. **8**(3), 55-61.

Shenoy, P. P. 1989. A valuation-based language for expert systems. *International Journal of Approximate Reasoning*. **3**(5), 383–411.

Shenoy, P. P. 1991. Conditional independence in valuation-based systems. Working Paper No. 236. School of Business, University of Kansas, Lawrence, KS.

Shenoy, P. P. 1992a. Valuation-based systems: A framework for managing uncertainty in expert systems. In Zadeh, L. A. and J. Kacprzyk (eds.). *Fuzzy Logic for the Management of Uncertainty*. 83–104. John Wiley & Sons, New York, NY.

Shenoy, P. P. 1992b. Valuation-based systems for Bayesian decision analysis. *Operations Research*. **40**(3), 463–484.

Shenoy, P. P. 1992c. Conditional independence in uncertainty theories. In D. Dubois, M. P. Wellman, B. D'Ambrosio and P. Smets (eds.). *Uncertainty in Artificial Intelligence: Proceedings of the Eighth Conference*. 284–291. Morgan Kaufmann, San Mateo, CA.

Shenoy, P. P. 1993a. A new method for representing and solving Bayesian decision problems. In Hand, D. J. (ed.). *Artificial Intelligence Frontiers in Statistics: AI and Statistics III*. 119–138. Chapman & Hall, London.

Shenoy, P. P. 1993b. A comparison of graphical techniques for decision analysis. *European Journal of Operational Research*. To appear.

Shenoy, P. P. 1993c. Valuation networks and conditional independence. Working Paper No. 238. School of Business, University of Kansas, Lawrence, KS.

Shenoy, P. P. 1993d. Game trees for decision analysis. Working Paper No. 239. School of Business, University of Kansas, Lawrence, KS.

Smith, J. E., S. Holtzman, and J. E. Matheson. 1989. Structuring conditional relationships in influence diagrams. Unpublished manuscript. Fuqua School of Business, Duke University.

Tatman, J. A. and R. D. Shachter. 1990. Dynamic programming and influence diagrams. *IEEE Transactions on Systems, Man, and Cybernetics*. **20**(2), 365–379.

von Neumann, J. and O. Morgenstern. 1944. *Theory of Games and Economic Behavior*. 1st edition. John Wiley & Sons, New York, NY.

Watson, S. R. and D. M. Buede. 1987. *Decision Synthesis: The Principles and Practice of Decision Analysis*. Cambridge University Press, Cambridge, U.K.

Wermuth, N. and S. L. Lauritzen. 1983. Graphical and recursive models for contingency tables. *Biometrika*. **70**, 537–552.

Zermelo, E. 1913. Uber eine anwendung der mengenlehre auf die theorie des schachspiels. *Proceedings of the Fifth International Congress of Mathematics*. **2**, 501–504. Cambridge, U.K.

Uncertainty in Intelligent Systems
B. Bouchon-Meunier et al. (Editors)

Propagation of Convex Sets of Probabilities in Directed Acyclic Networks[1]

José Cano, Serafín Moral and J.F. Verdegay López

Departamento de Ciencias de la Computación e I.A. Universidad de Granada. 18071 - Granada - Spain.

Abstract

This paper presents a generalization of Pearl's model of probability propagation to convex sets of probability distributions. It is based on an axiomatic framework for the propagation of general valuations in directed acyclic graphs, developed in Cano, Delgado, Moral [1]. In the first part of this work, this axiomatic system and the resulting propagation formulas are given. After, it is particularized to the case of valuations that are convex sets of probabilities. Special attention is given to the problems of knowledge acquisition. It is also shown the different nature of conditional information as part of the initial knowledge and 'a posteriori' conditional information, obtained after doing a restriction of initial probabilities from given observations.

1. INTRODUCTION

Shenoy, Shafer, [2], have shown that propagation of probabilities on graphs can be applied to different uncertainty formalisms or other problems as dynamic programming and integer optimization. They have proposed a set of axioms about the operations we have to perform. If these axioms are verified and if our original knowledge can be decomposed in an appropriate way then the calculus may be locally carried out.

Shafer, Shenoy work does not consider the idea of conditional information in general and focuses on the problem of distributed calculus. There are some aspects of knowledge representation and acquisition that are not addressed in this general framework. In fact in Probability Propagation, Lauritzen, Spiegelhalter,[3], consider a directed acyclic network to represent information and after, this is transformed in an undirected graph in which computation is performed.

In Cano, Delgado, Moral, [1], a generalization of Pearl,[4], model of representing and propagating information to general valuations has been developed. It is proposed an axiomatic which is based in Shenoy, Shafer, [2], axiomatic framework for propagation of uncertainty in hypertrees. However it is a bit more restrictive. The existence of two special elements, the contradiction and neutral valuations, is postulated in the new model. These elements will be essential to define conditional information and to determine propagation formulas.

[1]This work has been supported by the Commission of the European Communities under ESPRIT BRA 3085: DRUMS.

This model will be applied here to convex sets of probability distributions. The operators of combination are based on the methods proposed in, [5]. In this paper it is discriminated between 'a priori' information given for all the population and evidential information coming from restrictions for observations in a particular case of the population. The combination of 'a priori' and evidential information is the 'a posteriori' information. What is obtained is a generalization of Bayes rule. The underlying conditioning method is different of the known Dempster conditioning, [6], and upper and lower probabilities conditioning, [7, 8].

The propagation procedure that is obtained here is different of the generalization by Tessem, [9], of Pearl's propagation algorithms to interval probabilities, because the conditioning used by Tessem is upper and lower conditioning, different from the one used in this paper.

In the first section of this paper we introduce the calculus with convex set of probabilities. Then, in the following section we introduce axiomatic sytem for the propagation of Uncertainty in Directed Acyclic Graphs. This axiomatic framework to the particular case of convex sets of probabilities, in the last section.

2. CALCULUS WITH CONVEX SETS OF PROBABILITIES

Assume that we have a population Ω and a variable X taking its values on a finite set $U = \{u_1, ..., u_n\}$. About this variable we may have a piece of information consisting on a probability distribution p, that is a mapping, $p : U \longrightarrow [0,1]$ verifying

$$\sum_{i=1}^{n} p(u_i) = 1 \tag{1}$$

Under an objective interpretation, this probability informs about the relative frequencies with which X takes the different elements of U, in the population Ω.

If the knowledge about these frequencies is not so exact, we may have a set H^X of possible probability distributions. We shall consider that H^X is always a convex set with a finite set of extreme points $p_1, ..., p_k$. In case H^X is not convex, we shall always consider its convex hull (see [5]),

$$\mathrm{CH}(H^X) = \{\sum_i \alpha_i.p_i \mid p_i \in H^X, \alpha \in [0,1], \sum_i \alpha_i = 1\} \tag{2}$$

Reasons for this transformation may be found in [5, 10].

In the same way as probability convex set, we define a conditional piece of information, $H^{Y|X}$, as a convex set of conditional probabilities of Y given X with a finite number of extreme points.

If we have a convex set H^X for variable X with extreme points $\{p_1, \ldots, p_k\}$ and a conditional piece of information, $H^{Y|X}$, with extreme points $\{h_1, \ldots, h_l\}$, then we shall define a global information about the pair of variables (X, Y), as the convex set, $H^{X,Y}$, of probability distributions p on $U \times V$ generated by points:

$$H^{X,Y} = \mathrm{CH}\{p_1.h_1, \ldots, p_1.h_l, p_2.h_1, \ldots, p_2.h_l, \ldots, p_k.h_1, \ldots, p_k.h_l\} \tag{3}$$

There are two important remarks about this definition,

- In general, Not every point $p_i.h_j$ where p_i is an extrem point of H^X and h_j is an extreme point of $H^{Y|X}$ is an extreme point of $H^{Y,X}$. Then to calculate a minimal representation of this set, Ext$(H^{Y,X})$, we have to apply some convex hull algorithm to the all these points, to remove the non-extreme ones (see [11] for efficient algorithms to do this operation).

- This set $H^{X,Y}$ is not equal to the set $\{p.h \mid p \in H^X, h \in H^{Y|X}\}$ because this set is not convex, in general. However $H^{X,Y}$ is equal to the convex hull of above set. And, in the worst case, we are adding some extra probabilities for the sake of simplicity: convex sets may be represented by the extreme points (see [5, 9]).

2.1. Total Probability Theorem

First, we are going to describe the process of obtaining marginal information for a variable from some piece of information relating a wider set of variable. The following notation will be used: If $U_i, i = 1, \ldots, n$, are all finite sets, $I \subseteq \{1, \ldots, n\}$ and $u \in U_1 \times \ldots \times U_n$, then by $u^{\downarrow I}$ or $u^{\downarrow \prod_{i \in I} U_i}$ we shall denote the element $v \in \prod_{i \in I} U_i$ such that, $v_i = u_i, \forall i \in I$. That is the point obtained from u, by dropping the extra coordinates, that is, those non in I.

Now, in the same conditions as above, if h is a mapping, $h : U_1 \times \ldots \times U_n \longrightarrow [0, 1]$, then the mapping $\overline{h} : \prod_{i \in I} U_i \longrightarrow [0, 1]$ defined by

$$\overline{h}(v) = \sum_{u^{\downarrow I} = v} h(u) \tag{4}$$

will be denoted as $h^{\downarrow I}$ or $h^{\downarrow \prod_{i \in I} U_i}$.

In the probabilistic case, from a global information, $p^{X,Y}$, we may obtain a probability for X and a probability for variable Y, in the following way,

$$p^X(u) = \sum_{v \in V} p^{X,Y}(u, v) = (p^{X,Y})^{\downarrow U}(u) \tag{5}$$

$$p^Y(v) = \sum_{u \in U} p^{X,Y}(u, v) = (p^{X,Y})^{\downarrow V}(v) \tag{6}$$

If to build the global information for (X, Y) we have started from a probability on U, $\bar{p}^X$, and a conditional probability, $p^{Y|X}$. Then $\bar{p}^X$ is equal to the above calculated p^X. For p^Y we get the well known *Total Probability Theorem*,

$$p^Y(v) = \sum_{u \in U} p^X(u).p^{Y|X}(u, v) \tag{7}$$

that is,

$$p^Y = (p^X.p^{Y|X})^{\downarrow V} \tag{8}$$

In our model, if we have a global convex set $H^{X,Y}$ with extreme points $\{h_1, \ldots, h_n\}$ of possible bidimensional probabilities for variables (X, Y), then we may get marginal convex sets, H^X and H^Y, for variables X and Y, on an analogous way. H^X is the convex set generated by the points $p_i, i = 1, \ldots, n$, where p_i is a mapping of U on $[0,1]$ given by,

$$p_i(u) = \sum_{v \in V} h_i(u,v) \tag{9}$$

that is,

$$H^X = \mathrm{CH}\{h_1{}^{\downarrow U}, \ldots, h_n{}^{\downarrow U}\} \tag{10}$$

H^Y is the convex set generated by points $q_i, i = 1, \ldots, n$, where

$$q_i(v) = \sum_{u \in U} h_i(u,v) \tag{11}$$

that is,

$$H^Y = \mathrm{CH}\{h_1{}^{\downarrow V}, \ldots, h_n{}^{\downarrow V}\} \tag{12}$$

As before, it is important to remark that not every mapping on the set generating H^X or H^Y is an extreme point and, to get a minimal representation we have to remove the non-extreme ones.

For the case of the Total Probability Theorem, we start from a convex set H^X with extreme points $\{p_1, \ldots, p_n\}$ for variable X, and a convex set of conditional probabilities of Y given X, $H^{Y|X}$, with extreme points $\{h_1, \ldots, h_k\}$, then by combining H^X with $H^{Y|X}$ and marginalizing on V we get the *Total Probability Theorem* for convex sets of probabilities,

$$H^Y = \mathrm{CH}\{(p_1.h_1)^{\downarrow V}, .., (p_1.h_k)^{\downarrow V}, (p_2.h_1)^{\downarrow V}, .., (p_2.h_k)^{\downarrow V}, .., (p_n.h_1)^{\downarrow V}, .., (p_n.h_k)^{\downarrow V}\} \tag{13}$$

Example 1 *Assume that $U = \{u_1, u_2\}$ and $V = \{v_1, v_2, v_3\}$ and that H^X has as extreme points*

	u_1	u_2
p_1	0.2	0.8
p_2	0.6	0.4

$H^{Y|X}$ is a convex set of conditional probabilities of Y given X with extreme points,

	(u_1,v_1)	(u_1,v_2)	(u_1,v_3)	(u_2,v_1)	(u_2,v_2)	(u_2,v_3)
h_1	0.1	0.5	0.4	0.6	0.4	0.0
h_2	0.8	0.0	0.2	0.0	0.1	0.9

The combination of H^X and $H^{Y|X}$ is a bidimensional convex set of possible probability distributions with extreme points,

	(u_1,v_1)	(u_1,v_2)	(u_1,v_3)	(u_2,v_1)	(u_2,v_2)	(u_2,v_3)
$p_1.h_1$	0.02	0.10	0.08	0.48	0.32	0.00
$p_1.h_2$	0.16	0.00	0.04	0.00	0.08	0.72
$p_2.h_1$	0.06	0.30	0.24	0.24	0.16	0.00
$p_2.h_2$	0.48	0.00	0.12	0.00	0.04	0.36

To complete the application of the Total Probability Theorem, we marginalize these bidimensional probabilities on V, obtaining for variable Y the convex set, H^Y, *generated by points,*

	v_1	v_2	v_3
q_1	0.50	0.42	0.08
q_2	0.16	0.08	0.76
q_3	0.30	0.46	0.24
q_4	0.48	0.04	0.48

In these case, all these points are extreme and none of them may be eliminated.

2.2. Conditioning

Finally, in this section, we consider the problem of conditioning. Assume that $H^X = \text{CH}\{p_1, \ldots, p_n\}$ is a convex set for variable X and that we have observed *'X belongs to A'*, then the result of conditioning is the convex set, $H^X|A$, generated by points $\{p_1.l_A, \ldots, p_n.l_A\}$ where l_A is the likelihood associated with set A ($l_A(u) = 1, if\ u \in A; l_A(u) = 0, otherwise$).

It is important to remark that $H^X|A$ is a convex set of differently normalized functions. If we call $r_i = \sum_{u \in U} p_i(u).l_A(u) = P_i(A)$, then by calculating $(p_i.l_A)/r_i$ we get the conditional probability distribution $p_i(.|A)$. The set $H' = \{p(.|A) \mid p \in H\}$ was propossed by Dempster, [6], as the set of conditioning, and has been widely used. However, this set produces very large intervals, the reason being that, by normalizing each probability, we loose the information provided by values r_i.

To associate probability intervals with $H^X|A$ we have proposed, [12], the following procedure:

- Consider the extreme points of $H|A$: $\{p_1.l_A, \ldots, p_n.l_A\}$
- Normalize each extreme point, calculating $p_i(.|A)$, and assigning it, at the same time, a possibility value, $\pi(p_i(.|A))$, equal to $r_i/(\text{Max } r_k)$.
- If Π is the possibility measure defined by above possibility values and N its dual necessity measure, then lower and upper intervals are calculated as,

  $$P_*(B|A) = \text{I}(P_i(B|A) \mid N) \qquad P^*(B|A) = \text{I}(P_i(B|A) \mid \Pi)$$

  and I stands by the Choquet's integral, [13].

Example 2 *If* $U = \{u_1, u_2, u_3\}$ *and* $H^X = \text{CH}\{p_1, p_2\}$ *where*

	u_1	u_2	u_3
p_1	0.5	0.5	0.0
p_2	0.1	0.0	0.9

If $A = \{u_1, u_2\}$ *then* $H^X|A$ *has as extreme points,*

	u_1	u_2	u_3
$p_1.l_A$	0.5	0.5	0.0
$p_2.l_A$	0.1	0.0	0.0

	u_1	u_2	u_3	π
$p_1(.\|A)$	0.5	0.5	0.0	1.0
$p_2(.\|A)$	1.0	0.0	0.0	0.1

The intervals without taking into account the possibilities are,

$u_1 \to [0.5, 1] \qquad u_2 \to [0, 0.5] \qquad u_3 \to [0.0, 0.0]$

and by using Choquet's integral, we obtain,

$u_1 \to [0.5, 0.55] \qquad u_2 \to [0.45, 0.5] \qquad u_3 \to [0.0, 0.0]$

3. AN AXIOMATIC SYSTEM FOR THE PROPAGATION OF UNCERTAINTY IN DIRECTED ACYCLIC NETWORKS

Let $X = (X_1, ..., X_n)$ an n-dimensional variable such that each X takes its values on a finite set U. A valuation is a primitive concept representing the mathematical support for information in the corresponding theory. For each $I \subseteq \{1, \ldots, n\}$ there is a set $\mathcal{V}_I$ of valuations defined on the cartesian product $\prod_{i \in I} U_i$. $\mathcal{V}$ is the set of all valuations $\mathcal{V} = \cup_{I \subseteq \{1,...,n\}} \mathcal{V}_I$.

Two basic operations are necessary (see Zadeh, [14], Shenoy, Shafer, [2]):

- *Marginalization.-* If $J \subseteq I$ and $V_1 \in \mathcal{V}_I$ then the marginalization of V_1 to J is a valuation $V_1^{\downarrow J}$ defined on $\mathcal{V}_J$.
- *Combination.-* If $V_1 \in \mathcal{V}_I$ and $V_2 \in \mathcal{V}_J$, then its combination is a valuation $V_1 \otimes V_2$ defined on $\mathcal{V}_{I \cup J}$

In [2] it is considered that valuations verify the following three first axioms. We have added Axioms 4-6, [1].

1. $V_1 \otimes V_2 = V_2 \otimes V_1, \quad (V_1 \otimes V_2) \otimes V_3 = V_1 \otimes (V_2 \otimes V_3)$.
2. If $I \subseteq J \subseteq K$, and $V \in \mathcal{V}_J$ then $(V^{\downarrow J})^{\downarrow I} = V^{\downarrow I}$.
3. If $V_1 \in \mathcal{V}_I$, $V_2 \in \mathcal{V}_J$, then $(V_1 \otimes V_2)^{\downarrow I} = V_1 \otimes V_2^{\downarrow (J \cap I)}$.
4. *Neutral Element.-* There exits one and only one valuation V_0 defined on $U_1 \times \ldots \times U_n$ such that $\forall V \in \mathcal{V}_I, \forall J \subseteq I$, we have $V_0^{\downarrow J} \otimes V = V$.
5. *Contradiction.-* There exits one and only one valuation, V_c, defined on $U_1 \times \ldots \times U_n$, such that $\forall V$, $V_c \otimes V = V_c$.
6. $\forall V \in \mathcal{V}_\emptyset$, if $V \neq V_c^{\downarrow \emptyset}$, then $V = V_0^{\downarrow \emptyset}$.

V_0 represents the neutral valuation, V_c, the contradictory information. V_0 is used to define conditional information, V_c to define absorbent valuations. $(V_0)^{\downarrow I}$ and $(V_c)^{\downarrow I}$ are said to be the neutral and the contradictory valuations on $\mathcal{V}_I$, respectively. When there is not problem of confusion they are denoted as V_0 and V_c, simply.

Definition 1 *A valuation $V \in \mathcal{V}_I$ is said to be absorbent if and only if is different from the contradictory valuation and $\forall V' \in \mathcal{V}_I$, $(V \otimes V' = V)$ or $(V \otimes V' = V_c)$.*

If a valuation V represents an information about the values of variables $(X_i)_{i\in I}$, then an absorbent valuation represents a perfect knowledge about these values: it can not be consistently refined by combination with other valuation. We may obtain only the same information or the contradiction.

Definition 2 *If $V \in V_{I\cup J}$, it is said to be a valuation on $\prod_{i\in I} U_i$ conditioned to $\prod_{j\in J} U_j$, if and only if $V^{\downarrow J} = V_0 \in \mathcal{V}_I$, the neutral element on $\mathcal{V}_I$. The subset of $\mathcal{V}_{I\cup J}$ of valuations on $\prod_{i\in I} U_i$ conditioned to $\prod_{j\in J} U_j$ will be denoted as $\mathcal{V}_{I|J}$.*

The idea underlying conditional information is that if V is a valuation on $\prod_{i\in I} U_i$ conditioned to $\prod_{j\in J} U_j$ (sometimes said on I given J), then it inform us about $(X_i)_{i\in I}$ and its relationships with variables $(X_j)_{j\in J}$, but non about variables $(X_j)_{j\in J}$.

The fact that a valuation $V \in \mathcal{V}_{I\cup J}$ is conditional does not mean that is a valid conditional information. This will be considered below in the definition of systems of information. We only know that from a mathematical point of view is a possible conditional information.

For the definitions of dependence and conditional independence we follow the Pearl's idea of considering them primitive concepts, [15].

Definition 3 *Given a family of variables $(X_1, ..., X_n)$ a dependence structure on it is a mapping $D : \wp(\{1, ..., n\}) \times \wp(\{1, ..., n\}) \times \wp(\{1, ..., n\}) \longrightarrow \{0, 1\}$ where if $D(I,J,K) = 0$ is said that $(X_i)_{i\in I}$ is independent of $(X_k)_{k\in K}$ given $(X_j)_{j\in J}$ and verifying the following axioms (see Geiger, Pearl,[15])*

- Symmetry.- *If $D(I, J, K) = 0$ then $D(K, J, I) = 0$ and viceverse.*
- Decomposition.- *If $D(I, J, K \cup L) = 0$ then $D(I, J, K) = 0$ and $D(I, J, L) = 0$*
- Weak union.- *If $D(I, J, K \cup L) = 0$ then $D(I, J \cup L, K) = 0$*
- Contraction.- *If $D(I, J, K) = 0$ and $D(I, J \cup K, L) = 0$ then $D(I, J, K \cup L) = 0$*

An intuitive interpretation of these axioms can be found in Pearl, [4]. They are always verified for the dependences associated with a probability distribution and for the structure of dependences represented by a directed acyclic graph, or an undirected graph. A valuation on $\prod_{i\in I} U_i$ is the mathematical representation of an information about how $(X_i)_{i\in I}$ takes its values. The following properties establish how to build more complex pieces of information from elemental ones.

Definition 4 *Let $(X_1, \ldots, X_n)$ a n-dimensional variable taking values on $U_1 \times \cdots \times U_n$, and D an associated dependence structure. A system of information about this variable with respect to D is a family $Q \subseteq \mathcal{V}$ and a mapping $q : Q \rightarrow \wp(I) \times \wp(I)$ verifying the following properties:*

1. *$\forall V \in Q$, if $q(V) = (I, J)$ then $V \in \mathcal{V}_{I\cup J}$ and $I \cap J = \emptyset$. It is said that V is a valuation about variables $(X_i)_{i\in I}$ conditioned to variables $(X_j)_{j\in J}$. IF $J = \emptyset$, it is said that V is an information about $(X_i)_{i\in I}$, simply.*

2. *If* $V_1, V_2 \in Q$ *and* $q(V_1) = (J, \emptyset)$ *and* $q(V_2) = (I, J)$ *then* $V_1 \otimes V_2 \in Q$ *with* $q(V_1 \otimes V_2) = (I \cup J, \emptyset)$.

3. *If* $V \in Q$ *with* $q(V) = (I, J)$ *and* $D(I, J, K) = 0$ *then* $V \otimes (V_0)^{\downarrow K} \in Q$ *with* $q(V \otimes (V_0)^{\downarrow K}) = (I, J \cup K)$.

4. *If* $V \in Q$ *with* $q(V) = (I, J)$, *then if* $K \subseteq I$, $V^{\downarrow (K \cup J)} \in Q$ *with* $q(V^{\downarrow K}) = (K, J)$.

Given a system of information, an element from it will be called a valid information. There is not an axiom saying that the contradiction V does not belong to Q. However, it is clear than in a system with the contradiction there is some piece of information that can not be certain. In that case it would be interesting to determine methods to solve the inconsistency, changing the minimum number of valuations.

Definition 5 *A system of information (Q,q) is said to be complete and deterministic if and only if there exists one and only one valuation* $V \in Q$ *such that* $q(V) = (\{1, \ldots, n\}, \emptyset)$.

Graphical structures are very appropriate to represent dependence relationships among variables. In the following we consider the case of directed acyclic graphs. A directed acyclic graph is said to be associated with n-dimensional variable $(X_1, \ldots, X_n)$ if the set of vertices is $\{X_1, \ldots, X_n\}$. That is, we have a vertex for each variable. These graphs may be used to represent a dependence structure for these variables, as considered by Pearl 1989,[4] (D-separation criterion).

An acyclic dependence graph may be the basis to define a system of information.

Definition 6 *If* $(X_1, \ldots, X_n)$ *is a n-dimensional variable and* (T, E) *is a graph* $T = \{X_i\}_{i \in \{1,\ldots,n\}}$, *we say that* (Q, q) *is a system of information defined on* (T, E) *if and only if* Q *is the minimum information system generated by a family* $\{V_i\}_{i \in \{1,\ldots,n\}}$ *where* $q(V_i) = (i, P(i))$ *and* $P(i) = \{j \mid (X_j, X_i) \in E\}$, *that is the set of parents of* X_i, *and where the dependency structure is the one associated with* (T, E).

System of information defined on directed acyclic graphs are very appropriate because they are always complete and deterministic. We are sure that introducing for each variable a valuation conditioned to its parents, then we are able to obtain a global information for all the variables and this information is unique as is stated in the following theorem (a proof can be found in [1]).

Theorem 1 *If* (Q, q) *is a system of information defined on a graph* (T, E) *with* $T = \{X_i\}_{i \in \{1,\ldots,n\}}$ *then it is complete and deterministic about* $(X_1, \ldots, X_n)$.

To complete this abstract description of information and conditional information we need two additional definitions.

Definition 7 *A family of observations about a n-dimensional variable* $(X_1, \ldots, X_n)$ *is a set of valuations* $\{O_i\}_{i \in I}$, *where* $I \subseteq \{1, \ldots, n\}$, *and* O_i *is an absorvent valuation on* U_i, $\forall i \in I$.

Definition 8 *If (Q,q) is a complete deterministic system of information about $(X_1, \ldots, X_n)$ and $\{O_i\}_{i\in I}$ a family of observations about these variables, we call 'a posteriori' information to the family of valuations $((\otimes_{i\in I}O_i)\otimes V)^{\downarrow J}$, where $J \subseteq \{1,\ldots,n\}$ and V is the only global information about $(X_1,\ldots,X_n)$.*

Here is important to remark that 'a posteriori' information is considered different from given conditional pieces of information. In Probability Theory 'a posteriori' information are the result of conditioning and are of the same nature than original conditional information. However, in other fields like in upper and lower probabilities this is not the case.

In general the problem of calculus on graphs is the following: Let (Q,q) a system of information defined on a graph (T,E) and $\{O_i\}_{i\in I}$ a system of observations, the we have to calculate on an efficient way the 'a posteriori' valuations,

$$PS_j = ((\otimes_{i\in I}O_i)\otimes V)^{\downarrow\{j\}} \qquad (14)$$

for each $j \in \{1,\ldots,n\}$.

This may be done on a distributed way if the graph does not have undirected cycles (also called loops) and if the global information is not the contradiction. The following algorithms are a direct generalization of Pearl's ones. A proof may be found in [1]. In these algorithms we are going to calculate for each node X_j two valuations π_j and λ_j. PS_j will be equal to the combination of π_j and λ_j. For this, it will be necessary calculate for each node X_j and for each node, X_k, children of X_j ($k \in C(j)$), a valuation π_k^j, which are called messages from nodes X_j to its children. Analogously for each node X_i parent of X_j ($i \in P(j)$), we have to calculate a valuation λ_i^j, which are called messages from a node to its parents.

The first of the following algorithms says how to calculate PS_j when the set of observations is empty ($I = \emptyset$).

- For every $j = 1, ..., n$ calculate
  - $\pi_j = [V_j \otimes (\otimes_{k\in P(j)}\pi_j^k)]^{\downarrow\{j\}}$
  - $\lambda_j = V_o^{\downarrow j}$
  - $PS_j = \pi_j \otimes \lambda_j$
  - $\forall k \in C(j),\ \ \pi_k^j = \pi_k$
  - $\forall i \in P(j),\ \ \lambda_i^j = V_o^{\downarrow\{i\}}$

The following algorithm is to recalculate these values when we change from a set of observations I to the set $I \cup \{l\}$, that is, we add an observation for variable X_l. Let J be the set of pairs (j_1, j_2) where j_1 is a node to update and j_2 the incoming node. Then the algorithm is as follow:

- $J = \{(l,-1)\}$
- While $J \neq \emptyset$
  - Choose $(j_1,j_2) \in J$, $J \longleftarrow J - \{(j_1,j_2)\}$

- calculate
  - $\pi_{j_1} \longleftarrow [V_{j_1} \otimes (\otimes_{i \in P(j_1)} \pi^k_{j_1})]^{\downarrow\{j_1\}}$
  - $\lambda_{j_1} \longleftarrow (\otimes_{K \in C(j_1)} \lambda^k_{j_1}) \otimes O'_{j_1}$
  - $PS_{j_1} \longleftarrow \pi_{j_1} \otimes \lambda_{j_1}$
- For every $k \in C(j_1)$, $k \neq j_2$ calculate
  - $\pi^{j_1}_k \longleftarrow [\pi_{j_1} \otimes O'_{j_1} \otimes (\otimes_{i \in C(j_1)_{i \neq k}} \lambda^i_{j_1})]$
  - $J \longleftarrow J \cup \{(k, j_1)\}$
- For every $k \in P(j_1)$, $k \neq j_1$ calculate
  - $\lambda^{j_1}_k \longleftarrow [\lambda_{j_1} \otimes O'_{j_1} \otimes V_{j_1} \otimes (\otimes_{i \in P(j_1)_{i \neq k}} \pi^i_{j_1})]^{\downarrow\{k\}}$
  - $J \longleftarrow J \cup \{(k, j_1)\}$

$$\textit{where } O'_j = \begin{cases} O_j & \textit{if } j \in I \\ V_o^{\downarrow\{j\}} & \textit{otherwise} \end{cases}$$

4. PROPAGATION OF CONVEX SETS OF PROBABILITIES

Valuations on $\prod_{i \in I} U_i$ in the case of imprecise probabilities are convex sets, H, of non-necessarily normalized functions of $\prod_{i \in I} U_i$ on $\Re_0^+$, that is, vectors in $\Re^{\prod_{i \in I} n_i}$ (n_i is the number of elements of U_i), with non-negative components. Two valuations V_1 and V_2, given by convex sets, H_1 and H_2 are considered equivalent, if adding the null function, $h_c(u) = 0$, $\forall u \in U_1 \times \ldots \times U_n$, its convex hull are proportionals, that is, $\exists \alpha \in \Re^+$, such that

$$\mathrm{CH}(H_1 \cup \{h_c\}) = \alpha.\mathrm{CH}(H_2 \cup \{h_c\}) \tag{15}$$

The reasons for this equivalence are given in Section 2.

The operations of combination and marginalization are based on the ones defined on former section:

- If V_1, defined on $\prod_{i \in I} U_i$ is given by $\{h_1, \ldots, h_n\}$ and V_2, defined on $\prod_{i \in I} U_i$, is given by $\{f_1, \ldots, f_m\}$, then $V_1 \otimes V_2$ is the convex set, defined on $\prod_{i \in (I \cup J)} U_i$ generated by functions,

  $$\{h_1.f_1, \ldots, h_1.f_m, \ldots\ldots, h_n.f_1, \ldots, h_n.f_m\} \tag{16}$$

  where $(h_k.f_l)(u) = h_k(u^{\downarrow I}).f_l(u^{\downarrow J})$.

- If V is defined on $\prod_{i \in I} U_i$ given by functions $\{h_1, \ldots, h_n\}$ and $J \subseteq I$ then $V^{\downarrow J}$ is the convex set generated by functions

  $$\{h_1^{\downarrow J}, \ldots, h_n^{\downarrow J}\} \tag{17}$$

The neutral element is the convex set with only one element, the function h_0, where $h_0(u) = 1, \forall u \in U_1 \times \ldots \times U_n$. The contradiction has an only point: the null function, h_c.

Next theorem shows that these valuations verify Axioms 1-6.

Theorem 2 *The above defined valuations with the corresponding operations of combination and marginalization verify Axioms 1-6.*

We omit the proof here because is a bit long for the restriction of this paper.

As conclusion of this theorem, we can specify the uncertainty associated to a set of variables with a dependence structure given by a directed acyclic networks giving, for each variable, a convex set of possible conditional probabilities about the values of this variable conditioned to its parents on the graph. Then by theorem 1, we are sure that there is one and only one global valuation for all the variables. Furthermore , we can apply above algorithms to propagate convex sets of probabilities.

5. CONCLUSIONS

In this paper we have developed the theoretical results allowing to propagate upper and lower probabilities or, more concretely, convex sets of probability distributions in directed acyclic graphs. We have considered also different problems associated with the acquisition of information.

This method has been implemented as part of Project DRUMS (ESPRIT II B.R.A. 3085) in DECSAI (Department of Computer Science and Artificial Intelligence). It runs on a SUN Sparc-1 Workstation and needs SunView as Graphical Interface. To eliminate the non-extreme points after the operations of combination and marginalization it has been used the gift wrapping algorithm (see [11]). However an important problem of efficiency remains. It is possible to multiply the number of extreme functions when we perform a combination of valuations.

REFERENCES

1. Cano J.E., M. Delgado, S. Moral (1991) An axiomatic system for the propagation of uncertainty in directed acyclic networks. Submitted to the International Journal of Approximate Reasoning.
2. Shenoy P.P., G. Shafer (1990) Axioms for probability and belief-function propagation. In: Uncertainty in Artificial Intelligence, 4 (Shachter, Levitt, Kanal, Lemmer, eds.) 169-198.
3. Lauritzen S.L., D.J. Spiegelharter (1988) Local computation with probabilities on graphical structures and their application to expert systems. J. of the Royal Statistical Society, B 50, 157-224.
4. Pearl J. (1989) Probabilistic Reasoning with Intelligent Systems. Morgan & Kaufman, San Mateo.
5. Cano J.E., S. Moral, J.F. Verdegay-López (1991) Combination of Upper and Lower Probabilities. Proceedings of the 7th Conference on Uncertainty in A.I., Los Angeles 1991, 61-68.
6. Dempster A.P. (1967) Upper and lower probabilities induced by a multivalued mapping. Annals of Mathematical Statistics 38, 325-339.
7. Fagin R., J.Y. Halpern (1990) A new approach to updating beliefs. Research Report RJ 7222. IBM Almaden Research Center.

8. Campos L.M. De, M.T. Lamata, S. Moral (1990) The concept of conditional fuzzy measure. International Journal of Intelligent Systems, 5, 237-246.
9. Tessen B. (1989) Interval Representation of Uncertainty in Artificial Intelligence. Ph. D. Thesis, Department of Informatics, University of Bergen, Norway.
10. Walley P. (1991) Statistical Reasoning with Imprecise Probabilities. Chapman and Hall, London.
11. Edelsbrunner H. (1987) Algorithms in Combinatorial Geometry. Springer Verlag, Berlin.
12. Moral S., L.M. de Campos (1990) Updating uncertain information. Proceedings 3rd. IPMU Conference, Paris 1990, 452-454.
13. Choquet G. (1953/54) Theorie of capacities. Ann. Inst. Fourier 5, 131-292.
14. Zadeh L.A. (1979) A Theory of Approximate Reasoning. In: Machine Intelligence, 9 (J.E. Hayes, D. Mikulich, eds.) Elsevier, Amsterdam, 149-194.
15. Geiger D., J. Pearl (1988) Logical and algorithmic properties of conditional independence. Technical Report R-97, Cognitive Systems Laboratory, University of California, Los Angeles.

Uncertainty in Intelligent Systems
B. Bouchon-Meunier et al. (Editors)

Local propagation of information on directed Markov trees

Sandra Aparecida Sandri

Instituto Nacional de Pesquisas Espaciais (INPE)
C.P. 515, 12201-970 São José dos Campos S.P., Brazil*

We present here a strategy capable of reducing the computational cost of the application of the local propagation of information process to a knowledge base. Even though taking into account all the information in the knowledge base, the strategy performs combinations only on the part of the base that contains relevant information in relation to the query. This strategy draws together the usual rule-based system inference, and the local propagation paradigms. As answer to a query it yields a result that can be proved to be correct in relation to the uncertainty model adopted.

1. INTRODUCTION

One of the main concerns in the field of Artificial Intelligence is to endow computer systems with the capability of manipulating knowledge furnished by experts. In the last two decades we have seen the development of a large class of knowledge-based systems, in which we distinguish the so-called rule-based systems. In these systems all the knowledge about a given domain is coded in the form of production rules, and the inference is made through the propagation of information between the rules. In this way, an extreme easyness on the acquisition of an expert's knowledge is combined with an important computational power. On the other hand, the mechanisms for the propagation of information usually employed in these systems base themselves on the false hypothesis that the pieces of knowledge yielded by an expert are independent between themselves. This blind propagation of information usually leads to results that are not necessarily correct when the data furnished by the expert is imperfect, ie pervaded with uncertainty or imprecision.

No matter which is its underlying uncertainty model (including ad hoc ones), a knowledge base can be seen as a set of relations between groups of variables. In rule-based systems for instance, each rule (or fact) represents a relation between the various attributes referenced in its clauses. Seen this way, a knowledge base is implicitly modeled by a structure known as a hypergraph.

A hypergraph is composed by a set of hyperedges, each one of them being formed by the group of variables referenced in a piece of knowledge (e.g. the variables used in a rule). The hypergraph structure reveals the dependencies existing between the pieces of knowledge in a knowledge base $\mathcal{K}$: if $\mathcal{H}$ is the hypergraph derived from $\mathcal{K}$, then two hyperedges h and g in $\mathcal{H}$ are related if the pieces of knowledge relative to h and g have at least one variable in common. The blind propagation of information in $\mathcal{K}$, as commonly used in in rule-based systems, may not respect these dependencies and may lead to loss of precision of the final result if the pieces of knowledge are represented inside any uncertainty model. Moreover, it

---

* This work was developed at the Institut de Recherches en Informatique de Toulouse (IRIT), in Toulouse, France. It has been supported by IRIT, INPE, the Brazilian National Council for Research and Development (CNPq), and the São Paulo State Research Support Foundation (FAPESP).

may lead to errors caused by the over/under-estimation of hypotheses when the knowledge is represented inside uncertainty models that employ non-idempotent methods in the combination of information (e.g. Probability and Evidence Theories, Mycin's uncertainty model).

The same problem occurs when blind propagation is used in systems whose knowledge bases are defined as sets of valuations [1] in a given uncertainty model. In this case, the propagation process is made through the use of the marginalization, extension and combination operators defined inside the uncertainty model. The correct results of propagation are those obtained by what we call here the global propagation of information : all valuations are extended to the union of their reference domains, all the extended valuations are combined, and finally the result is marginalized to the variable of interest. This procedure leads however to a computational explosion, and is therefore not feasible in practice.

In recent years, one of the most important subjects in uncertainty management research has been the determination of tools that allow us to obtain correct results through the propagation of information, that do not however lead to a computational explosion. One important result in this field is the proof that, if the adopted uncertainty model obeys the local propagation of information axioms proposed in [1] (see also [2]), then the propagation of information on *Markov trees* - structures derivable from simple manipulations on $\mathcal{H}$ - produces correct results inside that model. This process using Markov trees is known as the *local propagation of information* and is extensively exposed in [1],[2],[3],[4]. From a hypergraph $\mathcal{H}$ underlying a knowledge base $\mathcal{K}$ many Markov tree representatives of $\mathcal{H}$ can be derived. The application of the local propagation of information process on any of them will yield the same results as with global computation, but with lower costs.

Here we show that, if we phocus our attention on the information needed to answer a given query, we can further reduce the cost of propagation, and still obtain the correct results. The most straightforward way to do this consists on applying the local propagation process only on the part of a Markov tree representative of $\mathcal{H}$, that contains relevant information in relation to the query. In other words, it suffices to prune away all the irrelevant information from a Markov Tree (representative of $\mathcal{H}$), and apply the local propagation process on it. We show here that we can further reduce this cost if, instead of just taking one of these Markov trees, we carefully construct it based on the query itself.

This work is organized as follows. In Section 2 we identify which are the nodes in a Markov tree that can proved to be irrelevant in relation to a given query. Then in Section 3 we use this result to propose a procedure that will minimize the propagation process. In Section 4 we present an example illustrating the ideas presented here and Section 5 brings the conclusion.

2. LOCAL PROPAGATION OF INFORMATION ON DIRECTED MARKOV TREES

Let $\mathcal{X}$ be a finite set of variables x_i on Ω_{x_i}, and $h \subset \mathcal{X}$ be a n-dimensional variable on $\Omega_h = \Omega_{x_{h1}} \times ... \times \Omega_{x_{hn}}$, $x_{hj} \in h$. Let $\mathcal{K}$ be a knowledge base whose pieces of knowledge are modeled through the use of variables of $\mathcal{X}$. Then $\mathcal{H} = (S, \mathcal{X})$, $S \subset 2^{\mathcal{X}}$, is the *hypergraph* underlying $\mathcal{K}$, where each element $h \in S$, called a *hyperedge*, is the set of variables appearing together in a piece of knowledge in $\mathcal{K}$.

Associated to each $h \in S$, there exists a set V_h, called a *valuation* on h. Valuations extend the concept of the value of a variable, in order to accommodate the imperfection of information. When there is no uncertainty and/or imprecision pervading the available information on a given variable h, then V_h has one single element, the value of h. In frameworks modeling uncertainty, each $V_h(A)$, $A \in 2^{\Omega_h}$, represents the accumulated evidence that the real value of h lies in A. This function is an application $V_h : 2^{\Omega_h} \rightarrow [0, 1]$, such that $V_h(\Omega_h) = 1$, $V_h(\emptyset) = 0$, and $\forall A \in 2^{\Omega_h}$, $V_h(A) + V_h(\overline{A}) \leq 1$. In Evidence Theory, for instance, a valuation is a belief function $Bel : 2^{\Omega_h} \rightarrow [0, 1]$, defined by $Bel(A) = \sum \{m(B) / B \subset A, B \neq \emptyset\}$, where $m : 2^{\Omega_h} \rightarrow [0, 1]$ is a mass distribution function that allocates the evidence concerning strictly A and not any subset of A. In this theory we can thus model a valuation V_h by a set of pairs (A, $m_h(A)$), where $A \in 2^{\Omega_h}$ and m_h is the mass allocation function on h.

A Markov Tree T = (N, E) on $\mathscr{X}$, with $N \subset 2^{\mathscr{X}}$ is an acyclic graph, such that if f and g are two distinct vertices of N and $x \in \mathscr{X}$ belongs to both f and g, then x also belongs to every vertex in the path from f to g [2], [4]. In other words, a Markov Tree is a structure whose edges are such that, if a variable x_i is contained in any two vertices h, $g \subset N$, then there exists one and only one path $p \subset E$ linking h and g, and x_i belongs to all the nodes along p. A Markov Tree T = (N, E), $N \subset 2^{\mathscr{X}}$, $E \subset N \times N$, is said to be a representative of a hypergraph $\mathscr{H} = (S, \mathscr{X})$ if for every hyperedge $s \in S$, there exists a node $n \in N$ such that $s \subset n$.

Let V_m and V_p represent valuations (in a given uncertainty model) on variables m and p respectively. Let $V_m{\downarrow^l}$, $l \subset m$, represent the marginalization of V_m to a referential of smaller cardinality, $V_m{\uparrow^p}$, $m \subset p$, represent the extension of V_m to a referential of larger cardinality, and $V_m \otimes V_p$ represent the combination of V_m and V_p. Let us suppose we want to derive the value of a variable $k \subset \mathscr{X}$ from $\mathscr{K} = \{V_f \,/\, f \subset \mathscr{X}\}$. Let T = (N, E) be a Markov tree representative of hypergraph $\mathscr{H}$, and h be a node containing k. Let f be a node of T with n neighbours, and g_i, $1 \le i \le n - 1$ be neighbours of f not contained in the path between f and h. The local propagation of information process consists on taking h as root, and applying the cycle composed of the following three steps, from the leaf nodes towards h :

1) *Marginalization* : We marginalize each V_{gi} to $V_{gi}{\downarrow^{gi \cap f}}$.
2) *Extension* : Each valuation $V_{gi}{\downarrow^{gi \cap f}}$ is extended to $(V_{gi}{\downarrow^{gi \cap f}}){\uparrow^f}$.
3) *Combination* : V_f on node f, and all the valuations $(V_{gi}{\downarrow^{gi \cap f}}){\uparrow^f}$ are combined altogether.

When node h is finally reached V_h is marginalized to variable k. The inference process consists thus in performing the combination operation on the smallest possible universes and propagating a result from a node to another through the intersection of their universes.

It has been shown [1][2] that, given a hypergraph $\mathscr{H} = (S, \mathscr{X})$, the overall valuation $V_S = V_{n1}{\uparrow^S} \otimes \dots \otimes V_{nk}{\uparrow^S}$, is obtained by performing local computation on a Markov tree representative of hypergraph $\mathscr{H}$, if the projection and combination operators of the subsiding mathematical framework M satisfy the following axioms :

Axiom A1 (*Commutativity and associativity of combination*) :
Suppose V_g, V_h, V_k are valuations on g, h, and k respectively.
Then $V_g \otimes V_h = V_h \otimes V_g$ and $V_g \otimes (V_h \otimes V_k) = (V_g \otimes V_h) \otimes V_k$.
Axiom A2 (*Consonance of marginalization*) :
Suppose V_h is a valuation on h, and suppose $f \subset g \subset h$.
Then $V_h{\downarrow^f} = (V_h{\downarrow^g}){\downarrow^f}$
Axiom A3 (*Distributivity of marginalization over combination*) :
Suppose V_g is a valuation on g and V_h is a valuation on h.
Then $(V_g \otimes V_h){\downarrow^g} = V_g \otimes (V_h{\downarrow^{g \cap h}})$

In other words, if the uncertainty and imprecision pervading a knowledge base are described by a model satisfying the axioms above, then the same result of global propagation of information on $\mathscr{K}$ is obtained by local propagation on a Markov Tree representative of the hypergraph underlying $\mathscr{K}$. In this way, we will eliminate both the loss of marginalization, and the problems due to the lack of idempotence that might arise if the information was propagated blindly, and at the same time reduce the cost of global propagation. This scheme can be used when valuations in $\mathscr{K}$ are modeled by Evidence Theory [2], Possibility Theory [5], or Probability Theory taking the Bayesian approach [4]. Similar schemes to local propagation of information have been proposed in [6] for the framework of Evidence Theory, in [6], [7], [8] for the framework of Possibility Theory or Fuzzy Sets Theory, and in [9], [10] for the Bayesian probabilistic framework.

A Markov tree T = (N, E) is a non-directed graph structure, i.e. each edge $(f, g) \in E$ can be used to propagate information from f to g, or from g to f. However, in the local propagation of information we make use in fact of *directed Markov trees* [4], since in this process we use $T_h = (N, E_h)$, which is the directed tree derived from giving an orientation to T when node h is set as root.

It has been shown in [11] that the derivation of the marginals on all the nodes in T is twice the cost of the derivation of the marginal on h in T_h. It suffices to transform each node in T in a processor that stores, processes and transmit information and we obtain the marginals for all the variables in $\mathcal{K}$. However, most of the time we are interested on the value of only a small set of variables. Moreover, one of the advantages of knowledge-based systems is the flexibility in the acquisition and modification of information ; it is probably useless to store the marginalized values of all variables, since they will have to be recalculated from time to time anyways. Last but not least, in real world applications knowledge bases are usually very large, and the queries may concern only a small part of it. In this case, it may be reasonable to give up the cost for the derivation of the marginals of all variables in the base if in exchange we are able to obtain a smaller cost in the derivation of only our variables of interest. For that we propose to prune directed Markov trees by disregarding the information in the base that is proved to be completely irrelevant in relation to the query.

Part of the information in a directed Markov tree may be discarded, if the uncertainty model underlying $\mathcal{K}$ obeys the following axiom, proposed in an early version of this work [12], and independently in [13] :

Axiom B (*Existence of identity*) : Let V_f denote a valuation on f. Then there exists a valuation V_{fe} on f such that $\forall\ V_f$, $V_f \otimes V_{fe} = V_{fe} \otimes V_f = V_f$. V_{fe} is called the identity valuation on f, or a vacuous valuation on f.

V_{fe} is completely uninformative, or in other words, from V_{fe} all we know about f is that its value is certainly contained in Ω_f. Axiom B is respected in the framework of Evidence Theory : for $f \subset \mathcal{X}$, V_{fe} is represented by the body of evidence $(\mathcal{F}, m)$, with $\mathcal{F} = \{\Omega_f\}$ and $m(\Omega_f) = 1$, where m is a mass allocation function.

Let T = (N, E) be a Markov tree. For each valuation V_g, $g \in N$, we can define the set of vacuous marginalizations that can be obtained from f :

Vacuous Marginalization : $\beta(V_g{}^{\downarrow f})$ denotes the proposition "the marginalization of valuation V_g on f is vacuous", which is true when $V_g{}^{\downarrow f} = V_{fe}$.

Vacuous Marginalization Set : $B(V_g) = \{f \,/\, f \subset g,\ V_g{}^{\downarrow f} = V_{fe}\}$ is the set of variables contained in g, to which the marginalization of V_g yields a vacuous valuation, i.e. $f \in B(V_g) \Leftrightarrow \beta(V_g{}^{\downarrow f})$ is true. Note that $f \in B(V_f) \Leftrightarrow V_f = V_{fe}$, and that $f \in B(V_g) \Leftrightarrow \forall\ h \subset f,\ h \in B(V_g)$. We assume here that $\varnothing \in B(V_f),\ \forall\ f \subset \mathcal{X}$.

Let $T_r = (N, E_r)$ be a directed Markov tree obtained from T as we set a node r in N as root. When we want to propagate the information in a leaf node g to its father node h we will marginalize the valuation in g to $g \cap h$, and then combine it with the information in h. However, if the marginalization is vacuous, i.e. if $g \cap h = f$ such that $V_g{}^{\downarrow g \cap h} = V_g{}^{\downarrow f} = V_{fe}$, then this propagation step as a whole can be considered as useless, since its final result will be the information already in h. Therefore, if the uncertainty model underlying $\mathcal{K}$ obeys axioms A1, A2, A3, and B, then we can prune a leaf node g with father h from T_r if $g \cap h \in B(V_g)$. The successive use of this result on the leaf nodes of T_r will allows us to prune away all the information that is irrelevant to r, and thus reduce the total cost of the process of propagation.

Let $\mathcal{K}$ be a knowledge base on $\mathcal{X} = \{x_i \,/\, 1 \leq i \leq 10\}$, such that $\mathcal{H} = (S, \mathcal{X})$, with $S = \{x_3,$

x_6, x_8, x_1x_2, x_3x_7, x_3x_8, x_6x_9, x_7x_8, x_8x_{10}, $x_1x_3x_4$, $x_1x_3x_7\}$, $B(V_{x8}) = \{x_8\}$ and $\forall f \in S$, $f \neq x_8$, $B(V_f) = 2^f - f$ (the hyperedges in S are here represented by concatenations in order to simplify the notation). Let us suppose that in the creation of a Markov tree T representative of $\mathcal{H}$ nodes x_3x_7, x_3x_8, and x_7x_8 are clashed together forming node $f = x_3x_7x_8$ ($V_f = V_{x3x7} \otimes V_{x3x8} \otimes V_{x7x8}$), with $B(V_f) = \{x_3, x_7\}$. Figure 1.a shows the Markov tree T derived from $\mathcal{K}$ (the symbol β on top of an arrow indicates that the projection in the direction of the arrow is vacuous). Let us now suppose that we want to derive the value of variable x_8. Figure 1.b shows T_{x8}, a directed Markov tree with root x_8 derived from T, already pruned. Nodes x_1x_2 and x_4x_5 can be discarded since they are proved to bring no information in relation to the root node x_8. Note that an internal node propagating a vacuous valuation can only be discarded if all the nodes preceding it also propagate vacuous valuations.

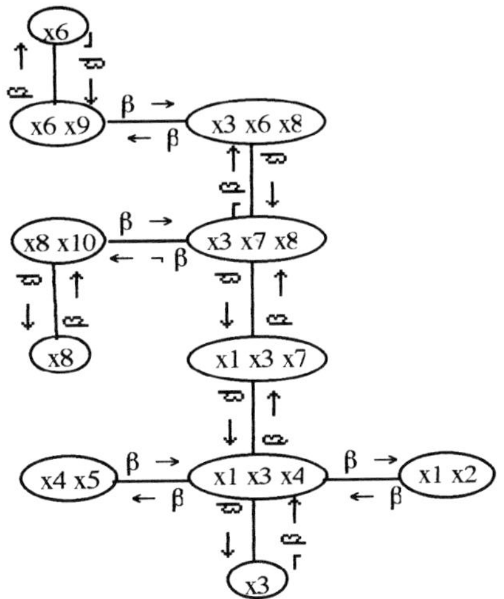

Figure 1.a. Markov tree T

Figure 1.b. Directed Markov tree T_{x8} already pruned.

It is important to note that the vacuous valuations set $B(V_f)$, $f \subset \mathcal{X}$, do not have to be recalculated every time the knowledge base is used. Moreover, sometimes, it does not even have to be explicitly calculated. For instance, let V_f be a valuation derived from the application of the principle of minimum specificity [16] on a rule such as if <x = a> then <y = b> (Pl, Bel), where Pl and Bel are respectively a plausibility and a credibility function defined in the framework of Evidence Theory [14] (see also [17] for details on transformations of rules into valuations). In this case, $B(V_f) = 2^f - f$ if Pl = 1.

3 - PRUNING DIRECTED MARKOV TREES

The choice of a Markov tree can be very important in the derivation of our goal variables. This is made clear when we compare the trees in Figures 1 and 2. In Figure 2.a we depict Markov tree T', also derived from the hypergraph $\mathcal{H}$ underlying $\mathcal{K}$. Figure 2.b shows the directed Markov tree T'_{x8} already pruned. We see that tree T in Figure 1.a differs from tree T' in Figure 2.a only by the different placing of the unary hyperedges. This small difference is however responsible for the large difference in the pruned directed trees shown in Figures 1.b and 2.b. In fact, even before constructing a Markov tree representative of $\mathcal{H}$, we know that these unary hyperedges will represent the nodes in the tree that are known to be informed . This comes from the fact that part of information in knowledge bases can be considered as general, as the case of those derived from rules. Based on this and on the fact that, for a given knowledge base $\mathcal{K}$ with hypergraph $\mathcal{H}$, the result of the application of the local propagation of information is the same on no matter which Markov tree derived from $\mathcal{K}$, we may further reduce the propagation process this process if we separate the general and the specific information in the knowledge base and construct a convenient Markov tree therefrom.

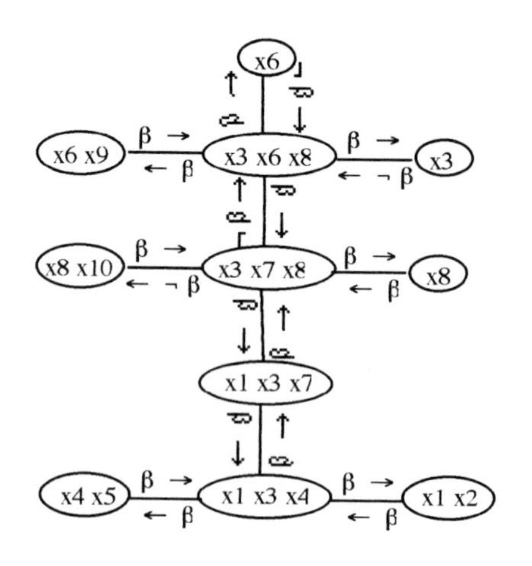

Figure 2.a. Markov tree T'

Figure 2.b. Directed Markov tree T'_{x8} already pruned

Here we propose a strategy to reduce the cost of the local propagation of information process which consists in the following steps :

. a Markov tree T is constructed only for the pieces of knowledge regarding general information, ie those whose hyperedges propagate vacuous valuations,

. a directed Markov tree T_r related to the query variable r is determined,

. the pieces of knowledge regarding specific information, ie those whose hyperedges are informative, are inserted into T_r, the closest possible to r,

. the resulting tree is pruned, and finally the information is propagated.

We divide the process in 6 phases as follows.

Phase 1 : Knowledge Base Separation and Markov tree Construction

We divide $\mathcal{K}$ in two parts : the non-informed knowledge base $\mathcal{NK}$, = $\{V_f \in \mathcal{K} \,/\, |f| > 1,$ $B(V_f) = 2^f - f\}$, and the informed knowledge base $\mathcal{IK} = \mathcal{K} - \mathcal{NK} - \{V_f \in \mathcal{K} \,/\, |f| = 1,$ $f \in B(V_f)\}$. The set $\mathcal{K} - \mathcal{NK} - \mathcal{IK}$ is discarded since it only contains the valuations $\{V_f \in \mathcal{K} \,/\, V_f = V_{fe}\}$

In order to guarantee that the structure that will be derived by our strategy is consistent with the definition of Markov trees we have to impose three restrictions on our pieces of knowledge :

. Let h and g be any two variables in $\mathcal{IK}$. Then $h \cap g = \varnothing$.
Otherwise V_h and V_g are substituted in $\mathcal{IK}$ by $V_{h \cup g} = V_h \otimes V_g$.
. Let h and g be any two variables in $\mathcal{NK}$. Then $h \not\subset g$.
Otherwise V_f and V_g are substituted in $\mathcal{NK}$ by $V_{h \cup g} = V_h \otimes V_g$.
. For every valuation V_g in $\mathcal{IK}$, there must exist a valuation V_h in $\mathcal{NK}$, such that $g \subset h$.

The second step of this phase consists in the construction of a Markov tree T = (N, E) for the hypergraph underlying $\mathcal{NK}$ using any cover algorithm presented in the literature (see [7], [18], [19], [20], [21]).

Phase 2 : Determination of vacuous edges

In phase 2 we take each edge e = (f, g) ∈ E, and label both its directions with either $\beta(V_f^{\downarrow f \cap g})$ or $\neg\beta(V_f^{\downarrow f \cap g})$, i.e. we verify the vacuous marginalizations on T. Note that the only edges labeled with $\neg\beta$ part from nodes that did not exist originally in $\mathcal{NK}$. These new nodes are created by the clash of nodes in $\mathcal{NK}$, during the construction of T .

Phase 3 : Markov Tree Partitioning

Let l(h, g) be the number of nodes lying in the path between h and g in T, and $l(h, G) = \inf_{g \in G} l(h, g)$ be the size of the shortest path from f to any node of G. Let x be the goal variable. Let then $R = \{r \in N / x \in r\}$ be the set of nodes in T containing x. We partition the set of nodes N - R by the shortest distance between each node $h \in N - R$ and a node in R. Formally, partition V_c is composed of classes $c_i = \{h \in N - R / l(h, R) = i\}$, $1 \leq i \leq |N - R|$. Similarly, we partition set R by the shortest distance between each node $f \in R$ and a node in N - R. Partition V_d is composed of the classes $d_i = \{f \in R / (h, N - R) = i\}$, $1 \leq i \leq | R |$. Class d_1 is called the roots class.

In this phase we generate the largest part of a directed Markov tree $T_r = (N, E_r)$; only the root $r \in d_1$ is left undefined for the moment. We use partitions V_1 and V_2 to give an orientation to all the edges of T except those in level d_1 : we create an edge $(f, g) \in E_r$, for each node f in c_i, $2 \leq i \leq |N - R|$, and a node g in c_{i-1}, such that (f, g) ∈ E. In the same way, we create the edges pointing from nodes in c_1 to nodes in d_1, and from nodes in d_i, $2 \leq i \leq | R |$, to nodes in d_{i-1}.

Phase 4 : Determination of indispensable nodes

When we propagate information from f to its father-node g, we first of all marginalize the valuation in f to $f \cap g$. If edge (f, g) has label $\neg\beta(V_f^{\downarrow f \cap g})$, the information in f cannot be discarded, because it is possibly informative. These nodes are therefore indispensable in the propagation process.

A node in N - R is indispensable if it is either informative or lay between an informative node and a node in the roots level d_1. In phase 4 we label the nodes in classes c_i, $i \geq 2$ as dispensable or indispensable . We start this process from the leaves in N - R, and progress up in the tree until we reach the nodes in class c_1. At this point, all the nodes which do not contain the goal variable x have been visited.

We now verify which nodes in R are indispensable. First of all, we mark as indispensable all the nodes h in R such that V_h has a non-vacuous marginalization in the direction of at least one neighbour node in R. Then, starting from the leaves until we reach d_2, we mark as indispensable all the nodes that are informative or that lay between an informative node and a node in d_1. Then every node in d_1 that has an indipensable node as neighbour on level c_1 or in level d_2 is marked as indispensable.

Finally, every node n in d_1 laying between two indipensables nodes is marked as indispensable. We denote by N_I the set of the indispensable nodes in T ; these are the nodes that must be taken into account in the propagation process, independently of the information stored in $\mathcal{IK}$.

Phase 5 : Insertion of informed nodes
Any variable in base $\mathcal{IK}$ is informative, and becomes therefore an indispensable node as it is incorporated in tree T. Moreover, it makes indispensable all nodes between it and the root. We should thus attach each valuation in $\mathcal{IK}$ to an already indispensable node in T, or else to a dispensable node the closest possible to the root.

Let $Y(f) = \{g / g \in N, f \subset g\}$ be the set of nodes in T containing f and let $i = \inf_{g \in Y(f)} l(g, d_1)$, be the shortest distance between Y(f) and the roots class. First of all, we try to attach f to an indispensable node in T. However, if there is no node h in N_I, such that $Y(f) \cap N_I = \varnothing$, we attach f to any node g in $N - N_I$ the closest possible to the roots class. That is, we find the lowest index i such that $Y(f) \cap c_i$ and the lowest index j such that $Y(f) \cap d_j$, and attach f a node g contained either in c_i if $i < j$, or in d_j otherwise. We then mark f and g as indispensable, as well as all the nodes in the path from g to a node on d_1.

Phase 6 : Propagation of information
We now finish characterizing our Markov Tree T_r, by choosing a node r in $d_1 \cap N_I$ as root of our directed Markov tree, and orienting the nodes in d_1 in the direction of r. We then prune T_r by eliminating all the nodes that are not marked as indispensable, i.e. we make $N = N_I$.

After applying the strategy presented above, we propagate the information in the resulting pruned directed Markov tree, starting from the leaf nodes until node r. If r does not represent the goal variable itself, we marginalize V_r to that variable.

This strategy thus constructs a Markov tree for the non-informed hyperedges of $\mathcal{K}$ (those in $\mathcal{NK}$), transforms it in a directed Markov tree, and then inserts the informed hyperedges of $\mathcal{K}$ (those in $\mathcal{IK}$). Since most of the time, the informed nodes in a knowledge base concerns the specific knowledge related to a query, the sectioning of the base in a non-informed part and informed one is equivalent to the division existing in rule-based systems, where the general pieces of information are contained in the rule-base and the query specific ones in the fact-base. In rule-based systems, only one part of the rule-base is explored, namely that concerned with the query. Here, we may also not explore (in the sense of combining) the whole knowledge base, but we make sure that the information discarded would really not interfere with the final result.

4. EXAMPLE

The following example is based on the one presented in [21], which is itself inspired from Lewis Carroll works. It deals on the characteristics that a person should have if he ever wanted to be a member of parliament. Table 1 below brings a knowledge base with the pieces of knowledge not yet characterized in any mathematical model. In Table 2 we describe the variables that can be used to model the pieces of knowledge in Table 1.

Let us suppose that the pieces of information in Table 1 have been transformed in valuations inside in the framework of evidence theory, using the variables described in Table 2. Table 3 brings a knowledge base $\mathcal{K}$ derived that way. Each valuation V_f is a body of evidence $(\mathcal{F}, m)$, represented by the set $\{(g, m(g)) / g \subset f, m(g) > 0\}$.

The hypergraph underlying $\mathcal{K}$ is $\mathcal{H} = (\{H_i, 0 \leq i \leq 12\}, \mathcal{X})$. In Phase 1 the set of empty marginalizations corresponding to each valuation are determined, which yields $B(V_{H0}) = \{x_8\}$, and $B(V_{Hi}) = 2^{Hi} - H_i$, $1 \leq i \leq 10$. Knowledge base $\mathcal{K}$ is then divided into the informed base $\mathcal{IK} = \{V_{H11}, V_{H12}\}$, containing specific information about the query, and the non-informed base $\mathcal{NK} = \{V_{Hi}, 0 \leq i \leq 10\}$, containing general information (V_{H0} is here included in $\mathcal{NK}$ only to simplify the visualization of the process in the figures). From the hypergraph associated to $\mathcal{NK}$ we construct a Markov tree $T = (N, E)$, with $N = \{H_i, 4 \leq i \leq 10\} \cup \{H_0, H_{13}\}$, and $V_{H13} = V_{H1} \otimes V_{H2} \otimes V_{H3}$ described by :

$V_{x3x7x8} = V_{H13} = \{(\{(S_c, F_{ge}, F_{ms}), (M_c, T_{ge}, T_{ms})\}, .36)$
$(\{(M_c, T_{ge}, T_{ms}), (M_c, T_{ge}, F_{ms}), (S_c, T_{ge}, F_{ms}), (S_c, F_{ge}, F_{ms})\}, .24)$
$(\{(M_c, T_{ge}, T_{ms}), (S_c, T_{ge}, T_{ms}), (S_c, F_{ge}, T_{ms}), (S_c, F_{ge}, F_{ms})\}, .24)$
$(\{T_{ge}\} + \{S_c\}, .16)\}$

with the empty marginalizations set $B(V_{H13}) = \{\{x_3\},\{x_7\}\}$.

Table 1
Informal Knowledge Base modeling a logical puzzle

Pieces of Knowledge	
P0 :	We do not know which are the chances of Mr Whatshisname becoming a M.P.
P1 :	A person who is not capable of expressing himself well has little chances of becoming a M.P.
P2 :	Sometimes a person who does not make speaches is so because he cannot express himself well.
P3 :	A person who is always making speaches has some chances of becoming a M.P.
P4 :	A person who possesses a clear mind and who has received a good education expresses himself well.
P5 :	Persons who have complicated minds usually do not know what they want.
P6 :	A person who has received a good education has usually good manners.
P7 :	A person who is always making speaches and who expresses himself well has most probably had a good education.
P8 :	A person who is not rich does not smoke cigars.
P9 :	A person who has very good chances of becoming a M.P. has often a minimum of influence.
P10 :	A rich person who expresses himself well has good chances of becoming a M.P.
P11 :	Mr Whatshisname has a very good expression.
P12 :	Mr Whatshisname is rich.

Table 2
Variables modeling concepts on a logical puzzle

Variable		Reference Domain
x_1 :	education	{none (N_e), medium (M_e), good (G_e)}
x_2 :	manners	{none (N_m), medium (M_m), good (G_m)}
x_3 :	good_expression	{true (T_{ge}), false (F_{ge})}
x_4 :	clear_mind	{true (T_{cm}), false (F_{cm})}
x_5 :	decided	{true (T_d), false (F_d)}
x_6 :	fortune	{small (S_f), medium (M_f), great (G_f)}
x_7 :	make_speaches	{true (T_{ms}), false (F_{ms})}
x_8 :	chances _to_become_MP	{small (S_c), medium (M_c), great (G_c)}
x_9 :	smoke	{nothing (N_s), cigars (CI_s), cigarettes (CT_s)}
x_{10} :	has_influence	{true (T_{hi}), false (F_{hi})}

Table 3
Formal Knowledge Base $\mathcal{K}$ modeling a logical puzzle

Valuations		
V_{x8} =	V_{H0} =	$\{((\Omega_{x8}, 1.)\}$
V_{x3x8} =	V_{H1} =	$\{(\{T_{ge}\} + \{S_c\}, .1)\}$
V_{x3x7} =	V_{H2} =	$\{(\{T_{ms}\} + \{F_{ge}\}, .6), (\Omega_{x3} \times \Omega_{x7}, .3)\}$
V_{x7x8} =	V_{H3} =	$\{(\{F_{ms}\} + \{M_c\}, .6), (\Omega_{x7} \times \Omega_{x8}, .4)\}$
V_{x1x3x4} =	V_{H4} =	$\{(\{F_{cm}\} + \{N_e, M_e\} + \{T_{ge}\}, 1)\}$
V_{x4x5} =	V_{H5} =	$\{(\{T_{cm}\} + \{F_d\}, .7), (\Omega_{x4} \times \Omega_{x5}, .3)\}$
V_{x1x2} =	V_{H6} =	$\{(\{N_e, M_e\} + \{G_m\}, .8), (\Omega_{x1} \times \Omega_{x2}, .2)\}$
V_{x1x3x7} =	V_{H7} =	$\{(\{F_{ms}\} + \{F_{ge}\} + \{G_e\}, .85), (\Omega_{x1} \times \Omega_{x3} \times \Omega_{x7}, .15)\}$
V_{x6x9} =	V_{H8} =	$\{(\{M_f, G_f\} + \{N_s, CT_s\}, 1)\}$
V_{x8x10} =	V_{H9} =	$\{(\{S_c, M_c\} + \{T_{hi}\}, .9), (\Omega_{x8} \times \Omega_{x10}, .1)\}$
V_{x3x6x8} =	V_{H10} =	$\{(\{S_f, Mf\} + \{F_{ge}\} + \{M_c, G_c\}, 1.)\}$
V_{x3} =	V_{H11} =	$\{(\{T_{ge}\}, .8), (\Omega_{x3}, .2)\}$
V_{x6} =	V_{H12} =	$\{(\{G_f\}, 1.)\}$

In Phase 2 we label the edges in T ; this corresponds to the trees shown in Figures 1.a and 2.a. with the exclusion of the unary nodes from them. In Phase 3 we partition the nodes in tree T according to the distance to the nodes containing variable x_8. We have $c_1 = \{H_7, H_8\}$, $c_2 = \{H_4\}$, $c_3 = \{H_5, H_6\}$, and $d_1 = \{H_0, H_9, H_{10}, H_{13}\}$. In Phase 4 we mark node V_{H13} in the tree, since it is the only one which is indispensable in the propagation process even before the addition of the specific information fom $\mathcal{IK}$. In Phase 5 we insert nodes V_{H11} as son of node V_{H13} and node V_{H12} as the son of node V_{H10} in the tree and mark node V_{H10} as indispensable. In Phase 6 we choose node $V_{H0} = V_{x8}$ as the root of the tree and propagate the information through the indispensable nodes from the lowest levels up to V_{H0} (see Figure 2b). We obtain valuation $V_{x8} = \{(\{M_c\}, .78), (\Omega_{x8}, .22)\}$ as the final result, which corresponds to the assertion "Mr Whatshisname has good chances ($\{M_c, G_c\}$) of becoming a MP with plausibility Pl = 1 and Bel = .78". Note that this result takes all the information in $\mathcal{K}$ into account, but the propagation process in itself is performed only on a subset of $\mathcal{K}$. In other words, using the strategy proposed here, we have been able to propagate the information in a reduced structure, and at the same time guaranteeing the consistance of the results. It is important to note that a good part of the work done in this specific case can be used with other specific pieces of information and with other goal nodes. Indeed, if the information in $\mathcal{NK}$ is not modified, there is no need to determine a new Markov tree or to recalculate the sets of empty valuations.

5. CONCLUSION

We presented a strategy for reducing the cost of the process of local propagation of information. This approach is based in the idea that the use of oriented structures allows us to disregard information that is proved to be irrelevant to a given query. It fits any uncertainty theory where the local propagation of information scheme can be implemented, and in which it can be proved that for any hyperedge h, there exists a vacuous valuation V_{fe} that combined to any other valuation V_f on f yields valuation V_f itself. This last requirement means that information in V_{fe} is irrelevant and can thus be discarded. The strategy presented here, based on the use of directed Markov trees, draws together the local propagation of information scheme, to the inference scheme used in rule-based systems. It guarantees that the results are correct and complete in relation to the uncertainty model adopted, examines the whole information contained in the base, without nevertheless requiring the combination of all the pieces of knowledge.

The scheme showed here can be profitably used when we implement an important characteristic of directed Markov Trees, namely the extension of its representation in such a way as to obtain a close expressibility power as that of rule-based systems. Information such as that modeled by rules of the kind if <x is a> then <y is f(a)>, $f : \Omega_X \rightarrow \Omega_Y$, can be incorporated in the local propagation of information scheme through the use a of relation on xy, if node xy is non-informed and if we can find a directed Markov tree such that the information is propagated from xy through y. In the same way, we can incorporate the capability of dealing with <attribute, object, value> triples modeled by general rules of the kind if <x(ob) is a> then <y(ob) is b>, which means that if attribute x has value a for object ob, then we assign the value b for attribute y on the same object.

REFERENCES

1. Shafer G. and Shenoy P.P., "Propagating Belief Functions with Local Computation", *IEEE Expert*, 1(3), pp 43-51, 1986.
2. Shafer G. and Shenoy P.P., *Local Computation in Hypertrees*, Working Paper No. 201, School of Business, The University of Kansas, Lawrence, 1988.
3. Shafer G. and Logan R., "Implementing Dempster's rule for Hierarchical Evidence", *Artificial Intelligence*, 33, pp 271-298, 1987.
4. Shafer G. and Shenoy P.P., *Bayesian and Belief-Function Propagation*, Working Paper No. 192, School of Business, The University of Kansas, Lawrence, 1988.
5. Dubois D., Prade H., "Inference in Possibilistic Hypergraphs", *Proc. of the 3rd Int. Conf. in Information Processing and Management of Uncertainty in Knowledge-Based Systems (IPMU '90)*, Paris, pp 228-230, 1990.
6. Chatalic P., Dubois D., Prade H., "A System for Handling Relational Dependencies in Approximate Reasoning", *Proc. of the 3rd Int. Expert Systems Conference*, London, 1987.
7. Kruse R. and Schwecke E., "Fuzzy Reasoning in a Multidimensional Space of Hypotheses", *Int. J. of Approximate Reasoning*, 4, pp 47-68, 1990.
8. Fonck P., "Building Influence Networks in the Framework of Possibility Theory", *Proc. of the 1st DRUMS RP2 WORKSHOP, Basic Research Action Esprit Project DRUMS*, Albi, France, pp 263-270, 1990.
9. Pearl J., *Fusion, Propagation, and Structuring in Bayesian Networks*, Tech. Rep. No. CSD-850022, Computer Science Department, University of California, Los Angeles, 1985.
10. Lauritzen S.L. and Spiegelhalter D.J., "Local Computations with Probabilities on Graphical Structures and their Application to Expert Systems", *J. R. Statist. Soc. B, 50(2)*, pp 157-224, 1988.
11. Shenoy P.P. and Shafer G., "Axioms for Probability and Belief-function Propagation", In *Uncertainty in Artificial Intelligence 4*, (Schachter R.D., Levitt T.S., Kanal L.N., Lemmer J.F., eds), Elsevier Science Publishers B.V. (North Holland), 1990.
12. Sandri S.A., "Local Propagation of Information on Directed Markov Trees", *DRUMS RP2 WORKSHOP, Basic Research Action Esprit Project DRUMS*, Blanes, Spain, 1991.
13. Cano J.E., Delgado M. & Moral S., "Propagation of Uncertainty in Dependence Graphs, Symbolic and Quantitative Approaches to Uncertainty", *Proc. European Conf. ECSQAU*, Marseille, 1991.
14. Shafer G., *A Mathematical Theory of Evidence*, Princeton University Press, 1976.
15. Dubois D., Prade H. (with the collaboration of Farreny H., Martin-Clouare R., Testemale C.), *Possibility Theory : An Approach to the Computerized Processing of Uncertainty*, Plenum Press, 1988.
16. Dubois D., Prade H., "The Principle of Minimum Specificity as a Basis for Evidentional Reasoning". In *Uncertainty in Knowledge-Based Systems*, (B. Bouchon, R.R. Yager, eds), Springer Verlag, 1986.

17. Chatalic P., Dubois D., Prade H., "An Approach to Approximate Reasoning Based on the Dempster Rule of Combination", *Int. J. of Expert Systems*, 1(1), pp 67-85, 1987.
18. Lianwen Z., *Studies on Finding Hypertree Covers for Hypergraphs*, Working Paper No. 198, School of Business, The University of Kansas, Lawrence, 1988.
19. Mellouli K., *On the Propagation of Belief in Networks Using the Dempster-Shafer Theory of Evidence*, Working Paper No.196, School of Business, The University of Kansas, Lawrence, 1988.
20. Kong A., *Multivariate Belief Functions and Graphical Models*, Doctoral Dissertation, Department of Statistics, Harvard University, 1986.
21. Chatalic P., *Raisonnement Déductif en Présence de Connaissances Imprécises et Incertaines : Un Système Basé sur la Théorie de Dempster-Shafer*, Doctoral Thesis, Université Paul Sabatier, Toulouse, France, 1986.

Uncertainty in Intelligent Systems
B. Bouchon-Meunier et al. (Editors)
1993 Elsevier Science Publishers B.V.

Approximations of probability distributions by three types of graph models

R. Jiroušek[a] [1]

[a]Department of Decision–Making Theory, Institute of Information Theory and Automation, Academy of Science of the Czech Republic, P.O. Box 18, 182 08 Praha 8, Czech Republic

Abstract

Graph modelling is a modern branch of probability theory in which a special interest is paid to classes of distributions advantageous from several reason. Dependence structure of these distributions usually makes possible to represent the distribution as a product of some functions which allows to process even high-dimensional distributions with the help of comparatively small number of parameters. Also computational problems are substantially reduced. The other advantage concerns the fact that dependence structures of such distributions are easier to explain in a human-adopted way, which can be used to advantage when designing explanatory modules determined for users without any probabilistic education.

Unfortunately, one can hardly expect that a distribution describing dependencies in the area of his/her interest belongs into one of those classes. Nevertheless, it is usually possible to modify it slightly (to approximate it) so that the resulting distribution still describes required relationships with a sufficient accuracy. The present paper brings a theoretical background for designing algorithms constructing approximations of distributions in three classes of graph models.

1. INTRODUCTION

Main problems connected with application of probabilistic methods in AI are those arising from high computational complexity of algorithms. Problems of practical importance are of high dimensionality (tens, hundreds or even thousands of random variables) which brings necessity to cope with a question how to handle such multidimensional probability distributions. The answer lies generally in considering only a class of distributions convenient from some point of view. Rather often log-linear, graphical or decomposable models are used. A common feature of these models is the fact that the respective distributions have dependence structures of convenient properties which make possible to reconstruct the distributions from a reasonable number of parameters (probabilities).

Several types of graph representation of dependence structures are commonly used. Let us mention mainly the representation by unoriented graphs [2, 4, 8–11], by directed acyclic graphs [7, 10, 11] and by decision trees [5]. It is not only the efficiency of computational

[1]The research was partly financially supported by the grant ČSAV no. 27 510.

algorithms what makes these approaches important. A graph representation, if simple enough, is probably the best way of presenting complex information to users, experts or even to authors (when debugging the system). It is reasonable to take full advantage of graph representation of dependence structures also when designing explanatory modules of probabilistic AI systems [4].

In practical situations neither the probability distribution nor its dependence structure is known exactly. Usually, knowledge is expressed in the form of a system of pieces of partial knowledge, or, in the form of a file of statistical data (or both). Choice of a distribution for a probabilistic model is therefore connected with a certain level of freedom. In fact, usually the whole set of equally acceptable distributions comes into consideration. Therefore, it makes good sense to look for a distribution meeting the required properties and simultaneously possessing the simplest dependence structure. In the present paper, the problem will be studied in a more specific version, which appears, for example, when designing explanatory modules: given a probability distribution, find its approximation with the simplest possible graph representation.

The problem is solved for three types of graph representation:

- graphical models,
- decomposable models,
- decision trees models.

All these approaches are based on the idea that a respective graph (unoriented, triangulated or a decision tree, respectively) uniquely describes the specific dependence structure of a class of probability distributions.

2. KULLBACK–LEIBLER DIVERGENCE

In the sequel, the following notation will be used. $X_1, \ldots, X_N$ will denote discrete (finite-valued) random variables. For a probability distributon $P(X_1, \ldots, X_N)$ and $C \subset V = \{1, \ldots, N\}$, P^C will denote the marginal distribution of P for variables $\{X_i\}_{i \in C}$ only. Similarly, let $\mathbb{X}^V$ denote the joint sample space (i.e. $(x_1, \ldots, x_N) \in \mathbb{X}^V$ is an N-tuple of values of all variables $\{X_i\}_{i \in V}$), then $\mathbb{X}^C$ (for $C \subset V$) denote the respective subspace corresponding to variables $\{X_i\}_{i \in C}$.

In this paper, we are interested in an approximation of a probability distribution P by a probability distribution Q with a simpler dependence structure. The "distance" between P and its approximation Q will be measured with the help of the *Kullback-Leibler divergence*

$$I(P||Q) = \sum_{(x_1, \ldots, x_N) \in \mathbb{X}^V} P(x_1, \ldots, x_N) \log \frac{P(x_1, \ldots, x_N)}{Q(x_1, \ldots, x_N)} \tag{1}$$

which is always nonnegative, turns to 0 iff $Q = P$, and to ∞ iff Q does not *dominate* P (i.e. there exists $(x_1, \ldots, x_N) \in \mathbb{X}^V$ for which $Q(x_1, \ldots, x_N) = 0$ and $P(x_1, \ldots, x_N) > 0$).

Having a class of "potential approximations", i.e. a class of distributions of required dependence structures, the goal is to choose that one which is the closest to the distribution being approximated. For this purpose, the following Lemma is very useful.

Lemma: If $I(P\|Q)+I(Q\|R)<\infty$ then $I(P\|Q)+I(Q\|R)=I(P\|R)$ iff

$$I(Q\|R)=\sum_{(x_1,\dots,x_N)\in\mathbb{X}^V} P(x_1,\dots,x_N)\log\frac{Q(x_1,\dots,x_N)}{R(x_1,\dots,x_N)}. \tag{2}$$

Proof.

$$\begin{aligned}
I(P\|Q)+I(Q\|R) &= \sum_{(x_1,\dots,x_N)\in\mathbb{X}^V} P(x_1,\dots,x_N)\log\left[\frac{P(x_1,\dots,x_N)}{R(x_1,\dots,x_N)}\cdot\frac{R(x_1,\dots,x_N)}{Q(x_1,\dots,x_N)}\right]+\\
&\quad+\sum_{(x_1,\dots,x_N)\in\mathbb{X}^V} Q(x_1,\dots,x_N)\log\frac{Q(x_1,\dots,x_N)}{R(x_1,\dots,x_N)}=\\
&= I(P\|R)-\sum_{(x_1,\dots,x_N)\in\mathbb{X}^V} P(x_1,\dots,x_N)\log\frac{Q(x_1,\dots,x_N)}{R(x_1,\dots,x_N)}+\\
&\quad+\sum_{(x_1,\dots,x_N)\in\mathbb{X}} Q(x_1,\dots,x_N)\log\frac{Q(x_1,\dots,x_N)}{R(x_1,\dots,x_N)}.
\end{aligned}$$

The extension of the ratio $[P(x_1,\dots,x_N)/Q(x_1,\dots,x_N)]$ by $[R(x_1,\dots,x_N)/R(x_1,\dots,x_N)]$ is possible because the assumption implies

$$R(x_1,\dots,x_N)=0\Rightarrow P(x_1,\dots,x_N)=Q(x_1,\dots,x_N)=0 \tag{3}$$

□

3. I - PROJECTION

In [1] Cziszár introduced the concept of an I-projection. Though we interchange the arguments in his definition we shall use his terminology.

If P is a probability distribution and Π a family of them over the same sample space then *I-projection* of P into Π is a distribution $Q\in\Pi$ for which

$$I(P\|Q)=\min_{R\in\Pi} I(P\|R). \tag{4}$$

Recall that, because of the convexity of the Kullback-Leibler divergence, if Π is a convex set with at least one distribution dominating P then there exists a unique I-projection of P into Π.

To characterize I-projections in a class of graphical models the concept of an extension of a set of marginal distributions will be useful. Consider a system $(C_1,\dots,C_K)$ of subsets of the index-set $\{1,\ldots,N\}: \ C_k \subset V = \{1,\ldots,N\}, k = 1,\ldots,K$. Let $P_k \ \ (k = 1,\ldots,K)$ be probability distributions over $\mathbb{X}^{C_k}$. Define a family of distributions

$$\Pi = \{Q : Q^{C_k} = P_k \ \text{ for all } \ k = 1,\ldots,K\}. \tag{5}$$

Any distribution $Q \in \Pi$ is called *an extension* of the system $\{P_1,\ldots,P_K\}$. The distribution $\hat{Q}$ maximizing Shannon entropy in Π is called the *maximum entropy extension* of the given system of marginal distributions [2]:

$$\hat{Q} = \arg \max_{Q\in\Pi} \left(- \sum_{(x_1,\ldots,x_N)\in\mathbb{X}^V} Q(x_1,\ldots,x_N) \log Q(x_1,\ldots,x_N) \right). \tag{6}$$

4. GRAPHICAL MODELS

Consider a simple (i.e. unoriented, without loops and multiple edges) graph $G = (V,E)$. Let $C_1,\ldots,C_K$ be the system of its *cliques* (maximal complete subgraphs). A probability distribution $Q((X_i)_{i\in V})$ is said to *factorize* [2, 12] with respect to G if there exist functions $\psi_1,\ldots,\psi_K$ defined on $\mathbb{X}^{C_1},\ldots,\mathbb{X}^{C_K}$ respectively such that

$$Q(X_1 = x_1,\ldots,X_N = x_N) = \prod_{k=1}^{K} \psi_k((X_i = x_i)_{i\in C_k}). \tag{7}$$

For any graph $G = (V,E)$, let us denote the set of all probability distributions on $\mathbb{X}^V$ factorizing with respect to G by Π_G.

Theorem 1: Let P be a probability distribution defined for $(X_i)_{i\in V}$ and $G = (V,E)$ be a graph with cliques $C_1,\ldots,C_K$. If $\hat{P}$ is the maximum entropy extension of the system of marginal distributions $(P^{C_1},\ldots,P^{C_K})$ then $\hat{P} \in \Pi_G$ and for any distributon $Q \in \Pi_G$

$$I(P||Q) = I(P||\hat{P}) + I(\hat{P}||Q). \tag{8}$$

and therefore $\hat{P}$ is an I-projection of P into Π_G.

Proof. To make the notation more transparent, denote $C_k = \{k_1,\ldots,k_{n(k)}\}$ for $k = 1,\ldots,K$. It was proven by Czisz'ar [1] that the maximum entropy extension, which can be got by the *Iterative Proportional Fitting Procedure*, of any system of distributions $R^{C_1},\ldots,R^{C_K}$ belongs to Π_G. Therefore, denoting $\hat{P}$ the maximum entropy extension of $(P^{C_1},\ldots,P^{C_K})$, there exist functions φ_k defined on $\mathbb{X}^{C_k}$ (for $k = 1,\ldots,K$) such that

$$\hat{P}(X_1 = x_1,\ldots,X_N = x_N) = \prod_{k=1}^{K} \varphi_k(x_{k_1},\ldots,x_{k_{n(k)}}). \tag{9}$$

The assumption $Q \in \Pi_G$ guarantees the existence of functions ψ_k defined on $\mathbb{X}^{C_k}$ for which

$$Q(X_1 = x_1,\ldots,X_N = x_N) = \prod_{k=1}^{K} \psi_k(x_{k_1},\ldots,x_{k_{n(k)}}). \tag{10}$$

Compute $I(\hat{P}\|Q)$:

$$
\begin{aligned}
I(\hat{P}\|Q) &= \sum_{(x_1,\ldots,x_N)\in\mathbb{X}^V} \hat{P}(x_1,\ldots,x_N)\log\frac{\hat{P}(x_1,\ldots,x_N)}{Q(x_1,\ldots,x_N)} = \\
&= \sum_{(x_1,\ldots,x_N)\in\mathbb{X}^V} \hat{P}(x_1,\ldots,x_N)\sum_{k=1}^{K}\log\frac{\varphi_k(x_{k_1},\ldots,x_{k_{n(k)}})}{\psi_k(x_{k_1},\ldots,x_{k_{n(k)}})} = \\
&= \sum_{k=1}^{K}\sum_{(x_{k_1},\ldots,x_{k_{n(k)}})\in\mathbb{X}^{C_k}} \hat{P}(x_{k_1},\ldots,x_{k_{n(k)}})\log\frac{\varphi_k(x_{k_1},\ldots,x_{k_{n(k)}})}{\psi_k(x_{k_1},\ldots,x_{k_{n(k)}})} = \\
&= \sum_{k=1}^{K}\sum_{(x_{k_1},\ldots,x_{k_{n(k)}})\in\mathbb{X}^{C_k}} P(x_{k_1},\ldots,x_{k_{n(k)}})\log\frac{\varphi_k(x_{k_1},\ldots,x_{k_{n(k)}})}{\psi_k(x_{k_1},\ldots,x_{k_{n(k)}})} = \\
&= \sum_{k=1}^{K}\sum_{(x_1,\ldots,x_N)\in\mathbb{X}^V} P(x_1,\ldots,x_N)\log\frac{\varphi_k(x_{k_1},\ldots,x_{k_{n(k)}})}{\psi_k(x_{k_1},\ldots,x_{k_{n(k)}})} = \\
&= \sum_{(x_1,\ldots,x_N)\in\mathbb{X}^V} P(x_1,\ldots,x_N)\log\frac{\hat{P}(x_1,\ldots,x_N)}{Q(x_1,\ldots,x_N)}.
\end{aligned}
$$

Replacement of $\hat{P}(x_{k_1,\ldots,x_{k_{n(k)}}})$ by $P(x_{k_1,\ldots,x_{k_{n(k)}}})$ is possible because $\hat{P}$ is an extension of $(P^{C_1},\ldots,P^{C_k})$ and therefore $P^{C_k} = \hat{P}^{C_k}$ for all $k = 1,\ldots,K$.

Regarding the previous Lemma, the proof is finished as we have proven that

$$
I(\hat{P}\|Q) = \sum_{(x_1,\ldots,x_N)\in\mathbb{X}^V} P(x_1,\ldots,x_N)\log\frac{\hat{P}(x_1,\ldots,x_N)}{Q(x_1,\ldots,x_N)}. \tag{11}
$$

□

Notice the importance of Theorem 1. It reads that for any distribution P and any graph $G = (V, E)$ there always exists an I-projection of P into the class Π_G of distributions factorizing with respect to G. The existence of this I-projection follows from the fact that there must exist a maximum entropy extension of $(P^{C_1},\ldots,P^{C_K})$ because P belongs among extensions of this system and therefore the set of these extensions is nonempty. Moreover, it provides (at least theoretically) the guideline how to construct this I-projection. One can use the *Iterative Proportional Fitting Procedure* or *Lagrange Multipliers Method* [2, 3] . Nevertheless, both these procedures are of high algorithmical complexity and therefore they can be used for low dimensions only. The situation differs substantially when the graph $G = (V, E)$ in question is triangulated.

5. DECOMPOSABLE MODELS

The distribution $P((X_i)_{i\in V})$ is said to be *decomposable* if it is factorizes with respect to a graph $G = (V, E)$ that is *triangulated* (i.e. with no chordless cycle of length greater than 3). For the system of cliques of a triangulated graph it is known that they can be ordered to meet the *running intersection property* [2, 6]:

$$\forall\, k = 2, \ldots, K \;\; \exists i,\; 1 \le i < k \;\; (C_k \cap (C_1 \cup \ldots \cup C_{k-1}) \subset C_i). \tag{12}$$

As decomposable distributions form a subclass of distributions factorizing with respect to graphs, Theorem 1 holds for them, too. In this case, however, it is simple to compute all values $I(P\|Q)$, $I(P\|\hat{P})$ and $I(\hat{P}\|Q)$ appearing in the statement of Theorem 1. Moreover, the distribution $\hat{P}$ can be computed without using algorithmically complex procedures like the Iterative Proportional Fitting. The respective well-known direct formula, presented in the following assertion, can be found (along with its proof) e.g. in [6].

Theorem 2: Let P and G be as in Theorem 1. If the graph G is triangulated and the ordering $C_1, \ldots, C_K$ of its cliques meets the running intersection property than the I-projection $\hat{P}$ of P into Π_G can be computed according to the following formula

$$\hat{P} = P^{C_1} \prod_{k=2}^{K} P^{C_k|C_k \cap (C_1 \cup \ldots \cup C_{k-1})}, \tag{13}$$

where $P^{C|D}$ denotes the conditional probability of variables $\{X_i\}_{i \in C-D}$ given variables $\{X_i\}_{i \in D}$ (for $C \subseteq D$, $P^{C|D} = 1$.

Corollary. Let P and G be a probability distribution and a triangulated graph as in Theorem 2. If $(C_1, \ldots, C_K)$ is a sequence (of all the cliques of G) meeting the running intersection property then

$$\min_{Q \in \Pi_G} I(P\|Q) = \sum_{k=2}^{K} I(P^{C_k|C_1 \cup \ldots \cup C_{k-1}} \| P^{C_k|C_k \cap (C_1 \cup \ldots \cup C_{k-1})}). \tag{14}$$

Remark. The symbol $I(Q^{C|D}\|Q^{C|B})$ for $B \subset D$ should be understood as follows: denote $D = \{X_{i_1}, \ldots, X_{i_n}\}$, $B = \{X_{i_1}, \ldots, X_{i_m}\}$ $(m < n)$, and $C - D = \{X_{j_1}, \ldots, X_{j_l}\}$, then

$$\begin{aligned} I(Q^{C|D}\|Q^{C|B}) &= \sum_{(x_{i_1}, \ldots, x_{i_n})} Q(x_{i_1}, \ldots, x_{i_n}) I(Q^{C|(x_{i_1}, \ldots, x_{i_n})} \| Q^{C|(x_{i_1}, \ldots, x_{i_m})}) = \\ &= \sum_{(x_{j_1}, \ldots, x_{j_l}, x_{i_1}, \ldots, x_{i_n})} Q(x_{j_1}, \ldots, x_{j_l}, x_{i_1}, \ldots, x_{i_n}) \log \frac{Q(x_{j_1}, \ldots, x_{j_l} | x_{i_1}, \ldots, x_{i_n})}{Q(x_{j_1}, \ldots, x_{j_l} | x_{i_1}, \ldots, x_{i_m})}. \end{aligned}$$

Proof of Corollary. Using Theorem 1 one gets $\min I(P\|Q) = I(P\|\hat{P})$. The Corollary is an immediate consequence of Theorem 2. As

$$\hat{P} = P^{C_1} \prod_{k=2}^{K} P^{C_k|C_k \cap (C_1 \cup \ldots \cup C_{k-1})} \tag{15}$$

and any distribution P can be expressed in the following form

$$P = P^{C_1} \prod_{k=2}^{K} P^{C_k|C_1\cup\ldots\cup C_{k-1}}, \tag{16}$$

the asserion is proven by a simple substitution of expressions (15) and (16) into the definition of $I(P\|\hat{P})$. □

Let us express this divergence for a positive distribution P in another way

$$\begin{aligned} I(P\|\hat{P}) &= \sum_{(x_1,\ldots,x_N)\in\mathbb{X}^V} P(x_1,\ldots,x_N)\left[\sum_{k=1}^{K}\log\frac{P^{C_1\cup\ldots\cup C_k}}{P^{C_{k-1}}} - \log\sum_{k=2}^{K}\frac{P^{C_1\cup\ldots\cup C_k}}{P^{C_k\cap(C_1\cup\ldots C_{k-1})}}\right] = \\ &= \sum_{k=1}^{K} I(P^{C_1\cup\ldots\cup C_k}\|P^{C_k}) - \sum_{k=1}^{K-1} I(P^{C_1\cup\ldots\cup C_k}\|P^{C_{k+1}\cap(C_1\cup\ldots\cup C_k)}) = \\ &= \sum_{k=1}^{K-1}\left[I(P^{C_1\cup\ldots\cup C_{k+1}}\|P^{C_{k+1}}) - I(P^{C_1\cup\ldots\cup C_k}\|P^{C_{k+1}\cap(C_1\cup\ldots\cup C_k)})\right], \end{aligned}$$

which gives an alternative intuitive view on it.

6. DECISION TREE MODELS

To simplify the notation as much as possible we shall not speak about decision tree models only but about a more general situation. For explanation why these models are called decision tree models the reader is refered to [5].

A partition $\{\mathbb{A}_w\}_{w\in W}$ of $\mathbb{X}^V$ (i.e. $\mathbb{A}_w \cap \mathbb{A}_{w'} = \emptyset$ pro $w \neq w'$ and $\bigcup_{w\in W} \mathbb{A}_w = \mathbb{X}^V$) is said to be *a cylinder partition* if any set $\mathbb{A}_w$ is a cylinder of $\mathbb{X}^V$ in the following sense:

$$\mathbb{A}_w = \mathbb{Y}_1 \times \ldots \times \mathbb{Y}_N \subseteq \mathbb{X}_1 \times \ldots \times \mathbb{X}_N, \tag{17}$$

where for all $n = 1,\ldots,N$, $\mathbb{Y}_n$ is either the whole set $\mathbb{X}_n$ of all values of the variable X_n or a singleton $\{x_n\}$ (one-element set for some $x_n \in \mathbb{X}_n$). Therefore, for each component $\mathbb{A}_w$ of such a partition the set of indices $\{1,\ldots,N\}$ of all variables can be split into two subsets:

$$\begin{aligned} \chi^-(w) &\quad - \quad \text{indices of the variables whose values in } \mathbb{A}_w \text{ are fixed}, \\ \chi^+(w) &= \{1,\ldots,N\} - \chi^-(w). \end{aligned}$$

Remark. To keep the idea of graph representation, let us mention that not all cylinder partitions correspond to decision trees. These are only the partitions which can be obtained by the following recursive process:

(i) One-set partition $\{\mathbb{A}\}$ for $\mathbb{A} = \mathbb{X}_1 \times \ldots \times \mathbb{X}_N = \mathbb{X}^V$ corresponds to a decision tree.

(ii) If $\{\mathbb{A}_w\}_{w\in W}$ is a partition corresponding to a decision tree such that $n \in \chi^+(w_o)$ for $w_o \in W$ then

$$\{\mathbb{A}_w\}_{w\in W-\{w_o\}} \cup \{\mathbb{A}_j : \mathbb{A}_{w_o} \cap \{\mathbb{X}_1 \times \ldots \times \mathbb{X}_{n-1}, \{x_n^j\}, \mathbb{X}_{n+1}, \ldots, \mathbb{X}_N\}, x_n^j \in \mathbb{X}_n\} \quad (18)$$

is a partition corresponding to a decision tree.

For more details see [5].

A probability distribution Q is said to be *representable by a cylinder partition* $\mathbb{A} = \{\mathbb{A}_w\}_{w\in W}$ if

$$Q(X_1, \ldots, X_N) = \sum_{w\in W} Q(\mathbb{A}_w) \prod_{n\in\chi^+(w)} Q(X_n|\mathbb{A}_w). \quad (19)$$

The set of all distributions representable by $\mathbb{A}$ will be denoted by $\Pi_{\mathbb{A}}$. In other words, a probability distribution is representable by a cylinder partition if all variables are conditionally independent on each component of the partition. We need not restrict ourself for variables $\{X_n\}_{n\in\chi^+(w)}$ as the remaining variables $\{X_n\}_{n\in\chi^-(w)}$ are constant on $\mathbb{A}_w$ and therefore also independent.

An analogon to Theorem 1 and Theorem 2 showing how to find the best approximation (i.e. the I-projection) is the following assertion whose simple but rather technical proof can be found in [5].

Theorem 3: For any probability distribution P defined for $(X_i)_{i\in V}$ and any cylinder partition $\mathbb{A} = \{\mathbb{A}_w\}_{w\in W}$ of the joint sample space (of the variables in question) $\mathbb{X}^V$, the I-projection $\hat{P}$ of P into $\Pi_{\mathbb{A}}$ can be computed according to the following formula

$$\hat{P}(X_1, \ldots, X_N) = \sum_{w\in W} P(\mathbb{A}_w) \prod_{n\in\chi^+(w)} P(X_n|\mathbb{A}_w). \quad (20)$$

Corollary: Let P and $\mathbb{A}$ be a probability distribution and a cylinder partition as in Theorem 3. Then

$$\min_{Q\in\Pi_{\mathbb{A}}} I(P||Q) = \sum_{w\in W} P(\mathbb{A}_w) I(P((X_n)_{n\in\chi^+(w)}|\mathbb{A}_w)), \quad (21)$$

where $I(P((X_n)_{n\in\chi^+(w)}|\mathbb{A}_w))$ is an information content of the conditional distribution $P((X_n)_{n\in\chi^+(w)}|\mathbb{A}_w)$ defined

$$I(P((X_n)_{n\in\chi^+(w)}|\mathbb{A}_w)) = \sum_{(x_n)_{n\in\chi^+(w)}\in\mathbb{X}^{\chi^+(w)}} P((x_n)_{n\in\chi^+(w)}|\mathbb{A}_w) \log \frac{P((x_n)_{n\in\chi^+(w)}|A_w)}{\prod_{n\in\chi^+(w)} P(x_n|A_w)}. \quad (22)$$

7. CONCLUSIONS

Three graph representations of probability distributions have been introduced each with its advantages and shortcomings. Decomposable models form proper subclass of graphical models, nevertheless, the latter are of less practical use because of the high algorithmical complexity of methods for computation of I-projections.

Application of decomposable models and decision tree models is made possible by Theorem 2, Theorem 3 and their corollaries. An extensive discussion in [5] shows that none of them is superior the other. The type of representation should be accommodated to the problem in question. Though we can hardly assume that a polynomial algorithm finding the best graph representation (i.e. the simplest decomposable, or decision tree approximation) for a given probability distribution will be found, the presented assertions gives a good theoretical basis for a number of heuristics which guarantee the suboptimal solution. Moreover, for limited dimensions, the optimal solution can be constructed using some dynamic programming techniques.

The suggested aproaches show new way of approximating dependence structures of probability distribution applicable in the field of AI. It reflects the fact that not all dependencies are of the same intensity and sometimes when neglecting weak dependence relations, incomparably simpler structures can be obtained.

REFERENCES

1. I. Csiszár, *I-divergence geometry of probability distributions and minimization problems.* Ann. Probab., 3 (1975), 146-158.
2. P. Hájek, T. Havránek and R. Jiroušek, Uncertain Information Processing in Expert Systems (CRC Press, Inc., Boca Raton, 1992).
3. R. Jiroušek, *A survey of methods used in probabilistic expert systems for knowledge integration.* Knowledge-Based Systems, 3 (1990), 1, 7-12.
4. R. Jiroušek, *Reasoning and derivation of knowledge in probabilistic expert systems.* Trans of the 11th Prague Conference on Inf. Theory, etc., Vol. B., Academia, Prague, 1992, 31-40.
5. Jiroušek R., 1991. *Decision-tree modelling of probability distributions.* To appear in: Int. J. on General Systems.
6. H. G. Kellerer, *Verteilungsfunktionen mit gegeben Marginalverteilungen.* Z. Warhsch. Verw. Gebiete 3 (1964), 247-270.
7. S. L. Lauritzen, A. P. David, B. N. Larsen and H. G. Leimer, Independence Properties of Directed Markov Fields. Tech. report R 88-32, University of Aalborg, Aalborg (1988), Denmark.
8. J. Pearl and T. S. Verma, *The logic of representing dependencies by directed graphs.* Proc. 6th Natl. Conf.on AI, Seatle (1987), 374-79.
9. J. Pearl and A. Paz, GRAPHOIDS: A Graph-Based Logic for Reasoning about Relevance Relations. TR850038 (R-53) UCLA Computer Science Dept. (1986).
10. J. Pearl, Probabilistic Reasoning in Intelligence Systems: Networks of Plausible Inference. (Morgan-Kaufmann, San Mateo, CA, 1988).
11. S. Ur and A. Paz, *The representation power of probabilistic knowledge by undirected*

graphs and directed acyclic graphs. A comparison. To appear in: Int. J. on General Systems.

12. J. Whittaker, Graphical Models in Applied Multivariate Statistics. (J. Wiley, New York, 1990).

Uncertainty in Intelligent Systems
B. Bouchon-Meunier et al. (Editors)

Using Bayesian Algorithms for Learning Causal Networks in Classification Problems

Rafael Molina, Luis M. de Campos, Javier Mateos

Departamento de Ciencias de la Computación e I.A. Universidad de Granada. 18071. Granada. España

Abstract

In this paper we describe CASTLE, a software developed at the University of Granada for the Esprit project StatLog and study its performance as classifier on two datasets. We also compare the results obtained by CASTLE and decision tree algorithms and describe how CASTLE can be used to select important attributes in classification problems.

1. INTRODUCTION

Causal networks ([5]) are powerful graphical knowledge representation tools able to efficiently represent relationships of relevance or dependency. Once a network has been built, it constitutes an efficient device to perform (probabilistic) inferences ([5],[4]). Obviously, there remains the previous problem of building such a network, that is, to provide the structure and conditional probabilities necessary for characterising the network. A very interesting task is then to develop methods able to learn the net directly from raw data, as an alternative to the method of eliciting opinions from the experts. Recently, several methods of automatic learning of networks have been proposed (see [5], [6], and [7] for references). Some members of the Department of Computer Science at the University of Granada are developing for the ESPRIT project StatLog CASTLE ([1],[2]) a tool that implements inductive learning algorithms and has propagation capabilities for causal networks.

The aim of this paper is to describe CASTLE and how it can be used in classification problems. In section 2, we describe the current facilities that CASTLE incorporates in order to learn network structures from raw data and to propagate knowledge through out the obtained networks; we also explain the approach we will adopt to tackle classification tasks using CASTLE. Section 3 is devoted to the study of the different techniques used by CASTLE for measuring (in)dependencies and for solving conflicts. In section 4 we apply CASTLE to two different real problems. Finally, in section 5 we study how CASTLE can be used as a complementary tool for decision trees (see [3] for references).

2. CLASSIFICATION PROBLEMS AND CASTLE

CASTLE can be used so far to learn singly-connected causal networks (polytrees) from raw data, propagate knowledge through out polytrees, simulate and also edit polytree dependent discrete distributions. In its learning mode, CASTLE basically estimates, from a file of examples, the (in)dependencies among the variables involved in the examples, in order to build a polytree displaying such (in)dependencies. First, CASTLE estimates the skeleton (the undirected graph) of the polytree using a maximum weight spanning tree algorithm, with different kinds of dependency measures, then, CASTLE uses several criteria for directing the edges based on different independency tests, and incorporates also methods for dealing with the inconsistencies that could appear when directing the skeleton. Finally, CASTLE attaches to each node the corresponding estimated marginal or conditional probabilities. Once the network has been learned, CASTLE can also exploit the net by fixing observed cases on some nodes, propagating these pieces of information through out the network, and consulting the (posterior) probabilities of the nodes (see [1], [2]).

After this brief view of CASTLE, now let us see how to use it to build a classifier. In any classification problem, we have a set of variables $V = \{X_i, i = 1, \ldots, n\}$ that (possibly) have influence on a distinguished classification variable C. The problem is, given a particular instantiation of these variables, to predict the value of C, that is, to classify this particular case in one of the possible categories of C. For this task, we need a set of examples and their correct classification, acting as training sample. In this context, the use of CASTLE is clear: in learning mode CASTLE will extract from this training sample a network structure displaying the causal relationships among the variables $\{X_i, i = 1, .., n\} \cup C$; next, in propagation mode, given a new case with unknown classification, CASTLE will instantiate and propagate the available information, showing the more likely value of the classification variable C.

A different variant of the classification task involves the use of costs of missclassification: we have a cost matrix (c_{ij}) where each c_{ij} represents the cost associated to classify an example as belonging to the jth category of the variable C, given that the true category is the ith. In this case, we look for the classification with minimum cost instead of the more likely classification. For those kind of problems, the posterior probability of the variable C calculated by CASTLE is used to obtain the average cost of classifying an example in each category, thus selecting the classification of minimum cost. Let us mention that maximum probability classification corresponds to minimum cost for a 1-0 cost function.

It is important to remark that this classifier can be used even when we do not know the exact value of all the variables in V. Moreover, the network shows which are the variables in V that directly have influence on C, in fact the parents of C, the children of C and the other parents of the children of C (the knowledge of these variables makes C independent of the rest of variables in V). So, the rest of the network could be pruned, thus reducing the complexity and increasing the efficiency of the classifier.

3. LEARNING WITH CASTLE

In order to evaluate the various techniques that CASTLE uses to reveal dependencies, directing edges and dealing with conflicts, in the next section we will perform some experiments on real data. So now, we are going to describe these techniques in some detail.

As we said previously, CASTLE estimates the skeleton of the polytree using a maximum weight spanning tree algorithm (Kruskal's algorithm, to be precise), where the weight of the branch connecting any two nodes X_i and X_j, $Dep(X_i, X_j)$, is a measure of the dependence degree between these two nodes, being zero if X_i and X_j are independent, and maximum if X_i and X_j are functionally dependent. In addition to the usual Kullback-Leibler information measure, CASTLE also incorporates other *Dep* functions: one of them (Rajski) is a normalized version of the Kullback-Leibler measure, and the rest are different distance measures between the bidimensional distribution and the product of the marginal distributions of X_i and X_j. The complete list of *Dep* functions follows:

1. Information
   - Kullback–Leibler
   $$Dep(X_i, Y_j) = \sum_{x_i, y_j} P(x_i, y_j) \log \frac{P(x_i, y_j)}{P(x_i)P(y_j)}$$
   - Rajski
   $$Dep(X_i, Y_j) = -\frac{\sum_{x_i, y_j} P(x_i, y_j) \log \frac{P(x_i, y_j)}{P(x_i)P(y_j)}}{\sum_{x_i, y_j} P(x_i, y_j) \log P(x_i, y_j)}$$
2. L1–norm
   - Unweighted
   $$Dep(X_i, Y_j) = \sum_{x_i} \sum_{y_j} |P(x_i, y_j) - P(x_i)P(y_j)|$$
   - Weighted
   $$Dep(X_i, Y_j) = \sum_{x_i} \sum_{y_j} P(x_i, y_j)|P(x_i, y_j) - P(x_i)P(y_j)|$$
3. L2–norm
   - Unweighted
   $$Dep(X_i, Y_j) = \sum_{x_i} \sum_{y_j} (P(x_i, y_j) - P(x_i)P(y_j))^2$$
   - Weighted
   $$Dep(X_i, Y_j) = \sum_{x_i} \sum_{y_j} P(x_i, y_j)(P(x_i, y_j) - P(x_i)P(y_j))^2$$
4. L∞

$$Dep(X_i, Y_j) = \max_{x_i} \max_{y_j} |P(x_i, y_j) - P(x_i)P(y_j)|$$

After obtaining the skeleton, CASTLE tries to recover the directions of the branches; given the subgraph X_i–X_k–X_j of the skeleton, we can only distinguish between two patterns of independence:

1. Marginal independence and conditional dependence, with associated graph

   $X_i \rightarrow X_k \leftarrow X_j$,

2. Marginal dependence and conditional independence, giving rise to any of the three graphs

   $X_i \leftarrow X_k \leftarrow X_j$, $X_i \leftarrow X_k \rightarrow X_j$ *or* $X_i \rightarrow X_k \rightarrow X_j$.

To distinguish between these two patterns we can use the following facts:

For the head–to–head pattern $X_i \rightarrow X_k \leftarrow X_j$, any marginal *Dep* function must be zero and any conditional *Dep* function must be greater than zero:

$Dep(X_i, X_j) = 0$ and $Dep(X_i, X_j|X_k) > 0$.

For the other pattern the opposite conditions hold:

$Dep(X_i, X_j) > 0$ and $Dep(X_i, X_j|X_k) = 0$.

The conditional *Dep* function $Dep(X_i, X_j|X_k)$ is defined as the mean with respect to the marginal distribution of X_k, $P(x_k)$, of the functions $Dep(X_i, X_j|X_k = x_k)$, which are analogous to $Dep(X_i, X_j)$ but each distribution being replaced by the corresponding conditional distribution given that X_k is equal to x_k.

The problem is that usually we do not have an exact distribution but we estimate it from a sample. Therefore, conditions like $Dep(X_i, X_j) = 0$ or $Dep(X_i, X_j|X_k) = 0$ are too hard for real cases. For that reason CASTLE may detect a head–to–head pattern $X_i \rightarrow X_k \leftarrow X_j$ using three different tests:

1. A marginal *Dep* test:

   $Dep(X_i, X_j) < \epsilon$,

   being ϵ a threshold chosen by the user to detect independencies.

2. A conditional *Dep* test:

   $Dep(X_i, X_j) < Dep(X_i, X_j|X_k)$.

3. A standard chi–square test of independence between X_i and X_j.

Another problem that must be taken into account is the possibility of finding inconsistencies. In that case, the Rebane and Pearl algorithm ([5]) does not provide any criterion to decide the dependencies or independencies that should be preserved (the selection depends on the order in which the nodes are examined). So, CASTLE also offers a more elaborate mean to give directions to the branches, based on the same previous methods to distinguish head–to–head patterns, but providing several criteria to decide which dependencies or independencies should be preserved in the case of a conflict. The basic idea is to take into account some measure of the strength of the dependencies and independencies, in order to include the more important first (using a local search with priorities).

The idea is simple: suppose we have the two subgraphs X_i–X_k–X_j and X_l–X_h–X_m.

(a) If we are using the marginal *Dep* criterion to detect the head-to-head patterns, then

If $Dep(X_i, X_j) < Dep(X_l, X_m) < \epsilon$

then the independence between X_i and X_j has more priority than the independence between X_l and X_m and should be preserved first.

If $\epsilon < Dep(X_i, X_j) < Dep(X_l, X_m)$

then the dependence between X_l and X_m has more priority than the dependence between X_i and X_j and should be preserved first.

In that case, given two nodes Y and Z, the strength of the independence between Y and Z is measured as $\epsilon - Dep(Y, Z)$, and the strength of the dependence between Y and Z is measured by mean of $Dep(Y, Z)$ (the nearer to zero is the value of *Dep*, the greater is the priority of the independence; the greater is the value of *Dep*, the greater is the priority of the dependence).

(b) If we are using the conditional *Dep* criterion, then

If $0 < Dep(X_l, X_m|X_h) - Dep(X_l, X_m) < Dep(X_i, X_j|X_k) - Dep(X_i, X_j)$

then the independence between X_i and X_j has also more priority than the independence between X_l and X_m.

If $0 < Dep(X_i, X_j) - Dep(X_i, X_j|X_k) < Dep(X_l, X_m) - Dep(X_l, X_m|X_h)$

then the dependence between X_l and X_m has again more priority than the independence between X_i and X_j.

In that case the strength of the independence (dependence respectively) is measured as the difference between the values of the conditional *Dep* and marginal *Dep* (marginal *Dep* and conditional *Dep* respectively) functions: the greater is the difference, the greater is the priority.

The previous rules apply to rank by priority the independencies on the one hand and the dependencies on the other hand. However an additional criterion to compare independencies and dependencies must be supplied. CASTLE incorporates three of these criteria:

1. To give more priority to the independencies.

2. To give more priority to the dependencies.

3. To put the priorities of both independencies and dependencies in a common dimension (the same range of variation) and then to preserve first the relationships with more priority regardless whether they are independencies or dependencies.

Finally, the strengths of the independencies and dependencies can be measured in a normalized or unnormalized way.

4. TEST EXAMPLES

In this section we are going to describe the experiments carried out with CASTLE on databases extracted from two real problems: the classification of satellite images of different types of soil, and the classification of patients with severe head injury.

As we stated in the previous section, CASTLE has different options to learn the skeleton of the polytree, to give directions to its branches and to solve inconsistencies. When we combine all of these options, we get a total of 84 different ways to learn the network. The aim of the experiments is to perform a comparative study of these techniques when applied to classification tasks. So, after running CASTLE using all its options, we will give a report of the results. Each database was randomly divided in a training set used for learning the network, and a test set, for testing the classification predicted by CASTLE against the actual classification. The comparison was based on the overall success rate of classification for the test set.

The first database we will try CASTLE on has been provided by the University of Strathclyde in Glasgow. Each sample in this dataset consists of the 4–spectral values of the pixels in 3x3 neighbourhoods in a satellite image and the classification associated with the central pixel. So each observation consists of 36 attributes and the class variable. Two of the spectral bands are in the visible region (corresponding approximately to green and red regions of the visible spectrum) and two are in the (near) infra-red. For each band, the attribute in each pixel is a 7-bit binary word, with 0 corresponding to black and 127 to white. The variable class can take the following values corresponding to different types of soil: *red soil, cotton crop, grey soil, damp grey soil, soil with vegetation and stubble* and *very damp grey soil.* In each line of data the four spectral values for the top-left pixel are given first followed by the four spectral values for the top-middle pixel and then those for the top-right pixel, and so on with the pixels read out in sequence left-to-right and top-to-bottom. Thus, the four spectral values for the central pixel are given by attributes 17,18,19 and 20.

The dataset consists of 6435 samples, divided in training set with 4435 samples and test set with 2000 samples. To perform the classification task the range of each of the 36 attributes was divided in four classes with equal probability.

The main results obtained when CASTLE is run on this dataset are:

1. The only thing that affects the success rate is the choice of the *Dep* function: once we fix a *Dep* function, the polytree obtained by changing the options relative to orientation and conflict resolution remains unchanged.

   This suggests that, at least for classification problems, the skeleton of the polytree is more important than the orientation. Another possible explanation is that the underlying true network structure is a tree, and therefore the directions we give to the arrows does not matter (as long as we do not introduce head-to-head patterns).

2. The success rates obtained by the different *Dep* functions are:

Success rate

	Kullback	*Rajski*	*L1+*	*L1-*	*L2+*	*L2-*	*L∞*
%	80.5	70.5	50.4	70.5	50.5	50.4	75.4

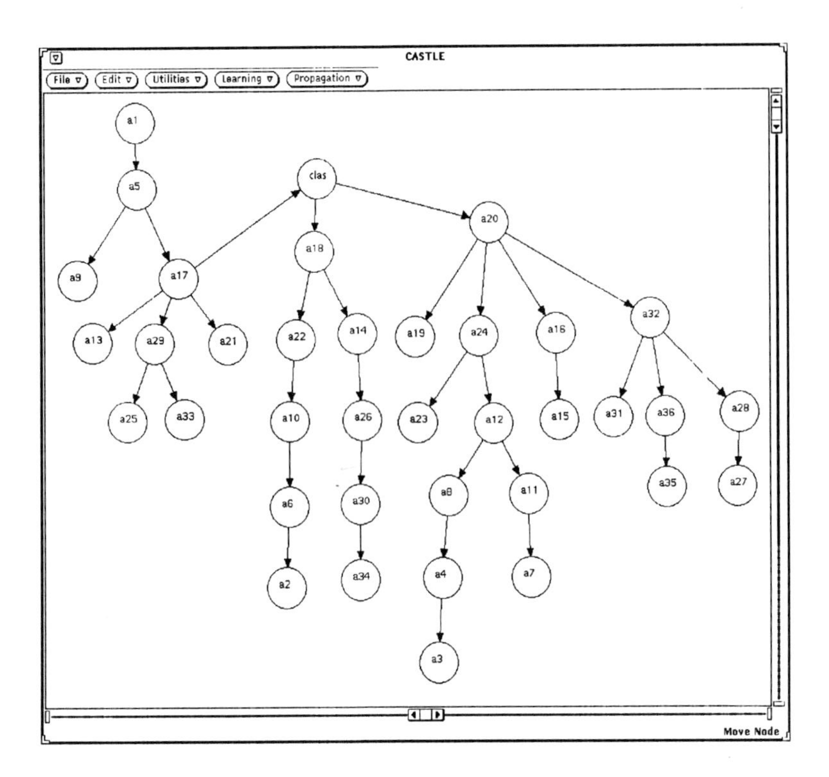

Figure 1. Satellite polytree

In this case the Kullback–Leibler measure of information gives the best results. Rajski measure, L1 unweighted and L∞ distances provide also good results. However, both versions of L2 and L1 weighted perform rather poorly.

3. The polytree we obtained using CASTLE with Kullback–Leibler measure on this dataset is displayed in figure 1. The most important conclusion to be drawn from this polytree is that, according to CASTLE, the attributes $a17, a18$ and $a20$ are the only needed to perform the classification task. These attributes correspond to three of the values observed on the central pixel. This pixel is the one we want to know of its land type.

4. The following confusion matrix was obtained over the test set. The row of each entry represents the actual classification and the column represents the predicted classification. The classes have been numbered in order from 1 to 6. It suggests that the classes *damp grey soil* and *very damp grey soil* are probably not very distinguishable: a 34.6% of all the errors are due to confusion between these two classes.

CONFUSION MATRIX

	1	*2*	*3*	*4*	*5*	*6*
1	416	0	10	13	14	8
2	5	200	5	1	10	3
3	1	0	353	40	0	3
4	1	0	29	117	1	63
5	21	15	3	4	142	52
6	0	0	8	72	8	382

The second data set consists of 900 records of patients with severe head injury, (source D.M. Titterington, University of Glasgow, [8]). The problem is to predict the degree of recovery individual patients would attain, using data colleted shortly after injury. Each observation consists of the values of age and five indicators of brain damage for a given patient together with his/her degree of recovery. So the seven variables are: *age*, grouped into eight decades, *EMV*, a combined measure of eye, motor and verbal response, coded from 1=nil to 7=normal, *MRP*, motor response in all four limbs (from 1-7), *Change*, which measures change in neurological function over 24 hours and is coded as 1(deteriorating), 2(static) and 3(improving), *Eye performance* which measures the eye performance coded as 1(bad), 2(impaired) and 3(good), *Pupils reaction to light* coded as 0 (non-reacting) or 1 (reacting) and the classification variable taking the values dead/vegetative, severe and moderate/good.

The data set was randomly divided into a training set with 600 records and a test set with 300 records. For this problem past experience provides the following cost matrix.

	d/v	*sev*	*m/g*
d/v	0	10	75
sev	10	0	90
m/g	750	100	0

After running every possible option of CASTLE on this dataset we have obtained the following results:

1. On this dataset the success rate does not only depend on the *Dep* function but on the independence test as well. Using searching with priorities to direct the skeleton improves the results very slightly.

2. Marginal independence test performs always better than any other tests.

3. The following table displays the success rate of the obtained polytrees without costs,

	Kullback	*Rajski*	*L1+*	*L1-*	*L2+*	*L2-*	*L∞*
Chi-test	69.3	72.3	71.7	71.7	71.7	71.7	71.7
Marginal test	69.3	72.3	73.3	73.3	73.3	73.3	73.3
Conditional test	67.7	71.3	71.7	71.7	71.7	71.7	73.3

The best results were obtained for L∞, L1 and L2 norms. Rajski performs slightly worse, and finally Kullback provides the worst results. In any case the differencies are very small.

4. Using the cost matrix given above the average classification cost is given in the following table

	Kullback	*Rajski*	*L1+*	*L1-*	*L2+*	*L2-*	$L\infty$
Chi-test	26.617	21.350	27.267	27.267	27.267	27.267	27.267
Marginal test	26.617	21.350	26.317	26.317	26.317	26.317	26.317
Conditional test	33.167	30.217	27.267	27.267	27.267	27.267	26.317

   As can be observed *Rajski* provides the minimum average cost and the Marginal dependence gives again the best results.

5. The following table displays the confusion matrix for the test set for the cost matrix provided above when using *Rajski* and marginal dependence,

	d/v	sev	m/g
d/v	66	67	21
sev	3	14	27
m/g	0	17	85

   The obtained polytree is displayed in figure 2.

5. COMBINING CASTLE AND DECISION TREES

Let us now briefly compare the results obtained by CASTLE with the results obtained on the same datasets by the IND package, a software developed by Buntine ([3]) and that among other things builds decision trees.

Let us start with the satellite dataset. The success rate of the decision tree obtained when only attributes $a17$, $a18$, and $a20$, the ones suggested by CASTLE , were used was 83.5 %, when the 36 attributes were used the success rate was 84.9%.

With regard to the head injure dataset, for the decision tree with the attributes selected by CASTLE, *Age*, *Change*, *Eye Performance* and *Pupils Reaction to Light*, the mean loss was 23.800 while the mean loss was 25.566 when all the six attributes were used. Again, CASTLE has been able to detect the most important variables but in this case it has also helped to improve the performance of the decision tree.

6. CONCLUSIONS

The results obtained on these two datasets show that CASTLE could be a useful and versatile tool for classification tasks. Furthermore, these results also suggest that CASTLE could be used to simplify the work of other classification methods.

Finally, we would like to mention that the process to build the network does not take into account the fact that we are only interested on classifying, and so the process does not specialize on the class node, expecting then as classifier a poorer performance than other classification oriented methods. However, as we have already mentioned, the built

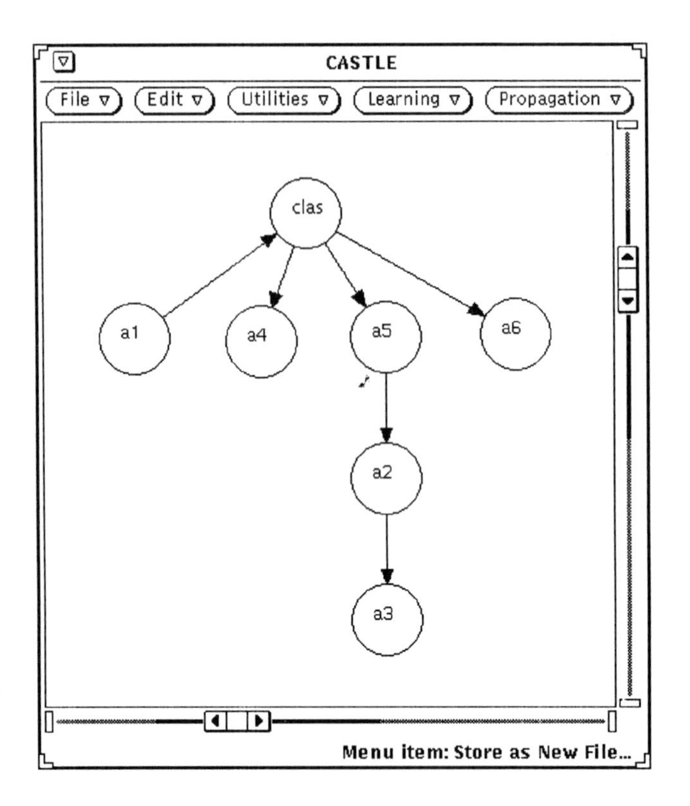

Figure 2. Injure Polytree

networks are able to display insights on the classification problem that other methods lack.

Acknowledgement

This work has been supported by the Commission of the European Communities under ESPRIT project no 5170: Comparative Testing of Statistical and Logical Learning, StatLog.

One of the authors (J.M.) has been supported by the Junta de Andalucia under its scheme 'Iniciación a la Investigación y la Docencia'.

REFERENCES

1. Acid, S., Campos, L.M.de, González, A., Molina, R., Pérez de la Blanca, N. (1991), Learning with CASTLE, in Symbolic and Quantitative Approaches to Uncertainty, Lecture Notes in Computer Science 548, R. Kruse, P. Siegel (Eds.), Springer Verlag, 99-106.

2. Acid, S., Campos, L.M.de, González, A., Molina, R., Pérez de la Blanca, N. (1991), CASTLE: A tool for Bayesian learning, Proc. of the 1991 ESPRIT Conference, 363-377.
3. Buntine, W. (1991). Introduction to IND and Recursive Partitioning. NASA Ames Research Center.
4. Lauritzen, S.L. and Spiegelhalter, D. J. (1988), Local computations with probabilities on graphical structures and their application to expert systems, J. R. Statist. Soc. Ser. B 50, 157-224.
5. Pearl, J. (1988), Probabilistic Reasoning in Intelligent Systems: Networks of Plausible Inference, Morgan and Kaufmann.
6. Spirtes, P., Glymour, C. and Scheines, R. (1991), Casuality, Statistics and Search, Unpublished manuscript.
7. Spiegelhalter, D.J., Dawid, A.P., Lauritzen, S.L and Cowell, R.G. (1992) Bayesian Analysis in Expert Systems. *BAIES report BR-27.*
8. Titterington, D.M., Murray, G.D., Murray, L.S., Spiegelhalter, D.J., Skene, A.M., Habbema, J.D.F., Gelpke, G.J. (1981) Comparison of Discrimination Techniques Applied to a Complex Data Set of Head Injured Patients (with discussion). J. Royal Statist. Soc., **A**, 144, 145-175.

Uncertainty in Intelligent Systems
B. Bouchon-Meunier et al. (Editors)

Convergence of Causal Probabilistic Networks

Ralph Matthes and Ulrich G. Oppel
Mathematisches Institut der Ludwig-Maximilians-Universität
Theresienstr. 39, D 8000 München 2, Germany

Key words: Causal probabilistic network, belief net, Markov composition, expert system; convergence, sensitivity analysis.

Complex technical, biological, economical, and psychological systems consist of a finite number of components. Such a system V and its components are subject to randomness, uncertainty, imprecision, incompleteness, and vagueness due to the variability between and within its components. The components v of such a system may be characterized by qualitative or quantitative random variables X_v assuming their values in rather general state spaces S_v with σ-algebras $\mathfrak{S}_v$.

From the probabilistic point of view the qualitative and quantitative aspects of the total system are completely determined by the common distribution of the random variables representing its components. The common distribution is a probability measure $\mathbb{P}: \mathfrak{S} \to [0,1]$ on the product σ-algebra $\mathfrak{S}$ on the product state space S belonging to the family $((S_v,\mathfrak{S}_v): v\in V)$ of state spaces. The system is completely determined by the probability space $(S,\mathfrak{S},\mathbb{P})$. However, to find $\mathbb{P}$ is often very difficult; usually $\mathbb{P}$ can be determined at most only approximately by theoretical or statistical procedures.

If V is finite or at most countably infinite, one way of finding $\mathbb{P}$ is to use a causal probabilistic network (CPN); e.g. see Pearl (9) or Lauritzen-Spiegelhalter (5). Such a CPN is a directed graph $G := (V,E)$ with $E \subset V\times V$ and $(u,v)\notin E$ for $(v,u)\in E$ and a family $\mathcal{P} := (\mathcal{P}_v: v\in V)$ of Markov kernels

$$\mathcal{P}_v: S(Pa(v)) \times \mathfrak{S}_v \to [0,1] \quad \text{with} \quad ((x_u: u\in Pa(v)), B) \to \mathcal{P}_v((x_u: u\in Pa(v); B)$$

where $Pa(v) := \{u\in V: (u,v)\in E\}$ is the set of the parents of v in G, $S(U) := \prod_{u\in U} S_u$ for $\emptyset \neq U \subset V$ is the product of the state spaces of the family $(X_u: u\in U)$ of random variables, and $S(\emptyset) \times \mathfrak{S}_v := \mathfrak{S}_v$. Every node $v\in V$ represents the random variable X_v with the state space $(S_v,\mathfrak{S}_v)$, E describes the dependency of $(X_v: v\in V)$ qualitatively, and $\mathcal{P}$ describes this dependency quantitatively. $\mathcal{P}_v$ is (or is supposed to be) the conditional distribution of X_v given $(X_u: u\in Pa(v))$. Associating a CPN to $\mathbb{P}$ is a localization procedure which reduces the problem of finding the high dimensional distribution $\mathbb{P}$ to finding the family $\mathcal{P}$ of lower dimensional conditional distributions.

For any finite CPN without cycles there is a probability measure $\mathbb{P}$ which is the common distribution of the random variables X_v represented by the nodes $v\in V$, which has the given Markov kernels $\mathcal{P}_v$ as conditional distributions of X_v given the values of the random variables associated to the parent nodes of v, and which is a directed Markov field. *Mathematically*, this can be shown by iterative integration:

Given any causal probabilistic network $G := (V,E)$ with finite V and Markov kernels $(\mathcal{P}_v: v\in V)$ without cycles it is possible to find an enumeration $V := \{v_i: i=0, \ldots, n\}$ of V such that no descendant is enumerated before any of his ancestors; such an enumeration is a well-ordering in the sense of Lauritzen-Dawid-Larsen-Leimer (3).

Let $S^{(i)} := S(\{v_j: j=0, \dots, i\})$ be the product set and $\mathfrak{S}^{(i)} := \bigotimes_{j=0}^{i} \mathfrak{S}_{v_j}$ be the product σ-algebra for $0 \le i \le n$. Defining associated Markov kernels

$P_i: S^{(i-1)} \times \mathfrak{S}_{v_i} \to [0,1]$ for $i = 0, \dots, n$ (where $S^{(-1)} \times \mathfrak{S}_{v_0} := \mathfrak{S}_{v_0}$) by
$((x_0, \dots, x_{i-1}), B) \to P_i(x_0, \dots, x_{i-1}; B) := \mathcal{P}_{v_i}((x_u: u \in Pa(v_i)); B)$

we may construct a probability measure $\mathbb{P}: \mathfrak{S} \to [0,1]$ by iterative integration:

$$(I) \quad \mathbb{P}(A) := \int_{S_{v_0}} \int_{S_{v_1}} \dots \int_{S_{v_n}} 1_A(x_0, \dots, x_n) \; P_n(x_0, \dots, x_{n-1}; dx_n) \dots P_1(x_0; dx_1) \, P_0(dx_0)$$

for $A \in \mathfrak{S} = \mathfrak{S}^{(n)}$.

This probability measure $\mathbb{P}$ does not depend on the chosen enumeration of this kind. (If V is countably infinite and if there is a well-ordering, such a probability measure $\mathbb{P}$ may constructed by applying a generalization of a theorem of C. Ionescu-Tulcea; e.g. see (7).)

In practice, however, it is very hard (if not impossible) to calculate $\mathbb{P}$ for large CPNs completely. Only for some situations (finite state spaces or certain mixtures of finite and Gaussian state spaces; e.g. see Lauritzen and Wermuth (4)) $\mathbb{P}$ can be calculated at least partially by an algorithm which is a mixture of local and global calculations.

In the case of finite and discrete state spaces S_v this probability measure $\mathbb{P}$ given by (I) is identical to the ones constructed using the procedures proposed by Pearl (9), Lauritzen-Spiegelhalter (5) and Jensen-Lauritzen-Oleson (2) and the procedures implemented in the software package HUGIN for the construction of expert systems based on causal probabilistic networks. For large CPNs $\mathbb{P}$ cannot be calculated totally because of storage problems: More than 10^{300} bytes would be needed for 1000 nodes with 2 states each. HUGIN calculates marginal distributions from the given conditional probabilities without making the detour to $\mathbb{P}$.

The probability measure $\mathbb{P}: \mathfrak{S} \to [0,1]$ given by (I) is a directed Markov field in the sense of Lauritzen-Dawid-Larsen-Leimer (3); see (6).

Under mild assumptions on the state spaces (e.g. for polish spaces) there are n! possibly different CPNs without cycles associated with $\mathbb{P}$, if V has n elements. This can be shown by iterative desintegration (conditioning), at least in principle; see (6).

Subsequently, we shall only consider finite CPNs without cycles. Furthermore, we shall assume that every state space is a polish space S_v, and $\mathfrak{S}_v$ is its σ-algebra of Borel subsets. Endowed with the product topology, the common state space S is also polish, and $\mathfrak{S}$ is its σ-algebra of Borel subsets. A topological space is called polish if it is a separable and completely metrizable space; e.g. a countable and discrete space, the Euclidean space $\mathbb{R}^n$ or any separable Hilbert and Banach space.

Hence, the difficult problem of finding the probability measure $\mathbb{P}$ characterizing the total system is *mathematically* equivalent to the problem of finding such a CPN with the directed graph $G := (V, E)$ without cycles and the Markov kernels $(\mathcal{P}_v: v \in V)$. However, in many applications it seems to be possible (or at least easier) to find and to quantify dependency relations between the components of the system by using theoretical, intuitive or statistical expert knowledge.

Even the best experts will not know these dependency relations between the components of the system exactly, but only approximately. Therefore it is necessary to consider proper concepts of convergence of CPNs. From what we said above, every CPN is associated to a probability measure $\mathbb{P}$ on the common state space $(S, \mathfrak{S})$. For probability measures, however, there are many concepts of convergence.

Subsequently, we shall study two extreme kinds of topologies (and hence of convergence): We shall study the topology of uniform convergence of probability measures and the topology of weak convergence of probability measures. First, we shall extend these topologies for spaces of probability measures in natural ways to topologies for spaces of Markov kernels. Then we shall relate the convergence of the Markov kernels of a sequence of CPN's to the convergence of the probability measures associated to these CPN's via (I). We shall prove continuity theorems and give examples of discontinuity of this iterative construction method (I).

For a (nonempty) set A and a σ-algebra $\mathfrak{A}$ on A let $\mathfrak{M}^b(A,\mathfrak{A})$ denote the linear space of signed real valued (and hence bounded) measures $\mu\colon \mathfrak{A} \to \mathbb{R}$, $\mathfrak{M}^{1+}(A,\mathfrak{A})$ denote the set of probability measures $\mathbb{P}\colon \mathfrak{A} \to [0,1]$, and $\mathfrak{F}_b(A,\mathfrak{A})$ denote the linear space of real valued and bounded measurable functions $f\colon (A,\mathfrak{A}) \to \mathbb{R}$. With the bilinear form

$$\langle f,\mu\rangle := \int_A f(x)\,\mu(dx) \quad \text{for } f\in \mathfrak{F}_b(A,\mathfrak{A}) \text{ and } \mu\in \mathfrak{M}^b(A,\mathfrak{A})$$

$\mathfrak{F}_b(A,\mathfrak{A})$ and $\mathfrak{M}^b(A,\mathfrak{A})$ is a separating dual pair. The topology of uniform convergence on $\mathfrak{M}^b(A,\mathfrak{A})$, denoted by $\beta(\mathfrak{M}^b,\mathfrak{F}_b)$, is induced by the norm of total variation:

$$\|\mu\| := \sup\{\,|\langle f,\mu\rangle|\colon f\in \mathfrak{F}_b(A,\mathfrak{A}) \text{ with } |f|\le 1\,\} \quad \text{for } \mu\in \mathfrak{M}^b(A,\mathfrak{A}).$$

If μ^+ denotes positive and μ^- the negative part of the Hahn-Jordan decomposition of μ, then $\|\mu\| = \mu^+(A)+\mu^-(A)$. Furthermore, the norm $\|\,.\,\|$ is equivalent to the norm $|||\mu|||$ $:= \sup\{|\mu(K)|\colon K\in\mathfrak{A}\}$. A sequence $(\mu_n\colon n\in\mathbb{N})$ in $\mathfrak{M}^b(A,\mathfrak{A})$ converges uniformly to $\mu\in$ $\mathfrak{M}^b(A,\mathfrak{A})$ if and only if the sequence $(\|\mu_n-\mu\|\colon n\in\mathbb{N})$ converges to 0.

For another measurable space $(B,\mathfrak{B})$ let $\mathcal{MK}(A,\mathfrak{A};B,\mathfrak{B})$ denote the set of Markov kernels $P\colon A\times\mathfrak{B} \to [0,1]$ with $(x;L) \to P(x;L)$ from $(A,\mathfrak{A})$ to $(B,\mathfrak{B})$; i.e. $P(x;.)\in\mathfrak{M}^{1+}(A,\mathfrak{A})$ for $x\in A$ and $P(.;L)\in\mathfrak{F}_b(A,\mathfrak{A})$ for $L\in\mathfrak{B}$. On the linear space $\mathcal{K}(A,\mathfrak{A};B,\mathfrak{B})$ of kernels $M\colon A\times\mathfrak{B} \to \mathbb{R}$ from the measurable space $(A,\mathfrak{A})$ to the measurable space $(B,\mathfrak{B})$ spanned by $\mathcal{MK}(A,\mathfrak{A};B,\mathfrak{B})$ we define the norm of uniform total variation by

$$\|M\| := \sup\{\,\|M(x;.)\|\colon x\in A\}$$

From $\mathfrak{F}_b(B,\mathfrak{B}) = \{\,g(\xi,.)\colon g\in\mathfrak{F}_b(A\times B,\mathfrak{A}\otimes\mathfrak{B}),\ \xi\in A\}$ we get

$$\begin{aligned}\|M\| &= \sup\{\,|\langle f,M(x;.)\rangle|\colon f\in\mathfrak{F}_b(B,\mathfrak{B}) \text{ with } |f|\le 1,\ x\in A\}.\\ &= \sup\{\,|\langle g(x,.),M(x,.)\rangle|\colon g\in\mathfrak{F}_b(A\times B,\mathfrak{A}\otimes\mathfrak{B}) \text{ with } |g|\le 1,\ x\in A\}\\ &= \sup\{\,|\langle g(\xi,.),M(x,.)\rangle|\colon g\in\mathfrak{F}_b(A\times B,\mathfrak{A}\otimes\mathfrak{B}) \text{ with } |g|\le 1,\ \xi\in A,\ x\in A\}\end{aligned}$$

We also call the topology $\beta(A,\mathfrak{A};B,\mathfrak{B})$ induced by this norm on $\mathcal{K}(A,\mathfrak{A};B,\mathfrak{B})$ and $\mathcal{MK}(A,\mathfrak{A};B,\mathfrak{B})$ the topology of uniform convergence. It is a very fine topology.

Let us now have a look at a topology which is much coarser, the weak topology. We assume that Y is a polish space, $\mathfrak{Y}$ is its Borel σ-algebra (generated by the open subsets of Y), and $\mathfrak{C}_b(Y)$ is the linear space of real valued and bounded continuous functions on Y. $\mathfrak{M}^b(Y,\mathfrak{Y})$ and $\mathfrak{C}_b(Y)$ with the bilinear form $\langle f,\mu\rangle$ is a separating dual pair. Hence, the weak topology on $\mathfrak{M}^b(Y,\mathfrak{Y})$, denoted by $\sigma(\mathfrak{M}^b, \mathfrak{C}_b)$, is a Hausdorff topology. Since Y is polish, i.e. separable and completely metrizable, also the restriction of this weak topology on $\mathfrak{M}^{1+}(Y) := \mathfrak{M}^{1+}(Y,\mathfrak{Y})$ is separable and completely metrizable; e.g. see Billingsley (1) or Parthasarathy (8). The weak topology $\sigma(\mathfrak{M}^b, \mathfrak{C}_b)$ is the coarsest topology such that the linear functional $\langle f,.\rangle\colon \mathfrak{M}^b(Y,\mathfrak{Y}) \to \mathbb{R}$ with $\mu \to \langle f,\mu\rangle$ is continuous for every $f\in\mathfrak{C}_b(Y)$. A sequence $(\mu_n\colon n\in\mathbb{N})$ in $\mathfrak{M}^b(Y,\mathfrak{Y})$ converges weakly to $\mu\in\mathfrak{M}^b(Y,\mathfrak{Y})$ if and only if for every $f\in\mathfrak{C}_b(Y)$ the sequence $(\langle f,\mu_n\rangle\colon n\in\mathbb{N})$ converges to $\langle f,\mu\rangle$ in $\mathbb{R}$. The weak convergence on $\mathfrak{M}^{1+}(Y)$ is characterized in many ways by the Portemanteau Theorem, e.g. see Billingsley (1) or Parthasarathy (8).

Let $\mu_n,\mu\in\mathfrak{M}^{1+}(\mathbb{R}^k)$ and $F_n\colon \mathbb{R}^k \to [0,1]$ with $y \to F_n(y) := \mu_n(H_y)$ for $H_y := \{z\in\mathbb{R}^k\colon z\le y\}$ and $F\colon \mathbb{R}^k \to [0,1]$ with $y \to F(y) := \mu(H_y)$ be the cumulative distribution functions of μ_n

and μ, respectively. Then the sequence $(\mu_n: n\in\mathbb{N})$ converges weakly to μ if and only if $(F_n(y): n\in\mathbb{N})$ converges to $F(y)$ for every $y\in\mathbb{R}^k$ in which F is continuous.

In general, the weak topology $\sigma(\mathfrak{M}^b,\mathfrak{C}_b)$ is much coarser than the topology $\beta(\mathfrak{M}^b,\mathfrak{F}_b)$ of uniform convergence on $\mathfrak{M}^b(Y,\mathfrak{Y})$ or on $\mathfrak{M}^{1+}(Y)$. For example, if $(y_n: n\in\mathbb{N})$ is a sequence of points y_n in Y converging to a point $y\in Y$ with $y\neq y_n$ for all $n\in\mathbb{N}$, then the normed Dirac measures $\delta_{y_n}: \mathfrak{Y}\to[0,1]$ (with $\delta_{y_n}(L):=1$ for $y_n\in L$ and $\delta_{y_n}(L):=0$ for $y_n\notin L$) are converging weakly to δ_y; because of $\|\delta_{y_n}-\delta_y\|=2$ for all $n\in\mathbb{N}$, this sequence is not converging uniformly.

However, if Y is finite (and hence endowed with the discrete topology) or Y is countably infinite and endowed with the discrete topology, then the two topologies coincide.

Let $\lambda: \mathfrak{Y}\to[0,\infty]$ be a σ-finite measure and $\mu_n,\mu\in\mathfrak{M}^{1+}(Y)$ which are absolutely continuous with respect to λ. If the sequence $(d\mu_n/d\lambda: n\in\mathbb{N})$ of the Radon-Nikodym derivatives of μ_n with respect to λ converges λ-almost everywhere to the Radon-Nikodym derivative $d\mu/d\lambda$, then according to a theorem due to Scheffé (e.g. see Billingsley (1)) the sequence $(\mu_n: n\in\mathbb{N})$ converges uniformly to μ and, hence, also weakly. Using this result, it is possible to establish convergence of sequences of probability measures and Markov kernels with respect to the topology of uniform convergence and to the weak topology. This can be done for one-dimensional and higher-dimensional Gaussian distributions which are of special interest here. For example, a sequence of one-dimensional Gaussian distributions $N(\alpha_n,\sigma_n^2)$ converges uniformly to $N(\alpha,\sigma^2)$ if and only if $\alpha_n\to\alpha$ and $\sigma_n\to\sigma$. The same is true for weak convergence. Also here the two topologies coincide. A first result about the continuity of the iterative construction of a probability measure $\mathbb{P}$ associated to a CPN via (I) is the following

Theorem 1: Let $(\mathbb{P}_m: m\in\mathbb{N})$ be the sequence of probability measures $\mathbb{P}_m$ associated to the CPN with the acyclic finite graph $G:=(V,E)$ and the family of Markov kernels $(\mathcal{P}_v^m: v\in V)$ via (I). Furthermore, let $\mathbb{P}$ be the probability measure associated to the CPN with the graph $G:=(V,E)$ and the family of Markov kernels $(\mathcal{P}_v: v\in V)$ via (I). If the sequence $(\mathcal{P}_v^m: m\in\mathbb{N})$ converges uniformly to $\mathcal{P}_v$ for every $v\in V$, then $(\mathbb{P}_m: m\in\mathbb{N})$ converges uniformly to $\mathbb{P}$.

Proof of Theorem 1: Let P_i^m and P_i be the Markov kernels associated to $\mathcal{P}_{v_i}^m$ and $\mathcal{P}_{v_i}$, respectively. According to our assumption we have that for every $i=0,\ldots,n$ $\|\mathcal{P}_{v_i}^m-\mathcal{P}_{v_i}\|\to 0$ for $m\to\infty$. This implies $\|P_i^m-P_i\|\to 0$ for $m\to\infty$. Let be $\varepsilon>0$ and $i\in\{0,\ldots,n\}$. Because of the uniform convergence of the sequence $(P_i^m: m\in\mathbb{N})$ to P_i we can find an $M_i\in\mathbb{N}$ such that

$$\varepsilon \geq \|P_i^m - P_i\| =$$

$$= \sup\{|\langle G(x_0,\ldots,x_{i-1},.), P_i^m(x_0,\ldots,x_{i-1};.)-P_i(x_0,\ldots,x_{i-1};.)\rangle|: G\in\mathfrak{F}_b(S^{(i)},\mathfrak{S}^{(i)}) \text{ with } |G|\leq 1;\ (x_0,\ldots,x_{i-1})\in S^{(i-1)}\} \quad \text{for } m\geq M_i.$$

Let be $N:=\max\{M_i: i=0,\ldots,n\}$ and $g\in\mathfrak{F}_b(S,\mathfrak{S})$ with $|g|\leq 1$.

$$\langle g,\mathbb{P}_m-\mathbb{P}\rangle = \sum_{i=0}^{n}\Big(\int_{S_{v_0}}\cdots\int_{S_{v_i}}\int_{S_{v_{i+1}}}\cdots\int_{S_{v_n}} g(x_0,\ldots,x_n)\,P_n(x_0,\ldots,x_{n-1};dx_n)\ldots$$

$$\ldots\ P_{i+1}(x_0,\ldots,x_i;dx_{i+1})\,P_i^m(x_0,\ldots,x_{i-1};dx_i)\ldots P_0^m(dx_0)\ -$$

$$- \int_{S_{v_0}} \cdots \int_{S_{v_{i-1}}} \int_{S_{v_i}} \int_{S_{v_{i+1}}} \cdots \int_{S_{v_n}} g(x_0, \dots, x_n) P_n(x_0, \dots, x_{n-1}; dx_n) \dots$$

$$\dots P_{i+1}(x_0, \dots, x_i; dx_{i+1}) P_i(x_0, \dots, x_{i-1}; dx_i) P_{i-1}^m (x_0, \dots, x_{i-2}; dx_{i-1}) .. P_0^m (dx_0) \Big)$$

We define g_i: $S^{(i)} \rightarrow \mathbb{R}$ by

$$(x_0, \dots, x_i) \rightarrow \int_{S_{v_{i+1}}} \cdots \int_{S_{v_n}} g(x_0, \dots, x_n) P_n(x_0, \dots, x_{n-1}; dx_n) \dots P_{i+1}(x_0, \dots, x_i; dx_{i+1}).$$

Then $g_i \in \mathfrak{F}_b(S^{(i)}, \mathfrak{S}^{(i)})$ and $|g_i| \le 1$. Hence, for $m \ge N$,

$$\left| \langle g, \mathbb{P}_m - \mathbb{P} \rangle \right| \le \sum_{i=0}^{n} \int_{S_{v_0}} \cdots \int_{S_{v_{i-1}}} \left| \int_{S_{v_i}} g_i(x_0, \dots, x_i) \Big(P_i^m(x_0, \dots, x_{i-1}; dx_i) - \right.$$

$$\left. - P_i(x_0, \dots, x_{i-1}; dx_i) \Big) \right| P_{i-1}^m (x_0, \dots, x_{i-2}; dx_{i-1}) .. P_0^m (dx_0) \le (n+1)\, \varepsilon$$

Finally, we have

$$\| \mathbb{P}_m - \mathbb{P} \| = \sup \{ \left| \langle g, \mathbb{P}_m - \mathbb{P} \rangle \right| : g \in \mathfrak{F}_b(S, \mathfrak{S}) \text{ with } |g| \le 1 \} \le (n+1)\, \varepsilon \text{ for } m \ge N \text{ ; q.e.d.}$$

The topologies of uniform convergence $\beta(S^{(i-1)}, \mathfrak{S}^{(i-1)}; S_i, \mathfrak{S}_i)$ and $\beta(S, \mathfrak{S})$ on $\mathcal{MK}(S^{(i-1)}, \mathfrak{S}^{(i-1)}; S_i, \mathfrak{S}_i)$ and $\mathfrak{M}^{1+}(S, \mathfrak{S})$, respectively, which we considered in Theorem 1 are very fine topologies. Let us now weaken the topology on the spaces of Markov kernels to the topology of pointwise weak convergence:

The topology of pointwise weak convergence on the linear space $\mathcal{K}(Y, \mathfrak{Y}; Z, \mathfrak{Z})$ of kernels M: $Y \times \mathfrak{Z} \rightarrow \mathbb{R}$ from the polish space Y with the Borel σ-algebra $\mathfrak{Y}$ into the polish space Z with the Borel σ-algebra $\mathfrak{Z}$ is the coarsest topology such that the seminorms $M \rightarrow |\langle f, M(y;.)\rangle|$ are continuous for all $y \in Y$ and all $f \in \mathfrak{C}_b(Z)$. We denote this topology by $\pi\sigma(Y, \mathfrak{Y}; Z, \mathfrak{Z})$ and call a sequence $(M_m: m \in \mathbb{N})$ in $\mathcal{K}(Y, \mathfrak{Y}; Z, \mathfrak{Z})$ converging pointwise weakly to $M \in \mathcal{K}(Y, \mathfrak{Y}; Z, \mathfrak{Z})$ if and only if it converges with respect to the topology of pointwise weak convergence. Such a sequence converges if and only if the sequence $(M_m(y;.): m \in \mathbb{N})$ in $\mathfrak{M}^b(Z, \mathfrak{Z})$ converges weakly to $M(y;.)$ for every $y \in Y$.

If we reduce the assumption of the uniform convergence of the sequences of the Markov kernels in Theorem 1 to pointwise weak convergence, the assertion of this theorem needs not to be true. This is demonstrated by the following very simple example:

Example 1: Let be $V := \{0,1\}$, $S_0 := [0,1] := \{y \in \mathbb{R}: 0 \le y \le 1\}$ with the usual topology and with the Borel σ-algebra $\mathfrak{S}_0$, $S_1 := \{0,1\}$ with the discrete topology and with $\mathfrak{S}_1 := \mathfrak{P}(S_1)$, and $E := \{(0,1)\}$. Then $G := (V, E)$ is an acyclic network with only one possible well-ordering. Let us define Markov kernels

$\mathcal{P}_0$: $\mathfrak{S}_0 \rightarrow [0,1]$ with $A \rightarrow \mathcal{P}_0(A) := 1/2\, \lambda(A) + 1/2\, \delta_0(A)$ and

$\mathcal{P}_1$: $S_0 \times \mathfrak{S}_1 \rightarrow [0,1]$ with $(x_0; B) \rightarrow \mathcal{P}_1(x_0; B) := \delta_0(B)$ for $x_0 = 0$ and

$\mathcal{P}_1(x_0; B) := \delta_1(B)$ for $x_0 \neq 0$

where λ: $\mathfrak{S}_0 \rightarrow [0,1]$ is the Lebesgue measure. Then the network G and the family $(\mathcal{P}_v: v \in V)$ of Markov kernels are a CPN. The probability measure $\mathbb{P}$: $\mathfrak{S} \rightarrow [0,1]$ which is associated to this CPN via (I) is determined by

$$\mathbb{P}(A \times B) = 1/2\, \delta_1(B)\, \lambda(A) + 1/2\, \delta_{(0,0)}(A \times B) \quad \text{for } A \in \mathfrak{S}_0 \text{ and } B \in \mathfrak{S}_1 .$$

For $m \in \mathbb{N}$ we define the Markov kernels

$\mathcal{P}_0^m$: $\mathfrak{S}_0 \rightarrow [0,1]$ with $A \rightarrow \mathcal{P}_0^m(A) := 1/2\, \lambda(A) + 1/2\, \delta_{1/m}(A)$ and $\mathcal{P}_1^m := \mathcal{P}_1$.

The sequence $(\mathcal{P}_0 : m \in \mathbb{N})$ of Markov kernels (which are simply probability measures in this case) is converging (pointwise) weakly to $\mathcal{P}_0$. Obviously, also the sequence $(\mathcal{P}_1^m: m\in\mathbb{N})$ of Markov kernels is converging pointwise weakly to $\mathcal{P}_1$. For every $m\in\mathbb{N}$ the family $(\mathcal{P}_v^m: v\in V)$ of Markov kernels together with G is a CPN. The measure $\mathbb{P}_m: \mathfrak{S} \to [0,1]$ which associated to this CPN via (I) is determined by

$$\mathbb{P}_m(A\times B) = 1/2\ \delta_1(B)\ \lambda(A) + 1/2\ \delta_{(1/m,1)}(A\times B) \quad \text{for } A\in\mathfrak{S}_0 \text{ and } B\in\mathfrak{S}_1 .$$

Since the sequence $(\delta_{(1/m,1)}: m\in\mathbb{N})$ in $\mathfrak{M}^{1+}(S)$ converges weakly to $\delta_{(0,1)}\in\mathfrak{M}^{1+}(S)$ and not to $\delta_{(0,0)}$ the sequence $(\mathbb{P}_m: m\in\mathbb{N})$ in $\mathfrak{M}^{1+}(S)$ does not converge weakly to $\mathbb{P}\in\mathfrak{M}^{1+}(S)$. We should notice that the Markov kernel $\mathcal{P}_1$ is no Feller kernel.

A Markov kernel $P: Y\times\mathfrak{Z} \to [0,1]$ from the polish space Y with the Borel σ-algebra $\mathfrak{Y}$ to the polish space Z with the Borel σ-algebra $\mathfrak{Z}$ is called a *Feller kernel* if one of the following two equivalent conditions is fulfilled:

(F1) For every sequence $(y_m: m\in\mathbb{N})$ in Y converging to $y\in Y$ the sequence $(P(y_m;.): m\in\mathbb{N})$ in $\mathfrak{M}^{1+}(Z)$ converges weakly to $P(y;.)\in\mathfrak{M}^{1+}(Z)$.

(F2) $Pg\in\mathfrak{C}_b(Y)$ with $y\to Pg(y):=\int_Z g(z)\, P(y;dz)$ for every $g\in\mathfrak{C}_b(Z)$.

Subsequently, we consider a CPN with the fixed finite acyclic network $G:=(V,E)$ and the family $\mathcal{P}:=(\mathcal{P}_v: v\in V)$ of Markov kernels $\mathcal{P}_v: S(Pa(v))\times\mathfrak{S}_v \to [0,1]$. For sake of simplicity we assume that $V:=\{0, \dots, n\}$ is well ordered by the natural order.

Furthermore, we consider a sequence of families $(\mathcal{P}_v^m: v\in V)$ of Markov kernels $\mathcal{P}_v^m: S(Pa(v))\times\mathfrak{S}_v \to [0,1]$ and a family $(\mathcal{P}_v: v\in V)$ of Markov kernels $\mathcal{P}_v: S(Pa(v))\times\mathfrak{S}_v \to [0,1]$. We shall assume:

(F) All Markov kernels $\mathcal{P}_v^m$ and $\mathcal{P}_v$ are Feller kernels.

(C) The sequence $(\mathcal{P}_v^m : k\in\mathbb{N})$ is pointwise weakly converging to $\mathcal{P}_v$ for every $v\in V$.

Again, we associate to every Markov kernel

$$\mathcal{P}_v^m : S(Pa(v))\times\mathfrak{S}_v \to [0,1] \quad \text{with } ((x_u: u\in Pa(v)),B) \to \mathcal{P}_v^m((x_u: u\in Pa(v)); B)$$

and $\mathcal{P}_v: S(Pa(v))\times\mathfrak{S}_v \to [0,1]$ with $((x_u: u\in Pa(v)),B) \to \mathcal{P}_v((x_u: u\in Pa(v)); B)$ the Markov kernel

$$P_v^m : S^{(v-1)}\times\mathfrak{S}_v \to [0,1] \quad \text{with}$$
$$((x_0, \dots, x_{v-1}),B) \to P_v^m(x_0, \dots, x_{v-1};B) := \mathcal{P}_v^m((x_u: u\in Pa(v)); B)$$

and $P_v: S^{(v-1)}\times\mathfrak{S}_v \to [0,1]$ with

$$((x_0, \dots, x_{v-1}),B) \to P_v(x_0, \dots, x_{v-1};B) := \mathcal{P}_v(x_u: u\in Pa(v)); B),$$

respectively, and notice that the following properties (F*) and (C*) are equivalent to (F) and (C), respectively:

(F*) The Markov kernels P_v^m and P_v for $v\in V$ and $m\in\mathbb{N}$ are Feller kernels.

(C*) The sequence $(P_v^m: m\in\mathbb{N})$ is pointwise weakly converging to P_v for every $v\in V$.

At first we shall state and prove some propositions which show that in very simple and special situations the assertion of Theorem 1 remains true if we replace "uniform convergence" by "pointwise weak convergence" and demand that all Markov kernels are Feller kernels. Then we shall present an example which shows that even in still very simple situations of this kind this assertion is not true in general. Finally we shall state and discuss additional conditions, which we call "strong Feller convergence property" and "compact uniform tightness", such that a modified assertion of Theorem 1 remains true.

Necessarily, the formulation of the resulting convergence theorems is somewhat more complicated, they are our Theorem 2 and Theorem 3.

Proposition 1: Let be $V:=\{0,1\}$ and $\mathcal{P}_0^m = \mathcal{P}_0$ for all $m \in \mathbb{N}$. Under the assumption (C) the sequence $(\mathbb{P}_m : m \in \mathbb{N})$ of probability measures $\mathbb{P}_m \in \mathfrak{M}^{1+}(S,\mathfrak{S})$ associated to $(\mathcal{P}_v^m: v \in V)$ via (I) converges weakly to $\mathbb{P} \in \mathfrak{M}^{1+}(S,\mathfrak{S})$ which is associated to $(\mathcal{P}_v: v \in V)$ via (I).

Proof of Proposition 1: Let be $h \in \mathfrak{C}_b(S^{(1)})$ bounded by γ. We shall show that the sequence $(\mathbb{P}_m(h): m \in \mathbb{N})$ in $\mathbb{R}$ converges to $\mathbb{P}(h)$ where

$$\mathbb{P}_m(h) := \int_{S_0}\int_{S_1} h(x_0,x_1)\ P_1^m(x_0;dx_1)\ P_0(dx_0) = \int_{S_0} f_m(x_0)\ P_0(dx_0)$$

with $f_m(x_0) := \int_{S_1} h(x_0,x_1)\ P_1^m(x_0;dx_1)$ and $\mathbb{P}(h)$ is defined analogously.

Since every continuous function is Borel measurable, the function $f_m: (S_0,\mathfrak{S}_0) \to \mathbb{R}$ is measurable by a generalized version of Fubini's theorem. The absolute value of f_m is bounded by γ. Since for every $x_0 \in S_0$ the function $h(x_0,.): S_1 \to \mathbb{R}$ is continuous, and since because of (C) and hence (C^*) the sequence $(P_1^m(x_0;.): m \in \mathbb{N})$ in $\mathfrak{M}^{1+}(S_1,\mathfrak{S}_1)$ converges weakly to $P_1(x_0;.) \in \mathfrak{M}^{1+}(S_1,\mathfrak{S}_1)$, the sequence $(f_m: m \in \mathbb{N})$ of measurable functions converges pointwise to the bounded measurable function

$$f: (S_0,\mathfrak{S}_0) \to \mathbb{R} \quad \text{with} \quad f(x_0) := \int_{S_1} h(x_0,x_1)\ P_1(x_0;dx_1).$$

Since P_0 is a probability measure, we get from the theorem of dominated convergence that the sequence $(\int_{S_0} f_m(x_0)\ P_0(dx_0): m \in \mathbb{N})$ in $\mathbb{R}$ converges to

$$\int_{S_0} f(x_0)\ P_0(dx_0) = \int_{S_0}\int_{S_1} h(x_0,x_1)\ P_1(x_0;dx_1)\ P_0(dx_0) = \mathbb{P}(h)\ ; \text{ q.e.d.}$$

Proposition 2: Let be $V:=\{0,1\}$ and $\mathcal{P}_1^m = \mathcal{P}_1$ for all $m \in \mathbb{N}$. Under the assumptions (F) and (C) the sequence $(\mathbb{P}_m: m \in \mathbb{N})$ of probability measures $\mathbb{P}_m \in \mathfrak{M}^{1+}(S,\mathfrak{S})$ associated to $(\mathcal{P}_v^m: v \in V)$ via (I) converges weakly to $\mathbb{P} \in \mathfrak{M}^{1+}(S,\mathfrak{S})$ which is associated to $(\mathcal{P}_v: v \in V)$ via (I).

As we saw in Example 1, the assuption (F) in Proposition 2 is essential. As the following Example 2 will show, the assumption "$\mathcal{P}_1^m = \mathcal{P}_1$ for all $m \in \mathbb{N}$" in Proposition 2 is essential.

Proof of Proposition 2: Let $\mathfrak{U}_b(S^{(1)})$ be the set of real-valued, bounded and uniformly continuous functions on $S^{(1)}$ (with respect to some metric inducing the topology). For $h \in \mathfrak{U}_b(S^{(1)})$ bounded by γ we shall show that the sequence $(\mathbb{P}_m(h): m \in \mathbb{N})$ in $\mathbb{R}$ converges to $\mathbb{P}(h)$ where

$$\mathbb{P}_m(h) := \int_{S_0}\int_{S_1} h(x_0,x_1)\ P_1(x_0;dx_1)\ P_0^m(dx_0) = \int_{S_0} f(x_0)\ P_0^m(dx_0),$$

with $f(x_0) := \int_{S_1} h(x_0,x_1)\ P_1(x_0;dx_1)$, and $\mathbb{P}(h)$ is defined analogously.

According to the Portemanteau theorem this is sufficient. Because of (F) and hence (F^*) and the uniform continuity of h, the function $f: S_0 \to \mathbb{R}$ is continuous. Furthermore, the absolute value of f is bounded by γ. Because of (C) and hence (C^*) the sequence $(P_0^m: m \in \mathbb{N})$ in $\mathfrak{M}^{1+}(S_0,\mathfrak{S}_0)$ converges weakly to $P_0 \in \mathfrak{M}^{1+}(S_0,\mathfrak{S}_0)$. Since f is continuous and bounded, the sequence $(\int_{S_0} f(x_0)\ P_0^m(dx_0): m \in \mathbb{N})$ in $\mathbb{R}$ converges to

$$\int_{S_0} f(x_0)\ P_0(dx_0) = \int_{S_0}\int_{S_1} h(x_0,x_1)\ P_1(x_0;dx_1)\ P_0(dx_0) = \mathbb{P}(h);\ \text{q.e.d.}$$

Example 2: $S_0 := \{0\} \cup \{1/k\colon k \in \mathbb{N}\}$ and $S_1 := \{0,1\}$ with the usual topologies are polish, and $\mathfrak{S}_0$ and $\mathfrak{S}_1$ are the Borel σ-algebras, respectively. For $P_0 := \delta_0$ and $P_0^m := \delta_{1/m}$ for $m \in \mathbb{N}$ the sequence $(P_0^m\colon m \in \mathbb{N})$ in $\mathfrak{M}^{1+}(S_0,\mathfrak{S}_0)$ converges weakly to $P_0 \in \mathfrak{M}^{1+}(S_0,\mathfrak{S}_0)$. Define $P_1(x_0;.) := \delta_0$ for all $x_0 \in S_0$ and $P_1^m(x_0;.) := \delta_1$ for $x_0 = 1/m$ and $:= \delta_0$ otherwise. Then P_1 and P_1^m for $m \in \mathbb{N}$ are Feller kernels. For all $x_0 \in S_0$ the sequence $(P_1^m(x_0;.)\colon m \in \mathbb{N})$ in $\mathfrak{M}^{1+}(S_1,\mathfrak{S}_1)$ converges weakly to $P_1(x_0;.) \in \mathfrak{M}^{1+}(S_1,\mathfrak{S}_1)$. (P_0,P_1) and (P_0^m,P_1^m) have the properties (F^*) and (C^*). $g\colon S_1 \to \mathbb{R}$ with $g(0) := 0$ and $g(1) := 1$ is continuous and bounded.

Define $f_m(x_0) := \int_{S_1} g(x_1)\ P_1^m(x_0;dx_1) = 1$ for $x_0 = 1/m$ and $= 0$ otherwise,

$f(x_0) := P_1 g\,(x_0) := \int_{S_1} g(x_1)\ P_1(x_0;dx_1) = 0$ for all $x_0 \in S_0$.

Then $f_m\colon S_0 \to \mathbb{R}$ for $m \in \mathbb{N}$ and $f\colon S_0 \to \mathbb{R}$ are continuous and bounded. The sequence $(f_m(x_0)\colon m \in \mathbb{N})$ in $\mathbb{R}$ converges to $f(x_0)$ for all $x_0 \in S_0$. If $h(x_0,x_1) := g(x_1)$, then $h \in \mathfrak{C}_b(S_0 \times S_1)$,

$$\mathbb{P}_m(h) := \int_{S_0}\int_{S_1} h(x_0,x_1)\ P_1^m(x_0;dx_1)\ P_0^m(dx_0) = \int_{S_0} f_m(x_0)\ P_0^m(dx_0) = 1$$

for all $m \in \mathbb{N}$, and

$$\mathbb{P}(h) := \int_{S_0}\int_{S_1} h(x_0,x_1)\ P_1(x_0;dx_1)\ P_0(dx_0) = \int_{S_0} f(x_0)\ P_0(dx_0) = 0\ .$$

Hence, the sequence $(\mathbb{P}_m\colon m \in \mathbb{N})$ in $\mathfrak{M}^{1+}(S_0 \times S_1)$ does not converge weakly to $\mathbb{P} \in \mathfrak{M}^{1+}(S_0 \times S_1)$.

Example 2 shows that the concept of pointwise weak convergence is too weak for our purposes. We have to introduce additional assumptions on the kind of convergence of the sequences of kernels involved. Having Prohorov's Uniform Tightness Theorem (e.g. Billingsley (1) or Parthasarathy (8)) in mind, the first idea would be to demand some uniform (!) kind of uniform tightness of the sequences of kernels. However, Example 2 shows that even this would not be enough. We notice:

In Example 2 the product space $S_0 \times S_1$ is compact. Furthermore, the functions $g\colon S_1 \to \mathbb{R}$ and $h\colon S_0 \times S_1 \to \mathbb{R}$ are bounded and uniformly continuous. The sequence $(f_m\colon m \in \mathbb{N})$ of functions $f_m\colon S_0 \to \mathbb{R}$ converges only pointwise and not uniformly to the function $f\colon S_0 \to \mathbb{R}$. Obviously, the sets of probability measures $\{P_0^m\colon m \in \mathbb{N}\} \cup \{P_0\}$ and $\{P_1^m(x_0;.)\colon m \in \mathbb{N},\ x_0 \in S_0\} \cup \{P_1(x_0;.)\colon m \in \mathbb{N},\ x_0 \in S_0\}$ are uniformly tight.

Let Y and Z be polish spaces. A sequence $(M_m\colon m \in \mathbb{N})$ in $\mathcal{MK}(Y,\mathfrak{Y};Z,\mathfrak{Z})$ is said to have the *strong Feller convergence property* with respect to $M \in \mathcal{MK}(Y,\mathfrak{Y};Z,\mathfrak{Z})$ if it has the following property:

(SFCP) For every uniformly bounded sequence $(g_m\colon m \in \mathbb{N})$ in $\mathfrak{C}_b(Y \times Z)$ converging uniformly on compact sets to $g \in \mathfrak{C}_b(Y \times Z)$ the sequence $(f_m\colon m \in \mathbb{N})$ of functions $f_m\colon Y \to \mathbb{R}$ with $y \to f_m(y) := \int_Z g_m(y,z)\ M_m(y;dz) = \langle g_m(y,.), M_m(y;.)\rangle$ is converging uniformly on compact sets to $f\colon Y \to \mathbb{R}$ with $y \to f(y) := \langle g(y,.), M(y;.)\rangle$.

Also here we consider measures as degenerated kernels and define the SFCP for these degenerated kernels with the obvious changes of the definition. Obviously, a sequence

$(M_m\colon m\in\mathbb{N})$ in $\mathcal{MK}(Y,\mathfrak{Y};Z,\mathfrak{Z})$ which has the SFCP with respect to $M\in\mathcal{MK}(Y,\mathfrak{Y};Z,\mathfrak{Z})$ converges pointwise weakly to M. The following propositions show important properties of the SFCP:

Proposition 3: Let P be a Feller kernel from $(Y,\mathfrak{Y})$ to $(Z,\mathfrak{Z})$. Then the constant sequence $(P_m\colon m\in\mathbb{N})$ with $P_m := P$ has the SFCP, with respect to P.
(Here we omit the proof. It is based on Prohorov's uniform tightness theorem.)

Proposition 4: If the sequence $(\mu_m\colon m\in\mathbb{N})$ in $\mathfrak{M}^{1+}(Y)$ converges weakly to $\mu\in\mathfrak{M}^{1+}(Y)$, then it has the SFCP with respect to μ.

Proof of Proposition 4: Let $(g_m\colon m\in\mathbb{N})$ be a sequence in $\mathfrak{C}_b(Y)$ which is uniformly bounded by $\gamma\geq 1$ and which converges uniformly on compact sets to $g\in\mathfrak{C}_b(Y)$. Since $(\mu_m\colon m\in\mathbb{N})$ converges weakly to μ, according to Prohorov's uniform tightness theorem for $\varepsilon>0$ there is a compact set C such that $\mu_m(Y\setminus C)\leq\varepsilon$ for every $m\in\mathbb{N}$. Because of the uniform convergence there is an $M\in\mathbb{N}$ such that $|g_m(y)-g(y)|\leq\varepsilon$ for all $y\in C$ and $m\geq M$. Since $g\in\mathfrak{C}_b(Y)$ and $(\mu_m\colon m\in\mathbb{N})$ converges weakly to μ, there is an $N\in\mathbb{N}$ with $N\geq M$ such that $|\int_Y g(y)\,\mu_m(dy) - \int_Y g(y)\,\mu(dy)|\leq\varepsilon$ for all $m\geq N$. From this we get

$$|\int_Y g_m(y)\,\mu_m(dy) - \int_Y g(y)\,\mu(dy)| \leq$$

$$\leq |\int_Y g_m(y)\,\mu_m(dy) - \int_C g_m(y)\,\mu_m(dy)| + |\int_C g_m(y)\,\mu_m(dy) - \int_C g(y)\,\mu_m(dy)|$$

$$+ |\int_C g(y)\,\mu_m(dy) - \int_Y g(y)\,\mu_m(dy)| + |\int_Y g(y)\,\mu_m(dy) - \int_Y g(y)\,\mu(dy)| \leq$$

$\leq 4\,\gamma\,\varepsilon$; q.e.d.

Proposition 5: If $P\colon Y\times\mathfrak{Z}\to[0,1]$ is a Feller kernel from $(Y,\mathfrak{Y})$ to $(Z,\mathfrak{Z})$, $h\in\mathfrak{C}_b(Y\times Z)$, and $f\colon Y\to\mathbb{R}$ with $y\to f(y) := \langle h(y,.),P(y,.)\rangle$, then $f\in\mathfrak{C}_b(Y)$.

Proof of Proposition 5: Let $(y_m\colon m\in\mathbb{N})$ be a sequence in Y which converges to $y\in Y$. Since P is a Feller kernel, the sequence $(P(y_m;.)\colon m\in\mathbb{N})$ converges weakly to $P(y;.)$. Because of Proposition 4 this sequence has the SFCP with respect to $P(y;.)$. $h(y_m,.)\in\mathfrak{C}_b(Z)$ and $h(y,.)\in\mathfrak{C}_b(Z)$. We show that the sequence $(h(y_m,.)\colon m\in\mathbb{N})$ converges uniformly on compact sets to $h(y,.)$. Let $L\subset Z$ be compact and $\varepsilon>0$. For every $z\in L$ there is an open neighbourhood U_z of y in Y and an open neighbourhood V_z of z in Z such that

$$|h(\upsilon,\zeta)-h(y,z)|\leq\varepsilon \quad\text{for all } \upsilon\in U_z \text{ and } \zeta\in V_z.$$

Since L is compact, there is a $k\in\mathbb{N}$ and $z_1,\ldots,z_k\in L$ such that

$$L\subset\bigcup_{i=1}^k V_{z_i} \quad\text{and}\quad U := \bigcap_{i=1}^k U_{z_i} \text{ is an open neighbourhood of y in Y.}$$

Hence, there is an $N\in\mathbb{N}$ with $y_m\in U$ for all $m\geq N$. For $m\geq N$ and $\zeta\in L$ there is a $p\in\{1,\ldots,k\}$ such that $\zeta\in V_{z_p}$ and $|h(y_m,\zeta)-h(y,\zeta)|\leq|h(y_m,\zeta)-h(y,z_p)|+|h(y,\zeta)-h(y,z_p)|\leq 2\varepsilon$. From the SFCP we get $f(y_m)=\langle h(y_m,.),P(y_m,.)\rangle\to\langle h(y,.),P(y,.)\rangle = f(y)$ for $m\to\infty$; q.e.d.

Theorem 2: Let $(\mathbb{P}_m\colon m\in\mathbb{N})$ be the sequence of probability measures $\mathbb{P}_m$ associated via (I) to the CPN with the acyclic finite graph $G := (V,E)$ and the family of Feller kernels $(\mathcal{P}_v^m : v\in V)$ and $\mathbb{P}$ the probability measure associated to the CPN with the graph $G := (V,E)$ and the family of Feller kernels $(\mathcal{P}_v\colon v\in V)$. For convenience, let be $V := \{0,\ldots,n\}$ with respect to the natural order be well-ordered in the sense of Lauritzen-

Dawid-Larsen-Leimer. Finally, for $v \in V$ associate the kernels P_v^m: $S^{(v-1)} \times \mathfrak{S}_v \to [0,1]$ and P_v: $S^{(v-1)} \times \mathfrak{S}_v \to [0,1]$ to $\mathcal{P}_v^m$: $S(Pa(v)) \times \mathfrak{S}_v \to [0,1]$ and $\mathcal{P}_v$: $S(Pa(v)) \times \mathfrak{S}_v \to [0,1]$, respectively. If $(P_v^m: m \in \mathbb{N})$ has the strong Feller convergence property with respect to P_v for every $v \in V$, then $(\mathbb{P}_m: m \in \mathbb{N})$ converges weakly to $\mathbb{P}$.

Proof of Theorem 2: Let be $g \in \mathfrak{C}_b(S^{(n)})$ bounded by γ, and define $g_{n,m} := g =: g_n$. Obviously, the sequence $(g_{n,m}: m \in \mathbb{N})$ is uniformly bounded and converges uniformly on compact subsets of $S^{(n)}$. For $k = n-1, \ldots, 0$ define by recursion

$g_{k,m}$: $S^{(k)} \to \mathbb{R}$ with $(x_0, \ldots, x_k) \to \langle g_{k+1,m}(x_0, \ldots, x_k, .), P_k^m(x_0, \ldots, x_k; .)\rangle$ and

g_k: $S^{(k)} \to \mathbb{R}$ with $(x_0, \ldots, x_k) \to \langle g_{k+1}(x_0, \ldots, x_k, .), P_k(x_0, \ldots, x_k; .)\rangle$.

The sequence $(g_{k,m}: m \in \mathbb{N})$ is uniformly bounded by γ. Since all the Markov kernels are Feller kernels, because of Proposition 5 the functions $g_{k,m}$ and g_k are continuous. Since $(P_k^m: m \in \mathbb{N})$ has the SFCP with respect to P_k, the sequence $(g_{k,m}: m \in \mathbb{N})$ converges uniformly on compact sets to g_k.

$$\langle g, \mathbb{P}_m \rangle = \int_{S_0} \int_{S_1} \ldots \int_{S_n} g(x_0, \ldots, x_n)\, P_n^m\, (x_0, \ldots, x_{n-1}; dx_n) \ldots P_1^m\, (x_0; dx_1)\, P_0\, (dx_0)$$

$$= \ldots\ldots\ldots = \int_{S_0} g_{0,m}(x_0)\, P_0\, (dx_0) = \langle g_{0,m}, P_0 \rangle$$

Therefore, the sequence $(\langle g_{0,m}, P_0 \rangle: m \in \mathbb{N})$ converges to $\langle g_0, P_0 \rangle = \langle g, \mathbb{P} \rangle$; q.e.d.

In Theorem 2 we made the assumption "$(P_v^m: m \in \mathbb{N})$ has the strong Feller convergence property with respect to P_v for every $v \in V$". This assumption is stronger than the more natural and weaker assumption that $(\mathcal{P}_v^m: m \in \mathbb{N})$ has the strong Feller convergence property with respect to $\mathcal{P}_v$ for every $v \in V$. But if we want to avoid reference to the associated kernels, we have to introduce an additional assumption:

Let Y and Z be polish spaces. A subset $\mathfrak{P}$ of $\mathcal{MK}(Y,\mathfrak{Y};Z,\mathfrak{Z})$ is said to have the *compact uniform tightness property* if it has the following property:

(CUTP) For every compact subset C_Y of Y and for every $\varepsilon > 0$ there is a compact subset C_Z of Z such that $P(y; Z \setminus C_Z) \leq \varepsilon$ for every $y \in C_Y$ and every $P \in \mathfrak{P}$.

Theorem 3: Let $(\mathbb{P}_m: m \in \mathbb{N})$ be the sequence of probability measures $\mathbb{P}_m$ associated via (I) to the CPN with the acyclic finite graph $G := (V,E)$ and the family of Feller kernels $(\mathcal{P}_v^m: v \in V)$ and $\mathbb{P}$ the probability measure associated to the CPN with the graph G and the family of Feller kernels $(\mathcal{P}_v: v \in V)$. If $(\mathcal{P}_v^m: m \in \mathbb{N})$ has the strong Feller convergence property with respect to $\mathcal{P}_v$ and $\{\mathcal{P}_v^m: m \in \mathbb{N}\} \cup \{\mathcal{P}_v\}$ has the compact uniform tightness property for every $v \in V$, then $(\mathbb{P}_m: m \in \mathbb{N})$ converges weakly to $\mathbb{P}$.

Proof of Theorem 3: Using the notation of Theorem 2 we have to prove that from the assumption of Theorem 3 the assumption of Theorem 2 follows. But this is a consequence of the following proposition; q.e.d.

Proposition 6: Let X, Y and Z be polish spaces with Borel σ-algebras $\mathfrak{X}$, $\mathfrak{Y}$ and $\mathfrak{Z}$, respectively. Furthermore, let $(\mathcal{P}_m: m \in \mathbb{N})$ be a sequence of Markov kernels $\mathcal{P}_m$: $Y \times \mathfrak{Z} \to [0,1]$ with $(y; B_Z) \to \mathcal{P}_m(y; B_Z)$ from $(Y,\mathfrak{Y})$ to $(Z,\mathfrak{Z})$, $\mathcal{P}$: $Y \times \mathfrak{Z} \to \mathbb{R}$ with $(y; B_Z) \to \mathcal{P}(y; B_Z)$ be a Markov kernel from $(Y,\mathfrak{Y})$ to $(Z,\mathfrak{Z})$, and P_m: $X \times Y \times \mathfrak{Z} \to [0,1]$

with $P_m(x,y;.) := \mathcal{P}_m(y;.)$ and $P: X\times Y\times \mathfrak{Z} \to [0,1]$ with $P(x,y;.) := \mathcal{P}(y;.)$ be the associated (Markov) kernels. If $(\mathcal{P}_m: m\in\mathbb{N})$ has the strong Feller convergence property with respect to $\mathcal{P}$ and $\{\mathcal{P}_m: m\in\mathbb{N}\}\cup\{\mathcal{P}\}$ has the compact uniform tightness property, then $(P_m: m\in\mathbb{N})$ has the strong Feller convergence property with respect to P.

Proof of Proposition 6: Let $(g_m: m\in\mathbb{N})$ be a sequence in $\mathfrak{C}_b(X\times Y\times Z)$ which is uniformly bounded by $\gamma>0$ and which converges uniformly on compact sets to $g\in\mathfrak{C}_b(X\times Y\times Z)$. Hence, also $|g|\le\gamma$. Take any compact subset C_0 of $X\times Y$ and define $C_X := pr_X(C_0)$ and $C_Y := pr_Y(C_0)$ with the canonical projections pr_X and pr_Y onto X and Y, respectively. C_X and C_Y are compact. Because of (CUTP) for $\varepsilon>0$ there is a compact subset C_Z of Z such that for every $y\in C_Y$

$$\mathcal{P}_m(y;Z\setminus C_Z) \le \gamma^{-1}\varepsilon \quad \text{for all } m\in\mathbb{N} \quad \text{and} \quad \mathcal{P}(y;Z\setminus C_Z) \le \gamma^{-1}\varepsilon .$$

Because of the Stone-Weierstraß Theorem we can find an $n\in\mathbb{N}$, and for $1\le i\le n$ functions $\Phi_i\in\mathfrak{C}(C_X\times C_Y)$ and $\Psi_i\in\mathfrak{C}(C_Z)$ such that

$$\Big|\sum_{i=1}^n \Phi_i(x,y)\,\Psi_i(z) - g(x,y,z)\Big| \le \min\{\varepsilon,\gamma\} =: \eta \text{ for every } (x,y,z)\in C_X\times C_Y\times C_Z =: C.$$

Applying Tietze's extension theorem, we get for $1\le i\le n$ functions $\hat\varphi_i\in\mathfrak{C}_b(X\times Y)$ and $\hat\psi_i\in\mathfrak{C}_b(Z)$ with $\hat\varphi_i|C_X\times C_Y = \Phi_i$ and $\hat\psi_i|C_Z = \Psi_i$. Define $\hat G: X\times Y\times Z \to \mathbb{R}$ by $\hat G(x,y,z) := \sum_{i=1}^n \hat\varphi_i(x,y)\,\hat\psi_i(z)$. Obviously, $\hat G\in\mathfrak{C}_b(X\times Y\times Z)$.

Let be $(x,y)\in C_X\times C_Y$. Let be $z\in C_Z$. There is an open neighbourhood $U_{x,y,z}$ of (x,y) in $X\times Y$ and an open neighbourhood $V_{x,y,z}$ of z in Z such that $|\hat G(\tilde x,\tilde y,\tilde z) - \hat G(x,y,z)| \le \eta$ for every $(\tilde x,\tilde y,\tilde z)\in U_{x,y,z}\times V_{x,y,z}$, especially $|\hat G(\tilde x,\tilde y,\tilde z)| \le 2\eta+\gamma \le 3\gamma$ for those $(\tilde x,\tilde y,\tilde z)$. Since C_Z is compact, there is a $k\in\mathbb{N}$ and $z_1, \dots, z_k\in C_Z$ such that $V_{x,y} := \bigcup_{j=1}^k V_{x,y,z_j} \supset C_Z$. $U_{x,y} := \bigcap_{j=1}^k U_{x,y,z_j}$ is an open neighbourhood of (x,y) in $X\times Y$, and $|\hat G(\tilde x,\tilde y,\tilde z)| \le 3\gamma$ for every $(\tilde x,\tilde y,\tilde z)\in U_{x,y}\times V_{x,y}$. Since $C_X\times C_Y$ is compact, there is an $l\in\mathbb{N}$ and $(x_1,y_1), \dots, (x_l,y_l)\in C_X\times C_Y$ such that $U := \bigcup_{j=1}^l U_{x_j,y_j} \supset C_X\times C_Y$. $V := \bigcap_{j=1}^l V_{x_j,y_j} \supset C_Z$. U and V are open, $|\hat G(x,y,z)| \le 3\gamma$ for every $(x,y,z)\in U\times V$.

Because of Urysohn's lemma there are continuous functions $\varphi: X\times Y \to [0,1]$ and $\psi: Z\to[0,1]$ such that $\varphi(x,y)=1$ for every $(x,y)\in C_X\times C_Y$, $\varphi(x,y)=0$ for every $(x,y)\in(X\times Y)\setminus U$, $\psi(z)=1$ for every $z\in C_Z$, and $\psi(z)=0$ for every $z\in Z\setminus V$.

For $1\le i\le n$ let us define $\varphi_i := \varphi\,\hat\varphi_i \in\mathfrak{C}_b(X\times Y)$ and $\psi_i := \psi\,\hat\psi_i\in\mathfrak{C}_b(Z)$, and $G: X\times Y\times Z \to \mathbb{R}$ by $G(x,y,z) := \sum_{i=1}^n \varphi_i(x,y)\,\psi_i(z)$. Then we have $|G(x,y,z) - g(x,y,z)| \le \varepsilon$ for every $(x,y,z)\in C$, and $|G(x,y,z)| = \varphi(x,y)\,\psi(z)\,|\hat G(x,y,z)| \le 3\gamma$ for $(x,y,z)\in U\times V$, and $=0$ for $(x,y,z)\notin U\times V$. Therefore $|G|\le 3\gamma$ on $X\times Y\times Z$.

Choose $\Gamma>0$ such that $|\varphi_i(x,y)|\le\Gamma$ for $1\le i\le n$ and every $(x,y)\in X\times Y$. For $1\le i\le n$ because of (SFCP) there is an $N_i\in\mathbb{N}$ such that

$$\Big|\int_Z \psi_i(z)\,\mathcal{P}_m(y;dz) - \int_Z \psi_i(z)\,\mathcal{P}(y;dz)\Big| \le \Gamma^{-1} n^{-1}\varepsilon \quad \text{for every } y\in C_Y \text{ and } m\ge N_i .$$

Choose $N_0\in\mathbb{N}$ such that $|g_m(x,y,z) - g(x,y,z)|\le\varepsilon$ for every $(x,y,z)\in C$ and $m\ge N_0$. For $m\ge\max\{N_0,N_1, \dots, N_n\}$ and every $(x,y)\in C_0\subset C_X\times C_Y$ we get

$$\Big|\int_Z g_m(x,y,z)\,P_m(x,y;dz) - \int_Z g(x,y,z)\,P(x,y;dz)\Big| \le$$

$$\leq \left| \int_{Z \setminus C_Z} g_m(x,y,z)\, P_m(x,y;dz) \right| + \int_{C_Z} |g_m(x,y,z) - g(x,y,z)|\, P_m(x,y;dz)$$

$$+ \int_{C_Z} |g(x,y,z) - G(x,y,z)|\, P_m(x,y;dz) + \left| \int_{Z \setminus C_Z} G(x,y,z)\, P_m(x,y;dz) \right|$$

$$+ \left| \int_Z G(x,y,z)\, P_m(x,y;dz) - \int_Z G(x,y,z)\, P(x,y;dz) \right| + \left| \int_{Z \setminus C_Z} G(x,y,z)\, P(x,y;dz) \right|$$

$$+ \int_{C_Z} |G(x,y,z) - g(x,y,z)|\, P(x,y;dz) + \left| \int_{Z \setminus C_Z} g(x,y,z)\, P(x,y;dz) \right|$$

$$\leq \gamma\, \gamma^{-1}\, \varepsilon + \varepsilon + \varepsilon + 3\, \gamma\, \gamma^{-1}\, \varepsilon + \sum_{i=1}^{n} |\varphi_i(x,y)| \left| \int_Z \psi_i(z)\, \mathcal{P}_m(y;dz) - \int_Z \psi_i(z)\, \mathcal{P}(y;dz) \right|$$

$$+ 3\, \gamma\, \gamma^{-1}\, \varepsilon + \varepsilon + \gamma\, \gamma^{-1}\, \varepsilon \leq 6\, \varepsilon + n\, \Gamma\, \Gamma^{-1}\, n^{-1}\, \varepsilon + 5\, \varepsilon = 12\, \varepsilon \quad ; \text{q.e.d.}$$

References

(1) Billingsley, P.: Convergence of Probability Measures. J. Wiley: New York, 1968.

(2) Jensen, F.V.; Lauritzen, S.L.; Oleson, K.G.: Bayesian updating in causal probabilistic networks by local computations. Computational Statistics Quaterly 4 (1990), 269-282.

(3) Lauritzen, S.L.; Dawid, A.P.; Larsen, B.N.; Leimer, H.-G.: Independence properties of directed Markov fields. Networks 20 (1990), 491-505.

(4) Lauritzen, S.L.; Wermuth, N.: Graphical models for associations between variables, some of which are qualitative and some quantitative. Annals of Statistics 17 (1989), 31-57.

(5) Lauritzen, S.L.; Spiegelhalter, D.J.: Local computations with probabilities on graphical structures and their applications to expert systems. J. Royal Stat. Soc. B 50 (2) (1988), 157-224.

(6) Oppel, U.G.: Every complex system can be determined by a causal probabilistic network without cycles and every such network determines a Markov field. In: Kruse, R.; Siegel, P.: Symbolic and Quantitative Approaches to Uncertainty. Proceedings of the European Conference ECSQAU, Marseille, France, October 15-17, 1991. Lecture Notes in Computer Science 548. Springer: Berlin 1991.

(7) Oppel, U.G.: Kausal-probabilistische Expertensysteme. Vorlesungsskriptum. Mathematisches Institut der Ludwig-Maximilians-Universität: München, 1991.

(8) Parthasarathy, K.R.: Probability Measures on Metric Spaces. Academic Press. New York - London, 1967.

(9) Pearl, J.: Probabilistic Reasoning in Intelligent Systems: Networks of Plausible Inference. Morgan Kaufmann: San Mateo, CA, USA; 1988.

CHAPTER 2:

REASONING WITH UNCERTAINTY

Uncertainty in Intelligent Systems
B. Bouchon-Meunier et al. (Editors)

Deriving Dempster's rule

Petr Hájek

Institute of Computer Science, Academy of Sciences of the Czech Republic, 182 07 Prague, Czech Republic (e-mail:HAJEK@CSPGCS11.BITNET)

Abstract.

Smets's axioms characterizing uniquely Dempster's rule of combination are revised, modified and simplified; some axioms are shown to be redundant.

INTRODUCTION

Dempster-Shafer theory, even if created with no reference to AI, has become one of prominent approaches to dealing with uncertainty in expert systems; this increases the importance of the question, which kind of uncertainty and belief is covered by this theory. There has been a highly interesting discussion on this topic ([Shafer], [Pearl], [Smets1]). Smets stresses the role of belief functions as an alternative to probability rather than as a generalization of probability and the crucial role of Dempster's rule. Technically, he works with belief functions and basic belief assignments (bba) of Dempster-Shafer theory but does not assume $m(0)$ to be 0 (open world assumption). Consequently, he uses unnormalized Dempster's rule. In [Smets2] he shows that Dempster's rule is the only operation satisfying eight axioms (A1)-(A8). Here we present result of an analysis of his beautiful result. Our axioms (A1)-(A4) will be (practically) the same as Smets's (A1)-(A4). (Briefly, $\oplus$ is a total operation, is commutative and associative and conditioning is expressible by $\oplus$.) In Part I we introduce a new axiom (A5), which is a strengthening of Smets's (A5) and uses the notion of a meet epimorphism (it says that $\oplus$ commutes with meet epimorphisms). Consequences of (A5) are investigated and our main result is that our (A1,A2),(A4,A5) characterize uniquely Dempster's rule. In Part II we investigate Smets's (A6) autofunctionality axiom and show that also (A1-A4, A6-A7) characterize uniquely Dempster's rule. This is hoped to contribute to our understanding of Dempster's rule, necessary to answer the question above. Note that another analysis of Smets's axioms was presented by [Klawonn-Schwecke]; it goes in a similar direction as ours. Our central device, i.e. meet automorphisms, is not used by them.

0.1. Definitions

Ω, Ω' vary over finite Boolean algebras; a *basic belief assignment* (bba) is a mapping $m : \Omega \rightarrow [0,1]$ such that $\sum_{A\in\Omega} m(A) = 1$. m is *regular* if $m(0) = 0$. (The former zero is the least element of Ω, the latter is a real; the reader will easily distinguish such things. Pedantically we should write 0_Ω in the former case).

The *closed world assumption* means the restriction of interest to regular bba's only; the *open world assumption* admits all bba's. *Dempster's rule of combination* for the open world case associates with each pair m_1, m_2 of bba's on the same algebra Ω the bba m_{12} satisfying

$$m_{12}(A) = \sum_{B \wedge C = A} m_1(B).m_2(C)$$

for $A \in \Omega$.

Let m be a bba on Ω; the corresponding *commonality* is the function $Q(m) = q$ such that

$$q(A) = \sum_{B \geq A} m(B)$$

for all $A \in \Omega$. One can easily show that if $q_i = Q(m_i)$ $(i = 1, 2)$ then

$$q_{12} = Q(m_1 \oplus m_2) \text{ iff } q_{12}(A) = q_1(A).q_2(A) \text{ for all } (A).$$

0.2. Facts

The following are well-known properties of Dempster's rule:

(A1) $\oplus$ is an operation associating with each Boolean algebra Ω and each pair of bba's on Ω a bba $m_1 \oplus m_2$ on Ω (pedantically we can write $m_1 \oplus_\Omega m_2$).

(A2) $\oplus$ is commutative, thus $m_1 \oplus m_2 = m_2 \oplus m_1$.

(A3) $\oplus$ is associative, thus $(m_1 \oplus m_2) \oplus m_3 = m_1 \oplus (m_2 \oplus m_3)$.

If m is a bba on Ω and $A \in \Omega$ let $cond(m, A)$ (conditioning) be the bba m' such that

$$m'(B) = \sum_{C \wedge A = B} m(C)$$

for $B \in \Omega$ (in particular, $m'(B) = 0$ if $B \not\leq A$). Furthermore, let m_A be the bba such that $m_A(A) = 1$ (and thus $m_A(B) = 0$ for $B \neq A$).

(A4) $$cond(m, A) = m \oplus m_A.$$

0.3. Remark

The point is now to consider *any* operation $\oplus$ satisfying (A1-A4) (and possibly other properties of Dempster's rule taken as axioms) and try to prove that $\oplus$ must be Dempster's rule. Smets proves his result in two steps: first he proves, using (A1-A4) and additional axioms that there is a binary function $F : [0,1] \times [0,1] \to [0,1]$ such that for any pair m_1, m_2 of bba's on Ω, if $q_i = Q(m_i)$ and $q_{12} = Q(m_1 \oplus m_2)$ then $q_{12}(A) = F(q_1(A), q_2(A))$; and then, using further additional axioms, he shows that F is the product, i.e. $(\forall x, y)(F(x, y) = x.y)$. We shall do the same, but with carefully chosen axioms.

1. MEET EPIMORPHISMS

1.1. Definition

Let Ω, Ω' be Boolean algebras; denote the corresponding operations and constants by $\wedge, \vee, 0, 1$ and $\wedge', \vee', 0', 1'$ respectively. A mapping f of Ω onto Ω' is a *meet epimorphism of Ω onto Ω'* if $f(A \wedge B) = f(A) \wedge' f(B)$ for all $A, B \in \Omega$. f is a Boolean epimorphism if, in addition, $f(A \vee B) = f(A) \vee' f(B)$ for all A, B.

1.2. Lemma

(1) If $f : \Omega \to \Omega'$ is a meet epimorphism then $f(0) = 0'$, $f(1) = 1'$, and $x \leq y$ implies $f(x) \leq f(y)$.
(2) If f is Boolean then f also preserves complements, i.e. $f(x) = -f(-x)$.

Proof. (1) Assume first $x \leq y$; this is equivalent to $x \wedge y = x$. Then $f(x) \wedge' f(y) = f(x)$, thus $f(x) \leq' f(y)$. Now let $f(1) = e$ and let $z \in \Omega'$; then $z = f(x)$ for some $x \in \Omega$ and $x \leq 1$, thus $z = f(x) \leq f(1) = e$. Hence e is maximal in Ω which implies $e = 1'$. Similarly one proves $f(0) = 0'$.
(2) is routine.

1.3. Definition

If m is a bba on Ω and f is a meet epimorphism of Ω to Ω' then the bba $f(m) = m'$ *induced* by f in defined as follows:

$$m'(A) = \sum_{f(B)=A} m(B).$$

1.4. Lemma

Dempster's rule commutes with meet epimorphisms, i.e. if $\oplus_\Omega$ is Dempster's rule for bba's on Ω then $f(m_1 \oplus_\Omega m_2) = f(m_1) \oplus_{\Omega'} f(m_2)$; thus, for each $A \in \Omega'$,

$$f(m_1 \oplus_\Omega m_2)(A) = f(m_1) \oplus_{\Omega'} f(m_2)(A).$$

Proof. Let $A \in \Omega'$ then

$$\begin{aligned} f(m_1) \oplus f(m_2)(A) &= \sum_{E \wedge F = A} f(m_1)(E).f(m_2)(F) = \\ &= \sum_{E \wedge F = A} \Big(\sum_{f(C)=E} m_1(C)\Big).\Big(\sum_{f(D)=F} m_2(D)\Big) = \\ &= \sum_{E \wedge F = A} \sum_{f(C)=E, f(D)=F} m_1(C).m_2(D) = \\ &= \sum_{f(C) \wedge f(D) = A} m_1(C).m_2(D) = \sum_{f(C \wedge D) = A} m_1(C).m_2(D). \end{aligned}$$

On the other hand,

$$\begin{aligned} f(m_1 \oplus m_2)(A) &= \sum_{f(B)=A} (m_1 \oplus m_2)(B) = \sum_{f(B)=A} \sum_{C \wedge D = B} m_1(C).m_2(D) = \\ &= \sum_{f(C \wedge D)=A} m_1(C).m_2(D). \end{aligned}$$

Thus $f(m_1 \oplus m_2)(A) = (f(m_1) \oplus f(m_2))(A)$ for all $A \in \Omega'$.

1.5. Definition

We introduce a new axiom (A5) saying

(A5) $\oplus$ commutes with meet epimorphisms.

(Note that Smets's A5 only says that $\oplus$ commutes with Boolean automorphisms, i.e. one-one Boolean epimorphisms of Ω to Ω).

1.6. Theorem

(A1) and (A5) imply that there is a function $F : [0,1] \times [0,1] \to [0,1]$ such that, if m_1, m_2 are bba's on Ω, $q_i = Q(m_i)$ and $q_{12} = Q(m_1 \oplus m_2)$ then

$$q_{12}(A) = F(q_1(A), q_2(A))$$

for all $A \in \Omega$.

Proof. Let $A \in \Omega$ be given; let $\Omega' = \{B \leq A | B \in \Omega\}$ be the algebra of elements of Ω less than or equal to A, whose operations $\wedge', \vee'$ are restrictions of the operations of Ω; clearly, $0' = 0$ and $1' = A$. The mapping f defined by $f(C) = C \wedge A$ for $C \in \Omega$ is a Boolean epimorphism of Ω to Ω'. Observe that for each bba m on Ω, $f(m)(A) = \sum_{f(C)=A} m(C) = \sum_{C \wedge A = A} m(C) = \sum_{C \geq A} m(C) = q(A)$ (where q is $Q(m)$). In particular $q_{12}(A) = f(m_1 \oplus m_2)(A) = (f(m_1) \oplus' f(m_2))(A) = (m'_1 \oplus' m'_2)(1')$, where $m'_i = f(m_i)$ and $\oplus'$ is $\oplus_{\Omega'}$.

Now let $\Omega'' = \{0'', 1''\}$ be a copy of the unique two-element Boolean algebra; put $g(1') = 1''$ and $g(B) = 0''$ for $B \neq 1'$, $B \in \Omega'$. This is a meet epimorphism (but not a Boolean epimorphism). Put $m''_i = g(m'_i)$ and let $\oplus''$ be $\oplus''_{\Omega}$. Then

$$(m''_1 \oplus'' m''_2)(1'') = g(m'_1 \oplus' m'_2)(1') = (m'_1 \oplus' m'_2)(1') = q_{12}(A),$$

thus $q_{12}(A) = (m''_1 \oplus'' m''_2)(1'')$; $(m''_1 \oplus'' m''_2)(1'')$ is a function of m''_1 and m''_2, but m''_i is uniquely given by $(m''_i(1'') = m'_i(1') = q_i(A)$ (clearly, $(m''(0'') = 1 - m''(1''))$. Let $F(x,y)$ be this function; thus if m, $\hat{m}$ are any bba's on Ω'' then $(m \oplus'' \hat{m})(1'') = F(m(1''), \hat{m}(1''))$. We have proved $q_{12}(A) = F(q_1(A), q_2(A))$.

1.7. Theorem

(A1) and (A5) imply that the function F of 1.5 is non-decreasing in either argument.

Proof as [Smets2], Lemma 7: given $x_1 \leq y_1$, $x_2 \leq y_2$, take an Ω (with more than two elements), let $0 < A < 1$, let $q_i(A) = y_i$, $q_i(1) = x_i$ (thus e.g. $m_i(1) = x_i$, $m_i(A) = y_i - x_i$, $m_i(0) = 1 - y_i$, $m_i(C) = 0$ otherwise). We get $q_{12}(1) \leq q_{12}(A)$, thus $F(x_1, x_2) \leq F(y_1, y_2)$.

1.8. Theorem

Assuming (A1) and (A5), the axiom (A4) is equivalent to the assumption that $F(x, 1) = x$ and $F(x, 0) = 0$ for each x.

Proof. Let $q = Q(m)$; $q' = Q(cond(m, A))$ iff the following holds:

$$\begin{aligned} q'(B) &= q(B) \text{ for } B \leq A, \\ q'(B) &= 0 \text{ otherwise.} \end{aligned}$$

Furthermore $q'' = Q(m_A)$ iff

$$\begin{aligned} q''(B) &= 1 \text{ for } B \leq A, \\ q''(B) &= 0 \text{ otherwise.} \end{aligned}$$

Finally, let $\hat{q} = Q(m \oplus m_A)$, thus $\hat{q}(B) = F(q(B), q''(B))$.
Now it is evident that $(\forall x)(F(x, 1) = x \;\&\; F(x, 0) = 0)$ implies $q' = \hat{q}$, thus (A4). On the other hand, if we assume (A4) we get $q' = \hat{q}$ (for each q and A) and hence, by an appropriate choice of m, A and B,

$$\begin{aligned} F(x, 1) &= F(q(B), 1) = cond(q, A)(B) = q(B) = x \quad (\text{if } B \leq A), \\ F(x, 0) &= F(q(B), 0) = cond(q, A)(B) = 0 \qquad (\text{if } B \not\leq A). \end{aligned}$$

1.9. Theorem

Assuming (A1)-(A5), F is a t-norm.

Proof. Clearly, the axioms guarantee $F(x, y) = F(y, x)$, $F(x, F(y, z)) = F(F(x, y), z)$ for all x, y, z, monotonicity and $F(1, x) = x$ for all x.

1.10. Lemma

Assuming (A1,A2,A4,A5), F satisfies $F(x, y) \geq x + y - 1$, for all $x, y \in [0, 1]$.

Proof. Let Ω be the power set of $\Theta = \{a, b\}$ with the usual operation; let $m_1(a) = 1 - x$, $m_1(\Theta) = x$, $m_2(b) = 1 - y$, $m_2(\Theta) = y$. Then we get

	$\emptyset$	a	b	Θ
q_1	1	1	x	x
q_2	1	y	1	y
q_{12}	1	y	x	$F(x, y)$

Compute $m_{12}(\emptyset) = 1 - y - x + F(x,y)$; since $m_{12}(\emptyset) \geq 0$ we set $F(x,y) \geq x + y - 1$, as desired.

1.11. Discussion

We shall now analyze Smets's proof of his Lemma 10 and Theorem 3. Let $f_x = F(x,y)$ (the one-argument function resulting from F by fixing the first argument to x).

(i) First observe $f_x(y) = f_y(x)$ and $f_1(x) = x$; thus second derivate f_1'' of f_1 is identically to 0.

(ii) It follows from the proof of [Bernstein] (p. 190ff) and [Smets 2], Lemma 9,10 and Theorem 3, that for $0 \leq x < 1$, the function f_x has the second derivative $f_x''(y)$ for each $0 \leq y < 1$ and this derivative equals to 0.

(iii) Hence for $0 \leq x, y < 1$, using Taylor's formula, we get

$$f_x(y) = f_x(0) + y.f_x'(0) + y^2.f_x''(\eta) = y.f_x'(0) = y.h(x);$$

$$F(x,y) = y.h(x) = x.h(y);$$

$$(\frac{\partial}{\partial y}F)(x,y) = x.h'(y);$$

$$h(x) = f_x'(0) = (\frac{\partial}{\partial y}F)(x,0) = x.h'(0) = x.c;$$

we have proved $F(x,y) = cxy$ for some fixed c and all $0 \leq x, y < 1$.

(iv) Clearly, $0 \leq c \leq 1$ since $0 \leq F(x,y) \leq 1$. One the other hand, it follows by elementary algebra that $c < 1$ would violate Lemma 1.10 since if we took $0 < x, y < 1$ such that $y > \frac{1-x}{1-cx}$ we would get $cxy < x + y - t$. Thus $c = 1$. (A3) has not then used. Thus we have proved the following.

1.12. Theorem

Axioms (A1,A2,A4,A5) (as presented above) imply that $\oplus$ is Dempster's rule.

2. AUTOFUNCTIONALITY

2.1. Definition

We shall now leave our axiom (A5) and investigate consequences of (A1) - (A4), (A6) where (A6) is the following axiom (from [Smets2] corrected):

Given Ω and $A \in \Omega$, $A \neq 0$, $(m_1 \oplus m_2)(A)$ does not depend on $m_1(x)$ for $X \leq -A$, (i.e. $(m_1 \oplus m_2)(A)$ is a function of A and $m_i(X)$ for all X such that $A \wedge X \neq 0$.

Smets calls this axiom the *axiom of autofunctionality*. Since now we have no relation between $\oplus_\Omega$ and $\oplus_{\Omega'}$, Ω, Ω' being different algebras, we shall get a weaker result than 1.5.

2.2. Theorem

Assuming (A1) - (A4), (A6), there is a function $F_\Omega : \Omega \times [0,1] \times [0,1] \to [0,1]$ such that, if m_1, m_2 are bba's on Ω, $q_i = Q(m_i)$ and $q_{12} = Q(m_1 \oplus m_2)$ then

$$q_{12}(A) = F_\Omega(A, q_1(A), q_2(A)).$$

This strengthens Theorem 2 from [Smets2]. Our Theorem is proved below.

2.3. Remark

Recall two important results of [Smets]:

Theorem 1 says that given Ω and $\oplus$ satisfying (A1-A4) q_{12} is a function of A, $\{q_1(B)|0 < B \leq A\}$, $\{q_2(B)|0 < B \leq A\}$. *Lemma 3* says that assuming (A1-A4,A6) for all $A \in \Omega$, $q_{12}(A)$ (as a function of $m_i(X)$, $X \in \Omega$) does not depend on $m_1(X)$, $X \leq -A$.

2.4. Definition

For each bba m and each A, let $sh(m, A)$ be the bba m' defined as follows:

$$\begin{aligned} m'(0) &= \sum\{m(X)|X \leq A\}, \\ m'(X) &= 0 \text{ for } 0 < X \leq A, \\ m'(X) &= m(X) \text{ otherwise.} \end{aligned}$$

We may say that m' results by shifting the belief masses $m(X)$, $0 < X \leq A$ to $m(0)$. More generally, if U is a subset of Ω containing with each element all smaller elements we can define $sh(m, U)$ by replacing "$\leq A$" by "$\in U$". Observe that a function H of m_i $(i = 1, 2)$ defined for each pair of bba's m_i does not depend on $\{m_1(X)|X \in U\}$ iff for each pair m_1, m_2 $H(m_1, m_2) = H(sh(m_1, U), m_2)$; similarly for m_2. Note also that $sh(m, U \vee V) = sh(sh(m, U), V)$.

2.5. Lemma

q_{12} as a function of A and $m_i(X)$, $X \neq 0$, does not depend on values $m_i(X)$ for $X < A$; thus if $m'_i = sh(m_i, (< A))$ and q'_{12} is given by m'_i then $q_{12}(A) = q'_{12}(A)$.

2.6. Corollary

$q_{12}(A)$ is a function of A and all $q_i(X)$ such that not$(X < A)$.

Proof. From $\{q_i(X)|$ not $(X < A)\}$ you may compute $\{m_i(X)|$ not $(X \geq A)\}$; if $0 < X < A$ implies $m_i(A) = 0$ you can determine all $q_i(X), 0 < X \leq A$. Thus you know all $q_i(X)$. Thus $q_{12}(A) = f(A, \{q_i(X)|$ not $(X < A)\}$ as desired. (In more details: let A be fixed, let $m'_i = sh(m_i, < A)$. Then for X such that not $(X < A)$ we have $q_i(X) = q'_i(X)$ and we know $q_{12}(A) = q'_{12}(A)$. But we have shown how to determine $q'_i(X)$ for $X < A$ from the others; thus $q'_{12}(A)$ is a function of $q'_i(X)$ such that not$(X < A)$ and therefore the same holds for q_{12} and q_i.

Remark. The following can be proved in the same manner: If $Y < A$ and $q_{12}(A)$ does not depend on any $m_i(X)$, $X \leq Y$ then $q_{12}(A)$ is a function of all $q_i(X)$ such that not$(X \leq Y)$. This will be used later.

2.7. Corollary

Theorem 2 follows: by Corollary 1, $q_{12}(A)$ is a function of $q_i(X)$ for X not less than A and by Theorem 1 we may set all $q_i(X)$ for X not $\leq A$ equal to 0. Thus $q_{12}(A)$ is a function of A, $q_1(A)$, $q_2(A)$. It remains to prove the lemma.

2.8. Proof of Lemma 2.5.

We simulate the proof of Smets, checking that (A5) is not used. For $1 \leq j < i$ make the following definition: (i, j) is an IP if whenever $|A| = i$ then $q_{12}(A)$ as a function of A and all $m_i(X)$ does not depend on $m_i(X)$ for $X < A$, $|X| \leq j$.

(1) We prove that $(n, 1)$ is an IP. Let $|A| = n - 1$ where n is the number of atoms of Ω and let $B = -A$ (thus B is an atom). We show that if $m_i' = sh(m_i, B)$ and q_{12}' belongs to m_i' then $q_{12}(t) = q_{12}'(t)$. This follows from the fact that $q_{12}(1) = q_{12}(A) - m_{12}(A)$ and neither $q_{12}(t)$ nor $m_{12}(A)$ depend on $m_i(B)$ ($q_{12}(A)$ by Lemma 2.3, $m_{12}(A)$ by Axiom (A6)). Thus $q_{12}(t) = q_{12}'(t)$.

Now work with m_i', take another C such that $|C| = n - 1$, put $D = -C$ and put $m_i'' = sh(m_i', D) = sh(sh(m_i, B), D)$; let q_{12}'' belong to m_i''. We get $q_{12}''(1) = q_{12}'(1) = q_{12}(1)$. Iterating this over all atoms $B_0 = B$, $B_1 = D$, $B_2 = \ldots$ we get bba's $m_{i*} = sh(m_i, |X| \leq 1)$ such that $m_{i*}(X) = \theta$ for $|X| = 1$ and such that $q_{12*}(1) = q_{12}(1)$. Observe that (A5) has not been used.

(2) Now we proceed by induction following Smetss' pattern: let

$(n, 1)$,
$(n - 1, 1), (n, 2)$,
$\ldots$
$(n - k + 2, 1), (n - k + 3, 2), \ldots, (n, k - 1)$,
$(n - k + 1, 1), \ldots (n - k + i, i)$

be IP's and consider $(n - k + i + 1, i + 1)$. Let $|A| = n - k$, $Y \leq -A$, let $|Y| = i + 1 \leq k$. By induction we assume that for all X such that $X \leq A \vee Y$ and $0 < |X| \leq i$ $m_i(X) = 0$. Express $m_{12}(A)$ as a linear combination of $q_{12}(A \vee B \vee C)$ where $B \leq Y$, $C \leq -(A \vee Y)$ and $|B \vee C| \leq k$; this gives you an expression of $q_{12}(A \vee Y)$ from the others. Now $m_{12}(A)$ (as a function of all $m_i(X)$) does not depend on $m_i(Y)$ by (A6). To show that $q_{12}(A \vee B \vee C)$ as a function of all $m_i(X)$ does not depend on $m_i(Y)$ it suffices to show that $q_{12}(A \vee B \vee C)$ as a function of $\{q_i(Z) | Z \leq A \vee B \vee C\}$ does not depend on any $q_i(D)$, where $D \leq B$. To obtain this one first shows that if not($B = Y$ and $C = 0$) then $(|A \vee B \vee C|, |D|)$ is an IP by induction assumption, hence $q_{12}(A \vee B \vee C)$ as a function of all $m_i(X)$ does not depend on $m_i(D)$ for any $D \leq B$. Then use the remark following Corollary 2.6: $q_{12}(A \vee B \vee C)$ as a function of $q_i(X)$ does not depend on $q_i(X)$ for $X \leq B$. Now use 2.3 to eliminate all $q_i(X)$ for not($X \leq A \vee B \vee C$).

We have proved that $q_{12}(A \vee B \vee C)$ does not depend on $m_i(Y)$. Since $q_{12}(A \vee Y)$ has been expressed as a combination of functions not depending on $m_i(Y) q_{12}$ itself does not depend on $m_{12}(Y)$.

Now we may take $(i+1)$-element subsets Y_h of $A \vee Y$ one after the other (Y_0 being Y) and step by step replace $m_i(Y_h)$ by 0; we get a sequence of bba's $m_{i,h}$ and corresponding $q_{12,h}$ such that $m_{i+1,h} = sh(m_{i,h}, Y_h)$ and $q_{12,0}(A) = q_{12,1}(A) = \ldots q_{12,h}(A) = \ldots$ Finally we get that $(n - k + i + 1,\ i + 1)$ is an IP. Observe that (A5) has not been used.

The transition from $(n, k-1)$ to $(n-k+1, 1)$ is similar to the proof that $(n, 1)$ is an IP and is left to the reader (simulate again [Smets]).

We have proved that for each $Y < A$, $m_i(Y)$ can be shifted to $m_i(0)$ without affecting $q_{12}(A)$. This completes the proof of Lemma 2.5.

2.9. Remark

Thus Theorem 2.2 is proved. This shows that in [Smets2], his (A5) is redundant (unnecessary for Smets's Theorem 2 and not used further.) A discussion analogous to 1.10 is appropriate here, with the difference that since now there is no apriori relation between F_Ω for different Ω, sometime one has to assume that Ω is not too small (has at least four atoms, cf. [Smets2] axiom (A7) and Lemma 10). The result is similar to 1.12.

2.10. Theorem

Axioms (A1-A4), (A6-A7) imply that $\oplus_\Omega$ is Dempster's rule.

3. CONCLUSION

The notion of a meet epimorphism was shown to be of importance for Dempster-Shafer theory; our axiom (A5) appears entirely natural. On the contrary, (A6) is hard to motivate, but it evidently follows from (A1), (A5). A proof of the fact that Dempster's rule is the only operation on bba's satisfying some natural axioms is undoubtedly an argument for the claim that Dempster's rule itself is natural. (Needlless to say, this is not the only thing we need to know about it).

Let me close by saying that I agree with Smets that Dempster's rule as in the centre of the theory; on the other hand, his claim that his presentation of Dempster-Shafer theory (called transferrable belief model) is built "without ever introducing explicity or implicitly any concept of probability" seems open to discussion.

REFERENCES

1. S. Bernstein: Lecon sur les Propriétés Extrémales et la Meilleure Approximation des Fonetions Analytiques d'une Variable Réelle, Gauthier-Villars, Paris 1926

2. F.Klawonn, E.Schwecke: On the axiomatic justification of Dempster's rule of combination, Int. Journ. of Intelligent Systems 7 (1992) 469-478

3. J.Pearl: Reasoning with belief functions: an analysis of compatibility, Int.J.Approx. Reasoning 4 (1990), 363-389

4. G.Shafer: Belief functions and possibility measures, in: Analysis of fuzzy information vol. 1 (Mathematics and logic), CRC Press 1987, 51-84

5. P.Smets: Resolving misunderstandings about belief functions, Int. Journ. Approx. Reasoning, to appear

6. P.Smets: The combination of evidence in the transferable belief model, IEEE Trans. Pattern Anal. and Machine Int. 12 (1990), 447-458

Uncertainty in Intelligent Systems
B. Bouchon-Meunier et al. (Editors)

INDEPENDENCE CONCEPTS IN UPPER AND LOWER PROBABILITIES

Luis M. de Campos and Juan F. Huete

Departamento de Ciencias de la Computación e I.A.
Universidad de Granada,- 18071 - Granada, Spain

The aim of this paper is the study of different concepts of independence in non probabilistic frameworks. The previous topics necessary to define independence relationships, as marginal, conditional and product measures are considered. Next we propose several alternative definitions of independence and study some of their properties. Finally, all of these concepts are illustrated by means of a simple example.

1. INTRODUCTION

The concept of (ir)relevance or (in)dependence between facts, propositions or variables is essential for reasoning tasks. Dependence is a relationship stating a possible change in our current belief due to a specified change in our knowledge. In the framework of uncertain knowledge management, the concept of independence (and conditional independence) has been extensively studied only for probability measures (see for example [7], [11], [13]). However, if we consider more general or different formalisms for representing and manipulating the uncertainty (as possibility measures [14], belief-plausibility functions [8], [12], Choquet capacities of order two [6] or upper and lower probabilities), neither independence in itself is a clear concept nor there is a general agreement on its definition (some works studying this topic are [9], [12], [14]).

The aim of this paper is to investigate different ways to define independence relationships in a very general context: the theory of upper and lower probabilities, what includes, at least in a formal sense, probability, possibility, Dempster-Shafer and capacity theories.

In addition to the obvious theoretical interest, there are also practical reasons that make the study of the independence an important topic. Independence permits us to modularise the knowledge in such a way that we only need to consult the pieces of information having relevance to the question we are interested in, instead of having to explore a complete knowledge base. So, reasoning systems should take into account independence considerations in order to get an efficient performance. This is the case, for example, of causal networks based systems [11]. This kind of systems encodes direct causal relevance relationships in a

---

This work has been supported by the DGYCIT under Project nº. PS89-0152 and by the European Economic Community under Project DRUMS (Esprit b.r.a. 3085)

graph, and exploits the (in)dependencies displayed by the graph, in order to obtain correct inferences using only local methods.

The ability of detecting independence is crucial if we intend to build a causal network in an automatic fashion, directly from empirical observations, without resort to assessments from experts. Thus, methods and software tools to detect probabilistic independence and carry out inductive learning of probabilistic causal networks have been developed ([11], [1], [2]). Following this idea, the definition of independence within the upper and lower probability formalism, and subsequent study of methods to identify it, would make possible the development of learning algorithms for non-probabilistic networks.

The paper is divided in 6 sections. In section 2, the problem formulation and the methodology we will use to solve it are stated. The tools and previous concepts needed to define independence are considered in section 3. In section 4, we propose several alternative definitions of independence. An example illustrating the different concepts of independence is presented in section 5. Finally, some remarks are outlined in the last section.

2. PROBLEM FORMULATION AND METHODOLOGY

The problem we want to solve may be stated as follows: let X, Y be two variables taking their values on the finite sets $V_X=\{x_1,x_2,\ldots,x_n\}$ and $V_Y=\{y_1,y_2,\ldots,y_m\}$ respectively. Consider a global piece of knowledge about the variables X and Y, which is represented as a lower-upper probability, g, on the cartesian product $V=V_X\times V_Y$. Then we want to define a concept of independence between X and Y given the available information g.

In order to illustrate the tools we will need to use, let us consider first the case of probabilities (our definition of independence for upper and lower probabilities should be an extension of the corresponding concept in probability theory). There are several equivalent definitions of independence for probabilities: given a bidimensional probability measure P on V, the variables X and Y are said independent (with respect to P) if

i) $P(A\times B) = P_x(A)P_y(B),\ \forall A\subseteq V_X, \forall B\subseteq V_Y,$ **(1)**

where P_x and P_y are the marginal probabilities of P on V_X and V_Y respectively.

ii) $P_y(B/A) = P_y(B) \quad (P_x(A/B) = P_x(A)),\ \forall A\subseteq V_X, \forall B\subseteq V_Y,$ **(2)**

where $P_y(./A)$ (resp. $P_x(./B)$) is the conditional probability on V_Y (resp. V_X) given A (resp. B).

As we see, both definitions use the concept of marginal probability. First definition establishes the independence when the original probability measure coincides with a "product" measure obtained by multiplication of the marginals. Second definition asserts the independence when all the conditional probabilities are equal to the marginal probability.

After these comments it is clear that, in order to define independence in upper and lower probabilities, we must define first the concepts of upper and lower marginal probability and either upper and lower product probability or conditional upper and lower probability. Next, we should compare in some sense either the original and product measures or the conditional and marginal measures.

3. MARGINAL, CONDITIONAL AND PRODUCT MEASURES

The concept of *marginal measure* that we will adopt is the following (see [9]):

Definition 3.1.
Given a bidimensional fuzzy measure g on V, the marginal measures g_x on V_X and g_y on V_Y are defined as

$$g_x(A) = g(A \times V_Y), \ \forall A \subseteq V_X, \qquad g_y(B) = g(V_X \times B), \ \forall B \subseteq V_Y, \tag{3}$$

This definition seems us natural and it is obvious that it reproduces the usual definition when g is a probability.

Proposition 3.1.
The set functions g_x and g_y of definition 3.1 are fuzzy measures on their respective universes V_X and V_Y.
Definition 3.1 preserves also duality (a pair of fuzzy measures (l,u) are said dual if $l(C)=1-u(\neg C) \ \forall \ C$):

Proposition 3.2.
If (l,u) is a pair of bidimensional dual fuzzy measures on V then (l_x,u_x) and (l_y,u_y) are also pairs of dual fuzzy measures on V_X and V_Y respectively.

Moreover, the next proposition shows that the marginal measures belong to the same class that the bidimensional (possibilities, evidences, capacities or upper and lower probabilities) (see [9]):

Proposition 3.3.
Let g be a bidimensional fuzzy measure defined on V.
i) If g is a lower or upper probability measure, then g_x and g_y are also lower or upper probability measures.
ii) If g is a Choquet capacity of order two, then g_x and g_y are also Choquet capacities of order two.
iii) If g is a evidence measure (belief or plausibility), then g_x and g_y are also evidence measures.
iv) If g is a possibility or necessity measure, then g_x and g_y are also possibility or necessity measures.
v) If g is a crisp measure focused on a set C, then g_x and g_y are also crisp measures.

On the other hand, the concept of *product measure* is not clear enough. Given two measures g_x and g_y defined on V_X and V_Y respectively, we are trying to define a product measure g_p on V (assuming that the corresponding variables X and Y are independent). Several definitions are possible depending on the particular class of measures that we consider. For example, for possibilities, one could construct a product π by means of

$$\pi_p(x,y) = \pi_x(x) * \pi_y(y), \tag{4}$$

with * being a t-norm, usually the minimum or the product. Note that using the minimum as t-norm, we obtain the definition of non-interaction for possibility measures (see [14]).

For evidence measures, we can build the product measure as follows:

Definition 3.2.

Let (bel_x,Pl_x) and (bel_y,Pl_y) two pairs of evidence measures with basic probability assignments (b.p.a.) m_x and m_y respectively. Then we define the product measure as the pair (bel_p,Pl_p) with b.p.a.

$$\begin{aligned} m_p(A\times B) &= m_x(A)m_y(B) \quad \forall A\subseteq V_X, \forall B\subseteq V_Y, \\ m_p(C) &= 0 \qquad \textit{otherwise} \end{aligned} \tag{5}$$

Using this definition of product measure, we can prove the following result:

Proposition 3.4.

The product measure of definition 3.2 verifies the following properties:

i) $bel_p(A\times B) = bel_x(A)bel_y(B)$

ii) $Pl_p(A\times B) = Pl_x(A)Pl_y(B)$

iii) $bel_p(A\times Y \cup X\times B) = bel_x(A) + bel_y(B) - bel_x(A)bel_y(B)$

iv) $Pl_p(A\times Y \cup X\times B) = Pl_x(A) + Pl_y(B) - Pl_x(A)Pl_y(B)$

Proof:

i) $bel_p(A\times B)=bel_x(A)\ bel_y(B)$

$$bel_p(A\times B) = \sum_{C\subseteq A\times B} m(C) = \sum_{A'\times B'\subseteq A\times B} m(A'\times B') = \sum_{A'\times B'\subseteq A\times B} m_x(A')m_y(B')$$

$$= \sum_{A'\subseteq A;B'\subseteq B} m_x(A')m_y(B') = \sum_{A'\subseteq A} m_x(A')\sum_{B'\subseteq B} m_y(B') = bel_x(A)bel_y(B)$$

ii) $Pl_p(A\times B)=Pl_x(A)Pl_y(B)$

$$Pl_p(A\times B) = \sum_{C\cap A\times B\neq\emptyset} m(C) = \sum_{A'\times B'\cap A\times B\neq\emptyset} m(A'\times B') = \sum_{A'\times B'\cap A\times B\neq\emptyset} m_x(A')m_y(B')$$

since $A'\times B'\cap A\times B <> \emptyset$ if and only if $A'\cap A <> \emptyset$ and $B'\cap B <> \emptyset$

$$\sum_{A'\cap A\neq\emptyset, B\cap B'\neq\emptyset} m_x(A')m_y(B') = \sum_{A'\cap A\neq\emptyset} m_x(A')\sum_{B'\cap B\neq\emptyset} m_y(B') = Pl_x(A)Pl_y(B)$$

iii) $bel_p(A\times Y \cup X\times B)=bel_x(A) + bel_y(B) - bel_x(A)bel_y(B)$

$$bel_p(A\times Y\cup X\times B) = 1-Pl_p(\overline{A\times Y\cup X\times B}) = 1-Pl_p(\overline{A}\times\overline{B}) = 1-Pl_x(\overline{A})Pl_y(\overline{B})$$

$$= 1-[\ (1-bel_x(A))\ (1-Bel_y(B))\ ] = bel_x(A) + bel_y(B) - bel_x(A)bel_y(B)$$

iv) $Pl_p(A\times Y \cup X\times B)=Pl_x(A) + Pl_y(B) - Pl_x(A)Pl_y(B)$

equal to (iii), but replacing "bel" by "Pl". □

In the more general context that we are considering, a definition of product of upper and lower probabilities is the following (it does not coincide with the previous definitions when we restrict the measure to the appropriate subclasses):

Definition 3.3.

Given two upper and lower probabilities (l_x,u_x) and (l_y,u_y), defined on V_X and V_Y, let $\{P^1_x,...,P^r_x\}$ and $\{P^1_y,...,P^s_y\}$ be the extreme points of the convex set of probabilities associated to (l_x,u_x) and (l_y,u_y) respectively. Then we define the product measure (l_p,u_p) as the upper and lower probability associated to the convex hull of the probabilities $\{P^{i,j}, i=1,...r; j=1,...,s\}$, where $P^{i,j}$ is the probability product of P^i_x and P^j_y.

This product measure verifies the following properties:

Proposition 3.5.

The product measure of definition 3.3 verifies:

i) $l_p(A\times B) = l_x(A)l_y(B)$

ii) $u_p(A\times B) = u_x(A)u_y(B)$

iii) $l_p(A\times Y \cup X\times B) = l_x(A) + l_y(B) - l_x(A)l_y(B)$

iv) $u_p(A\times Y \cup X\times B) = u_x(A) + u_y(B) - u_x(A)u_y(B)$

v) The marginal measures of (l_p,u_p) are again (l_x,u_x) and (l_y,u_y).

Proof:

i) $l_p(A\times B) = l_x(A)l_y(B)$

Let $\mathcal{P}_{XY}$ the convex set of probabilities $\{P^{i,j}, i=1..r, j=1..s\}$ where $P^{i,j}$ is the probabilistic product of P^i_x and P^j_y. In this case, l_p is defined as

$$l_p(A\times B) = \inf_{P\in\mathcal{P}_{xy}} P(A\times B)$$

$$\forall P\in\mathcal{P}_{xy};\quad P=\sum_{i,j}\lambda_{i,j}P^{i,j}\ ;\quad \lambda_{i,j}\geq 0\ ;\quad \sum_{i,j}\lambda_{i,j}=1$$

Then

$$\forall P\in\mathcal{P}_{xy}\ ;\quad P(A\times B) = \sum_{i,j}\lambda_{i,j}P^{i,j}(A\times B) = \sum_{i,j}\lambda_{i,j}P^i_x(A)P^j_y(B)\geq$$

$$\geq\sum_{i,j}\lambda_{i,j}(\min_i P^i_x(A))\ (\min_j P^j_y(B)) = \min_i P^i_x(A)\ \min_j P^j_y(B) = l_x(A)l_y(B)$$

so

$$\inf_{P\in\mathcal{P}_{xy}} P(A\times B) = l_p(A\times B) \geq l_x(A)l_y(B)$$

Moreover, given any A and B, there exist i_A and j_B such $l_x(A)=P_x^{iA}(A)$ and $l_y(B)=P_y^{jB}(B)$. Then $l_x(A)l_y(B) = P_x^{iA}(A)P_y^{jB}(B)$ and $P_x^{iA}P_y^{jB}\in\mathcal{P}_{xy}$ $(\lambda_{iA,jB}=1\ ;\ \lambda_{ij}=0\ \forall\ i,j \neq i_Aj_B.)$. Therefore $l_p(A\times B) = l_x(A)l_y(B)$.

ii) $u_p(A\times B) = u_x(A)u_y(B)$

The proof is similar to (i).

iii) $l_p(A\times Y \cup X\times B) = l_X(A) + l_Y(B) - l_X(A)l_Y(B)$

$$l_p(A\times Y\cup X\times B) = 1-u_p(\overline{A\times Y\cup X\times B}) = 1-u_p(\overline{A}\times\overline{B}) = 1-u_p(\overline{A})u_p(\overline{B})$$
$$= 1-[\ (1-l_x(A))\ (1-l_y(B))\ ] = l_x(A) + l_y(B) - l_x(A)l_y(B)$$

iv) $u_p(A\times Y \cup X\times B) = u_X(A) + u_Y(B) - u_X(A)u_Y(B)$
The proof is equal than in (iii).

v) The marginal measures of (l_p,u_p) are again (l_x,u_x) and (l_y,u_y)

$$l_p^{\downarrow x}(A) = l_p(A\times Y) = l_x(A)l_y(Y) = l_x(A)$$
$$u_p^{\downarrow x}(A) = u_p(A\times Y) = u_x(A)u_y(Y) = u_x(A)$$

□

An interesting relation between both definitions of product measure is the following:

Proposition 3.6.
Let (bel_x,Pl_x) and (bel_y,Pl_y) be two pairs of evidence measures; let (bel_2,Pl_2) the corresponding product measure obtained with definition 3.2 and let (bel_3,Pl_3) be the product measure of definition 3.3 .Then

$bel_2(C) \le bel_3(C)$ and $Pl_2(C) \ge Pl_3(C)$ $\forall\ C \subseteq V$

that is to say, the product measure of definition 3.2 is always included (see [3]) in that one of definition 3.3.

Proof:
bel_2 has the b.p.a $m_2(A\times B)=m_X(A)m_Y(B)$. Given an evidence measure m, for any probability P belonging to $\mathcal{P}$ there exists a function (see[5])

$$\lambda:\ \bigcup_{B\subseteq X-\{\emptyset\}} \{(B,Z)\,|Z\in B\}\rightarrow\mathbb{R}$$

such that λ is positive or null and

$$\sum_{Z\in B}\lambda(B,Z) = 1\ \forall B \quad and \quad P(Z)=\sum_{B|Z\in B} m(B)\lambda(B,Z)$$

Let P_x a probability belonging to $\mathcal{P}_X$ and P_y belong to $\mathcal{P}_Y$; then there exist λ_X and λ_Y such that

$$P_x(x) = \sum_{A|x\in A} m_x(A)\lambda_x(A,x) \qquad P_y\ (y)= \sum_{B|y\in B} m_y(B)\lambda_y(B,Y)$$

and moreover

$$P_x(x)P_y(y) = \sum_{\substack{A\times B\\(x,y)\in A\times B}} m_x(A)m_y(B)\lambda_x(A,x)\lambda_y(B,y) = \sum_{\substack{A\times B\\(x,y)\in A\times B}} m_2(A\times B)\lambda_{xy}(A\times B,(x,y))$$

where $\lambda_{XY}(\ A\times B,(x,y))=\lambda_X(A,x)\lambda_Y(B,y)\ge 0$, with

$$\sum_{(x,y)\in A\times B} \lambda_{xy}(A\times B,(x,y)) = \sum_{x\in A}\lambda_x(A,x)\sum_{y\in B}\lambda_y(B,y) = 1$$

In the sets of X×Y that are not rectangles (so, they are not focal sets) definition of lambda is irrelevant. So P_xP_y is a probability that belongs to the convex set associated to bel_2 , that is $\forall P_x;\ \forall P_y\quad P_xP_y\ge bel_2$ *and therefore* $bel_3\ge bel_2$ ($Pl_2\ge Pl_3$ *is obtained by duality*)

□

Finally, the concept of *conditional measure* neither is clear enough. There are many different definitions, some of them valid only for particular subclasses of measures (see [4], [8], [10], [12], among others). Three of them will be considered here:

-Shafer's conditioning (see [12]) for evidence measures (although it can be also used for general upper and lower probabilities):

$$Pl_x(A/B) = \frac{Pl(A\times B)}{Pl_y(B)} = \frac{Pl(A\times B)}{Pl(X\times B)} \tag{6}$$

$$bel_x(A/B) = 1-Pl_x(\overline{A}/B) = \frac{Pl(X\times B)-Pl(\overline{A}\times B)}{Pl(X\times B)} \tag{7}$$

-Campos/Moral's conditioning (see [4]), defined for every fuzzy measure:

$$g_x(A/B) = \frac{g(A\times B)}{g(A\times B)+g^*(\overline{A}\times B)} \tag{8}$$

where g* is the dual measure of g.

-Dempster's conditioning (see [8]) for upper and lower probabilities:

$$\begin{aligned} l_x(A/B) &= \inf_{P\in\mathcal{P}} P(A/B) \\ u_x(A/B) &= \sup_{P\in\mathcal{P}} P(A/B) \end{aligned} \tag{9}$$

where $\mathcal{P}$ is the convex set of probabilities associated to the lower and upper probability pair (l,u).

Note that Dempster and Campos/Moral's conditionings coincide for Choquet capacities of order two (see [4]).

Next, we are going to show several properties relating marginal and conditional measures:

Proposition 3.7

Let (bel_x,Pl_x) and (bel_y,Pl_y) be two pairs of evidence measures, and let (bel_p,Pl_p) the corresponding product measure obtained with either definition 3.2 or definition 3.3; then

i) The conditional measures of (bel_p,Pl_p) coincide with the marginal ones, if we use Shafer's conditioning, that is

$bel_p(A/B) = bel_x(A)$ and $bel_p(B/A) = bel_y(B)$
$Pl_p(A/B) = Pl_x(A)$ and $Pl_p(B/A) = Pl_y(B)$

ii) The conditional measures of (bel_p,Pl_p) are included in the marginal ones using Campos/Moral's conditioning. , that is

$bel_p(A/B) \leq bel_x(A)$, and $bel_p(B/A) \leq bel_y(B)$
$Pl_p(A/B) \geq Pl_x(A)$, and $Pl_p(B/A) \geq Pl_y(B)$

Proof:

i) In the case of Pl measure we have that

$$Pl_p(A/B) = \frac{Pl_p(A\times B)}{Pl_p(X\times B)} = \frac{Pl_x(A)Pl_y(B)}{Pl_y(B)} = Pl_x(A)$$

and for the bel measure

$$bel_p(A/B) = 1 - Pl_p(\overline{A}/B) = 1 - \frac{Pl_p(\overline{A}\times B)}{Pl_p(X\times B)} = 1 - \frac{Pl_x(\overline{A})Pl_y(B)}{Pl_y(B)} = 1 - Pl_x(\overline{A}) = bel_x(A)$$

ii) In the case of bel measure

$$bel_p(A/B) = \frac{bel_p(A\times B)}{bel_p(A\times B) + Pl_p(\overline{A}\times B)} = \frac{bel_x(A)bel_y(B)}{bel_x(A)bel_y(B) + Pl_x(\overline{A})Pl_y(B)}$$

$$= \frac{bel_x(A)bel_y(B)}{bel_x(A)bel_y(B) + [1 - bel_x(A)]Pl_y(B)} \leq \frac{bel_x(A)bel_y(B)}{bel_x(A)bel_y(B) + [1 - bel_x(A)]bel_y(B)} = bel_x(A)$$

The proof for the Pl measure is similar □

4. DEFINITIONS OF INDEPENDENCE

As we said before, one way to define an independence relationship is to compare the original bidimensional measure g and the product measure g_p obtained from the marginal measures of g. This can be done in several ways. As the claim of equality between g and g_p seems us unrealistic, we will impose a weaker condition:

Definition 4.1.

Given two variables X and Y taking their values in the sets V_X and V_Y respectively, and given a bidimensional upper and lower probability (l,u) on $V=V_X\times V_Y$, X and Y are said independent (given (l,u)) if (l,u) and the product measure of its marginals, (l_p,u_p), are compatible (in the sense that their corresponding convex sets of probabilities $\mathcal{P}$ and $\mathcal{P}_p$ have non empty intersection, see [3]).

This is a very weak definition of independence: it rather points out only the possibility of independence. As soon as there exists one probability in the convex set $\mathcal{P}$ that coincides with the product of its marginals probabilities, definition 4.1 asserts the independence of X and Y. For example, if the bidimensional information on X and Y is a necessity-possibility pair, then X and Y are always independent using definition 4.1.

Another alternative is to define a graded ε-independence by imposing to some kind of distance between the bidimensional and the product measures to be lesser than a non negative threshold ε; for example:

Definition 4.2.

In the same conditions of definition 4.1, X and Y are said independent (given (l,u)) if

$$\max_{P_p\in\mathcal{P}_p} \min_{P\in\mathcal{P}} d(P,P_p) \leq \varepsilon \qquad \textbf{(9)}$$

where d is a distance measure for probabilities, $\mathcal{P}$ and $\mathcal{P}_p$ are the convex sets of probabilities associated to (l,u) and (l_p,u_p) respectively.

In order to obtain independence, definition 4.2 states that, for each probability in $\mathcal{P}_p$, we can find a "similar enough" probability in $\mathcal{P}$ (similarity being measured through a distance measure and a threshold ε), that is to say, definition 4.2 represents a weakening of the inclusion condition of $\mathcal{P}_p$ in $\mathcal{P}$. Although definition 4.2 is usually stronger than definition 4.1, the former implies the later only for ε equal to zero.

A different approach would be obtained if we compare the conditional and marginal measures instead of the bidimensional and the product measures. So, another condition of independence is to impose to the conditional measures to be included (in the sense of being less informative, see [3]) in the marginal ones, more precisely:

Definition 4.3.

Given two variables X and Y taking their values in the sets V_X and V_Y respectively, and given a bidimensional upper and lower probability (l,u) on $V=V_X \times V_Y$, X and Y are said independent (given (l,u)) if

$$l_y(B/A) \leq l_y(B) \text{ and } l_x(A/B) \leq l_x(A), \ \forall A \subseteq V_X, \forall B \subseteq V_Y, \tag{10}$$

or equivalently

$$u_y(B/A) \geq u_y(B) \text{ and } u_x(A/B) \geq u_x(A), \ \forall A \subseteq V_X, \forall B \subseteq V_Y, \tag{11}$$

where (l_x,u_x) and (l_y,u_y) are the marginals of (l,u), and (l_x(./B),u_x(./B)) and (l_y(./A),u_y(./A)) are the conditionals of (l,u), given B and A respectively.

This definition matches our intuition that independence holds if we do not get additional information after conditioning.

Note that in all of these definitions there is a degree of freedom with respect to the choice of the type of product measure or the kind of conditioning to be used. Nevertheless, we can show some relationships between the definitions of independence with different conditional measures:

Proposition 4.1.

Let (l,u) be a pair of bidimensional Choquet capacities of order two. If X and Y are independent in the sense of definition 4.3, using Shafer's conditioning, then X and Y are also independent in the sense on definition 4.3 using Campos/Moral's conditioning.

Proof:

Let (l,u) be a pair of Choquet capacities, let X and Y independent variables (using Shafer`s conditioning). We denote Shafer`s and Campos/Moral`s conditioning by u_{DS} and u_{CM} respectively; in this case $u_{DS_y}(B|A) \geq u_y(B)$ and $u_{DS_x}(A|B) \geq u_x(A)$. For capacities of order two, we have that

$$u(A \times B) + l(A \times \overline{B}) \leq u(A \times Y)$$

so, we can see that

$$u_{CM_y}(B/A) = \frac{u(A \times B)}{u(A \times B) + l(A \times \overline{B})} \geq \frac{u(A \times B)}{u(A \times Y)} = u_{DS_y}(B|A)$$

and this does imply that $u_{CM_y}(B|A) \geq u_{DS_y}(B|A) \geq u_y(B)$; that is, X and Y are independent.□

Some connections between definitions 4.2 and 4.3 can also be established:

Proposition 4.2.
The independence in the sense on definition 4.2 for $\varepsilon=0$ implies independence in the sense of definition 4.3 for Shafer's conditioning (and therefore also implies independence using Campos/Moral's conditioning, for Choquet capacities of order two).

5. EXAMPLE

Some of the topics we have seen before can be illustrated with the following example.

Suppose that we have two variables X and Y, which take their values in $V_X=\{x_1,x_2\}$ and $V_Y=\{y_1,y_2\}$ respectively, and the following b.p.a. over the set $V_X \times V_Y$:

$m(x_1,y_1) = 0.25$; $m(x_1,y_2) = 0.10$;
$m(x_2,y_1) = 0.45$; $m(x_1,y_1) = 0.15$;
$m(\{(x_1,y_1),(x_2,y_1)\}) = 0.05$

The associated evidence measures are (we only list those values that will be needed after):

$bel(x_1,y_1) = 0.25$; $bel(x_1,y_2) = 0.10$ $\quad$ $Pl(x_1,y_1) = 0.30$; $Pl(x_1,y_2) = 0.10$
$bel(x_2,y_1) = 0.45$; $bel(x_2,y_2) = 0.15$ $\quad$ $Pl(x_2,y_1) = 0.50$; $Pl(x_2,y_2) = 0.15$

Using definition 3.1, we can obtain the marginal measures over V_X and V_Y:

$bel_X(x_1) = 0.35$, $Pl_X(x_1) = 0.40$ $\quad$ $bel_Y(y_1) = Pl_Y(y_1) = 0.75$
$bel_X(x_2) = 0.60$, $Pl_X(x_2) = 0.65$ $\quad$ $bel_Y(y_2) = Pl_Y(y_2) = 0.25$

As proposition 3.3 asserts, these marginal measures are also evidence measures with associated b.p.a.

$m_X(x_1) = 0.35$; $m_X(x_2) = 0.60$ $\quad$ $m_X(x_1,x_2) = 0.05$
$m_Y(y_1) = 0.75$; $m_Y(y_2) = 0.25$

From these measures, we can get the extreme points that define the convex set of probabilities associated to V_X and V_Y, i.e. for V_X we obtain two extreme probabilities and one probability for V_Y:

$P_1(x_1) = 0.35$, $P_1(x_2) = 0.65$; $P_2(x_1) = 0.40$, $P_2(x_2) = 0.60$,

$P(y_1) = 0.75$, $P(y_2) = 0.25$.

The product measures can be obtained from the b.p.a. m_X and m_Y (definition 3.2), or by the product of the extreme points of the convex set of probabilities associated with m_X and m_Y (definition 3.3). In this example, both definitions give rise to the same result, which is :

	(x_1,y_1)	(x_1,y_2)	(x_2,y_1)	(x_2,y_2)
l = bel	0.2625	0.0875	0.4500	0.1500
u = Pl	0.3000	0.1000	0.4875	0.1625

Finally, the conditional measures that we are considering take the following values:
Using Shafer`s conditioning :

	$Pl(./y_1)$	$bel(./y_1)$	$Pl(./y_2)$	$bel(./y_2)$
x_1	0.4000	0.3333	0.4000	0.4000
x_2	0.6666	0.6000	0.6000	0.6000

	$Pl(./x_1)$	$bel(./x_1)$	$Pl(./x_2)$	$bel(./x_2)$
y_1	0.7500	0.7500	0.7693	0.7693
y_2	0.2500	0.2500	0.2307	0.2307

Using Campos/Moral`s conditioning:

	$Pl(./y_1)$	$bel(./y_1)$	$Pl(./y_2)$	$bel(./y_2)$
x_1	0.4000	0.3333	0.4000	0.4000
x_2	0.6666	0.6000	0.6000	0.6000

	$Pl(./x_1)$	$bel(./x_1)$	$Pl(./x_2)$	$bel(./x_2)$
y_1	0.7500	0.7143	0.7693	0.7500
y_2	0.2857	0.2500	0.2500	0.2307

Now, we have got the previous values that we need to test the independence between variables. If we use definition 4.1, there exists only one probability in the convex set $\mathcal{P}$ that coincides with the product of its marginal probabilities, which is

$$P(\{x_1,y_1\}) = 0.30, \quad P(\{x_1,y_2\}) = 0.10, \quad P(\{x_2,y_1\}) = 0.45, \quad P(\{x_2,y_2\}) = 0.15$$

Therefore X and Y are independent using definition 4.1. However, definition 4.2 could be used instead. In this case we take as distance measure between probabilities the L1-norm:

$$d(P,P_p) = \sum_{i,j} |P(x_i,y_j) - P_p(x_i,y_j)|$$

The minimun threshold, ε, that makes X and Y independent is 0.05. So we can say that these variables are independent at any degree greater than or equal to 0.05.

If we use the approach which compares the conditional and marginal measures, following the definition 4.3, we can say that X and Y are dependent, because with the two conditioning we have, for example, that $Pl(./y_2)$ is not greater than or equal to $Pl_X(.)$.

6. CONCLUDING REMARKS

We have studied independence concepts in non probabilistic frameworks, pointing out different ways and tools necessary to tackle the problem. However there remain a lot of work to be done in this field: the relations between the different definitions of independence must be studied in more detail; even there are more possible definitions of independence that should be explored (for example, that the extreme points of the associated convex set of probabilities $\mathcal{P}_p$ coincide with the product of their marginal probabilities). The relationships among concepts such as information or entropy and independence should be also considered. Finally, a formal study of the properties that a non probabilistic independence relationship should verify is of great interest.

REFERENCES

1. Acid, S., Campos, L.M.de, González, A., Molina, R., Pérez de la Blanca, N. (1991), Learning with CASTLE, in Symbolic and Quantitative Approaches to Uncertainty, Lecture Notes in Computer Science 548, R. Kruse, P. Siegel (Eds.), Springer Verlag, 99-106.
2. Acid, S., Campos, L.M.de, González, A., Molina, R., Pérez de la Blanca, N. (1991), CASTLE: A tool for bayesian learning, Proc. of the 1991 ESPRIT Conference, 363-377.
3. Campos, L.M.de, Lamata, M.T., Moral, S. (1988), Logical connectives for combining fuzzy measures, in Methodologies for Intelligent Systems 3, Z.W. Ras, L. Saitta (Eds.), 11-18, North-Holland.
4. Campos, L.M.de, Lamata, M.T., Moral, S. (1990), The concept of conditional fuzzy measure, International Journal of Intelligent Systems 5, 237-246.
5. Chateauneuf, A. , Jaffray J., (1989) Some characterizations of lower probabilities and other monotone capacities through the use of mobius inversion., Math. Social Sciences 17, 263-283
6. Choquet, G. (1953), Theory of capacities, Ann. Inst. Fourier 5, 131-295.
7. Dawid, A.P. (1979), Conditional independence in statistical theory (with discussion), Journal of the Royal Statistical Society, Ser. B, 41, 1-31.
8. Dempster, A.P. (1967), Upper and lower probabilities induced by a multivaluated mapping, Ann. Math. Stat. 38, 325-339.
9. Lamata, M.T. (1985), Modelos de decisión con información general, Ph. D. Thesis, Universidad de Granada.
10. Moral, S., Campos, L.M.de (1991), Updating uncertain information, in Uncertainty in Knowledge Bases, Lecture Notes in Computer Science 521, B. Bouchon-Meunier, R.R. Yager, L.A. Zadeh (Eds.), Springer Verlag, 58-67.
11. Pearl, J. (1988), Probabilistic reasoning in intelligent systems. Networks of plausible inference, Morgan and Kaufmann.
12. Shafer, G. (1976), A mathematical theory of evidence, Princeton University Press.
13. Spohn, W. (1980), Stochastic independence, causal independence and shieldability, Journal of Philosophical Logic 9, 73-99.
14. Zadeh, L.A. (1978), Fuzzy sets as a basis for a theory of possibility, Fuzzy Sets and Systems 1, 3-28.

Uncertainty in Intelligent Systems
B. Bouchon-Meunier et al. (Editors)

A MASS ASSIGNMENT THEORY AND MEMORY BASED REASONING

J. F. BALDWIN*
University of Bristol
Bristol BS8 1TR, England

ABSTRACT
A mass assignment theory for handling both fuzzy and probabilistic uncertainties is reviewed and extended to provide a means of handling IF ... THEN rules involving conditional incomplete probability specifications and fuzzy sets.
The theory is applied to memory and case based reasoning. An evidential support logic is given which can form part of a general theory of inductive reasoning. Applications are modeled as rules in which the body represents a list of features. If a weighted sum of these features are satisfied by some object then a support is given for the body satisfying the head of the rule. Supports from various rules are combined using the mass assignment theory.

KEYWORDS
Fuzzy Logic, Evidential Reasoning, Logic Programming, Decision Theory, Case Based Reasoning, Neural Nets

1. INTRODUCTION

A new mass assignment theory for handling both fuzzy and probabilistic uncertainties has been given [Baldwin 1989, 1990a,b,c, 1991a,b, 1992a,b,c,d]. This theory allows incompleteness in the specification of the probability distributions and also unifies the fuzzy and probabilistic approaches to uncertainty. In this paper we illustrate the theory and show its application to memory based reasoning. An evidential support logic, [Baldwin 1992e], is also given for case based reasoning.

2. FUZZY SETS AND MASS ASSIGNMENTS

Definition of complete mass assignment
A mass assignment, MA, over a finite frame of discernment F is a function m

m : P(F) → [0, 1] where P(F) is the power set of F, such that

$$\sum_{A \in P(F)} m(A) = 1$$

$m(A) \geq 0$ for any $A \subseteq F$
If $m(\emptyset) > 0$ then the mass assignment is said to be incomplete. Otherwise it is complete.

This is similar to the basic probability assignment function of the Dempster / Shafer theory of evidence, [Shafer 1976], [Smets 1988]. The theory presented here is not that of the Shafer

* Professor Baldwin is a S.E.R.C Senior Research Fellow

theory of Evidence since the method of combining mass assignments is different and also we allow a mass to be assigned to the empty set.

Every subset, A, of F for which m(A) > 0 is called a focal element. If M is the set of focal elements of *P*(F) for m then the mass assignment can be represented by $\{L_i : m_i\}$ where $L_i \in$ M and $m(L_i) = m_i$. Thus we can write $m = \{L_i : m_i\}$.

Total ignorance is then represented by the mass assignment {F : 1}. In this case the only focal element is F so that m(F) = 1 and m(A) = 0 for all subsets, A, of F other than F.

We illustrate the main ideas of the theory with simple examples. Consider that evidence for diagnosing the illness of a patient provides the following diagnosis
D = a / 1 + b / 1 + c / 0.3
expressed as a fuzzy set, [Zadeh 1965, 1975, 1978, 1983], which provides a possibility distribution over the set of diseases.

We can use a voting model to interpret this fuzzy set. A representative group of experts accept or reject each of the possible diseases given the evidence. Everyone accepts a and b but only 30% accept c. We assume in the voting model that anyone who accepts an element x will accept all elements y with larger membership values in the fuzzy set. The pattern of voting for a group of 10 is

a	a	a	a	a	a	a	a	a	a
b	b	b	b	b	b	b	b	b	b
c	c	c							

The additional assumption is not required for this pattern but in general it is necessary to provide a unique pattern from the recorded voting information. The concept of the group can be replaced by one person voting at different times. The membership 0.3 then would mean that only 30% of the time does he accept c as a possible disease.

Suppose we wish to give a probability distribution over the set of diseases. We imagine choosing one of the voters or one time for the single voter. This gives a probability of 0.7 that the disease is one of {a, b} and 0.3 probability of it being one of {a, b, c}. This corresponds to a family of probability distributions over the set of diseases. We represent this by the mass assignment
m1 = {a, b} : 0.7, {a, b, c} : 0.3

More generally, let **f** be a normalized fuzzy set, $\mathbf{f} \subseteq X$, such that

$$\mathbf{f} = \sum_{x_i \varepsilon X} x_i / \chi_f(x_i) \quad ; \quad \chi_f(x_1) = 1, \ \chi_f(x_k) \le \chi_f(x_j) \text{ for } k > j$$

where $\chi_f(x)$ is the membership level, $x \varepsilon X$, of x in **f**

The mass assignment associated with **f** is

$x_1 : 1 - \chi_f(x_2)$, $\{x_1, ..., x_i\} : \chi_f(x_i) - \chi_f(x_{i+1})$ for i = 2, ... ; with $\chi_f(x_k) = 0$ for $x_k \notin X$
This defines a family of probability distributions over F for the instantiation of variable X, given the statement X is **f**

This definition is consistent with the voting model given above.

Let **f** be an unnormalised fuzzy set, $\mathbf{f} \subseteq X$, such that $\max_i\{\chi_f(x_i)\} = \beta < 1$ where

$$\mathbf{f} = \sum_{x_i \in X} x_i / \chi_f(x_i) \quad ; \quad \chi_f(x_1) = \beta,\ \chi_f(x_k) \leq \chi_f(x_j) \text{ for } k > j$$

then the mass assignment associated with **f** is

$$x_1 : \beta - \chi_f(x_2),\ \{x_1, ..., x_i\} : \chi_f(x_i) - \chi_f(x_{i+1}),\ \varnothing : 1 - \beta \text{ for } i = 2, ... ;\text{ with } \chi_f(x_k) = 0 \text{ for } x_k \notin X$$

Restrictions

If m = {Li : mi} then

$$m' = \left(\begin{array}{l} \{Li : mi \mid Li \neq Lk, L'1, L'2\} \cup \{L'1 : m(L'1) + x\} \cup \{L'2 : m(L'2) + y\} \\ \qquad \text{if } x + y = mk \\ \{Li : mi \mid Li \neq Lk, L'1, L'2\} \cup \{L'1 : m(L'1) + x\} \cup \{L'2 : m(L'2) + y\} \\ \qquad \cup \{Lk : mk - x - y\} \\ \qquad \text{if } x + y < mk \end{array} \right)$$

is a <u>restriction of</u> m and is denoted by m' ≤ m.

Mass assignments s1 and s2 are said to be **orthogonal** if one cannot be obtained from the other by restriction, ie $\neg(s1 \leq s2) \wedge \neg(s2 \leq s1)$.

complements

The complement of a mass assignment (m, M), denoted by $\overline{(m, M)}$, is (m', F') where the set of focal elements F' = {L'i} where L'i is the complement of Li with respect to F and the mass associated with L'i is m(L'i) = m(Li). If for m a mass is associated with F then this mass will be associated with Ø for $\overline{m}$ and the complement is an incomplete mass assignment. If m is an incomplete mass assignment then the mass associated with Ø is associated with F for $\overline{m}$

3. UPDATING

If we are given the prior information that a, b, and c are equally likely then we can use this prior to "fill in" the mass assignment to give a single distribution. Thus the 0.7 given to {a, b} is equally split between a and b and 0.3 is equally split between a, b and c giving the distribution

a : 0.45, b : 0.45, c : 0.1

This filling in process represents updating the prior with the specific mass assignment m1 and generalizes the Bayesian updating method. In the theory of mass assignments this is termed the iterative assignment method, [Baldwin 1989, 1990b, 1991a] .

We denote the updating of a prior mass assignment m with the specific mass assignment n by

$m \uparrow n$

We will not include more details in this paper since for the application to evidential support logic and case based reasoning as defined in this paper the updating process is a special case and is equivalent to the inference rule of FRIL, [Baldwin et al 1987].

4. Combining Mass Assignments (Meet and Join)

The meet of two mass assignments m1 and m2, denoted by $m1 \wedge m2$, is the least restricted mass such that

$m1 \wedge m2 \leq m1$; $\quad m1 \wedge m2 \leq m2$

if this is unique. In general there will be several orthogonal masses {s1, s2, ...} satisfying these conditions. The meet is then Σ $\alpha i.si$ where the weights αi sum to 1.

In fact $m1 \wedge m2 = s1 \vee s2 \ldots$

The join of m1 and m2, denoted by $m1 \vee m2$, is the most restricted mass such that

$m1 \vee m2 \geq m1$; $\quad m1 \vee m2 \geq m2$

if this is unique. In general there will be several orthogonal masses {s1, s2, ...} satisfying these conditions. The meet is then Σ $\alpha i.si$ where the weights αi sum to 1.

Consider the two mass assignments m1 = {Xi : xi} for i = 1, 2, ..., n1 and m2 = { Yj : yj} for j = 1, 2, ..., n2 where each Xi, Yj are subsets of F for which xi, yj $\neq$ 0.
Let m = {Zk : zk}, where Zk are subsets of F for which zk $\neq$ 0, be the meet of m1 and m2.
Write M = $\{Xi \cap Yj : m_{ij}\}$, called the meet matrix, and let

$$zk = \sum_{i,j : Xi \cap Yj = Zk} m_{ij}$$

then for the meet to correspond to the intersection of distributions of m1 and m2 the following constraints, called the constraint set C, must be satisfied

$$\sum_j m_{ij} = xi \; ; i = 1, \ldots, n1 \quad ; \quad \sum_i m_{ij} = yj \; ; j = 1, \ldots, n2$$

Example
Consider the example above for which
m1 = {a, b} : 0.7, {a, b, c} : 0.3
Suppose we were also presented with a second piece of evidence which provided the mass assignment
m2 = a : 0.1, {a, b} : 0.5, {b, c} : 0.4

We require to combine the two masses m1 and m2 to give the final specific mass assignment. This combined mass assignment is given by the meet of m1 and m2, which by inspection is

$m1 \wedge m2$ = m say = a : 0.1, b : 0.1, {a, b} : 0.5, {b, c} : 0.3 as given in the tableau below.

	0.1 a	0.5 {a, b}	0.4 {b, c}
0.7 {a, b}	a 0.1	{a, b} 0.5	b 0.1
0.3 {a,b,c}	a 0	{a, b} 0	{b, c} 0.3 max

m is equivalent to the family of probability distributions obtained by intersecting the family of distributions corresponding to m1 and that for m2. This is an example of the general assignment method for determining the mass of the conjunction of two evidences, each expressed as a mass assignment. The rules for finding the combination are : labels in cells are intersections of the corresponding row and column sets in the two evidences, the masses in the cells are chosen such that the row cell masses add up to the corresponding evidence row mass and similarly for the column cell masses.

Methods for finding the meet and join of two mass assignments have been discussed in [Baldwin 1989, 1991a, b, c, 1992a]. Additional assumptions can be made to ensure a unique meet and these are discussed in [Baldwin 1992b].

We can use the **multiplication model** for the meet. In this case the masses of the cells are obtained by multiplying the corresponding row and column masses of the given assignments. This corresponds to an expectation model in which all possible equally likely voting patterns for m1 and m2 are considered. Combining these in all possible ways and taking an expected solution gives the multiplication model. We will denote this multiplication meet as m1 ∧. m2. If the information sources giving rise to m1 and m2 are independent then the meet of m1 and m2 is given by the multiplication meet.

If m1 and m2 are derived from fuzzy sets **f1** and **f2** and we require the meet of m1 and m2 to correspond to a fuzzy set, then the meet is given by

$m1 \wedge_f m2 = m_{f1 \cap f2}$

where the MIN rule is used for fuzzy intersection.
In [Baldwin 1991a, b, c] this is referred to as the least commitment model for the meet.

Algebra of Mass Assignments

Let **M** represent all MA's over F. <**M**, ∨, ∧> is an algebra with idempotence, commutativity, associativity, absorption, distributivity properties of a Boolean algebra. Full complementation properties are not satisfied. The algebra is a pseudo Boolean Algebra

5. IF ... THEN RULES

IF ... THEN ... statements involving both fuzzy sets and probabilities can be represented in mass assignment form and used for many applications such as fuzzy and probabilistic control and databases, memory based reasoning, non-monotonic and probabilistic logics and expert systems.

Consider the simple case
IF A = a THEN B = b with probability 0.7, B = ¬b with probability 0.3
i.e. IF A with {a _} : 1 THEN B with {_ b} : 0.7, {_ ¬b} : 0.3
corresponding to Pr(b | a) = 0.7.
The statement "IF A = a THEN B =b" is represented by the {ab, ¬a _} : 1 and similarly the statement "IF A = a THEN B =¬b" by {a¬b, ¬a _} : 1. The first occurs with a probability of 0.7 and the second with probability 0.3, so that the combined mass assignment for the rule is
{ab, ¬a _} : 0.7, {a¬b, ¬a _} : 0.3
which is equivalent to

$\overline{\{a\ \_\} \wedge \overline{\{\_\ b\} : 0.7, \{\_\ \neg b\} : 0.3}}$

It should be noted that this is not the same as the mass assignment for Pr{a ⊃ b} = 0.7 which is {ab, ¬a _} : 0.7, a¬b : 0.3.
Consider the rule

IF X = {a, b} THEN Y with {α : 0.3, β : 0.5, {α, β} : 0.2 where X can take values from {a, b, c} and Y from {α, β, χ} and X = {a, b} means value of X is a or b, exclusive or.
This is equivalent to the mass assignment

{aα, bα, c_} : 0.3, {aβ, bβ, c _} : 0.5, {aα, aβ, bα, bβ, c_} : 0.2
More generally the mass assignment associated with the IF ... THEN statement of the form IF A with m_A THEN B with m_B is given by

$\overline{m_A \wedge \overline{m_B}}$

For example, the statement

IF X with m1 = a : 0.4, {a, b} : 0.6 THEN Y with m2 = α : 0.3, {α, β} : 0.7, where X can take values from {a, b, c} and Y from {α, β, χ}, is equivalent to

IF X = {a, b} THEN Y = α with mass m1 $\wedge$ m2({aα,bα})

IF X = {a, b} THEN Y = {α , β} with mass m1 $\wedge$ m2({aα, aβ, bα, bβ})

IF X = a THEN Y = α with mass m1 $\wedge$ m2({aα})

IF X = a THEN Y = {α , β} with mass m1 $\wedge$ m2({aα, aβ})

A multiplication model meet could be assumed but care should be taken in knowing exactly what is meant by a statement of this form.

This form of definition can be extended to include fuzzy sets. We can then derive mass assignments for such statements as

IF someone is tall THEN the person will wear large shoes

IF someone is tall THEN the person will wear large shoes with probability 0.8 and medium shoes with a probability 0.2.

For the first case, the definition of the fuzzy IF ... THEN rule is used to derive a fuzzy relation, [Dubois, Prade 1984], which can be converted to a mass assignment or the fuzzy sets can first be converted to mass assignments and the above IF ... THEN mass assignment rule used. The additional assumptions used for the meet will correspond to different fuzzy IF ... THEN rule interpretations. This requires more study. The second case can be treated similarly, weighting the mass assignments to take account of the probabilities in a similar manner to above for the non-fuzzy cases.

More generally we will denote a random variable X with probability distribution belonging to a family of probability distributions defined by the mass assignment m as "X with m".

The rules

IF X with m_{iX} THEN Y with m_{iY} for i = 1, ..., n

where {m_{iX}} and {m_{iY}} are sets of mass assignments over F_I and F_O respectively, provide the general mass assignment

$$m_{X \times Y} = \bigwedge_i (\overline{m_{iX}} \vee m_{iY}) = \bigwedge_i \overline{(m_{iX} \wedge \overline{m_{iY}})}$$

so that, for a given specific input m_X, we obtain the output

$$m_Y = \mathrm{Proj}_Y[m_{X \times Y} \uparrow m_X]$$

where $m_1 \uparrow m_2$ is the mass assignment resulting from updatingm_1 with m_2 using the iterative assignment algorithm and $\mathrm{Proj}_Z m$ represents the projection of m onto the space Z.

A special case is that for the fuzzy rules

IF x is $\mathbf{f_i}$ THEN y is $\mathbf{g_i}$ where $\mathbf{f_i} \subseteq_f F_I$ and $\mathbf{g_i} \subseteq_f F_O$ for i = 1, ..., n

which are replaced by

IF X with m_{iX} THEN Y with m_{iY} for i = 1, ..., n

where m_{iX} is the mass assignment associated with the fuzzy set $\mathbf{f_i}$ and m_{iY} is the mass assignment associated with the fuzzy set $\mathbf{g_i}$.

6. MEMORY BASED REASONING

This represents reasoning from examples, i.e. case reasoning. In its simplest form it represents the classification of an object with known features in terms of a set of relevant classified objects with known features. A classified object provides a statement of the form
If feature F1 has value f1 and ..., and feature Fn has value fn then classification C is **c**, where **c** is a fuzzy set over the set of possible classifications. Generalization is obtained by fuzzifying this statement to the following:
If feature F1 has value **g1** and ..., and feature Fn has value **gn** then classification C is **c** where **g1** and ..., and **gn** are fuzzy sets on G1, ..., Gn and Gi is a set of fuzzy labels for feature i.
This statement has the associated mass assignment

$m_i = \overline{(m_{\mathbf{g1}} \wedge \ldots \wedge m_{\mathbf{gn}}) \wedge m_{\mathbf{c}}}$ where $m_{\mathbf{gi}}$ is the mass assignment corresponding to the fuzzy set **gi**..
We consider all relevant examples for the classification of the given object. Let the mass assignments for these statements be given by $\{m_k\}$. A relevant object is one that is prescribed as having near features to that of the given object and can be defined in several ways. The mass assignment for all relevant objects is then given by

$$m = \bigwedge_i m_i$$

This is combined with a prior mass assignment over C. Suppose this prior mass assignment, which is independent of m, is m0. The combined mass assignment is then given by

$$n = m \wedge_{.} m0$$

The mass assignment for the feature values in terms of fuzzy sets on Gi of the given object, namely {**g'i**} is given by

$$n' = m_{\mathbf{g1}} \wedge \ldots \wedge m_{\mathbf{gn}}$$

The classification for the given object is then given by

$$\mathrm{Proj}_C \{n \uparrow n'\}$$

This represents a mass assignment over C from which a decision on the classification can be made.

It is important to note that the generalization of the knowledge of the examples to include knowledge about the given unclassified object results from expressing the feature values of the examples in terms of fuzzy sets over sets of labels for each feature.

7. SEMANTIC UNIFICATION OF FUZZY SETS

We use the theory of mass assignments to provide the semantic unification of **g** with respect to **g'**, denoted by **g | g'** where **g** and **g'** are fuzzy sets on G. We illustrate this with an example. More details can be found in [Baldwin 1992e]

Example
g = a / 1 + b / 0.4 + c / 0.1 ; **g'1** = a / 0.8 + b / 1 + c / 0.1 ; G = {a, b, c, d}

The mass assignment for **g** and **g'** are given by
$m_{\mathbf{g}}$ = a : 0.6, {a, b} : 0.3, {a, b, c} : 0.1 ; $m_{\mathbf{g'}}$ = b : 0.2, {a, b} : 0.7, {a, b, c} : 0.1 so that

m_g \ $m_{g'}$	0.2 b	0.7 {a, b}	0.1 {a, b, c}
0.6 a	f 0.12	u 0.42	u 0.06
0.3 {a, b}	t 0.06	t 0.21	u 0.03
0.1 {a, b, c}	t 0.02	t 0.07	t 0.01

T(**g1** | **g'1**)

The symbolic entries in the cells correspond to the truth of the associated set of m_g given the associated set of $m_{g'}$, in terms of t, f, u representing true, false, uncertain.

$m_{T(\mathbf{g1} \mid \mathbf{g'1})}$ = t : 0.37, f : 0.12, u : 0.51

i.e. **g1** | **g'1** : [0.37, 0.88]

Numerical entries in cells given by the multiplication model.

8. EVIDENTIAL SUPPORT LOGIC

The head of FRIL rule can represent a classification and the body of the rule a list of features with their importances. A function, S, of the weighted sum of the degrees of match of the features determines the support for the body of the rule. A family of S functions is given. Members of this family can be interpreted as conjunctions of various degrees of optimism, disjunctions of various degrees of optimism and mixtures of these. The features in the body are compared with those of the body to be classified. A FRIL semantic unification is performed to obtain a support pair for each clause in the body of the rule.
The FRIL rule is of the form

((conclusion y for X is **f**)
(ev_body ((feature 1 of importance θ1 has value for X of **g1**) . . .
(feature n of importance θn has value for X of **gn**)))) : (α β)

This is interpreted as head of rule, (conclusion y for X is **f**), is supported to a degree lying in [α β] if the body of the rule, (ev_body ((feature 1 of importance θ1 has value for X of **g1**) . . . (feature n of importance θn has value for X of **gn**))), is true.

Feature facts about the object to be classified are given by

((feature i of importance θi has value for x of **g'i**))

for i = 1, ..., n, where **f**, **gi**, **g'i** are fuzzy sets defined on F, Gi, Gi respectively. An inference of the form

((conclusion y of x is **f**)) : (γ1, γ2)

where γ1 and γ2 are given by $\gamma 1 = \alpha\rho$; $\gamma 2 = \beta\rho + (1 - \rho)$ and ρ is the necessary support for the body of the rule given the facts calculated according to the rule

$$\rho = S\left(\sum_{i=1}^{n} \theta_i \alpha_i\right)$$

where {αi} are determined by semantic unifications {**gi** | **g'i**} = [αi, βi]} and

$$S(x) = \begin{pmatrix} 0 \text{ for } 0 \leq x \leq a \\ \frac{1}{b - a} x - \frac{a}{b - a} \text{ for } a < x \leq b \\ 1 \text{ for } x > b \end{pmatrix}$$

for a, b $\in$ [0, 1) ; $a \leq b$

In the limiting case as a→1 then S(x) = 1 if x =1 and 0 otherwise.
The limiting case corresponds to interpreting the combination in a "hard logic" sense.

The inference uses the mass assignment theory of updating to derive the support pair for the head of the rule given the support pair for the body of the rule. This is equivalent to the FRIL inference rule. S(x) is a fuzzy set membership function defined on [0, 1]. If we choose a = 0.2 and b = 0.8, then the body behaves like a disjunction for large values of x and as a conjunction for small values of x. This is explained more fully in [Baldwin 1992f]

Even if nothing is known about certain features of the object to be classified, some support can be given to the head of the rule.

The mass assignment theory combines solutions from different rules.

We can give this a neural net like interpretation as seen in Figure below.

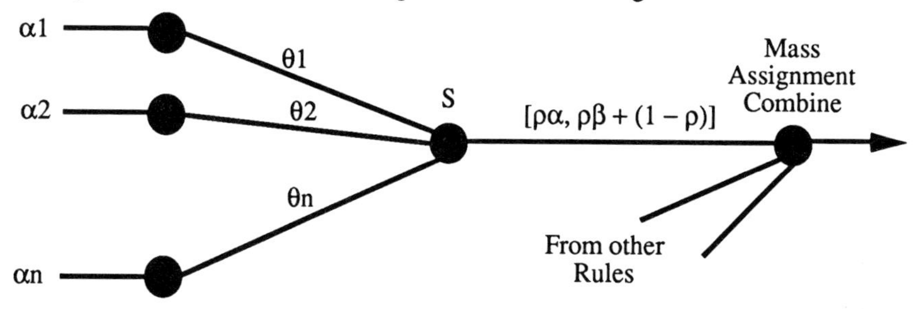

An application can be modeled with rules of the form above. The importances {θi} and the rule support pair (α β) can be obtained using statistical methods or neural net type learning rules or be provided by the user.

Example.

Rule 1
((conclusion y for X is **f**)
(ev_body ((feature 1 of importance 0.3 has value for X of **g1**)
(feature 2 of importance 0.7 has value for X of **g2**)))) : (1 1)
where
f = u / 1 + v / 0.4 ; F = {u, v, w}
g1 = a / 1 + b / 0.4 + c / 0.1 ; G1 = {a, b, c, d} , **g2** = x / 1 + y / 0.3 ; G2 = {x, y, z}

((feature 1 of importance 0.3 has value for x of **g'1**))

((feature 2 of importance 0.7 has value for x of **g'2**))
where
g'1 = a / 0.8 + b / 1 + c / 0.1 , **g'2** = x / 1 + y / 0.5

The multiplication model for semantic unification given above gives
g1 | **g'1** : [0.37, 0.88] ; **g2** | **g'2** : [0.65, 1]

Therefore ρ = (0.37)(0.3) + (0.65)(0.7) = 0.566 so that the conclusion is given as
(conclusion y for x is **f**)) : (0.65, 1)
which corresponds to the mass assignment
m1 = { u : 0.6, {u, v} : 0.4 } : 0.65, { {u, v, w} : 0.35 }
= u : 0.39, {u, v} : 0.26, {u, v, w} : 0.35
This actually corresponds to the fuzzy set
$\mathbf{f_{c1}}$ = u / 1 + v / 0.61 + w / 0.35

Rule 2
((conclusion y for X is **f2**)
(ev_body ((feature 1 of importance 0.7 has value for X of **g3**)
(feature 2 of importance 0.3 has value for X of **g4**)))) : (1 1)
where
f2 = w / 1 + v / 0.6 ; F = {u, v, w}
g3 = d / 1 + b / 0.6 + c / 0.9 ; G1 = {a, b, c, d} , **g4** = z / 1 + y / 0.7 ; G2 = {x, y, z}

((feature 1 of importance 0.3 has value for x of **g'1**))
((feature 2 of importance 0.7 has value for x of **g'2**))

The multiplication model for semantic unification gives
g3 | **g'1** : [0.12, 0.63] ; **g4** | **g'2** : [0, 0.35]

Therefore ρ = (0.12)(0.7) = 0.084 so that the conclusion is given
((conclusion y for x is **f2**)) : (0.084, 1)
which corresponds to the mass assignment
m2 = { w : 0.4, {u, w} : 0.6 } : 0.084, { {u, v, w} : 0.916 }
= w : 0.0336, {u, w} : 0.0504, {u, v, w} : 0.916
This actually corresponds to the fuzzy set
$\mathbf{f_{c2}}$ = w / 1 + u / 0.9664 + v / 0.916

Finding the meet of the two inferences gives
m = m1 $\wedge$ m2 = u : 0.39, w : 0.0336, {u, v} : 0.26, {u, v, w} : 0.3164

If we add the restriction that we want the conclusion to correspond to a fuzzy set of F then we obtain using the mass assignment theory
m = m1 $\wedge_f$ m2 = Ø : 0.0336, u : 0.3564, {u, v} : 0.26, {u, v, w} : 0.35
corresponding to the fuzzy set
$\mathbf{f_c}$ = u / 0.9664 + v / 0.61 + w / 0.35
If we use the multiplication model for the meet to combine the inferences then we obtain
m = m1 $\wedge.$ m2
= Ø : 0.0218, w : 0.0118, u : 0.39, {u, v} : 0.2382, {u, w} : 0.0176, {u, v, w} : 0.3206

Any sensible decision rule to make the final classification would in all these cases give classification = u

In applications of case based reasoning we can form a rule for each example in the example set. Alternatively a rule for a typical case, a centroid of a cluster of example points, can be used. The rules can be more complicated than those given here allowing for conjunction of features, relations between features etc. to be considered.

REFERENCES

Baldwin J.F. , Pilsworth B.W, Martin T, (1987), "FRIL Manual", Fril Systems Ltd, St Anne's House, St Anne's Rd, Bristol BS4 4A, UK

Baldwin J.F., (1989), A new approach to combining evidences for evidential reasoning,ITRC 151 Univ. of Bristol Report.

Baldwin J.F., (1990a), Computational models of uncertainty reasoning in expert systems, Computers Math. Applic., **19**, 105-119.

Baldwin J.F., (1990b), Towards a general theory of intelligent reasoning, 3rd Int. Conf IPMU, Paris, July 1990

Baldwin J, F., (1990c), Inference under uncertainty for expert system rules, ITRC 152 Univ. of Bristol Report, To Appear

Baldwin J.F., (1991a), Combining evidences for evidential reasoning, Int. J. of Intelligent Systems, 6, 569-617

Baldwin J. F. (1991b), Fuzzy and probabilistic databases with automatic reasoning, ITRC 160, University of Bristol Report, To Appear.

Baldwin J, F, (1992a), Evidential reasoning under probabilistic and fuzzy uncertainties, in An introduction to Fuzzy Logic Applications in Intelligent Systems, **R. R. Yager** and **L. A. Zadeh**, (Eds), Dordrecht: Kluwer

Baldwin J.F. (1992b), Fuzzy and probabilistic uncertainties, in Encyclopedia of Artificial Intelligence, Second edition, S. C. Shapiro, (Ed), New York: John Wiley.

Baldwin J.F. (1992c), Inference for information systems containing probabilistic and fuzzy uncertainties, in Fuzzy Logic for the Management of Uncertainty, . **L. A. Zadeh** and **J. Kacprzyk**, (Eds), New York: Wiley, To Appear

Baldwin J. F. (1992d), A calculus for Mass Assignments in Evidential Reasoning, To Appear in "Advances in the Dempster-Shafer Theory of Evidence, Ed **Mario Fedrizza, Janusz Kacprzyk and Ronald Yager**, Wiley & Sons, Inc. New York 1992

Baldwin J. F., (1992e), Evidential Support logic, Fril and Case Based Reasoning, ITRC Report, University of Bristol.

Baldwin J. F., (1992f), Evidential Support logic, ITRC Report, University of Bristol.

Dubois D., Prade H, (1984), Fuzzy Logics and the Generalized Modus Ponens, Cybernetics and Systems, **15**, 293-331.

Shafer G., (1976), A mathematical theory of evidence , Princeton Univ. Press

Smets P., (1988), Belief Functions, in Non-Standard Logics for Automated Reasoning, Eds **Smets et al**, Academic Press, pp 253-286

Yager R.R, (1991a), On Structures for Fuzzy Modeling and Control, Technical Report MII-1213, Iona College, New Rochelle.

Yager R.R, Filev D. P, (1991b)Generalizing the Modeling of Fuzzy Controllers by Parametrized Aggregation Operators, Technical Report MII-1219, Iona College, New Rochelle.

Zadeh L, (1965), Fuzzy sets, Information and Control, **8,** 338-353.

Zadeh L,. (1975), Fuzzy logic and approximate reasoning, Synthese, **30**, 407-428

Zadeh L, (1978), Fuzzy sets as a basis for a theory of possibility, Fuzzy Sets and Systems **1**, 3-28

Zadeh L., (1983), The role of fuzzy logic in the management of uncertainty in expert systems, Fuzzy Sets and Systems, **11**, 199-227

Uncertainty in Intelligent Systems
B. Bouchon-Meunier et al. (Editors)
1993 Elsevier Science Publishers B.V.

CONDITIONAL EVENTS AND SUBJECTIVE PROBABILITY IN MANAGEMENT OF UNCERTAINTY

Angelo Gilio

Dipartimento Metodi e Modelli Matematici, Universita'
"La Sapienza", Via A. Scarpa 10 - 00161 Roma, Italy.

Subjective probability is the natural theoretical tool to model the uncertain knowledge and to revise beliefs. Based on de Finetti's coherence principle, this theory, which does not require to assume strictly positive probabilities for all conditioning events, provides the most general framework for checking consistency of conditional probability assessments. In this paper it is argued that, through the concept of generalized atom, a natural representation of the truth values of conditional events can be given. Moreover, some results on probabilistic consistency are obtained.

1. INTRODUCTION

The (classical) probabilistic approach to the management of uncertainty has often been criticized. Some of the objections against the use of probability have been: the frequentistic approach is not always possible; there exist different kinds of uncertainties; probabilities are (sometimes) imprecise or not well known; the elicited probabilities are fuzzy; probability is (in some cases) ... highly subjective! And so on.

Nevertheless, despite the above objections, probability always plays a key role to suitably measure uncertainty. Furthermore, adopting the subjective approach to probability (de Finetti (1974), (1975)), many objections disappear and the need for alternative theories of uncertainty measures is greatly reduced (see Scozzafava (1991) for a discussion on subjective probability versus belief functions).

It is worthwhile to remark that probability is just a provisional and, depending on relevant information, non monotonic numerical representation of our beliefs. Moreover, whatever our information or uncertainty may be on a given event, the only *objective* fact (in the sense that it does not depend on individual opinions) is the *truth* or *falsity* (and *not* the probability) of the event at hand.

Therefore, in all circumstances, probability is a *subjective* and not an *objective* concept. Consequently, it is very doubtful to give any objective meaning to claims such as "probability is imprecise or not well known".

The subjective approach, besides distinguishing well founded problems from pseudo ones, has the following advantages:

(a) de Finetti's *coherence principle* can be used as the unique *axiom* by means of which probability theory can be developed;

(b) consistent probabilistic judgements on *arbitrary* families of events can be made and extensions of probabilistic assessments are always possible, using (linear and non linear) programming techniques (see Bruno and Gilio (1980), Nilsson (1986), Gilio and Scozzafava (1988), Paass (1988), Lad, Dickey and Rahman (1990), Andersen and Hooker (1990));

(c) if the assessment of numerical probabilities entails some difficulties, then a qualitative approach can be adopted (see Coletti (1991), Coletti, Gilio and Scozzafava (1990a), (1990b), (1991));

(d) a *direct* assessment of *conditional* probabilities (and checking for coherence) is allowed, i.e. the use of the *compound probability theorem* is not explicitly required.

(e) differently from the usual *axiomatic* approach, no theoretical problem arises when some (or possibly all) *conditioning* events have *zero* probability.

Thus, contrarily to what is claimed in Shenoy (1991), within the subjective approach, probability is perfectly adequate for revising beliefs. Also, subjective probability provides the natural framework for the probabilistic extension of terminological logics, as well as probabilistic consistency, proposed in Heinsohn (1991).

In this paper some of the above aspects are deepened. A concise review of de Finetti's coherence principle is given in Section 2. In Section 3 the concept of generalized atom is related to the representation of truth values of conditional events. Then, some results on probabilistic consistency are obtained in Section 4 .

2. DE FINETTI'S COHERENCE PRINCIPLE: A CONCISE REVIEW

De Finetti's coherence principle can be based on the *penalty criterion* or on the *betting scheme*. Here we examine, in the framework of the penalty criterion, a definition introduced in Gilio (1990), which slightly modifies that of de Finetti.

We use the symbol EH to shorten the *logical product* $E \wedge H$ of the events E and H. Moreover, given an event E, we denote by the same symbol its indicator. We denote by $\mathcal{K}$ an arbitrary family of conditional events and by P a real function defined on $\mathcal{K}$. Given a finite family $\mathcal{F} = \{E_1|H_1, \ldots, E_n|H_n\} \subseteq \mathcal{K}$, we put

$$P(E_i|H_i) = p_i \ , \quad i = 1,2,\ldots,n \ , \quad \mathcal{P} = (p_1,p_2,\ldots,p_n) \ .$$

As it is well known, based on de Finetti's penalty criterion, to the point $\mathcal{P}$ it is associated a loss

$$\mathfrak{L} = \sum_{i=1}^{n} H_i \left(E_i - p_i\right)^2 .$$

For each subscript i , it is $E_iH_i \vee E_i^cH_i \vee H_i^c = \Omega$, so that, developing the expression

$$\left(E_1H_1\vee E_1^cH_1\vee H_1^c\right)\wedge\left(E_2H_2\vee E_2^cH_2\vee H_2^c\right)\wedge\ldots\wedge\left(E_nH_n\vee E_n^cH_n\vee H_n^c\right),\quad(1)$$

we obtain a partition $\Pi = \{C_0, C_1, \ldots, C_s\}$ of Ω, where the atoms C_h are all the (not impossible) logical products $A_1A_2 \ldots A_n$, with $A_i \in \{E_iH_i,\ E_i^cH_i,\ H_i^c\}$. We denote by H_0 the event $H_1\vee\ldots\vee H_n$ and by C_0 the atom $H_1^cH_2^c \ldots H_n^c = H_0^c$. Then, for each $h = 1,\ldots,s$, it is $C_h \subseteq H_0$, with $s \le 3^n - 1$. Note that, if $H_0 = \Omega$, then $C_0 = H_1^cH_2^c\ldots H_n^c = \emptyset$ and $\Pi = \{C_1, \ldots, C_s\}$.

We denote by L_h the value of $\mathfrak{L}$ which corresponds to C_h. Obviously, $L_0 = 0$. Moreover, to the atoms $C_1,\ldots,C_s$, contained in H_o, we associate the *generalized atoms* $Q_1,\ldots,Q_s$, defined as

$$Q_h = (q_{h1}, q_{h2}, \ldots, q_{hn}),\qquad h = 1,2,\ldots,s,$$

where

$$q_{hi} = \begin{cases} 1, & \text{if } C_h \subseteq E_iH_i \\ 0, & \text{if } C_h \subseteq E_i^cH_i, \\ p_i, & \text{if } C_h \subseteq H_i^c \end{cases}\qquad \begin{array}{l} i=1,2,\ldots,n \\ h=1,2,\ldots,s \end{array}.$$

As it easy to verify, to each atom C_h there corresponds a value L_h of $\mathfrak{L}$ given by

$$L_h = \sum_{i=1}^{n}\left(q_{hi}-p_i\right)^2 = \overline{\mathcal{P}Q_h}^{\,2}.$$

Given a point $\mathcal{P}^* \ne \mathcal{P}$, we denote by $\mathfrak{L}^*$ the loss associated to $\mathcal{P}^*$ and by L_h^* the value of $\mathfrak{L}^*$ which corresponds to the atom C_h.

Then, we give the following definition (*C_o-coherence*).

(2.1). Definition. The real function P defined on the class $\mathcal{K}$ is a coherent conditional probability on $\mathcal{K}$ if, for each n and for each family $\mathcal{F} = \{E_1|H_1,\ E_2|H_2,\ \ldots,\ E_n|H_n\} \subseteq \mathcal{K}$, putting $P(E_i|H_i) = p_i$, $i = 1,\ldots, n$, $\mathcal{P} = (p_1, p_2, \ldots, p_n)$, for every point $\mathcal{P}^* \ne \mathcal{P}$, there exists a subscript h such that $L_h^* > L_h$.

In other words, P is coherent if a point $\mathcal{P}^*$, such that $\mathfrak{L}^* \le \mathfrak{L}$, $\mathfrak{L}^* \ne \mathfrak{L}$, does not exist.

Notice that if P is coherent, then for each $\mathcal{F} \subseteq \mathcal{K}$ the assessment $\mathcal{P}$, as a *restriction of* P to $\mathcal{F}$, is coherent too.

Conversely, if $\mathcal{P}$ is a coherent assessment on $\mathcal{F}$, given an arbitrary family $\mathcal{K} \supseteq \mathcal{F}$, there exists a coherent extension P of $\mathcal{P}$ on $\mathcal{K}$.

In Section 4 the following Theorems (see Gilio (1990), (1992)) will be useful to obtain some results on coherence of probability assessments.

(2.2) Theorem. If there exist s *positive* numbers $\lambda_1, \ldots, \lambda_s$, with $\sum_{h=1}^{s} \lambda_h = 1$, such that $\mathcal{P} = \sum_{h=1}^{s} \lambda_h Q_h$, then $\mathcal{P}$ is coherent.

Given a subset $J \subseteq \{1,2,\ldots,n\}$, denote by $\mathfrak{I}_J$ the convex hull of the generalized atoms relative to $\mathcal{F}_J = \{E_j|H_j : j \in J\} \subseteq \mathcal{F}$ and point $\mathcal{P}_J = (p_j : j \in J)$. Then, we have

(2.3) Theorem. The prevision point $\mathcal{P}$ is coherent *if and only if* $\mathcal{P}_J \in \mathfrak{I}_J$ for every $J \subseteq \{1,2,\ldots,n\}$.

(2.4) Remark. Assume $\emptyset \subset E_j H_j \subset H_j$. Then, for $J = \{j\}$, the condition $\mathcal{P}_J \in \mathfrak{I}_J$ amounts to $0 \le p_j \le 1$.

As shown by Theorems (2.2) and (2.3), the generalized atoms allow to extend the *geometrical characterization* of coherence, given by de Finetti for unconditional events, to the case of conditional events too.

3. TRUTH VALUES FOR CONDITIONAL EVENTS

Given a conditional event $E|H$ and denoting by T the *truth assignment* function, in Dubois and Prade (1991) it is defined $T(E|H) = 1$, or 0 , or $?$, according to whether E and H are *both true*, or E is *false* and H is *true*, or H is *false*.

In the same paper it is proved that $0 \le ? \le 1$; thus, the symbol ? is interpreted as a number in [0,1].

The subjective theory allows us to operatively recognize that, assessing $P(E|H) = p$, the suitable truth value of $E|H$ when H is false is $T(E|H) = ? = p$.

In fact, if we define $T(E|H) = p$ when H is false, then the loss function $\mathfrak{L}$ associated to $E|H$ can be represented as

$$\mathfrak{L} = H(E - p)^2 = \Big(T(E|H) - p\Big)^2 .$$

More in general, given an assessment $\mathcal{P} = (p_1, p_2, \ldots, p_n)$ on a family $\mathcal{F} = \{E_1|H_1, E_2|H_2, \ldots, E_n|H_n\}$, we have

$$\mathfrak{L} = \sum_{i=1}^{n} H_i \Big(E_i - p_i\Big)^2 = \sum_{i=1}^{n} \Big(T(E_i|H_i) - p_i\Big)^2 .$$

This interpretation of the truth value of a conditional event is naturally related to the concept of generalized atoms: for each h , the generalized atom Q_h is the vector of truth values, assumed by the conditional events $E_1|H_1, E_2|H_2, \ldots, E_n|H_n$, which correspond to the *possible world* C_h .

Remark. The above interpretation of the truth values of conditional events, considered in Goodman, Nguyen and Walker (1991, p. 162) too, is in agreement with the partial order, as proposed in Goodman and Nguyen (1988):

$B|A \le D|C \iff AB \subseteq CD$ and $CD^c \subseteq AB^c$.

In fact, as it will be shown in the next Section (cfr. Proposition (4.2)), we have

$$B|A \le D|C \implies T(B|A) \le T(D|C) . \qquad (2)$$

4. SOME RESULTS ON PROBABILISTIC CONSISTENCY

In the framework of de Finetti's theory the checking of coherence does not require to assume positive probabilities for the conditioning events. To deepen this aspect, some Propositions will be proved in the next Section.

To avoid trivial cases, when considering some conditional events $E_i|H_i$, $i = 1,2,\ldots,$ for each subscript i we will assume $\emptyset \subset E_iH_i \subset H_i$.

(4.1) Proposition. Given the events H and $E|H$, the assessment $P(H) = p_1$, $P(E|H) = p_2$ is coherent for all values $p_1,p_2 \in [0,1]$.

Proof. We use Theorem (2.3), with $\mathcal{F} = \{H|\Omega, E|H\}$, $\mathcal{P} = (p_1,p_2)$. Applying (1) to $\mathcal{F}$ we obtain, observing that $C_0 = \emptyset$,

$$(H \vee H^c) \wedge (EH \vee E^cH \vee H^c) = C_1 \vee C_2 \vee C_3 ,$$

where $C_1 = EH$, $C_2 = E^cH$, $C_3 = H^c$.

Then, the generalized atoms are

$$Q_1 = (1,1) , \quad Q_2 = (1,0) , \quad Q_3 = (0,p_2)$$

and the convex hull $\mathfrak{I}$ is represented by the triangle with vertices the points $(1,1)$, $(1,0)$, $(0,p_2)$. It is $\mathcal{P} = (p_1,p_2) \in \mathfrak{I}$ for every point $(p_1,p_2) \in [0,1] \times [0,1]$ and the assert follows from Remark (2.4) .

In particular, from Proposition (4.1) we obtain that the assessment $P(H) = 0$, $P(E|H) = p$ is coherent for all $p \in [0,1]$.

In the next Proposition we examine the compatibility between the partial order of Goodman and Nguyen and conditional probability.

(4.2) Proposition. Consider the family $\mathcal{F} = \{B|A, D|C\}$ and the assessment $\mathcal{P} = (p_1, p_2)$, where $P(B|A) = p_1$, $P(D|C) = p_2$, with $0 \le p_1, p_2 \le 1$. If $B|A \le D|C$, then the assessment (p_1, p_2) is coherent *if and only if* $p_1 \le p_2$.

Proof. Assume that $B|A \le D|C$ and apply (1) to $\mathcal{F} = \{B|A, D|C\}$. Then, the generalized atoms relative to $\mathcal{F}$ and $\mathcal{P}$ are

$$Q_1 = (1,1),\ Q_2 = (0,0),\ Q_3 = (0,1),\ Q_4 = (0,p_2),\ Q_5 = (p_1,1),$$

whose convex hull $\mathfrak{I}$ is the triangle with vertices the points Q_1,

Q_2, Q_3. If $\mathcal{P}$ is coherent, then $\mathcal{P} \in \mathfrak{I}$, which amounts to $p_1 \leq p_2$. Conversely, under the hypothesis $B|A \leq D|C$, from the condition $p_1 \leq p_2$ it follows $\mathcal{P} \in \mathfrak{I}$. Then, the coherence of $\mathcal{P}$ is obtained from Remark (2.4) and Theorem (2.3).

Obviously, from Proposition (4.2) it follows (2).

Remark. The compatibility between the partial order defined by Goodman and Nguyen and the conditional probability has been considered also in Dubois and Prade (1991, Proposition 1, p. 141), Goodman, Nguyen and Walker (1991, Lemma 2, p. 48), Calabrese (1991, Theorem 12, p. 90). But we observe that, differently from the quoted papers, our proof of Proposition (4.2) does not require to assume $P(A) > 0$, $P(C) > 0$.

Given the *rule* $A|B$ and the *material implication* $A \vee B^c$, it is $A|B \leq A \vee B^c$. Then, from Proposition (4.2) we obtain

(4.3) Corollary. The assessment $P(A|B) = p_1$, $P(A \vee B^c) = p_2$ is coherent if and only if $0 \leq p_1 \leq p_2 \leq 1$.

Using Theorem (2.2), the following Proposition can be proved.

(4.4) Proposition. Consider the family $\mathcal{F} = \{A|B, A \vee B^c, B\}$ and the assessment $\mathcal{P} = (p_1, p_2, p_3)$, where $p_1 = P(A|B)$, $p_2 = P(A \vee B^c)$, $p_3 = P(B)$. If there exist three positive numbers λ_1, λ_2, λ_3, such that $\lambda_1 + \lambda_2 + \lambda_3 = 1$ and

$$p_1 = \lambda_1 / (\lambda_1 + \lambda_2), \quad p_2 = \lambda_1 + \lambda_3, \quad p_3 = \lambda_1 + \lambda_2, \qquad (3)$$

then the assessment $\mathcal{P}$ is coherent.

Proof. Applying (1) to $\mathcal{F}$, we obtain

$$(AB \vee A^cB \vee B^c) \wedge [(A \vee B^c) \vee A^cB] \wedge (B \vee B^c) = C_1 \vee C_2 \vee C_3,$$

where $C_1 = AB$, $C_2 = A^cB$, $C_3 = B^c$. Then, the generalized atoms relative to $\mathcal{F}$ and $\mathcal{P}$ are the points

$$Q_1 = (1,1,1), \quad Q_2 = (0,0,1), \quad Q_3 = (p_1,1,0).$$

If (3) is satisfied for some positive quantities λ_1, λ_2, λ_3, with $\lambda_1 + \lambda_2 + \lambda_3 = 1$, then $\lambda_1 Q_1 + \lambda_2 Q_2 + \lambda_3 Q_3 = \mathcal{P}$ and from Theorem (2.2) the coherence of $\mathcal{P}$ follows.

On the basis of (3), to each choice of $\lambda_1, \lambda_2, \lambda_3$ it is associated a coherent assessment $\mathcal{P}$. In particular, given two positive constants k, δ, we can choose the values

$$\lambda_1 = k/(1+k+k\delta), \quad \lambda_2 = k\delta/(1+k+k\delta), \quad \lambda_3 = 1/(1+k+k\delta).$$

Then, from the closure property of coherence with respect to the limit, letting k tend to zero, we obtain the limiting coherent assessment $p_1 = 1/(1+\delta)$, $p_2 = 1$, $p_3 = 0$.

Therefore, for each $p_1 \in (0,1)$, the assessment $(p_1,1,0)$ is coherent. Finally, letting δ tend to zero or $+\infty$, we obtain, respectively, the limiting assessments $(1,1,0)$ or $(0,1,0)$.

Remark. The assessment $\mathcal{P} = (p_1, 1, p_3)$, with $p_3 > 0$, is coherent if and only if $p_1 = 1$. In fact, $(1,1,p_3)$ is coherent because it is obtained from (3), letting λ_2 and λ_1 tend respectively to zero and to $p_3 > 0$. Conversely, denoting by $\mathfrak{I}$ the convex hull of the points Q_h's, if $\mathcal{P}$ is coherent then $\mathcal{P} \in \mathfrak{I}$, i.e. there exist three non negative values λ_1, λ_2, λ_3, such that $\lambda_1 Q_1 + \lambda_2 Q_2 + \lambda_3 Q_3 = \mathcal{P}$, with $\lambda_1 + \lambda_2 + \lambda_3 = 1$. Then, it is

$$\lambda_1 + \lambda_3 p_1 = p_1 \ , \quad \lambda_1 + \lambda_3 = 1 \ , \quad \lambda_1 + \lambda_2 = p_3 \ ,$$

from which we obtain $\lambda_2 = 0$. Moreover, being $p_3 > 0$, we have $\lambda_1 > 0$, which implies $p_1 = 1$.

(4.5) Proposition. The assessment $\mathcal{P} = (p_1, p_2)$ on the family $\mathcal{F} = \{A|B,\ A^c|BC\}$, where $p_1 = P(A|B)$, $p_2 = P(A^c|BC)$, is coherent for all the values $p_1, p_2 \in [0,1]$.

<u>Proof</u>. The generalized atoms relative to $\mathcal{F}$ and $\mathcal{P}$ are the points

$$Q_1 = (1,0) \ , \quad Q_2 = (0,1) \ , \quad Q_3 = (1,p_2) \ , \quad Q_4 = (0,p_2)$$

and, denoting by $\mathfrak{I}$ their convex hull, it is easy to verify that, for all the values $p_1, p_2 \in [0,1]$, it is $\mathcal{P} \in \mathfrak{I}$. Then, from Remark (2.4) and Theorem (2.3) coherence of $\mathcal{P}$ follows.

Remark. Note that, as we cannot have the two "rules" $A|B$ and $A^c|BC$ simultaneously true, it could be "intuitively" argued that the judgements $P(A|B) = 1$, $P(A^c|BC) = 1$ are *incompatible*. In fact, as shown in Dubois and Prade (1991), the above assessment implies $P(A^cBC) = P(BC) = 0$ and, in the usual axiomatic theory, $P(A^c|BC)$ is *undefined*. But, as already noted, adopting the subjective approach no problem arises if some conditioning events have zero probability. A direct checking of the probabilistic consistency is allowed and, as shown by Proposition (4.5), the seemingly contradictory values $P(A|B) = 1$ and $P(A^c|BC) = 1$ are actually *compatible*!

(4.6) Proposition. Given, for the family $\mathcal{F} = \{A|B,\ A|BC\}$, the assessment $\mathcal{P} = (p_1,p_2)$, where $p_1 = P(A|B)$, $p_2 = P(A|BC)$, the point $\mathcal{P}$ is coherent for all the values $p_1, p_2 \in [0,1]$.

<u>Proof</u>. The generalized atoms relative to $\mathcal{F}$ and $\mathcal{P}$ are the points

$$Q_1 = (1,1) \ , \quad Q_2 = (0,0) \ , \quad Q_3 = (1,p_2) \ , \quad Q_4 = (0,p_2)$$

and, for all the values $p_1, p_2 \in [0,1]$, the point $\mathcal{P}$ belongs to the convex hull $\mathfrak{I}$ of points Q_h's. Then, from Remark (2.4) and Theorem (2.3) it follows that $\mathcal{P}$ is coherent.

Proposition (4.6) is in agreement with the following statement made in Dubois and Prade (1991):

"*there is no universal ordering between* $A|B$ *and* $A|BC$" .

Using Propositions (4.5) and (4.6) the following more general Proposition can be proved.

(4.7) Proposition. Given the family $\mathcal{F} = \{A|B,\ A^c|BC,\ A|BC\}$, the assessment $\mathcal{P} = (p_1,p_2,p_3)$, where $p_1 = P(A|B)$, $p_2 = P(A^c|BC)$, $p_3 = P(A|BC)$ is coherent *if and only if* $0 \le p_1, p_2, p_3 \le 1$ and $p_2 + p_3 = 1$.

<u>Proof</u>. If $\mathcal{P}$ is coherent, then, from Theorem (2.3), it follows $\mathcal{P}_J \in \mathfrak{I}_J$ for every $J \subseteq \{1,2,3\}$, so that, recalling Remark (2.4), it is $0 \le p_j \le 1$, $j = 1, 2, 3$. Moreover, as it can be easily verified, for $J = \{2,3\}$, condition $\mathcal{P}_J \in \mathfrak{I}_J$ amounts to $p_2 + p_3 = 1$. In fact, the generalized atoms relative to $\mathcal{F}_J = \{A^c|BC, A|BC\}$ and $\mathcal{P}_J = (p_2, p_3)$ are the two points $Q_1 = (0,1)$, $Q_2 = (1, 0)$, whose convex hull $\mathfrak{I}_J$ is the segment Q_1Q_2 . Therefore, $\mathcal{P}_J \in \mathfrak{I}_J$ if and only if $p_2 + p_3 = 1$.

Conversely, assume: 1. $0 \le p_1, p_2, p_3 \le 1$; 2. $p_2 + p_3 = 1$. Then, from property 1., using Remark (2.4) and Propositions (4.5), (4.6), we obtain $\mathcal{P}_J \in \mathfrak{I}_J$ for $J = \{1\}$, $J = \{2\}$, $J = \{3\}$, $J = \{1, 2\}$ and $J = \{1, 3\}$. Moreover, from property 2. it follows $\mathcal{P}_J \in \mathfrak{I}_J$ for $J = \{2, 3\}$. It remains to prove that $\mathcal{P} \in \mathfrak{I}$.

The generalized atoms relative to $\mathcal{F}$ and $\mathcal{P}$ are the points

$$Q_1 = (1,0,1)\ ,\ Q_2 = (0,1,0)\ ,\ Q_3 = (1,p_2,p_3)\ ,\ Q_4 = (0,p_2,p_3).$$

Then, condition $\mathcal{P} \in \mathfrak{I}$ holds if there exist four non negative numbers $\lambda_1, \lambda_2, \lambda_3, \lambda_4$, with $\lambda_1 + \lambda_2 + \lambda_3 + \lambda_4 = 1$, such that

$$\lambda_1 + \lambda_3 = p_1\ ,\quad \lambda_2 + p_2(\lambda_3 + \lambda_4) = p_2\ ,\quad \lambda_1 + p_3(\lambda_3 + \lambda_4) = p_3\ . \qquad (4)$$

We distinguish three cases:

(i) $p_2 = 0$. In this case, every vector $(\lambda_1, 0, p_1-\lambda_1, 1-p_1)$, with $0 \le \lambda_1 \le p_1$, is a solution of (4).

(ii) $p_2 = 1$. In this case, every vector $(0, \lambda_2, p_1, 1-p_1-\lambda_2)$, with $0 \le \lambda_2 \le 1-p_1$, is a solution of (4).

(iii) $0 < p_2 < 1$. In this case, defining the quantities

$\lambda_0 = (1-p_1)(1-p_2)/p_2$, $\delta = \min\left(p_1, \lambda_0\right)$,

for each $\lambda_1 \in [0, \delta]$, the vector $(\lambda_1, \lambda_1 p_2/p_3, p_1-\lambda_1, \lambda_0)$ is a solution of (4) .

Therefore, in all cases (4) is compatible, i.e. $\mathcal{P} \in \mathfrak{I}$.

Then, the condition $\mathcal{P}_J \in \mathfrak{I}_J$ is satisfied for every subset $J \subseteq \{1,2,3\}$ and the assessment $\mathcal{P} = (p_1, p_2, p_3)$ is coherent.

Propositions (4.5) and (4.6) make evident the complete freedom that we have, within the subjective approach, to revise beliefs when new information is gained. So, we can also discuss the claim contained in Shenoy (1991) that probability is inadequate for revising beliefs. In the Introduction of his paper Shenoy makes the following assertion:
"I believe A is true cannot be represented by P(A) = 1 because a probability of 1 is incorrigible, that is, P(A|B) = 1 for all B such that P(A|B) is well defined. However, plain belief is clearly corrigible. I may believe it is snowing outside but when I look out the window and observe that is has stopped snowing, I now believe that it is not snowing outside".

In the usual axiomatic framework, the above reasoning is correct: in fact, from $P(A) = 1$ and $P(B) > 0$ it follows $P(A^c) = P(A^cB) = 0$, $P(B) = P(AB) > 0$. Then $P(A|B) = P(AB)/P(AB) = 1$.

But, what happens if we evaluate $P(B) = 0$?

We prove a Proposition, which will be useful to answer the above question.

(4.8) Proposition. Given the family $\mathcal{F} = \{A, A|B, B\}$, consider the assessment $\mathcal{P} = (p_1,p_2,p_3)$, where $p_1 = P(A)$, $p_2 = P(A|B)$, $p_3 = P(B)$. If there exist three positive quantities $\lambda_1, \lambda_2, \lambda_3$ such that

$$p_1 = \lambda_1+\lambda_3 , \quad p_2 = \lambda_1 / (\lambda_1+\lambda_2) , \quad p_3 = \lambda_1+\lambda_2 , \tag{5}$$

with $\lambda_1 + \lambda_2 + \lambda_3 < 1$, then the assessment $\mathcal{P}$ is coherent.

<u>Proof</u>. The generalized atoms relative to $\mathcal{F}$ and $\mathcal{P}$ are the points

$Q_1 = (1,1,1)$, $Q_2 = (0,0,1)$, $Q_3 = (1,p_2,0)$, $Q_4 = (0,p_2,0)$.

If (5) is satisfied for some positive quantities $\lambda_1, \lambda_2, \lambda_3$, with $\lambda_1+\lambda_2+\lambda_3 < 1$, then $\sum_{h=1}^{4}\lambda_h Q_h = \mathcal{P}$, where $\lambda_4 = 1-(\lambda_1+\lambda_2+\lambda_3)$, and from Theorem (2.2) the coherence of $\mathcal{P}$ follows.

Remark. On the basis of (5), to each choice of $\lambda_1,\lambda_2,\lambda_3$ it is associated a coherent assessment $\mathcal{P} = \left(\lambda_1+\lambda_3, \lambda_1/(\lambda_1+\lambda_2), \lambda_1+\lambda_2\right)$.
As an example, given a positive constant k, we can choose

$$\lambda_1 = \left(\frac{k}{k+1}\right)^2 , \quad \lambda_2 = \frac{k}{(k+1)^2} , \quad \lambda_3 = \frac{1}{(k+1)^2} .$$

Then, letting k tend to 0, we obtain the limiting assessment

$$p_1 = P(A) = 1 \ , \quad p_2 = P(A|B) = 0 \ , \quad p_3 = P(B) = 0 \ , \qquad (6)$$

which can be applied, in particular, to the events:

A = "it is snowing outside",

B = "looking out of the window I observe that it has stopped snowing" .

Given a value $p \in (0,1)$ and an arbitrary positive constant k, a more general result can be obtained choosing the coefficients

$$\lambda_1 = \frac{kp}{p + (1+p)k} \ , \quad \lambda_2 = \frac{k(1-p)}{p + (1+p)k} \ , \quad \lambda_3 = \frac{p}{p + (1+p)k} \ .$$

For each $p \in (0,1)$, letting k tend to zero, we obtain the coherent assessment

$$p_1 = P(A) = 1 \ , \quad p_2 = P(A|B) = p \ , \quad p_3 = P(B) = 0 \ . \qquad (7)$$

Then, letting p tend to zero, from (7) we obtain (6). Moreover, letting p tend to one, from (7) we obtain

$$p_1 = P(A) = 1 \ , \quad p_2 = P(A|B) = 1 \ , \quad p_3 = P(B) = 0 \ .$$

Therefore, if $P(B) = 0$, from $P(A) = 1$ it does not necessarily follow $P(A|B) = 1$. The conclusion is that (at least in the subjective approach) probability is not inadequate to revise beliefs.

In Heinsohn (1991), to take into account uncertain knowledge, a probabilistic extension of terminological logics is proposed and the probabilistic consistency is studied. For example, given the events $C|A$, $A|C$, $B|A$, $A|B$, $C|B$, $B|C$, with positive probabilities p, p', q, q', r, r', Theorem 4 of the quoted paper asserts that if equation $r'\cdot p\cdot q' = r\cdot p'\cdot q$ is violated, then the given probabilities are inconsistent.

In the following Proposition the same result is obtained, without assuming positivity of probability values.

(4.9) Proposition. Given, for the family $\mathcal{F} = \{C|A, A|C, B|A, A|B, C|B, B|C\}$, the probability assessment $\mathcal{P} = (p,p',q,q',r,r')$, if $\mathcal{P}$ is coherent then

$$r'\cdot p\cdot q' = r\cdot p'\cdot q \ . \qquad (8)$$

<u>Proof</u>. If $\mathcal{P}$ is coherent, then there exists a probability P, consistent with $\mathcal{P}$, which is defined on the events $A|H_0$, $B|H_0$, $C|H_0$, where $H_0 = A \vee B \vee C$. Moreover, we have

$$\begin{cases} pP(A|H_0) = p'P(C|H_0) \\ q'P(B|H_0) = qP(A|H_0) \\ r'P(C|H_0) = rP(B|H_0) \end{cases} \ . \qquad (9)$$

Observing that $P(A|H_0) + P(B|H_0) + P(C|H_0) \geq 1$, if the values

$P(A|H_0)$, $P(B|H_0)$, $P(C|H_0)$ are all positive, then from (9) we immediately obtain (8). In all the other cases we can verify that (8) reduces to $0 = 0$.

For example, if $P(A|H_0) = 0$, $P(B|H_0) > 0$, $P(C|H_0) > 0$, then from (9) it follows $p' = q' = 0$ and (8) is verified.

Further results on consistency of conditional probability assessments are given in Gilio and Spezzaferri (1992).

5. CONCLUSIONS

The subjective approach to probabilistic treatment of uncertainty, distinguishing well founded problems from pseudo ones, allows to put in the right perspective all the relevant aspects to model uncertain knowledge and to revise beliefs. Then, as a consequence, the need for alternative theories of uncertainty measures is greatly reduced.

In this paper it has been shown that, using the concept of generalized atom, a natural representation of the truth values of conditional events can be given and that the subjective methodology provides the most general framework for checking probabilistic consistency. Moreover, some results on coherence of conditional probability assessments have been obtained.

REFERENCES

Andersen, K.A. and Hooker, J.N. (1990), *Probabilistic logic for belief nets*, Proc. of The 8th Int. Congr. of Cybernetics and Systems, Vol. 1, (C.N. Manikopoulos, Ed.), New York, The NJIT Press, 245-251.

Bruno, G. and Gilio, A. (1980), *Applicazione del metodo del simplesso al teorema fondamentale per le probabilita' nella concezione soggettiva*, Statistica, 40, 337-344.

Calabrese, P. G. (1991), *Deduction and inference using conditional logic and probability*, In: Conditional Logic in Expert Systems (I.R. Goodman, M.M. Gupta, H.T. Nguyen and G.S. Rogers,Eds.), North-Holland, Amsterdam, 71-100.

Coletti, G. (1991), *Comparative probabilities ruled coherence conditions and use in expert systems*, Intern. Journal of General Systems (in press).

Coletti, G., Gilio, A. and Scozzafava, R. (1990a), *Coherent qualitative probability and uncertainty in Artificial Intelligence*, Proc. of The 8th Int. Congr. of Cybernetics and Systems, Vol. 1, (C. N. Manikopoulos, Ed.), New York, The NJIT Press, 132-138.

Coletti, G., Gilio, A. and Scozzafava, R. (1990b), *Conditional events with vague information in expert systems*, Lecture Notes in Computer Science (B. Bouchon-Meunier, R. R. Yager, L. A. Zadeh, Eds.), n. 521, 106-114.

Coletti, G., Gilio, A. and Scozzafava, R. (1991), *Assessment of qualitative judgements for conditional events in expert systems*,

Lecture Notes in Computer Science (R. Kruse, P. Siegel, Eds.), n. 548, 135-140.

De Finetti, B. (1974, 1975), *Theory of probability*, 2 Volumes, A.F.M. Smith and A. Machì, trs. John Wiley, New York.

Dubois, D. and Prade H. (1991), *Conditioning, non-monotonic logic and non-standard uncertainty models*, In: Conditional Logic in Expert Systems (I.R. Goodman, M.M. Gupta, H.T. Nguyen and G.S. Rogers, Eds.), North-Holland, Amsterdam, 115-158.

Gilio, A. (1990), *Criterio di penalizzazione e condizioni di coerenza nella valutazione soggettiva della probabilita'*, Boll. Un. Mat. Ital., Vol. **4**-B, n. 3, Serie 7[a], 645-660.

Gilio, A. (1992), *C_o- coherence and extensions of conditional probabilities*, Bayesian Statistics 4 (Eds. J. M. Bernardo, J. O. Berger, A.P. Dawid and A.F.M. Smith), Clarendon Press, Oxford, 633-640.

Gilio, A. and Scozzafava, R. (1988), *Le probabilita' condizionate coerenti nei sistemi esperti*, Atti Giornate di lavoro A.I.R.O. su "Ricerca Operativa e Intelligenza Artificiale", Offset Grafica, Pisa, 5-6-7 Ottobre, 317-330.

Gilio, A. and Spezzaferri, F. (1992), *Knowledge integration for conditional probability assessments*, Uncertainty in Artificial Intelligence (Eds. D. Dubois, M. P. Wellman, B. D'Ambrosio, P. Smets), Morgan Kaufmann Publishers, San Mateo, California, 98-103.

Goodman, I. R. and Nguyen, H. T. (1988), *Conditional objects and the modeling of uncertainties*, Fuzzy Computing Theory, Hardware and Applications (M.M. Gupta, T. Yamakawa, Eds.), North- Holland, New York, 119-138.

Goodman, I. R., Nguyen, H. T. and Walker, E. A. (1991), *Conditional inference and logic for intelligent systems: a theory of measure-free conditioning*, North-Holland, Amsterdam.

Heinsohn, J. (1991), *A hybrid approach for modeling uncertainty in terminological logics*, Lecture Notes in Computer Science (R. Kruse, P. Siegel, Eds.), n. 548, 198-205.

Lad, F., Dickey, J.M. and Rahman M.A. (1990). *The fundamental theorem of prevision*, Statistica, 50, 19-38.

Nilsson, N. J. (1986). *Probabilistic logic*, Artificial Intelligence, **28**, 71-87.

Paass, G. (1988). *Probabilistic logic*, Non-standard logics for automated reasoning, eds. P. Smets, E. H. Mamdani, D. Dubois and H. Prade, New York: Academic Press, 213-51.

Scozzafava, R. (1991), *Subjective probability versus belief functions in artificial intelligence*, Intern. Journal of General Systems (in press).

Shenoy, P. P. (1991). *On Spohn's rule for revision of beliefs*, Int. Journal of Approximate Reasoning, 5, 149-181.

Uncertainty in Intelligent Systems
B. Bouchon-Meunier et al. (Editors)
1993 Elsevier Science Publishers B.V.

HOW TO SOLVE SOME CRITICAL EXAMPLES BY A PROPER USE OF COHERENT PROBABILITY

Romano Scozzafava

Dipartimento Metodi e Modelli Matematici, Università
"La Sapienza", Via Scarpa, 10 - 00161 Roma, Italy

This paper shows how to solve some real examples introduced in the literature on Artificial Intelligence to challenge the probabilistic approach. They are usually treated using different theories (such as belief functions, fuzzy sets, probability without product rule) : we single-out for each example the crucial point where a peculiar and restrictive interpretation of the probability concept is used and we give their solution in terms of a coherent subjective approach.

1. INTRODUCTION AND SUMMARY OF KNOWN RESULTS

A number of real examples used to (putatively) challenge the probabilistic approach are considered in this paper. These examples are treated in the literature on Artificial Intelligence by using different theories, such as belief functions, fuzzy sets, possibility theory, new rules for the logic of conditional events, etc.: we show for each example the crucial points where very peculiar and restrictive interpretations of the concept of event or of the probabilistic scenario come to the fore. Moreover we discuss how to solve these examples in terms of a coherent subjective approach, introduced for the first time in Artificial Intelligence by Gilio & Scozzafava (1988). To this purpose an useful tool will be a theorem given in Gilio (1990) and already used in Scozzafava (1991) to show that a classical example discussed in terms of belief functions can easily be dealt with by resorting to our framework.

In order to take into account the need of updating probability evaluations, all uncertain statements in the expert system are expressed by *conditional* events, which are the natural tool for a Bayesian methodology. A conditional event $E|H$ is looked upon as a 3-valued entity taking one of the three values 1, 0, p (denoting by p its probability), according to whether which of the events $E\,H$, or E^cH, or H^c, respectively, is true. Notice that in our framework an *event* can be singled-out by a (nonambiguous) *proposition* E, that is a statement that can be either *true* or *false*. Moreover E^c denotes the *contrary* of E, and to avoid cumbersome notation we dropped the conjunction operator $\cap$, so that an event such as $A \cap B$ is simply denoted by $A\ B$.

Our approach refers to an *arbitrary* (i.e., *with no underlying structure*) set $\mathcal{C} = \mathcal{E} \times \mathcal{H}$ of conditional events : it follows that *it is possible to assess the probability* P *only on a set of events of interest*. We require *coherence* of the function P on $\mathcal{C}$, based on mathematical conditions derived by the two equivalent classical criteria (bet or penalty method) introduced by de Finetti (1974). This entails that P is a *conditional probability distribution*. The converse in general is not true: sufficient conditions for the coherence of P concern the structure of $\mathcal{C}$, and so *the interest for a direct check of coherence is clear*.

If $\mathcal{F}$ is a family of n conditional events $E_1|H_1,\ldots,E_n|H_n$, introduce the *atoms* C_h ($h = 1,2,\ldots,r \le 2^{2n}$) generated by the 2n events $E_1,H_1,E_2,H_2,\ldots,E_n,H_n$ and contained in the union of the H_i's. As it is well-known, the atoms (different from the impossible event $\emptyset$) generated by k events $B_1,B_2,\ldots,B_k$ are obtained from the intersection $A_1\cap A_2\cap\ldots\cap A_k$ by putting in place of each A_i the event B_i or its contrary B_i^c in all possible ways (for $i = 1,2,\ldots,k$). A sufficient condition for the existence of a conditional probability on $\mathcal{F}$ can be found through a *coherence* condition, as in the following theorem, due to Gilio (1990). We give here a *direct* approach to the relevant procedure.

<u>Theorem</u> - Let $\mathscr{E}$ be an *arbitrary* family of conditional events. For *any* arbitrary finite subfamily $\mathcal{F} = \{E_1|H_1,\ldots,E_n|H_n\} \subseteq \mathscr{E}$, denote by C_h $(h = 1,2,\ldots,r)$ those atoms generated by the events E_i, H_i, and contained in the union of the H_i's ; given the evaluations $p_i = P(E_i|H_i)$, with $i = 1,2,\ldots,n$, and the atoms C_h, put

$$\alpha_{hi} = \begin{cases} 1 & \text{if } C_h \subseteq E_i \cap H_i \\ 0 & \text{if } C_h \subseteq E_i^c \cap H_i \\ p_i & \text{if } C_h \subseteq H_i^c \end{cases} . \tag{1}$$

If there exists a solution $(\lambda\ ,\lambda\ ,\ldots,\lambda\ ) = \lambda$ of the system

$$\sum_{h=1}^{r} \lambda_h = 1\ ,\ \lambda_h > 0\ (h = 1,\ldots,r)\ ;\ p_i = \sum_{h=1}^{r} \lambda_h \alpha_{hi}\quad (i = 1,\ldots,n), \tag{2}$$

then the vector $(p_1,p_2,\ldots,p_n)$ is a *coherent assessment* defining a *conditional probability* $P(\cdot|\cdot)$ on $\mathcal{F}$.

Remark 1 - The theorem gives a *sufficient* condition for an assessment $(p_1,p_2,\ldots,p_n)$ to be coherent (so that it defines a conditional probability). The above condition is not necessary: yet notice that it follows from it, for $n = 1$, that any $p_1 = p$, with $0 \le p \le 1$, can be chosen as a coherent assessment.

Remark 2 - In general, system (2) has a solution which *is not unique* : the corresponding ranges of values for the λ_h's lead to probability assignments $(p_1,p_2,\ldots,p_n)$ which are coherent for p_i $(i = 1,2,..n)$ belonging to a suitable closed interval.

The practical implementation of the theorem leads to a step-by-step assessment of probability : for a further discussion of these aspects, see Scozzafava (1992).

2. COHERENT PROBABILITY VERSUS FUZZY REASONING

The following example is taken from a paper by Nies and Camarinopoulos (1991), concerning the long term safety assessment of a radioactive waste repository in salt.

After the disposal of waste has been finished, "almost impermeable" dams are built at strategical positions within an underground gallery system in order to prevent the transport of fluid possibly intruding at later times. The problem is to predict the future development of the permeability of these dams for time periods of hundreds or thousands of years.

In the quoted paper, fuzzy set theory is presented as a different approach to manage data uncertainty in a complex system : available information about possible values of dams permeability is used to construct a subjective membership function of a fuzzy set (of "almost impermeable" dams) rather than a probability density function. For values of the permeability between 10^{-21} and 10^{-17} the membership function is put equal to 1, while it is put equal to 0 for values greater than 10^{-15} ; finally, the membership function is decreasing from 1 to 0 in the interval from 10^{-17} to 10^{-15}. The motivation given by the authors rests on the argument that, given the range of values from 0 to 1, there is no restriction for the definition of membership functions, in contrast to probability that, regardless of its subjective interpretation, obeys certain rules such as, for example, the axiom of additivity of the probability measure when applied to mutually exclusive events : it follows that, as soon as an expert assigns a subjective probability of (say) 0.4 to the statement that in the future the permeability of the dam will be between 10^{-17} and 10^{-16}, he inescapably assigns a degree of belief of 0.6 to the contrary, and he may not have for the latter fact any justification apart from the consistency argument represented by the additivity rule.

A second argument brought forward to contrast probabilistic methods with the fuzzy approach concerns the putative merits of the rules according to which the possibility of an object belonging to two fuzzy sets is obtained as the minimum of the possibilities that it belongs to either fuzzy set. The issue is the computation of the probability that the value of a safety parameter belongs to a given (dangerous) interval *for all four components* (grouped according to the similarity of their physico-chemical conditions) of the repository section. For each component this probability is computed as equal to 1/5, and the conclusion is that "*in terms of a safety assessment, the fuzzy calculus is more conservative*", since in the fuzzy calculus (interpreting those values as values of a membership function) the possibility of a value of the parameter in the given interval *for all components* is still 1/5 (which is the minimum taken over numbers all equal to 1/5), while the assumption of independence gives the same event the small probability $(1/5)^4$.

In our probabilistic framework the way-out from these putative difficulties is indeed very simple. Notice that the above choice of the membership function implies that dams whose permeability is less than 10^{-17} are "almost impermeable", while those with a permeability greater than 10^{-15} are not. So the real problem is that we are uncertain on being or not "almost impermeable" those dams having a permeability between 10^{-17} and 10^{-15} : then our interest is in fact directed toward the *conditional event* $E|H$, with

E = *the dam is "almost impermeable"* ,

H = *the permeability of the dam is between* 10^{-17} *and* 10^{-15}.

It follows that an expert may assign a subjective probability $P(E|H)$ equal to (say) 0.25 without any need to assign a degree of belief of 0.75 to the event E under the assumption H^c (i.e., the permeability of the dam is *not* between 10^{-17} and 10^{-15}), since an additivity rule with respect to the *conditioning* events *does not hold*. Nevertheless we are able, thanks to the Theorem in Sect. 1, to give suitable bounds concerning, for example, the probabilities $P(E|H)$, $P(E|H^c)$, $P(E)$. Denoting them, respectively, by p_1, p_2, p_3, and putting $E_1 = E_2 = E_3 = E$ and $H_1 = H$, $H_2 = H^c$, $H_3 = \Omega$, the Theorem gives

$$p_1 = \frac{\lambda_1}{\lambda_1 + \lambda_2} \quad , \quad p_2 = \frac{\lambda_3}{\lambda_3 + \lambda_4} \quad , \quad p_3 = \lambda_1 + \lambda_3 \quad ,$$

with

$$\sum_{h=1}^{4} \lambda_h = 1 \ , \ \lambda_h > 0 \ (h = 1, \dots, 4) \ .$$

Since both denominators are less than 1, we get in particular $p_1 + p_2 > p_3$. A better inequality can be obtained taking into account that

$$P(E) = P(H)P(E|H) + P(H^c)P(E|H^c) \ ,$$

so that

$$\min \{p_1, p_2\} \le p_3 \le \max \{p_1, p_2\} \ .$$

As far as the second issue is concerned, it is enough (to simplify calculations) referring only to two components. If A and B are the relevant events, with $P(A) = P(B) = 1/5$, to be able to say on $P(A \cap B)$ something that can be sensible (and different from the assignment of a *unique* value such as $(1/5)^2$ or $1/5$) there is no need to assume independence or fuzziness, respectively. Resorting again to the Theorem of Section 1 gives (taking $E_1 = A$, $E_2 = B$, $E_3 = A \cap B$, and $H_1 = H_2 = H_3 = \Omega$)

$$\lambda_1 + \lambda_3 = \lambda_1 + \lambda_2 = 1/5 \quad , \quad P(A \cap B) = \lambda_1 \quad ,$$

so that, putting $\lambda = \lambda_3 = \lambda_2$, we get $P(A \cap B) = 1/5 - \lambda$.

Clearly, since λ (with $0 < \lambda \le 1/5$) is arbitrary, *any* value less than (but as near as we please to) $1/5$ is admissible for $P(A \cap B)$; of course, the value $(1/5)^2$ is admissible too, as follows by taking $\lambda = 4/25$.

3. SLIPPERY STREETS, THERMOMETER AND BELIEFS

In Shafer (1987), the following example is considered : "*Is Fred, who is about to speak to me, going to speak truthfully, or is he, as he sometimes does, going to speak carelessy, saying whatever comes into his mind ?*". Shafer denotes *truthful* and *careless* as the possible answers to the above question : since he knows from previous experience Fred's announcements being truthful reports on what he knows 80% of the time and careless statements the other 20% of the time, he writes

$$P\{\text{truthful}\} = 0.8 \quad , \quad P\{\text{careless}\} = 0.2 \ . \tag{3}$$

If we introduce the event

E = *the streets outside are slippery*

and Fred announces that E is true (let us denote by A the latter event, i.e. *Fred's announcement*), the usual Shafer's belief function argument gives

$$\text{Bel}(E) = 0.8 \quad , \quad \text{Bel}(E^c) = 0 \ . \tag{4}$$

Then the putative merits of (4) are discussed with respect to what is called "*a Bayesian argument*" and to its presumed "*inability to model Fred when he is being careless*" and "*to fit him into a chance picture at all*" ; not to mention reasonings based on undefined and vague concepts such as the "quality of the evidence".

Successively, other evidence about whether the streets are slippery is considered, that is the event

T = *a thermometer shows a temperature of* $31^{o}F$.

It is known that streets are not slippery at this temperature, and there is a 99% chance that the thermometer is working properly ; moreover, Fred's behavior is independent of whether it is working properly or not. Then a belief function is obtained through the loose concept of "combination of independent items of evidence", getting a result which should reflect the fact that more trust is put on the thermometer than in Fred, i.e.

$$Bel(E) = 0.04 \quad , \quad Bel(E^c) = 0.95 \; . \tag{5}$$

Finally, other computations follow concerning the case of the so-called "dependent evidence".

Our probabilistic solution of the above example is very simple and fits and encompasses the solution obtained via the belief function approach. First of all, we challenge the possibility of defining the probabilities in (3), since "*truthful*" and "*careless*" cannot be considered *events* : in fact, their truth or falseness cannot be directly verified, while we can instead ascertain whether are true or false the *conditional* events $E|A$ and $E^c|A$; moreover, the equalities in (3) must be replaced by inequalities, since E *may be true also in the case that* A *is a careless statement*. So we have

$$P(E|A) \geq 0.8 \quad , \quad P(E^c|A) \leq 0.2 \tag{6}$$

The belief values (4) can be seen as particular choices of the two conditional probabilities satisfying (6) ; on the other hand, it is not unusual that a belief is interpreted as a lower probability (see the discussion in Spiegelhalter, 1987, and the example discussed in Scozzafava, 1991).

As far as eqs.(5) are concerned (for the sake of brevity, we will discuss only the first one), in a probabilistic context we are actually interested in coherently assessing $p = P(E|A \cap T)$, that should be consistent with the value of $p_1 = P(E|A)$ and that of $p_2 = P(E|T)$. Notice that, since there is a 99% chance that the thermometer is working properly, we have $p_2 = 1/100$. So we apply the general theorem given in Sect. 1, with $E_1 = E_2 = E_3 = E$, $H_1 = A$, $H_2 = T$, $H_3 = A \cap T$: there are six atoms contained in $A \cup T$ (that is, dropping the symbol $\cap$, ETA^c, ETA, ET^cA, E^cTA^c, E^cTA, E^cT^cA) so that, performing the relevant computations the system (2) becomes

$$\begin{cases} p_1\lambda_1 + \lambda_2 + \lambda_3 + p_1\lambda_4 = p_1 \\ \lambda_1 + \lambda_2 + p_2\lambda_3 + p_2\lambda_6 = p_2 \\ p\lambda_1 + \lambda_2 + p\lambda_3 + p\lambda_4 + p\lambda_6 = p \\ \lambda_1 + \lambda_2 + \lambda_3 + \lambda_4 + \lambda_5 + \lambda_6 = 1 \end{cases}$$

from which it follows easily

$$p_1 = \frac{\lambda_2 + \lambda_3}{\lambda_2 + \lambda_3 + \lambda_5 + \lambda_6} \;, \quad p_2 = \frac{\lambda_1 + \lambda_2}{\lambda_1 + \lambda_2 + \lambda_4 + \lambda_5} \;, \quad p = \frac{\lambda_2}{\lambda_2 + \lambda_5} \;. \qquad (7)$$

Notice that, given p_1, p_2, p, the system has (besides the conditions $\lambda_h > 0$) four equations and six unknowns : choosing, for example, 0.8 as the value of $p_1 = P(E|A)$ and recalling that $p_2 = 0.01$, can the first value given in (5), i.e. $Bel(E) = 0.04$, be interpreted as a coherent assessment of the conditional probability $p = P(E|A \cap T)$? Inserting the numerical values of p_1, p_2, p, the system takes the form

$$\begin{cases} 4\lambda_1 + 5\lambda_2 + 5\lambda_3 + 4\lambda_4 = 4 \\ 100\lambda_1 + 100\lambda_2 + \lambda_3 + \lambda_6 = 1 \\ \lambda_1 + 25\lambda_2 + \lambda_3 + \lambda_4 + \lambda_6 = 1 \\ \lambda_1 + \lambda_2 + \lambda_3 + \lambda_4 + \lambda_5 + \lambda_6 = 1 \end{cases}$$

so that, taking, for example, $\lambda_1 = 1/400$ and $\lambda_6 = 1/100$, one has that the relevant coefficient determinant Δ is different from 0 (that is, $\Delta = 780$) ; then standard computations lead to the solution

$$\lambda_2 = 0.00359 \ , \ \lambda_3 = 0.381 \ , \ \lambda_4 = 0.517 \ , \ \lambda_5 = 0.0859 \ .$$

In conclusion, not only the chosen values of p_1, p_2, p constitute a coherent assessment, but the system has clearly many other solutions that may be plugged into eqs. (7) giving other coherent evaluations.

4. ARE THERE "STRANGE" PROBABILITIES WITHOUT PRODUCT RULE ?

Given an election with three candidates A, B, C in which the probability of either one winning is 1/3 , denote by the same symbols also the corresponding events, so that

$$P(A) = P(B) = P(C) = \frac{1}{3} \ .$$

Moreover, $A \cup B \cup C = \Omega$, the certain event. Now, suppose that C withdraws and that then all his votes will go to B: according to Schay (1968), this situation involves probabilities for which the product rule

$$P(B \cap H) = P(B|H)\, P(H) \ , \qquad (8)$$

with $H = (A \cup B)$, *does not hold.* He argues as follows : since $P(A \cup B) = \frac{2}{3}$ and

$$P(B|(A \cup B)) = \frac{2}{3} \ , \qquad (9)$$

taking into account that $B \cap (A \cup B) = B$ gives for the left member of (8) the value $P(B) = 1/3$, while the right member of the product rule is equal to $(2/3)(2/3) = 4/9$.

Actually, a careful singling-out of the conditioning event in

(9) is needed : in our framework this can be done also outside the initial "space" $\{A,B,C\}$, introducing a suitable *proposition* (that is a statement that can be either *true* or *false*), which in this context is not $A \cup B$, but the event

E = C *withdraws and all his votes go to* B .

Notice that $E \subset A \cup B$ and that in fact (9) must be replaced by

$$P(B|E) = \frac{2}{3} . \tag{10}$$

Let us now find a coherent assessment of $P(B|(A \cup B)) = P(B|H)$ by means of the theorem given in Sect. 1. Consider the four conditional events

$$B|E \quad , \quad H|\Omega \quad , \quad B|\Omega \quad , \quad B|H \quad ,$$

with $p_1 = P(B|E) = 2/3$, $p_2 = P(H|\Omega) = 2/3$, $p_3 = P(B|\Omega) = 1/3$ and the conditional probability $p = P(B|H)$ to be determined *coherently*. We have the five atoms HEB^c, HE^cB^c, HEB, HE^cB, H^c and the system

$$\begin{cases} p_1\lambda_2 + \lambda_3 + p_1\lambda_4 + p_1\lambda_5 = p_1 \\ \lambda_1 + \lambda_2 + \lambda_3 + \lambda_4 = p_2 \\ \lambda_3 + \lambda_4 = p_3 \\ \lambda_3 + \lambda_4 + p\lambda_5 = p \\ \lambda_1 + \lambda_2 + \lambda_3 + \lambda_4 + \lambda_5 = 1 \end{cases}$$

From second and third equation we get

$$\lambda_1 + \lambda_2 = \lambda_3 + \lambda_4 = \frac{1}{3} \quad ;$$

hence fourth and fifth equation give

$$p = \frac{\lambda_3 + \lambda_4}{\lambda_1 + \lambda_2 + \lambda_3 + \lambda_4} = \frac{1}{2} \quad ,$$

that is $P(B|H) = 1/2$. It follows that the product rule (8) holds.

In conclusion, the thesis of this paper could be summarized by a statement such as : "*tell me which approach to the management of uncertainty you support in a given example, and I will tell you how to cope with it through a proper use of coherent probability*".

REFERENCES

B.DE FINETTI (1974), *Theory of probability*, Vols. 1 and 2, Wiley & Sons, Chichester.

A.GILIO (1990), *Criterio di penalizzazione e condizioni di coerenza nella valutazione soggettiva della probabilita'*, Bollettino Unione Matematica Italiana (7) 4-B, pp. 645-660.

A.GILIO, R.SCOZZAFAVA (1988), *Le probabilita' condizionate coerenti nei sistemi esperti*, Atti A.I.R.O. su "Intelligenza artificiale e Ricerca operativa", Pisa, pp. 317-330.

A.NIES, L.CAMARINOPOULOS (1991), *Application of fuzzy set and probability theory to data uncertainty in long term safety assessment of radioactive waste disposal systems*, P.S.A.M. (G.Apostolakis, Ed.), Elsevier, N.Y., pp. 1389-1394.

G.SCHAY (1968), *An Algebra of Conditional Events*, Journal of Mathematical Analysis and Applications 24, pp. 334-344.

R.SCOZZAFAVA (1991), *Subjective probability versus belief functions in Artificial Intelligence*, International Journal of General Systems (in press).

R.SCOZZAFAVA (1992), *Expert systems: a step-by-step coherent method of assessing subjective probabilities*, 3rd Conf. on "Practical Bayesian Statistics", Nottingham.

G.SHAFER (1987), *Probability judgement in Artificial Intelligence and Expert Systems*, Statistical Science 2, pp. 3-16.

D.J.SPIEGELHALTER (1987), Rejoinder, in *Probabilistic Expert Systems in Medicine : practical issues in handling uncertainty*, Statistical Science 2, pp.43-44.

Uncertainty in Intelligent Systems
B. Bouchon-Meunier et al. (Editors)

Handling Uncertainty and Incompleteness with multiple lines of reasoning[1]

Thomas Chehire[a] and Francesco Fulvio Monai[b]

[a]Thomson-CSF/SDC - 18, Avenue du Maréchal Juin, 92366 Meudon-La-Forêt Cedex, France

[b]Thomson-CSF/RCC - 160, Boulevard de Valmy, BP 82 92704 Colombes Cedex, France

ABSTRACT

Applications in complex domain need the ability of representing and reasoning about uncertain and somewhat contradictory informations. A classical first order inference engine is unable to deal properly with such kind of imperfect knowledge. A more powerful environment which allows to represent uncertain data and perform uncertain inferences must be developed together with a Reason Maintenance System (RMS) able to pursue the reasoning process in presence of partially inconsistent fact bases.
We propose an architecture integrating an OPS-like inference engine and a Possibilistic Assumption based Truth Maintenance System (ATMS).
After a brief recall of the basic concepts of ATMSs and Possibilistic Logic, the extension of ATMSs to manage uncertain knowledge in the framework of Possibilistic Logic is presented. Finally the coupling of a Possibilistic ATMS with a first order inference engine is outlined and the representation and the treatment of negation and defaults is investigated.

1. ATMS BACKGROUND

ATMSs [deKleer 86] [deKleer 88] make the distinction between *assumptions* and other data (or *facts*).
Assumptions are data which are presumed to be true, unless there is evidence of the contrary. Other data are primitive data always true, or that can be derived from other data or assumptions. The ATMS records such dependencies through *justifications*. It is then in charge of determining which combinations of choices (assumptions) are consistent, and which conclusions they enable to draw.
To achieve this, each datum is stamped with a *label* consisting of the list of *environments* (i.e. sets of assumptions) under which it holds. When a new justification for a datum is provided, its label is updated with the label of the left-hand-side of the justification (i.e. list of environments under which all facts or hypotheses supporting the datum through this justification are simultaneously true).
An environment is inconsistent if it enables to derive a special datum representing the contradiction. It is then called a *nogood*. When such an environment is discovered, it has to be removed from all the labels to guaranty their consistency.
The *context* of a consistent environment is the set of facts that can be derived from the assumptions of that environment. A problem with many possible solutions, will thus generate many contexts.
The main advantage of an ATMS is that all solutions are developed in parallel, and maximum work is shared between solutions.

---

[1]This work has been partially funded by the DRET (french MOD under the grant 89/568)

When dealing with uncertain knowledge one should be able to distinguish in the set of hypotheses the ones which are too uncertain to be considered as true, the ones which are almost certain and the ones which have an intermediary degree of certainty.
In the same manner justifications should be weighted by a degree of uncertainty in order to be able to graduate the certainty of the conclusions on the basis of the certainty of the premises.
Possibilistic Logic seems a good framework to formalise such kind of imperfect knowledge, that is to express the uncertainty in a relative manner (one believes in an event more than in another).

2. POSSIBILISTIC LOGIC

Possibilistic Logic [Dubois 88] [Lang 91] is an extension of Classical Logic where ground formulas are weighted by a number belonging to [0, 1] representing *lower bounds* of necessity or possibility degrees.
The possibility degree evaluates at which degree a proposition p is possible, that is coherent with the available knowledge.
The necessity degree evaluates at which degree a proposition p is certain, that is implied by the available knowledge.
A necessity measure **N** is defined by the following axioms :

$$\mathbf{N}(\bot)=0;\ \mathbf{N}(\mathrm{T})=1;\ \mathbf{N}(p\wedge q)=\min(\mathbf{N}(p),\mathbf{N}(q))$$

where $\bot$ and T represent, respectively, contradiction and tautology.

By duality, to a necessity measure a possibility measure Π is associated through the following relations :

$$\Pi(p)=1-\mathbf{N}(\neg p);\ \mathbf{N}(p)=1-\Pi(\neg p)$$

The Figure 1 summarises the meaning of the weights attached to a proposition p:

$\Pi(p)$ \ $N(p)$	$N(p)=0$	$0<N(p)<1$	$N(p)=1$
$\Pi(p)=0$	p is false	Contradiction	
$0<\Pi(p)<1$	p is somewhat false		
$\Pi(p)=1$	Ignorance	p is somewhat true	p is true

Figure 1

We note that when simultaneously $0<N(p)\leq 1$ and $0\leq\Pi(p)<1$ we have a (partial) contradictory situation since this means that both p and $\neg p$ are somewhat true.
Being able to pursue the reasoning in presence of partial inconsistency is one of the main issue of a Possibilistic ATMS.

3. POSSIBILISTIC ATMS

Possibilistic ATMSs [Dubois 90] (noted Π-ATMS in the following) are an extension of ATMSs to handle uncertainty in the framework of Possibilistic Logic.
In a Π-ATMS, hypotheses, facts and justifications can be uncertain.
This allows to obtain a weight for the generated environments, to evaluate the inconsistency degree of a set of hypotheses and to calculate the uncertain consequences of a set of uncertain hypotheses.
To represent the uncertainty we will consider in the following just lower bounds of a necessity measure **N**.

In fact the valuation $\mathbf{N}(p) \geq \alpha > 0$ is equivalent to $\Pi(\neg p) \leq 1-\alpha < 1$ and we can also obtain from the axioms $\Pi(p)=1$ and $\mathbf{N}(\neg p)=0$.

The lower bound of a possibility measure is a much poorer information since from $\Pi(p) \geq \beta > 0$ nothing can be said about $\mathbf{N}(p)$ or $\Pi(\neg p)$.
We will thus associate to each fact or assumption in the fact base a lower bound of a necessity measure representing its certainty.
In the same manner, a lower bound of a necessity measure is attached to the conclusion of a justification, representing to which extent it is sufficient to be certain in the premises to believe in the conclusion.

This means that from the premise p such that $\mathbf{N}(p) \geq \alpha$ and from the justification $p \rightarrow c$ such that $\mathbf{N}(p \rightarrow c) \geq \alpha'$ we can obtain the conclusion c such that $\mathbf{N}(c) \geq \min(\alpha, \alpha')$.
In a Π-ATMS the definitions of environment, label and context of a classic ATMS must be modified to take into account the uncertainty.
Let J be a set of weighted justifications, H a set of weighted hypotheses, E a subset of H and d a datum.
Then we have the following definitions:

Environments: $[E\ \alpha]$ is an *environment* of d iff d can be deduced from $J \cup E$ with a certainty degree α.

$[E\ \alpha]$ is an *α-environment* of d iff $[E\ \alpha]$ is an environment of d and $\forall\ \alpha' > \alpha$, $[E\ \alpha']$ is not an environment of d (α is maximal).

$[E\ \alpha]$ is an *α-contradictory environment*, or *α-nogood* iff $J \cup E$ is *α-inconsistent*, that is $\perp$ can be deduced from $J \cup E$ with α maximal (α is called the *inconsistency degree* of $J \cup E$).
The α-nogood $[E\ \alpha]$ is *minimal* iff there is no β-nogood $[E'\ \beta]$ such that $E \supset E'$ and $\alpha \leq \beta$.

Labels: The label of a datum d, noted $L(d)=\{[E_i\ \alpha_i],\ i \in I\}$, is the only subset of the set of environments which satisfies the following properties:

- *(weak) consistency* : $\forall\ [E_i\ \alpha_i] \in L(d)$, $J \cup E_i$ is β-inconsistent with $\beta < \alpha_i$ ($J \cup E_i$ has an inconsistency degree which is strictly less than the certainty degree obtained for d from $J \cup E_i$ (d is deduced using a consistence sub-base of $J \cup E_i$)).
- *soundness* : $\forall\ [E_i\ \alpha_i] \in L(d)$ is an environment of d.
- *completeness* : $\forall$ E' such that d can be deduced from $J \cup E'$ with a degree α', $\exists\ [E_i\ \alpha_i] \in L(d)$ such that $E' \supset E_i$ and $\alpha' \leq \alpha_i$ (all the minimal α-environments of d are in L(d)).
- *minimality* : L(d) doesn't contain two environments $[E_1\ \alpha_1]$ and $[E_2\ \alpha_2]$ such that $E_2 \supset E_1$ and $\alpha_2 \leq \alpha_1$ (L(d) contains only the most specific α-environments of d).

By ordering the environments in the labels on the basis of their weight, a Π-ATMS can determine the set of hypotheses which allow to deduce a given datum with the greatest degree of certainty.

Contexts: The context of a set of weighted hypotheses H is the set of pairs $(d, val_{[H]}(d))$, where d is a datum and $val_{[H]}(d)=\sup\{\alpha$ such that d can be deduced from $J \cup H$ with a degree $\alpha\}$.

Let's take as an example the weighted justification: $A \wedge B \rightarrow C \ \alpha_C$.
Then the label of the conclusion C will be updated by combining its previous label with the labels of the premises A and B.
The weight of the environments in this label will depend on the weights of the environments of the left-hand-side data and the weight of the rule conclusion, as shown in the Figure 2:

$L(A) = \{.. [E_A \ \alpha_A] ..\}$ A

$L(B) = \{.. [E_B \ \alpha_B] ..\}$ B

→ C α_C $L(C) = \{.. [E_A \cup E_B \ \min(\alpha_A, \alpha_B, \alpha_C)] ..\} \cup OLDL(C)$

with verification of (weak) consistency and minimality

Figure 2

4. COUPLING A Π-ATMS AND AN INFERENCE ENGINE

ATMSs draw inferences and build multiple contexts based on initial facts, hypotheses, and justifications.
Justifications could be viewed as simple rules without variables (Propositional Logic).
Most applications require more expressiveness, and OPS-like first order rules are needed.
Forward chaining inference engines rely usually on a *match-select-act* cycle.
The propositions contained in the fact base may match some rules condition parts, and thus instantiate one or more rules. Instantiated rules are queued in a so called conflict set for future selection and eventual firing. When a rule is fired, it may in turn create new facts.
In this context, a justification is a link between a fact created in the right-hand-side of a rule and the facts which instantiated this rule.
Let us illustrate this through an example in an OPS-like syntax:

(rule birds_fly (Bird ?x)→(assert (Fly ?x)))
(Bird Tweety)

The firing of rule birds_fly on fact (Bird Tweety) will produce the justification:

(Bird Tweety)→(Fly Tweety)

The proposed architecture integrates an inference engine and a Π-ATMS.
The role of the inference engine is to produce the weighted justifications whilst the role of the Π-ATMS is to manage the uncertainty pervading the Π-ATMS nodes (facts, assumptions and justifications) by updating weighted environments in labels, handling α-contradictions, etc.
The user can create uncertain hypotheses, facts, and rules. Facts and hypotheses are stored as Π-ATMS nodes. Corresponding working memory elements are created in the inference engine and can eventually match rule conditions.
When a rule is selected and then fired, its action part doesn't modify directly the working memory of the inference engine (as it normally does in the standard match-select-act cycle).

Instead, new uncertain facts and hypotheses can be created, or new justifications can be installed on existing facts and hypotheses.
A detailed description of efficient techniques that can be used to interface an ATMS and an inference engine can be found in [Morgue 91].
The interface provided between the inference engine and the Π-ATMS is represented in the Figure 3:

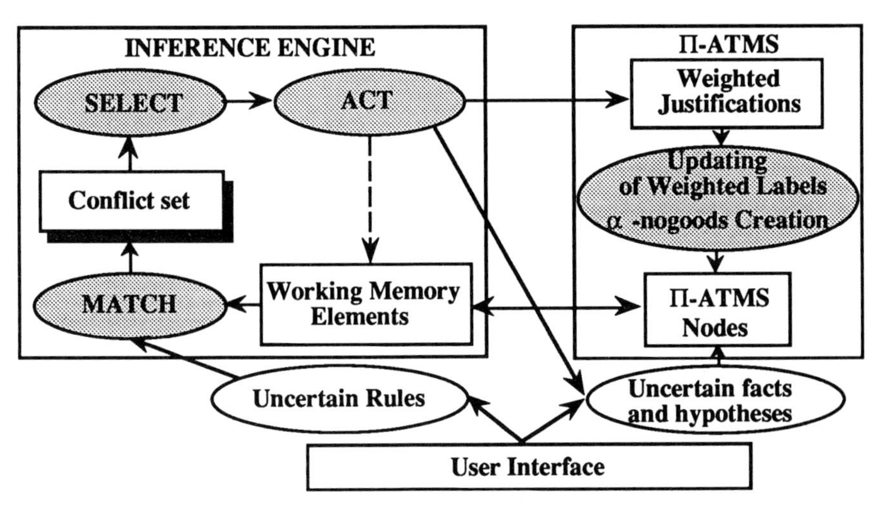

Figure 3

5. NON-MONOTONIC REASONING

Handling negation is an important requirement for many applications. Possibilistic Logic provides a convenient mean for representing the negation of a proposition, due to the fundamental relation of duality which links necessity and possibility measures (equivalence between $N(\neg p)\geq\alpha>0$ and $\Pi(p)\leq 1-\alpha<1$). Indeed to represent that we are certain to the degree α that $\neg p$ is true, we can say that the possibility of p is *upper bounded* by the weight $1-\alpha$.

So we can use *upper bounds* of a possibility measure Π to express the degree of certainty in the negation of a proposition p.
Moreover, extending the ATMS, it is possible to have both contexts in which a datum d is true and contexts in which $\neg d$ is true.

This leads to a new definition of the label L(d) for a given datum d. It will now contain two sub-labels, noted respectively $L_+(d)$ and $L_-(d)$.
$L_+(d)$ is the set of environments in which d is somewhat certain.
On the other hand $L_-(d)$ is the set of environments in which $\neg d$ is somewhat certain.
We will refer to the environments in L_+ as *α-valued* environments (the weights here are *lower bounds* of a necessity measure N) and to the environments in L_- as *β-valued* environments (the weights here are *upper bounds* of a possibility measure Π).

The truth table of a proposition p, as given in section 2, can be generalised. A truth table can be defined for each environment E, as shown in the Figure 4:

Is p true in the environment E?	$\nexists$ [A α]∈L+(p) / E ⊃ A	Maximize α / [A α]∈L+(p), E⊃A
Minimize β / [B β]∈L-(p), E⊃B	p is somewhat false in E N(p)=0, Π(p)≤β	E is a nogood with weight min(α, 1-β)
$\nexists$ [B β]∈L-(p) / E ⊃ B	Ignorance	p is somewhat true in E N(p)≥α, Π(p)=1

Figure 4

The previous Π-ATMS axioms are extended:

The property of *minimality in* L_+ is the following:

$\forall\ [E_1\ \alpha_1] \in L_+(d)\ \forall\ [E_2\ \alpha_2] \in L_+(d)$

If $E_2 \supseteq E_1$ and $\alpha_2 > \alpha_1$ then $L_+(d)$ is minimal;

If $E_2 \supseteq E_1$ and $\alpha_2 \leq \alpha_1$ then $[E_2\ \alpha_2]$ is removed from $L_+(d)$.

This means that in $L_+(d)$ we maximise the value of α for the minimal (in the classical sense) environments of d.

The property of *minimality in* L_- is the following:

$\forall\ [E_1\ \beta_1] \in L_-(d)\ \forall\ [E_2\ \beta_2] \in L_-(d)$

If $E_2 \supseteq E_1$ and $\beta_1 > \beta_2$ then $L_-(d)$ is minimal;

If $E_2 \supseteq E_1$ and $\beta_1 \leq \beta_2$ then $[E_2\ \beta_2]$ is removed from $L_-(d)$.

This means that in $L_-(d)$ we minimise the value of β for the minimal (in the classical sense) environments of d.

The property of *(weak) consistency in* L_+ is the following:

$\forall\ [E_i\ \alpha_i] \in L_+(d)$ and $\forall\ [E_\perp\ \alpha_\perp] \in$ α-nogoods set

If $E_i \supseteq E_\perp$ and $\alpha_i > \alpha_\perp$ then $L_+(d)$ is (weakly) consistent;

If $E_i \supseteq E_\perp$ and $\alpha_i \leq \alpha_\perp$ then $[E_i\ \alpha_i]$ is removed from $L_+(d)$.

This means that the inconsistency degree of E_i is strictly less then the degree of certainty with which d can be derived.

The property of *(weak) consistency in* L_- is the following:

$\forall\ [E_i\ \beta_i] \in L_-(d)$ and $\forall\ [E_\perp\ \alpha_\perp] \in$ α-nogoods set

If $E_i \supseteq E_\perp$ and $1-\beta_i > \alpha_\perp$ then $L_-(d)$ is (weakly) consistent;

If $E_i \supseteq E_\perp$ and $1-\beta_i \leq \alpha_\perp$ then $[E_i\ \beta_i]$ is removed from $L_-(d)$.

Moreover from the relation $d \wedge \neg d \rightarrow \perp$ we have:

$\forall\ [E_i\ \alpha_i] \in L_+(d)$ and $\forall\ [E_k\ \beta_k] \in L_-(d)$ with $0 < \alpha_i \leq 1$ and $0 \leq \beta_k < 1$

the environment $[E_i \cup E_k\ \min(\alpha_i, 1-\beta_k)\ ]$ is added to the α-nogoods set.

Since we can represent the negation of a proposition in the Π-ATMS, we can extend the rule syntax to test for the negation of a condition, thus obtaining non-monotonic justifications.

For example the rule:

(rule find_humans Object(x)$\wedge$Intelligent(x)$\wedge$OUT(Robot(x))$\rightarrow$Human(x) α)

means that in all the environments in which it is known that an object is intelligent and that it is not a robot, it can be deduced (with a certainty α) that it is a human being.

The problem here is that this rule cannot fire if we do not know if the object is a robot or not. Default Logic overcomes this well known limitation.

We propose to work under the *closed word assumption:* a datum which is not represented in the Π-ATMS nodes, will be assumed to be false.
This additional expressiveness power has of course some cost, that is an additional hypothesis is created for each datum.
We will denote with CWAH(d) the hypothesis that d is absent (Closed World Assumption Hypothesis for d).

An additional justification is also needed for each new hypothesis:

CWAH(d)$\rightarrow$OUT(d) 1

which means that d is certainly false under the hypothesis that it is absent.

In order to reduce the number of such hypotheses, and to avoid the computation of L_- labels if no default rules are specified, the inference engine is responsible for asking the Π-ATMS to initialise the L_+ and L_- labels, when needed, and to create the associated hypotheses.

In the previous example, let's suppose that we have both Object(O_999) and Intelligent(O_999) in the fact base but no fact Robot(O_999) exists.
Then the inference engine will ask the Π-ATMS to create the CWAH(Robot(O_999)) hypothesis.
The Π-ATMS also creates the fact Robot(O_999) with the following labels:

L+=not_initialised
L_-={[{CWAH(Robot(O_999))} 0]}

This fact is finally transmitted back to the inference engine which can instantiate the rule and put it in the conflict set.
When the rule is selected and fired, the L_+ label of the fact Human(O_999) is updated by combining the L_+ labels of Object(O_999) and Intelligent(O_999), and the L_- label of Robot(O_999).

In order to present the labels combinations and updatings that occur when rules are fired, we summarise the possible justifications in the case of two premises and one conclusion (the general case can be derived easily).

Let A and B be two Π-ATMS nodes with the following labels:

$L_+(A)=\{..[E_{A_+}\ \alpha_{A_+}]..\}$ $L_+(B)=\{..[E_{B_+}\ \alpha_{B_+}]..\}$
$L_-(A)=\{..[E_{A_-}\ \beta_{A_-}]..\}$ $L_-(B)=\{..[E_{B_-}\ \beta_{B_-}]..\}$

Then, depending on the type of the rule to be applied, we can have the following cases:

1. **$A \wedge B \rightarrow C\ \alpha C$**
   $L_+(C)=\{..[E_{A_+} \cup E_{B_+}\ \min(\alpha_{A_+},\alpha_{B_+},\alpha_C)]..\} \cup OLDL_+(C)$
   $OLDL_-(C) \otimes L_+(C) \rightarrow \bot$
2. **$A \wedge OUT(B) \rightarrow C\ \alpha C$**
   $L_+(C)=\{..[E_{A_+} \cup E_{B_-}\ \min(\alpha_{A_+},1-\beta_{B_-},\alpha_C)]..\} \cup OLDL_+(C)$
   $OLDL_-(C) \otimes L_+(C) \rightarrow \bot$
3. **$OUT(A) \wedge OUT(B) \rightarrow C\ \alpha C$**
   $L_+(C)=\{..[E_{A_-} \cup E_{B_-}\ \min(1-\beta_{A_-},1-\beta_{B_-},\alpha_C)]..\} \cup OLDL_+(C)$
   $OLDL_-(C) \otimes L_+(C) \rightarrow \bot$
4. **$A \wedge B \rightarrow OUT(C)\ \alpha C$**
   $L_-(C)=\{..[E_{A_+} \cup E_{B_+}\ 1-\min(\alpha_{A_+},\alpha_{B_+},\alpha_C)]..\} \cup \{[CWAH(C)\ 0]\} \cup OLDL_-(C)$
   $OLDL_+(C) \otimes L_-(C) \rightarrow \bot$
5. **$A \wedge OUT(B) \rightarrow OUT(C)\ \alpha C$**
   $L_-(C)=\{..[E_{A_+} \cup E_{B_-}\ 1-\min(\alpha_{A_+},1-\beta_{B_-},\alpha_C)]..\} \cup \{[CWAH(C)\ 0]\} \cup OLDL_-(C)$
   $OLDL_+(C) \otimes L_-(C) \rightarrow \bot$
6. **$OUT(A) \wedge OUT(B) \rightarrow OUT(C)\ \alpha C$**
   $L_-(C)=\{..[E_{A_-} \cup E_{B_-}\ 1-\min(1-\beta_{A_-},1-\beta_{B_-},\alpha_C)]..\} \cup \{[CWAH(C)\ 0]\} \cup OLDL_-(C)$
   $OLDL_+(C) \otimes L_-(C) \rightarrow \bot$

The $\cup$ operation in the above schemas is subject to the verification of the minimality and (weak) consistency properties defined previously. The $\otimes$ operator is the cartesian product of sets.
The updating of the L_+ (or L_-) label of a datum d entails the updating of the nogoods set due to the $d \wedge \neg d \rightarrow \bot$ justification.

It can be noted that if a data d is explicitly stated as "almost certainly false" (with a certainty $0<\alpha<1$) then we obtain $L_-(d)=\{[\{\}\ 1-\alpha][\{CWAH(d)\}\ 0]\}$ whereas if d is stated as "certainly false" then $L_-(d)=\{[\{\}\ 0]\}$ due to the minimality property.

6. EXAMPLE

In the following example we will illustrate the coupling between the inference engine and the Π-ATMS. A solution using possibilistic resolution can be found in [Dubois 91].
Let's consider the following knowledge base:

r1. If Bob attends a meeting, then Mary does not.
r2. It is certain that Bob comes to the meeting tomorrow.
r3. If Betty attends a meeting, then it is likely that the meeting will not be quiet.
r4. It is only somewhat certain that Betty comes to the meeting tomorrow.
r5. If Albert comes tomorrow and Mary does not, then it is almost certain that the meeting will not be quiet.
r6. It is likely that Mary or John will come tomorrow.
r7. If John does not come tomorrow, it is almost certain that the meeting will be quiet.
r8. If John comes tomorrow, it is rather likely that Albert will come.

This knowledge can be represented by the following rules, where ?meeting is a variable that can be bound to any meeting whilst m denotes the particular meeting under consideration ("tomorrow meeting" in the text):

(rule r1 Bob(?meeting)→(assume (OUT(Mary(?meeting))) 1))

(rule r2 →(assume (Bob(m)) 1))

(rule r3 Betty(?meeting)→(assert (OUT(quiet(?meeting))) 0.7))

(rule r4 →(assume (Betty(m)) 0.3))

(rule r5 Albert(?meeting)∧OUT(Mary(?meeting))→(assert (OUT(quiet(?meeting))) 0.8))

(rule r6 OUT(John(?meeting))→(assume (Mary(?meeting)) 0.7))

(rule r7 OUT(John(?meeting))→(assert (quiet(?meeting)) 0.8))

(rule r8 John(?meeting)→(assume (Albert(?meeting)) 0.6))

Rules r2 and r4, having no condition part, will fire first and will create the Π-ATMS nodes for the hypotheses Bob(m) and Betty(m) with the following labels:

L_+(Bob(m))={[{Bob(m)} 1]}
L_-(Bob(m))=not_initialised

L_+(Betty(m))={[{Betty(m)} 0.3]}
L_-(Betty(m))=not_initialised

Then, two working memory elements corresponding to these two hypotheses are created in the inference engine and match the condition part of rules r1 and r3 respectively.

The firing of rule r1 results in the creation of the Π-ATMS node for the hypothesis Mary(m) (and of the associated CWAH) with a L_- label computed using the L_+ label of the hypothesis Bob(m):

L_+(Mary(m))=not_initialised
L_-(Mary(m))={[{Bob(m)} 0] [{CWAH(Mary(m))} 0]}

The weighted justification Bob(m)→OUT(Mary(m)) 1 is installed in the Π-ATMS.

The firing of rule r3 results in the creation of the Π-ATMS node for the fact quiet(m) (and of the associated CWAH) with a L_- label computed using the L_+ label of the hypothesis Betty(m):

L_+(quiet(m))=not_initialised
L_-(quiet(m))={[{Betty(m)} 0.7] [{CWAH(quiet(m))} 0]}

The weighted justification Betty(m)→OUT(quiet(m)) 0.7 is installed in the Π-ATMS.

Then, the inference engine asks the Π-ATMS to create the CWAH(John(m)) hypothesis which will enable to pursue the reasoning process with rules r6 and r7.
The Π-ATMS node for the hypothesis John(m) is also created with the following labels:

L_+(John(m))=not_initialised
L_-(John(m))={[{CWAH(John(m))} 0]}

The creation in the inference engine of the working memory element for these new hypotheses causes the instantiation of rules r6, r7 and r8.

The firing of rule r6 results in the updating of the L_+ label of the hypothesis Mary(m) (after its initialisation to {[{Mary(m)} 1]}) using the L_- label of the hypothesis John(m):

L_+(Mary(m))={[{Mary(m)} 1] [{CWAH(John(m))} 0.7]}

The following weighted nogoods are also created by the Π-ATMS due to this updating:

Nogoods={[{Bob(m), Mary(m)} 1]
[{Bob(m), CWAH(John(m))} 0.7]
[{CWAH(Mary(m)), Mary(m)} 1]
[{CWAH(Mary(m)), CWAH(John(m))} 0.7]}

The weighted justification OUT(John(m))→Mary(m) 0.7 is installed in the Π-ATMS.

The firing of rule r7 results in the updating of the L_+ label of the fact quiet(m) (after its initialisation to empty) using the L_- label of the hypothesis John(m):

L_+(quiet(m))={[{CWAH(John(m))} 0.8]}

The nogoods set is also modified by the Π-ATMS due to this updating:

Nogoods=OLDNogoods ∪ {[{CWAH(John(m)), Betty(m)} 0.3]
[{CWAH(John(m)), CWAH(quiet(m))} 0.8]}

The weighted justification OUT(John(m))→quiet(m) 0.8 is installed in the Π-ATMS.

The firing of rule r8 results in the creation of the Π-ATMS node for the hypothesis Albert(m) with the L_+ label obtained from the L_+ label of the hypothesis John(m) (after its initialisation to {[{John(m)} 1]}):

L_+(Albert(m))={[{Albert(m)} 1] [{John(m)} 0.6]}
L_-(Albert(m))=not_initialised

The weighted justification John(m)→Albert(m) 0.6 is installed in the Π-ATMS.

The creation in the inference engine of the working memory element for this new hypothesis causes the instantiation and firing of rule r5.

This results in the updating of the L_- label of the fact quiet(m) using the L_+ label of the hypothesis Albert(m) and the L_- label of the hypothesis Mary(m):

L_-(quiet(m))={[{Betty(m)} 0.7]
[{CWAH(quiet(m))} 0]
[{Albert(m), Bob(m)} 0.2]
[{Albert(m), CWAH(Mary(m))} 0.2]
[{John(m), Bob(m)} 0.4]
[{John(m), CWAH(Mary(m))} 0.4]}

The nogoods set is also modified by the Π-ATMS due to this updating:

Nogoods=OLDNogoods ∪ {[{Albert(m), Bob(m), CWAH(John(m))} 0.8]
[{Albert(m), CWAH(Mary(m)), CWAH(John(m))} 0.8]}

The weighted justification Albert(m)∧OUT(Mary(m))→quiet(m) 0.8 is installed in the Π-ATMS.

The final set of α-nogoods is the following:

Nogoods={[{Bob(m), Mary(m)} 1]
[{Bob(m), CWAH(John(m))} 0.7]
[{CWAH(Mary(m)), Mary(m)} 1]
[{CWAH(Mary(m)), CWAH(John(m))} 0.7]
[{CWAH(John(m)), Betty(m)} 0.3]
[{CWAH(John(m)), CWAH(quiet(m))} 0.8]
[{Albert(m), Bob(m), CWAH(John(m))} 0.8]
[{Albert(m), CWAH(Mary(m)), CWAH(John(m))} 0.8]}

The final labels for all facts and hypotheses are listed in the Figure 5:

Data	Label+	Label-
Bob(m)	{[{Bob(m)} 1]}	not_initialised
Betty(m)	{[{Betty(m)} 0.3]}	not_initialised
Mary(m)	{[{Mary(m)} 1] [{CWAH(John(m))} 0.7]}	{[{Bob(m)} 0] [{CWAH(Mary(m))} 0]}
John(m)	{[{John(m)} 1]}	{[{CWAH(John(m))} 0]}
Albert(m)	{[{Albert(m)} 1] [{John(m)} 0.6]}	not_initialised
quiet(m)	{[{CWAH(John(m))} 0.8]}	{[{Betty(m)} 0.7] [{CWAH(quiet(m))} 0] [{Albert(m), Bob(m)} 0.2] [{Albert(m), CWAH(Mary(m))} 0.2] [{John(m), Bob(m)} 0.4] [{John(m), CWAH(Mary(m))} 0.4]}

Figure 5

By looking at the L_+ (respectively L_-) label of a given datum, it is possible to derive in which contexts the datum is somewhat true (respectively somewhat false).
For example, from the L_- label of the fact quiet, it can be deduced that it is rather likely that the meeting will not be quiet (with a certainty $\mathbf{N}(\neg\text{quiet(m)})\geq 1-0.4=0.6$) if John and Bob come to the meeting.
On the contrary, from the L_+ label of the fact quiet, it can be deduced that if John doesn't come to the meeting it is almost certain that the meeting will be quiet (with a certainty $\mathbf{N}(\text{quiet(m)})\geq 0.8$).

7. CONCLUSIONS

The integration of a Π-ATMS with an OPS-like first order inference engine into a common architecture has been outlined in this paper. The resulting environment allows to represent and reason about uncertain informations even if partial inconsistencies are discovered in the available knowledge.
The general ability of Possibilistic Logic for non-monotonic reasoning is captured in the L_+ and L_- labelling mechanism and associated non-monotonic justifications.
The proposed architecture explores concurrently both contexts in which a given datum d is (somewhat) true and contexts in which its negation $\neg d$ is (somewhat) true.
In order to pursue the reasoning process under incomplete knowledge, absent facts are assumed to be false. A special hypothesis (called Closed World Assumption Hypothesis) and an additional justification are created to support the negation of a fact which is absent.

However, labels are initialised by the Π-ATMS only if required by the inference engine.

There is thus no overhead, with L_- labels and closed world assumption hypotheses, in purely monotonic applications.

ACKNOWLEDGEMENTS

The work described in this paper has been partially funded by the DRET (French MOD).
The inference engine component of the proposed architecture was built on top of XIA which results partly from ESPRIT project P96 and from Thomson Strategic Project on AI.
We would like to thank Pr Didier Dubois, Pr Henri Prade, Dr Jerome Lang of IRIT (Paul Sabatier University - Toulouse - France) for providing the Possibilistic ATMS knowledge and Mme Geneviève Morgue for her precious collaboration.

BIBLIOGRAPHY

[deKleer 86] **de Kleer J.**, "*An Assumption based Truth Maintenance System*", AI 28, pp 127-224.

[deKleer 88] **de Kleer J.**, "*A general labeling algorithm for assumption-based Truth Maintenance*", Proceedings of AAAI 1988 Saint-Paul MN, pp 188-192.

[Dubois 88] **Dubois D., & Prade H.et al.**, "*Possibility Theory- an approach to computerized Processing of Uncertainty*", Plenum Press, NY 1988.

[Dubois 90] **Dubois D., & Lang J & Prade H..**, "*Handling uncertain knowledge in an ATMS using possibilistic logic*", ECAI-90, Workshop on Truth Maintenance Systems, Stockholm, Sweden, Aug 6-7 1990.

[Dubois 91] **Dubois D., & Prade H..**, "*Possibilistic logic, preferential models, non-monotonicity and related issues*", IJCAI-91, Sydney, Australia, Aug 24-30 1991.

[Lang 91] **Lang J.**, "*Logique possibiliste : aspects formels, déduction automatique et applications*", Thèse de l'université Paul Sabatier, Toulouse, Janvier 1991.

[Léa Sombé 90] **Léa Sombé**, "*Reasoning Under Incomplete Information in Artificial Intelligence*", Int. J. of Intelligent Systems, 5(4), pp 323-471, 1990.

[Morgue 91] **Morgue G., Chehire T.**, "*Efficiency of Production Systems when Coupled with an Assumption based Truth Maintenance System.* ", Proceedings of AAAI 1991, pp 268-274.

Uncertainty in Intelligent Systems
B. Bouchon-Meunier et al. (Editors)

AN ENTROPY FORMULATION OF EVIDENTIAL MEASURES AND THEIR APPLICATION TO REAL-WORLD PROBLEM SOLVING*

Leonard P. Wesley
Artificial Intelligence Center
SRI International
333 Ravenswood Ave.
Menlo Park, CA.

415-859-3368 (Business)
415-859-3735 (Fax)

Wesley@AI.SRI.COM

1 INTRODUCTION

The Dempster-Shafer (DS) theory of belief functions provides a powerful framework to support the mathematical representation of evidence and its use in probable reasoning [1-2]. A touted virtue of the theory is its ability to explicitly represent bounded ignorance which is not possible with pure probabilistic schemes. Despite an increasing number of applications that use belief functions, little work has focused on exploiting the representational strength of belief functions in order to extend the ability of knowledge-based systems (KBSs) to cope with real-world uncertainty. In this extended abstract, we describe our initial efforts to explore and test, more fully than has been previously conducted, the usefulness of the belief function representation for uncertain reasoning in real-world contexts.

We begin with a brief review of belief functions. Next we we introducing our notion of *evidential measures* which are extracted from the probability interval (commonly called an *evidential interval*) that is induced by a belief function. We posit that evidential measures provide a more useful characterization a KBSs situation knowledge than is possible with pure probabilistic representations. A benefit of being able to produce such a characterization is that KBSs will be better situated to subsequently choose the most appropriate problem solving action. Since these evidential measures can be viewed in an information-theoretic manner, next we provide

*This work was supported in part by DARPA contract No. F30 602-90-C-0086, SRI Project No. 1520.

an entropy-based formulation of these measures and describe, in an example mobile robot application, how they might be used to guide the information acquisition activities necessary to carry out locative reasoning from uncertain environmental information.

1.1 Belief Functions

The DS theory begins with the view that problem-solving activities can be described as a process of attempting to answer questions of interest. Within this theory, possible answers to a question are represented as a set of interrelated propositional sentences, called a frame of discernment (FOD), denoted Θ. Possible answers, for example, to a question about a mobile platform's current location may be represented by the following FOD which consists of locations under consideration:

$$\Theta = \{Location_1,\ Location_2, \ldots,\ Location_n\}.$$

Given a FOD, one and only one propositional sentence is possibly true at any instant. Discerning the correct answer involves determining which proposition is true. Accomplishing this involves forming a consensus about the truth of propositions based on opinions that bear upon the question of interest. In practice, this is done by first having distinct sources convey their opinions about which propositions they believe to be true based on environmental observations. Next, a consensus is formed by pooling distinct opinions with Dempster's Rule [1].

Beliefs derived from real world observations are usually partial, imprecise, and occasionally unreliable to varying degrees. Within the DS framework, each independent source conveys its belief via a mass distribution, m, which is defined as:

$$\begin{aligned} m : 2^{\Theta} &\mapsto [0,1], \text{ where} \\ \textstyle\sum_{p \subseteq \Theta} m(p) &= 1, \text{ and} \\ m(\emptyset) &= 0. \end{aligned}$$

When two independent opinions, m_1 and m_2, are expressed, Dempster's rule is used to combine them into a third mass distribution, m_3, that represents their consensus. Dempster's rule is defined to be:

$$\begin{aligned} m_3(c) &= m_1(a) \oplus m_2(b) \\ &= \frac{1}{1-\kappa} \sum_{a \cap b = c} m_1(a) m_2(b) \\ \kappa &= \sum_{a \cap b = \emptyset} m_1(a) m_2(b) \\ &< 1\ . \end{aligned}$$

Since Dempster's rule is both commutative and associative, multiple independent opinions can be combined in any order without affecting the result. If the initial opinions are independent, then the derivative opinions are independent as long as they share no common ancestors.

Thus, attention must be paid to the manner in which opinions are combined to guarantee the independence of the consensus at each combination.

Interpreting an opinion that is expressed as a mass distribution involves calculating the *Spt* and *Pls* values, called the *evidential interval* $[Spt, Pls] \subseteq [0,1]$, relative to propositions of interest as follows:

$$\begin{aligned} Spt(p) &= \sum_{q \subseteq p \subseteq \Theta} m(q) \\ Pls(p) &= 1 - Spt(\Theta - p) \quad . \end{aligned}$$

Propositional statements that are attributed nonzero mass are called the *focal elements* of the distribution. When a mass distribution's focal elements are all single element sets, the distribution corresponds to a classical *additive* probability distribution and the evidential interval, for any proposition discerned by the frame, collapses to a point: support is equivalent to plausibility. For any other choice of focal elements, some propositional statement discerned by the frame will have an evidential interval with support strictly less than plausibility. This reflects the fact that mass attributed to a set consisting of more than one element represents an incomplete assessment; if additional information were available, the mass attributed to this set of elements would be distributed over its single element subsets. Thus, an evidential interval with support strictly less than plausibility is indicative of incomplete information relative to the frame.

1.2 Evidential Measures

Systems rarely solve complex problems in one fell "swoop." Rather, problem solving is an iterative process that involves acquiring information, interpreting the information, and then deciding what action to pursue next. Deciding what action to take next requires being able to characterize and take into account the current situation in terms of what **is** and is **not** known. This is particularly true since the consequences of pursuing some actions can be undesirable and irrevocable. Within a belief function framework, this means it is important to not only know the degree to which we believe propositional sentences to be true or false, but to know the degree to which we are ignorant about the truth or falsity of propositional sentences. The evidential interval interpretation of belief functions provides a readily accessible means to extract this notion of ignorance as well as other useful measures such as certainty and dissonance that are associated with propositions of interest. A detailed discussion of the measures developed in previous work can be found in [3].

Having knowledge of the degree to which one is *ignorant*, *certain*, *dissonance*, or *ambiguous* about propositions is important for effective problem solving. For example, while moving through our environment, knowing that we are significantly ignorant about the presence or absence of particular objects in our environment might warrant taking actions to deliberate further or acquire and interpret specific information to better assess out current location before taking any action to alter our current position. Ignorance about propositional sentences is represented as an evidential interval, $[Spt, Pls]$ and the value $Igr = Pls - Spt$ in some sense captures the degree to which the evidence neither supports nor refutes the truth of a proposition.

The Spt measure captures the degree to which we are *certain* about the truth of propositions. However, knowing that the $Spt(p)$ is not much greater than the $Spt(\neg p)$ suggests the evidence does not support a *decisive* conclusion about the truth or falsity of p relative to $\neg p$. It may be desirable, therefore, to acquire additional information to arrive at a more conclusive opinion about the truth of p before choosing to take any alternative problem solving action. Decisiveness is defined to be:

$$Dec(p) = Spt(p) - Dbt(p),$$

where $Dbt(p) = 1 - Pls(p)$ is the dubiety of p and represents the degree to which p is false. A positive Dec measure means the evidence tends to support the truth of p. A negative Dec measure means the evidence tends to refute the truth of p. As Dec approaches 1 or -1, the strength of the suggestion increases proportionally.

Just as Igr represents the total amount of evidence that neither supports nor refutes a proposition, the total amount of information that both supports or refutes a proposition is captured by *informidness* which is defined to be:

$$Inf(p) = Spt(p) + Dbt(p).$$

Now if sources convey information that is significantly dissonant then taking actions to identify and then discount the errant source prior to forming a consensus might be warranted. We say a consensus opinion is consonant to the degree the consensus either tends to completely support or completely refute the proposition. That is, either the $Spt(p) \sim 1$ and $Dbt(p) \sim 0$, or $Spt(p) \sim 0$ and $Dbt(p) \sim 1$. Conversely, we say a proposition exhibits dissonance to the degree $Spt(p) - Dbt(p)$ approaches 0, i.e. $Spt(p) \approx Dbt(p)$. That is, there is relatively equal evidence both for and against the proposition, and as a consequence it is difficult to choose to believe p or $\neg p$. Notice that if $Inf(p)$ is the total amount of information about a proposition, then we can be no more decisive about choosing p than we are informed. What logically follows is that $Inf(p) - |Dec(p)|$ is the total amount of evidence that could potentially contribute to the dissonance of a proposition p. The portion of this potential evidence that actually contributes to dissonance is that evidence which does **not** support choosing a proposition over its negation, i.e., the complement of decisiveness, $1 - |Dec(p)|$. Taking the product of these two measures gives us the amount of the available evidence that actually contributes toward dissonance. Thus, one measure of the degree of dissonance Dis associated with proposition p is defined as:

$$Dis(p) = (Inf(p) - |Dec(p)|) * (1 - |Dec(p)|)$$

These and other evidential measures can help to form characterizations of a system's situational knowledge, i.e., knowledge about the certainty, ignorance, dissonance and so forth that is associated with propositions of interest.

In some of our previous work on knowledge-based high-level image interpretation we demonstrated how KBSs might benefit from using these evidential measures to construct better characterizations of knowledge about the current state of the interpretation process [3]. Our KB image interpretation system was able to correctly identify, for a specific interval of time, more objects in a partitioned image than was possible without using these measures. It accomplished this because being able to characterize what was **not** known as well as what **was** known improved the system's likelihood of pursuing the most appropriate information acquisition and interpretative actions. Furthermore, it accomplished this using fewer resources. An implication of this work is that evidential measures can provide systems with an improved ability to cope with the uncertainties inherent in the real world.

In the following section, we reformulate our *certainty*, *ignorance*, and *dissonance* measures within an entropy-based framework. Space limitations preclude us from discussing our reformulation of additional measures developed in our previous work. An advantage of reformulating these measures is that it provides access to mathematical machinery that can be used to conduct a dynamic and more thorough analysis of the benefits of using these measures. It also makes the subject material accessible to a wider audience since entropy-based information theoretic notions are widely understood.

1.3 Entropy Formulation Of Evidential Measures

Some previous work as involved relating entropy-based notions to aspects of the theory of belief functions. For example, Yeager has reformulated probabilistic notions of entropy to the theory of belief functions [4]. Philippe Smets also explored information-content and entropy notions of belief functions [5-6]. Stephanou and Lu have generalized specific entropy criterion to notions of characterizing the effectiveness of consensus formation using Dempster's rule [7]. The work presented here differs from previous related efforts in that we present an entropy formulation of some evidential measures that have not been defined before, and can be readily extracted from belief function representations.

An entropy-based characterization of the evidential measures discussed in the previous section proceeds as follows. The *certainty* entropy, H_C, of a mass distribution is defined as:

$$H_C = \sum_{p \subseteq \Theta} m(p) \log_2(m(p)).$$

Intuitively, H_C represents the degree of uncertainty about the portion of unit belief that should be attributed to $p \subseteq \Theta$. H_C approaches ∞ value as $m(p) \rightarrow 0$, and approaches its minimum value of 0 as $m(p) \rightarrow 1$.

The *dissonance* entropy, H_D, of a mass distribution is defined as:

$$H_D = -\sum_{p \subseteq \Theta} m(p) \log_2(f(p)), \text{ where}$$

$$f(p) = 1 + (\text{the number of focal elements disjoint from } p) \cdot 0.5$$

H_D reflects the degree to which an opinion is discordant, i.e., the extent to which a source of an opinion disagrees with itself. This entropy measure is maximum when the number of disjoint focal elements equals $|\Theta|$. It is minimum when there are no disjoint focal elements.

The *ignorance* entropy, H_I, of a mass distribution is defined as:

$$H_I = (1 - \sum_{p \subset \Theta} m(p)) \log_2 (1 - \sum_{p \subset \Theta} m(p)).$$

H_I reflects the degree of uncertainty about the portion of unit belief that should be attributed to any $p \subset \Theta$. H_I approaches its maximum value as $\sum_{p \subset \Theta} m(p) \rightarrow 0$, and reaches its maximum value when $\sum_{p \subset \Theta} m(p) \rightarrow 1$.

What logically follows is that the total entropy, based on the measures we have introduced here, associated with a mass distribution is:

$$H_T = H_C + H_D + H_I.$$

An objective, therefore, of any KBS that characterizes its situational knowledge in this manner is to reduce H_T by reducing one or more of H_C, H_D, and H_I.

If we consider two mass distributions m_1 and m_2, there are only four logical possible situations that can arise when focal elements p_1 and p_2, from m_1 and m_2 respectively, are pooled using Dempster's rule:

1. $p_1 \equiv p_2$.
2. $p_1 \subset p_2$ or $p_2 \subset p_1$, we consider this to be one possibility since which proposition subsumes the other is essentially immaterial.
3. $p_1 \cap p_2 \neq \emptyset$.
4. $p_1 \cap p_2 = \emptyset$.

Stephanou and Lu have shown ([7]) that for each of these possibilities, the total entropy H_{T3} associated with $m_3 = m_1 \oplus m_2$ is:

$$H_{T3}(c) \leq H_{T1}(a) + H_{T2}(b), \text{ where}$$
$$a, b, c \subseteq \Theta \text{ and } c = a \cap b \ .$$

This means, for each of the four possibilities above, Dempster's rule does not increase the entropy when forming a consensus. Furthermore, it is possible to predict the change in entropy for each possibility. Consider H_C, for the first possibility, $p_1 \equiv p_2$:

$$\begin{aligned} H_C(c) = \quad & -[1 - (1 - m_1(a))(1 - m_2(b))] \\ & \cdot \log_2[1 - (1 - m_1(a))(1 - m_2(b))] \\ & -(1 - m_1(a))(1 - m_2(b)) \log_2[(1 - m_1(a))(1 - m_2(b))] \ . \end{aligned}$$

For the second possibility, $p_1 \subset p_2$ or $p_2 \subset p_1$:

$$\begin{aligned} H_C(c) = & -m_1(a)(1-m_2(b))\log_2[m_1(a)(1-m_2(b))] - m_2(b)\log_2 m_2(b) \\ & -(1-m_1(a))(1-m_2(b)) \\ & \cdot\log_2[(1-m_1(a))(1-m_2(b))] \ . \end{aligned}$$

For the third possibility, $p_1 \cap p_2 \neq \emptyset$:

$$\begin{aligned} H_C(c) = & -m_1(a)(1-m_2(b))\log_2[m_1(a)(1-m_2(b))] \\ & -m_2(b)(1-m_1(a))\log_2(m_2(b)(1-m_2(ab)) \\ & -m_1(a)m_2(b)\log_2(m_1(a)m_2(b)) - ((1-m_1(a))(1-m_2(b))) \\ & \cdot\log_2[(1-m_1(a))(1-m_2(b))] \ . \end{aligned}$$

For the last possibility, $p_1 \cap p_2 = \emptyset$:

$$\begin{aligned} H_C(c) = & [m_1(a)(1-m_2(b))\log_2[m_1(a)(1-m_2(b))/(1-m_1(a)m_2(b))] \\ & +m_2(b)(1-m_1(a))\log_2[m_2(b)(1-m_1(a))/(1-m_1(a)m_2(b))] \\ & +(1-m_1(a))(1-m_2(b))\log_2[(1-m_1(a))(1-m_2(b))/(1-m_1(a)m_2(b))] \\ & \cdot(1-(m_1(a))m_2(b))) \ . \end{aligned}$$

Similar predictions can be made for the remaining entropy measures H_D, and H_I. The predicted $H_T(c)$ is then the sum of the predictions for H_C, H_D, and H_I.

1.4 Example Application

In previous work ([8]), we implemented, on a mobile platform, an evidential approach to extracting and matching linear perceptual cues characteristic of indoor office objects such as doors, walls, corridors and so forth to symbols in a map of an office environment. Objects are represented as distinct propositional sentences in a frame of discernment (FOD). A second FOD is used to represent possible locations in the office environment. Linear perceptual cues are extracted from two distinct types of sensor data, an array of sonar sensors and a laser range finder (i.e, structured-light (SL) sensor) mounted on the platform. Compatibility relations ([8]) were constructed that relate extracted perceptual cues to possible objects and their spatial relationships. A second set of compatibility relations relate possible objects and their spatial relationships to distinct locations in the environment. For example, LOCATION-1 might be the junction of two hallways which will exhibit a different set of perceptual cues than say, LOCATION-2, that is an office door in the middle of a corridor. Since every possible location in an office environment is not unique, knowledge about the platform's previous location was used to help constrain competing hypotheses.

To perform locative reasoning, it is necessary to infer the presence and absence of objects, their relative spatial positions, and then match these results to an environmental map. As linear cues were extracted from sensor data, a mass distribution was constructed over the corresponding FOD. The associated compatibility relation was then used to translate the separate sonar and

SL opinion to a common object FOD. Dempster's rule was then used to form a consensus about the perceived objects and their spatial relationships.

To navigate through the environment, the current location must be periodically identified. This was accomplished by construction an opinion in the form of a mass distribution, based on the presence and absence of recently identified objects and relationships as well as the previously known location, over an objects and spatial relations FOD and using compatibility relations to translate these mass distributions to a common locations FOD. Dempster's rule was then used to form a consensus opinion about the perceived location based on objects and their spatial relationships.

Since all perceptual processes and sensors are imperfect to some degree, and given the situational dynamics of the environment it must be anticipated that at times there will be significant ignorance, dissonance, and ambiguity associated with possible objects, their relative spatial location and as a consequence any hypotheses about the current location of the platform. Time and other resource limitations may preclude taking actions to acquire all of the information that might disambiguate the current situation. In addition, each type of sensor is not capable of perceiving the perceptual cues that can discern the correct object or location hypothesis. For example, while the sonar sensor data tends to be inaccurate for wall convex and concave wall junctions, the SL data is generally reliable. Conversely, the likelihood of error in the SL data is greater for darker surfaces,[1] however, sonar readings are unaffected by the color of surfaces. Thus, the most appropriate action to pursue next is not always obvious. Sonar data might provide information that is, say, only disjount or subsumes the location hypotheses of interest. Alternatively, the SL data might only provide information that is, say, equivalent or overlaps with location hypotheses of interest. What information, therefore, if any should be acquired and interpreted to refine current location hypotheses? What sensors should be used to acquire the desired information?

Suppose we are given a mass distribution that reflects a very uncertain consensus about the current location hypothesis, i.e., H_T is relatively large, it is possible to predict a decrease in entropy (as described above) for each of the four logical possibilities. In other words, we can straightforwardly compute the amount of decrease in H_T required to be decisive about a location hypothesis. Once this value is determined, then replacing the m_1 values in each entropy measure with that in the consensus mass distribution, and then solving algebraically for m_2 will yield the amount of support a sensor must attribute to propositions in the respective frame in order to be certain about the current location.

While it is impossible to always know what data and information a sensor will return, one strategy might be to task the most reliable sensor that is capable of providing the required *disjount*, *subset*, *intersecting*, and *equivalent* cases described above with the expectation that it is more likely to provide the minimum support than using another sensor to obtain data. Depending on available resources, other strategies might be appropriate as well. Clearly, acquiring additional information in this manner to infer the current location is not guaranteed to result in an optimal solution because sensor data and the information extracted from them is unpredictable. However, the entropy-based characterization of evidential measures and the notion of

[1] Dark surfaces tend to absorb light, hence not reflecting the light back to the sensor in order to make time-of-flight calculations

pursuing actions that minimize the entropy of a consensus is near optimal given the available information.

2 SUMMARY

In summary, this formal information theoretic characterization of evidential measures extends and facilitates the automated assessment of situation knowledge. As a consequence, KBSs are better situated to make critical and more informed decisions about how to focus their information acquisition activities and hence improve their problem solving capabilities. A logical next step in this work is to develop and characterize additional evidential measures and explore their impact on the ability of KBS to handle uncertainty.

3 ACKNOWLEDGEMENTS

The author is grateful to John Lowrance, Thomas Strat, Enrique Ruspini, and Thomas Garvey for helpful comments, insights, and suggestions. Any errors and omissions in this abstract are the sole responsibility of the author.

References

[1] Arthur P. Dempster, 'A Generalization of Bayesian Inference," *Journal of the Royal Statistical Society*, Vol 30, Series B, pp. 205-247, (1968).

[2] Glenn Shafer, *A Mathematical Theory of Evidence*, Princeton University Press, Princeton, NJ, (1976).

[3] Leonard P. Wesley, *Evidential-Based Control in Knowledge-Based Systems*, Department of Computer and Information Science, University of Massachusetts, Amherst, MA., (1988).

[4] Ronald R.Yeager, "Entropy and specificity in a mathematical theory of evidence," in *International Journal of General Systems*, Vol. 9, pp. 249-260, (1983).

[5] Philippe Smets and Paul Magrez, "Additive Structure of the Measure of Information Content," *Approximate Reasoning in Expert Systems*, Eds. M.M. Gupta, A.Kandel, W.Bandler, J.B.Kiszka, Elsevier Science Publishers B.V. (North Holland), pp. 195-197, (1985)

[5] Philippe Smets, "Information Content of an Evidence," *Int. Journal Man-Machine Studies*, Academic Press Inc. (London) Limited, pp. 33-43.

[7] Harry E. Stephanou and Shin-Yee Lu, "Measuring Consensus Effectiveness by Generalized Entropy Criterion," in *IEEE Transactions on Pattern Analysis and Machine Intelligence*, Vol. 10, No. 4, pp. 544-554, (1988).

[8] John D. Lowrance, Thomas M. Strat, Leonard P. Wesley, Thomas D. Garvey, Enrique H. Ruspini, David E. Wilkins, "The Theory, Implementation, and Practice of Evidential Reasoning," Final DARPA Project Report, DARPA Contract No. N00039-88-C-0248, ARPA Order No. 4783, SRI Internal Project No. 5701., (June 1991).

Uncertainty in Intelligent Systems
B. Bouchon-Meunier et al. (Editors)

Properties of probabilistic imprecision

Gernot D. Kleiter[a*]

[a]Institut für Psychologie,
Universität Salzburg,
Hellbrunnerstr. 34,
A-5020 Salzburg, Austria

The probability that a single case, described by a set of features, belongs to a given class is called 'member probability'. It is obtained from Bayes' theorem. The (first-order) member probability is treated as a statistical parameter and its (second-order) density function - called member distribution - is derived. The member distribution expresses the precision and amount of knowledge with which a diagnosis can be made. The paper discusses a series of general properties of the distribution and their implications for probabilistic single case inferences in belief nets.

1. INTRODUCTION

Consider an inference system that supports the probabilistic classification of objects into classes. Take, as a typical example, medical diagnosis: a patient shows a specific symptom pattern and is suspected to suffer from one of several not directly observable diseases. The system's information about the diseases and the symptoms may consist of objective statistics, subjective expertise, or both. The system calculates the probabilities that the new object is a member of one of the alternative classes. The objective statistics may be based on large or small samples. Similarily, the subjective assessments may have been obtained from highly experienced experts or from novices. In the first case, the system is based on a lot, in the second, on little knowledge only. Consequently, an inference system processing the information of large samples and of highly experienced experts should contribute more precise knowledge to a diagnosis than a system processing information of small samples and less experienced experts only.

We propose to express the amount of knowledge a system contributes to its inferences by *second-order probability density functions*. We call the function *member distribution*. The distribution tells us how sure we can be that the classification probability falls into any given region between zero and one. It expresses the *precision* the system provides for the specific single case classification at hand. The difference between a conventional classification probability and a member distribution is analogue to that between a point estimate and a probability distribution: The distribution communicates information about the favored values *and* about their precision. In the literature, several proposals were made to handle imprecision in dependency structures, including lower and upper bounds [5, 6], propagation of variances [16, 21], and second-order distributions [17, 20].

*This research was supported by the Fonds zur Förderung der wissenschaftlichen Forschung, Vienna.

2. BASIC MODEL

Fundamental to every classification problem is (i) the description $\boldsymbol{x}$ of a single case, (ii) a set C of classes, and (iii) the question what the probability is that, given its description $\boldsymbol{x}$, the case comes from the focus class c, $c \in C$. More formally, we have a $1 \times d$ random vector of discrete *features* $\underset{\sim}{\boldsymbol{x}} = (\underset{\sim}{x}_1, \ldots, \underset{\sim}{x}_d)$. Each feature is associated with a domain of k_i possible values, $X_i = \{x_{i,1}, \ldots, x_{i,k_i}\}$. The feature space $\mathcal{X}$ is given by the Cartesian product of the domains, $\mathcal{X} = \times_{i=1}^{d} X_i$. Its cardinality is *card*$(\mathcal{X}) = \prod_{i=1}^{d} k_i$. Next, we have a set of m, $2 \leq m < \infty$, exhaustive and disjoint *classes* (diseases, hypotheses). We assign an index c to every class and denote the class by its index c, $c \in C = \{1, 2, \ldots, m\}$. If the class is random, we write $\underset{\sim}{c}$, if it is instantiated, we write c.

Classifications are based on previously observed cases. For these cases, both, the class they came from and their features are known. We call $\boldsymbol{z} = (c, \boldsymbol{x})$ an *object record*, where c is the class the object comes from and $\boldsymbol{x}$ is the object's feature vector. We denote by $\mathcal{E}$ a sample of n of object records. $\mathcal{E}$ consists of a partition of m sub-samples, $\mathcal{E} = (\mathcal{E}_1, \ldots, \mathcal{E}_m)$, with the according sample sizes $N = n_1 + \cdots + n_m$. Furthermore, $\boldsymbol{x}_{n+1}$ is the feature vector of a *new object* for which c is unknown. Throughout, we write $\boldsymbol{x}$ instead of $\boldsymbol{x}_{n+1}$.

Let $\boldsymbol{\tau} = (\tau_1, \ldots, \tau_m)$ denote the $1 \times m$ vector of *base rates* (prevalences, arrival parameters) of the classes. We denote by $\boldsymbol{\theta_x} = (\theta_{\boldsymbol{x},1}, \ldots, \theta_{\boldsymbol{x},\mathrm{d}})$ the parameters of the class conditional feature distributions. $\theta_{\boldsymbol{x},\mathrm{j}}$ may be a scalar or a tupel of parameters. In the case of a univariate multinomial process we write $\boldsymbol{\pi}_x = (\pi_{x,1}, \ldots, \pi_{x,d})$ for the symptom probabilities. Where possible we consider, without loss of generality, univariate feature variables and write x instead of the more general bold notation $\boldsymbol{x}$.

Consider a new case with the instantiated feature value x, and the focus class c. The member probability that the new case comes from class c is denoted by $\mu_{c;x}$. In the elementary case, the base rates and the class conditional symptom probabilities are known constants and then the member probability is obtained from 'plain' Bayes' formula:

$$\mu_{c;x} = \frac{\tau_c \pi_{x,c}}{\sum_{h=1}^{m} \tau_c \pi_{x,h}} \tag{1}$$

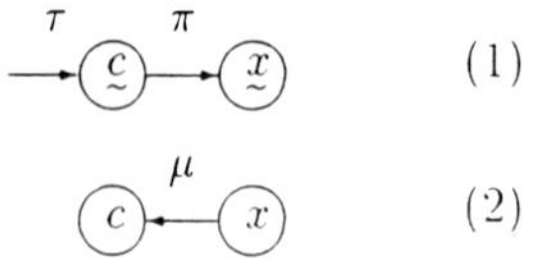

Figure 1. Simple classification problem; (1): single case data and classes are random variables; (2): member probability after instantiation of the data and class nodes and the application of Bayes' theorem.

Figure 1 shows the directed independence graphs associated with this elementary situation. We use two types of arcs: (a) initial arcs having no predecessors and (b) non-initial arcs having predecessors. Initial arcs represent marginal distributions, non-initial ones sets of conditional distributions. For each possible value of the predecessor, the conditional distribution specifies the probability (density) function for each of the random variables at the successor node. If two nodes are (conditionally) independent, they are *not* connected by an arc. A random variable is *instantiated*, if the variable is substituted by one of its possible values. The structure of an independence graph can be changed by the application of a few elementary rules. Two such rules are node removal and arc reversal [19]. Node removal is accomplished by averaging out (integrating out) the random variable associated with the node to be removed. Arc reversal is performed by Bayes' theorem. In the case of a parameter node, Bayes' theorem transforms the prior distribution and the likelihoods into the posterior distribution. We remove the prior and revert the arrow.

The graphs in Figure 1 contain two nodes. In diagram (1) they are associated with the random variables $\underset{\sim}{c}$ and $\underset{\sim}{x}$, and in diagram (2) with the according instantiated values. The initial arc represents the array of base rates $\boldsymbol{\tau} = (\tau_1, \ldots, \tau_m)$. The non-initial arc corresponds to a multiway array of the order $m \times card(\mathcal{X})$ containing the conditional probabilities $\boldsymbol{\pi}_x = (\pi_{x_1}, \ldots, \pi_{x_m})$ for $x \in \mathcal{X}$ and $c \in C$. The joint distribution of the two nodes in diagram (1) is

$$f(c, x) = f(c) f(x \mid c) = \tau_c \pi_{x,c} \,. \tag{2}$$

The instantiation of the two random variables and the application of Bayes' theorem transform diagram (1) into diagram (2). Bayes' theorem reverses the non-initial and deletes the initial arc. The arc from the data to the class node represents the member probability of c given x, that is $\mu_{c;x}$.

We now drop the assumption that $\boldsymbol{\tau}$ and $\boldsymbol{\pi}_x$ are precisely known constants. We assume, instead, that they are known probabilistically only, meaning that they become uncertain quantities or random variables $\underset{\sim}{\boldsymbol{\tau}}$ and $\underset{\sim}{\boldsymbol{\pi}}_x$. Accordingly, we introduce two additional nodes, one for each of the new random variables, and we change the interpretation of the arcs slightly (Figure 2, diagram (1)). Two arcs are now associated with probability density functions. We select Dirichlet distributions. The first arc is associated with the prior distribution of the base rates. The second arc τ corresponds to the conditional probabilities of the classes given the base rates. This is equivalent to the base rates. Obviously, the parent node inherits its values to the output arc. The third arc represents the class conditional symptom probabilities, again assumed to be Dirichlet. Without the incorporation of observational data, this is a prior distribution. The fourth arc, at the right side of the diagram, represents the single case probability distribution given the symptom probability. Again, the parent node inherits its values to the output arc. From diagram (1) in Figure 2 the following three pairwise independence assumptions can be read off: (1) $\underset{\sim}{\boldsymbol{\tau}} \perp\!\!\!\perp \underset{\sim}{\boldsymbol{\pi}}_x \mid c$, (2) $\underset{\sim}{\boldsymbol{\tau}} \perp\!\!\!\perp \underset{\sim}{x} \mid c, \pi_x$, and (3) $\underset{\sim}{c} \perp\!\!\!\perp x \mid \boldsymbol{\pi}_x$. The joint distribution for the nodes in diagram (1) of Figure 2 is obtained by factorization:

$$f(\underset{\sim}{\boldsymbol{\tau}}, \underset{\sim}{c}, \underset{\sim}{\boldsymbol{\pi}}_x, \underset{\sim}{x}) = f(\underset{\sim}{\boldsymbol{\tau}}) f(\underset{\sim}{c} \mid \tau) f(\underset{\sim}{\boldsymbol{\pi}}_x \mid c) f(\underset{\sim}{x} \mid \pi_x) \,. \tag{3}$$

Each factor corresponds to one arc.

The uncertainty at the parameter nodes $\underset{\sim}{\boldsymbol{\tau}}$ and $\underset{\sim}{\boldsymbol{\pi}}_x$ is reduced by the observation of sample data. We assume that samples are taken with a pre-fixed total number of observations N. Thus, the sub-sample sizes $\underset{\sim}{\boldsymbol{n}} = (\underset{\sim}{n}_1, \ldots, \underset{\sim}{n}_m)$ within each class are random,

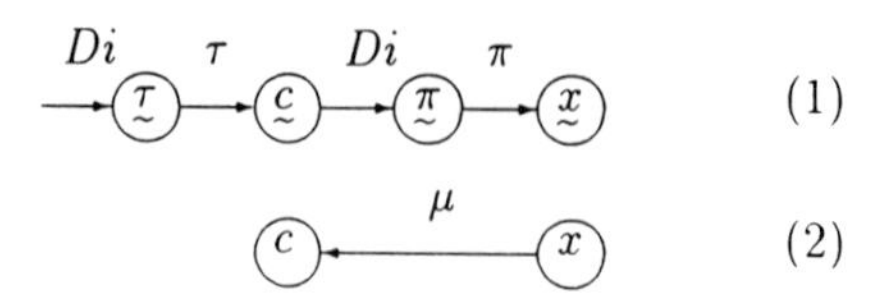

Figure 2. Classification problem with probabilistically known parameters. Diagram (1): base rates, classes, conditional symptom parameters and data are random variables. Diagram (2): member probability after instantiation of the data and class nodes, change of variable and integrating out parameters.

and, of couse, the sample statistics $\underset{\sim}{s} = (\underset{\sim}{s}_1, \ldots, \underset{\sim}{s}_m)$ are random variables. In the discrete example, the sample statistics are the frequencies of the symptoms, $\boldsymbol{s}_i = (s_{i,1}, \ldots, s_{i,d})$, $i = 1, \ldots, m$. Diagram (1) in Figure 3 contains two more nodes, one representing the sub-sample sizes and one the sample statistics. Bayes' theorem transforms diagram (1) into diagram (2). The transformation is accomplished by arc reversal and by the inheritance of parent nodes [19]. The arc from the base rate to the sample sizes and the arc from the symptom probabilities to the sample statistics are reverted. The involved nodes mutually inherit their parental input arcs.

Diagram (2) of Figure 3 can be simplified, if the model is specified for *natural conjugate distributions*. A prior distribution is natural conjugate if updating through Bayes' theorem leads to a posterior distribution belonging to the same family of distributions as the prior. Updating changes the parameters of the priors only, but does not change the distribution family. We assume that both input arcs to the base rate node are natural conjugate. Both input arcs are Dirichlet distributions. The arcs can be unified by updating the prior and removal of the sample size node. Similarily, we assume that input arcs to the symptom node are natural conjugate. Two input arcs are Dirichlet and the third one provides the information about the sample size. We update the conditional prior and remove the sample statistics node. We thus obtain diagram (3) in Figure 3. To emphasis the difference between the original priors and the updated posteriors, the posteriors are represented by double arcs. Diagram (3) in Figure 3 has the same structure as diagram (1) in Figure 2. This equivalence is an expression of the Bayesian philosophy underlying learning and the revision of probabilities in the light of new evidence. The final transformation – from diagram (3) to diagram (4) in Figure 3 – to obtain the member distribution is completely analogous to the 'no data' case in Figure 2. Finally, diagram (4) of Figure 3 represents the conditional probability distribution for a class given the single case data. This corresponds to the member distribution. In the following sections, we show how this transformation can be accomplished.

Diagram (1) in Figure 3 contains one initial and five non-initial arcs. Nine pairwise independence conditions can be read off. The joint distribution of the graph is obtained

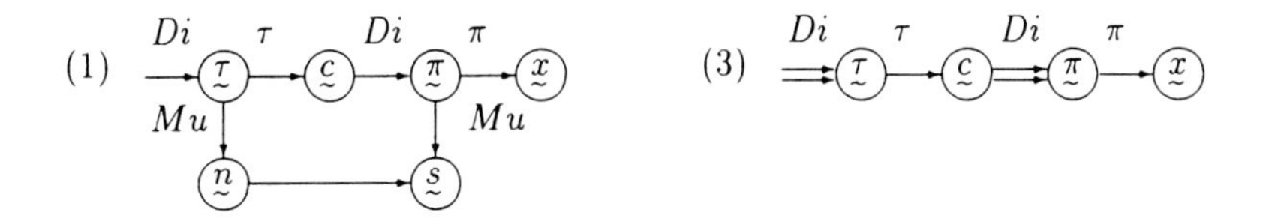

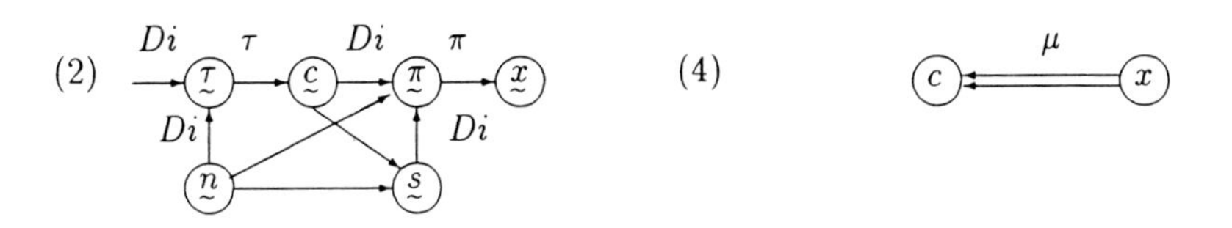

Figure 3. Classification problem with probabilistically known parameters and sample data. (1): Initial situation. (2): After applying Bayes' theorem. (3): After simplifying by natural conjugate distributions. (4): After instantiating, change of variable, and integrating-out parameters. Di: Dirichlet distribution, Mu: multinomial distribution.

by factorization of the six arc probability distributions:

$$f(\underset{\sim}{\tau}, \underset{\sim}{c}, \underset{\sim}{\pi}_x, \underset{\sim}{n}, \underset{\sim}{s}, \underset{\sim}{x}) = f(\underset{\sim}{\tau}) f(\underset{\sim}{c} \mid \boldsymbol{\tau}) f(\underset{\sim}{\pi}_x \mid c) f(\underset{\sim}{n} \mid \boldsymbol{\tau}, N) f(\underset{\sim}{s} \mid \boldsymbol{\pi}_x, \boldsymbol{n}) f(\underset{\sim}{x} \mid \boldsymbol{\pi}_x), \tag{4}$$

where $f(\underset{\sim}{\tau})$ is the prior distribution of the base rates, e.g., a Dirichlet distribution, $f(\underset{\sim}{c} \mid \boldsymbol{\tau})$ is the probability of the classes given the base rates, $f(\underset{\sim}{\pi}_x \mid c)$ are the prior distributions of the class conditional symptom parameters, e.g., Dirichlet distributions, $f(\underset{\sim}{x} \mid \boldsymbol{\pi}_x)$ is the single case probability, $f(\underset{\sim}{n} \mid \boldsymbol{\tau}, N)$ is the distribution of the sub-sample sizes, e.g., a multinomial distribution, $f(\underset{\sim}{s} \mid \boldsymbol{\pi}_x, \boldsymbol{n})$ is the probability of the sample statistics, e.g., a multinomial distribution.

3. MEMBER DENSITY DISTRIBUTION

Assume, for a moment, the base rate parameters $\tau_1, \ldots, \tau_m$ and the conditional feature parameters $\theta_1, \ldots, \theta_m$ to be known constants. Then, the probability that a new object with feature vector $\boldsymbol{x}$ belongs to the focus class c, generally, is given by Bayes' formula

$$\mu_{c;\boldsymbol{x}} = \frac{\tau_c l_{c;\boldsymbol{x}}(\theta_c)}{\sum_{k=1}^{m} \tau_k l_{k;\boldsymbol{x}}(\theta_k)} \qquad c = 1, \ldots, m. \tag{5}$$

$l_{k;\boldsymbol{x}}(\theta_k)$ denotes the value of the likelihood function of θ_k at the instantiated value of the feature vector $\boldsymbol{x}$ of the new object. The likelihood terms $l_{c;\boldsymbol{x}}$ in (3) are more general than the probability terms $\pi_{\boldsymbol{x},c}$ in (1). The latter apply to the special case of multinomial feature generating processes but not, say, to a Gaussian processes. The *member probability* $\mu_{c;\boldsymbol{x}}$ is a scalar-valued function of the base rates and the likelihoods.

If the parameters are not known exactly we express our knowledge about them by probability distributions that are based on our previous experience or on statistical data or on both. Once we have assessed the probability distributions for the parameters, we may derive the distribution for the 'inverse Bayes' probability' $\mu_{c;\boldsymbol{x}}$ given the previous experience $\mathcal{E}$. The member probability is a function of the parameters and we obtain its probability density function by a series of formal manipulations only - no new entities are introduced. The *member distribution* $f(\underset{\sim}{\mu}_{c;\boldsymbol{x}} \mid \mathcal{E})$ tells us, by a distribution defined over the interval $[0,1]$, how sure we can be that the new object comes from class c. If the distribution is flat the system has little to say about its diagnosis, if the distribution is tight it provides a substantial knowledge. From the distribution we may obtain confidence intervals for the classification probability. The width of the intervals is one way to express the (second order) precision about (first order) member probability. Figure 4 shows examples of member distributions.

To obtain the member density function we start from the joint posterior distribution of the parameters

$$f(\underset{\sim}{\tau}_1,\ldots,\underset{\sim}{\tau}_c,\ldots,\underset{\sim}{\tau}_m,\underset{\sim}{\theta}_1,\ldots,\underset{\sim}{\theta}_m \mid \mathcal{E})\,. \tag{6}$$

We change the metric of the variable τ_c

$$u:(\underset{\sim}{\tau}_1,\ldots,\underset{\sim}{\tau}_c,\ldots,\underset{\sim}{\tau}_m,\underset{\sim}{\theta}_1,\ldots,\underset{\sim}{\theta}_m \mid \mathcal{E}) \rightarrow (\underset{\sim}{\tau}_1,\ldots,\underset{\sim}{\mu}_{c;\boldsymbol{x}},\ldots\tau_m,\underset{\sim}{\theta}_1,\ldots,\underset{\sim}{\theta}_m \mid \mathcal{E}) \tag{7}$$

according to (5), multiply the result by the Jacobian $\mid \partial\underset{\sim}{\tau}_c/\partial\underset{\sim}{\mu}_{c;\boldsymbol{x}} \mid$, and finally integrate out all parameters except $\underset{\sim}{\mu}_{c;\boldsymbol{x}}$:

$$\begin{aligned} f(\underset{\sim}{\mu}_{c;\boldsymbol{x}} \mid \mathcal{E}) \;=\; & \int_{\tau_1}\cdots\int_{\tau_{c-1}}\int_{\tau_{c+1}}\cdots\int_{\tau_m}\cdots\int_{\theta_1}\cdots\int_{\theta_m} \\ & f(\underset{\sim}{\tau}_1,\ldots,\underset{\sim}{\tau}_{c-1},u^{-1}(\underset{\sim}{\mu}_{c;\boldsymbol{x}}),\underset{\sim}{\tau}_{c+1},\ldots,\underset{\sim}{\tau}_m,\underset{\sim}{\theta}_1,\ldots\underset{\sim}{\theta}_m) \\ & \left|\frac{\partial\underset{\sim}{\tau}_c}{\partial\underset{\sim}{\mu}_{c;\boldsymbol{x}}}\right| d\underset{\sim}{\tau}_1\cdots d\underset{\sim}{\tau}_{c-1}\underset{\sim}{\tau}_{c+1}\cdots d\underset{\sim}{\tau}_m d\underset{\sim}{\theta}_1\cdots d\underset{\sim}{\theta}_m\,, \end{aligned} \tag{8}$$

where u^{-1} is the inverse of the transformation u obtained from the rearrangement of (5):

$$u^{-1}(\mu_{c;\boldsymbol{x}}):\tau_c = \frac{\mu_{c;\boldsymbol{x}}\sum_{k\neq c}\tau_k l_{k;\boldsymbol{x}}(\theta_k)}{(1-\mu_{c;\boldsymbol{x}})l_{c;\boldsymbol{x}}(\theta_c)}\,. \tag{9}$$

For Dirichlet models the integrals were solved [12]. The solution is complex and for practical purposes an approximation is needed.

4. APPROXIMATION

The symptoms may all be discrete, all be continuous, or may be mixed. The member distribution $f(\mu_{c;x})$ for all three models can be approximated by a beta distribution. Space allows only to treat the discrete case in this paper. We denote a beta distribution by $Be(p,q)$. The shape of the distribution is determined by p and q; its mean is $p/(p+q)$ and its variance $pq/[(p+q)^2(p+q+1)]$. The distributions in Figure 4 are betas. We conceive p as the weigth of evidence in favor of a classification, q as the weigth of evidence against a classification, and $p+q$ as the total weigth of evidence. The approximation is based on the δ rule [4]. The rule states that the distribution of a variable that is a function of a normally distributed random variable is asymptotically normal with mean and variance being a function of the original distribution. Now, the member parameter is just a function of variables that follow beta distributions. As a beta can be approximated by a Normal, the δ rule provides a convenient method to approximate (8).

In the discrete case we assume (without loss of generality) that all the d symptoms are binary, $x_{i,j} \in \{0,1\}$, $i = 1,\ldots,m$, $j = 1,\ldots,d$. We denote the weigths (coefficients) of the base rates by $g_1,\ldots,g_m$ and the weigths of the symptoms by $a_{i,x_1,\ldots,x_d}$, $i = 1,\ldots,m$. The symptom weigths may be arranged in m multiway contingency tables each having 2^d cells. The posterior base rate weigths and the posterior symptom weigths comprise the actual starting point for the analysis.

We denote the mean of a distribution by $E(\cdot)$ and its approximation by $m(\cdot)$. Similarily, we denote the variance of a distribution by $V(\cdot)$ and its approximation by $var(\cdot)$. The approximation $m(\mu_{c;x})$ for the mean $E(\mu_{c;x})$ of the member distribution is obtained by the δ rule:

$$m(\underset{\sim}{\mu}_{c;x}) = \frac{E(\underset{\sim}{\tau}_c)E(\underset{\sim}{l}_{c;x})}{\sum_{k=1}^{m} E(\underset{\sim}{\tau}_k)E(\underset{\sim}{l}_{k;x})} \approx E_\mu(\underset{\sim}{\tau}_1,\ldots,\underset{\sim}{\tau}_{m-1},\underset{\sim}{l}_{1;x},\ldots,\underset{\sim}{l}_{m;x} \mid \mathcal{E}), \tag{10}$$

where $E(\underset{\sim}{\tau}_i) = g_i/G$ and $G = \sum_{k=1}^{m} g_k$. In the present model the likelihood function for the single case data is $l_{k;x}(\pi) = \pi_{k,x}$, so that the expected value of the function is equivalent to the expected value of the parameter (or the mean of the according posterior distribution):

$$m(\underset{\sim}{l}_{i;x}) = E(\underset{\sim}{l}_{i;x}) = \frac{a_{i,x}}{A_i} \qquad i = 1,\ldots,m, \tag{11}$$

where A_i is the sum of all symptom weigths in class i. In the terminology of [1], the approximate mean of the member distribution is equal to the diagnostic probability, and the likelihoods are obtained from the predictive probabilities.

The δ method approximates the variance of the distribution of a function of a set of random variables by a sum of variance components. Each component consists of a product of the variance of one of the random variables in the original set and the square of a partial derivative. We denote the approximate variance of the member distribution for the focus class c by $var(\underset{\sim}{\mu}_{c;x})$. Its approximation consists of the sum of $2m-1$ components: $m-1$ base rate components and m likelihood components:

$$var(\underset{\sim}{\mu}_{c;x}) = \sum_{k=1}^{m-1} var(\underset{\sim}{\mu}_{\tau_k}) + \sum_{k=1}^{m} var(\underset{\sim}{\mu}_{l_k}). \tag{12}$$

The $m-1$ terms in the first sum are a consequence of the fact that a Dirichlet distribution of m probabilities summing up to 1 has only $m-1$ dimensions. The mth dimension is dummy. Four different types of terms enter the sums: the variance components of (i) the focus class base rate, (ii) the non-focus class base rates, (iii) the focus class likelihoods and (iv) the non-focus likelihoods. We give the expressions for the four sources.

(i) The variance components for the *base rate of the focus class* c in the first sum of (12) is obtained from

$$var(\underset{\sim}{\mu}_{\tau_c}) = V(\underset{\sim}{\tau}_c)\left(\frac{\partial m(\underset{\sim}{\mu}_{c;x})}{\partial m(\underset{\sim}{\tau}_c)}\right)^2. \tag{13}$$

The variance of the base rate distribution $\underset{\sim}{\tau}_c \sim Be(g_c, G-g_c)$ is

$$V(\underset{\sim}{\tau}_c) = \frac{g_c(G-g_c)}{G^2(G+1)} \tag{14}$$

and the partial derivative

$$\frac{\partial m(\underset{\sim}{\mu}_{c;x})}{\partial m(\underset{\sim}{\tau}_c)} = \frac{m(\underset{\sim}{l}_{c;x}) \sum_{j \in I} m(\underset{\sim}{\tau}_i)[m(\underset{\sim}{l}_{j;x}) - m(\underset{\sim}{l}_{z;x})] + m(\underset{\sim}{l}_{c;x}) m(\underset{\sim}{l}_{z;x})}{D^2}, \tag{15}$$

z denotes the dummy Dirichlet dimension, $z \in \{1, \ldots, m\} \setminus \{c\}$, $D = \sum_{k=1}^{m} m(\underset{\sim}{\tau}_k) m(\underset{\sim}{l}_{k;x})$, and $I = \{1, ..., m\} \setminus \{c, z\}$.

A complication arises in the calculation of $var(\mu_{\tilde{\tau}_c})$. The approximation can be obtained in $m-1$ different ways depending upon which of the $m-1$ non-focus classes is treated as the dummy dimension. For example, for five classes and $c = 1$ we have the four index sets $\{1, 2, 3, 4\}$, $\{1, 2, 3, 5\}$, $\{1, 2, 4, 5\}$, and $\{1, 3, 4, 5\}$. We will come back to this problem below.

(ii) The variance components for the *base rates of the non-focus classes* $i \in \{1, \ldots, m\} \setminus \{c\}$, are obtained from

$$var(\underset{\sim}{\mu}_{\tau_i}) = V(\underset{\sim}{\tau}_i) \left(\frac{\partial m(\underset{\sim}{\mu}_{c;x})}{\partial m(\underset{\sim}{\tau}_i)} \right)^2. \tag{16}$$

The variance of the beta distribution $\underset{\sim}{\tau}_i \sim Be(g_i, G - g_i)$ is

$$V(\underset{\sim}{\tau}_i) = \frac{g_i(G - g_i)}{G^2(G+1)}, \quad i \in \{1, \ldots, m\} \setminus \{c\}, \tag{17}$$

and the partial derivative is given by

$$\frac{\partial m(\underset{\sim}{\mu}_{c;x})}{\partial m(\underset{\sim}{\tau}_i)} = -\frac{m(\underset{\sim}{\mu}_c)[m(\underset{\sim}{l}_{i;x}) - m(\underset{\sim}{l}_{z;x})]}{D}, \tag{18}$$

and z is the dummy dimension of the Dirichlet. Again, this variance component can be calculated in $m-1$ different ways. In our computer programs we calculate the weigths of the approximate member distribution for each z and use the mean of the $m-1$ weigths as the final approximation.

(iii) The variance component for the *likelihood term in the focus class c* in the second term of (12) is

$$var(\underset{\sim}{\mu}_{l_{c,x}}) = V(\underset{\sim}{l}_{c;x}) \left(\frac{\partial m(\underset{\sim}{\mu}_{c;x})}{\partial m(\underset{\sim}{l}_{c;x})} \right)^2. \tag{19}$$

The variance is given by

$$V(\underset{\sim}{l}_{c;x}) = \frac{a_{c,x}(n_x - a_{c,x})}{n_x^2(n_x+1)} \quad \text{where} \quad n_x = \sum_{k=1}^{m} a_{k,x} \tag{20}$$

and the partial derivative is

$$\frac{\partial m(\underset{\sim}{\mu}_{c;x})}{\partial m(\underset{\sim}{l}_{c;x})} = \frac{m(\underset{\sim}{\tau}_c)[D - m(\tau_c) m(\underset{\sim}{l}_{c;x})]}{D^2}. \tag{21}$$

(iv) The variance components for the likelihood terms in the *non-focus classes* are

$$var(\underset{\sim}{\mu}_{l_{i,x}}) = V(\underset{\sim}{l}_{i;x}) \left(\frac{\partial m(\underset{\sim}{\mu}_{i,x})}{\partial m(\underset{\sim}{l}_{i;x})} \right)^2. \tag{22}$$

The variance is obtained from

$$V(\underset{\sim}{l}_{i;x}) = \frac{a_{i,x}(n_x - a_{i,x})}{n_x^2(n_x+1)}, \quad i \in \{1,\ldots,m\} \setminus \{c\} \tag{23}$$

and the partial derivative from

$$\frac{\partial m(\underset{\sim}{\mu}_{c;x})}{\partial m(\underset{\sim}{\pi}_j)} = -\frac{m(\underset{\sim}{\mu}_{c;x})m(\underset{\sim}{\tau}_i)}{D}. \tag{24}$$

Finally, we fit a beta distribution $Be(p,q)$ to the mean $m(\underset{\sim}{\mu}_{c;x})$ and the variance $var(\underset{\sim}{\mu}_{c;x})$. With

$$T = \frac{m(\underset{\sim}{\mu}_{c;x})[1 - m(\underset{\sim}{\mu}_{c;x})]}{var(\underset{\sim}{\mu}_{c;x})} - 1 \tag{25}$$

we obtain

$$p = m(\underset{\sim}{\mu}_{c;x})T \qquad \text{and} \qquad q = T - p. \tag{26}$$

In summary: We approximate the member distribution by a beta distribution. The mean of the beta is proportional to the product of the estimate of the base rate and the density of the predictive distribution. The variance is represented as the sum of variance components. Each variance component corresponds to a model parameter.

5. PROPERTIES OF THE MEMBER DISTRIBUTION

We discuss several properties of the member distribution which are relevant for the treatment of precision in belief nets.

5.1. Natural sampling

The relationship between the amount of knowledge about the base rates and about the class conditional symptom probabilities may differ in three ways: (a) Little is known about the base rates: the sums of the weigths of the base rates are smaller than those of the symptoms; take, as an example, the case in which little is known about the frequency of a family of bacteria in a natural habitat, though, in the laboratory the *in vitro* tests are known to have a high diagnosticity. (b) Substantial knowledge about the base rates is available: the sums of the base rate weigths are greater than those of the symptoms; take, as an example, the case in which the epidemiology of a human disease is well documented, though, early clinical diagnosis is only poorly understood. (c) Natural sampling: This is the special case in which the base rate weigths are equal to the sums of the according symptom weigths. In statistics natural sampling occurs in contingency tables with random marginals. The symptom counts add up to the class counts and are used to estimate the base rates. We formulate the following result for the most elementary case:

Natural sampling. Consider $m = 2$ diseases, the focus class $c = 1$, and let X be a univariate binary symptom. Let the base rate distribution be beta, $\underset{\sim}{\tau}_1 \sim Be(g_1, g_2)$ and let the conditional symptom distributions be beta, $\underset{\sim}{\pi}_i \sim Be(a_{i,0}, a_{i,1}), i = 1, 2$. A patient shows symptom x. If the natural sampling condition holds, i.e., if $g_1 = a_{1,0} + a_{1,1}$ and $g_2 = a_{2,0} + a_{2,1}$ then the member distribution is the beta $\underset{\sim}{\mu}_{1,x} \sim Be(a_{1,x}, a_{2,x})$.

Table 1

Top: Natural sampling example with five classes, equal base rates g_i, and different symptom counts a_i and b_i; column m shows the means of the member distributions $Be(p, q)$ and column N the sum of $p + q$. N is approximately 50, or the sum of column a, in all classes. *Middle*: Same symptom information as on top, but assuming little epidemiological base rate knowledge (all $g_i = 2$) on the left side and assuming substantial epidemiological base rate knowledge (all $g_i = 40$) on the right side. *Bottom* Same symptom information as on top but class 3, 4, and 5 collapsed.

class	g	a	b	m	p	q	N
1	20	2	18	0.04	2.04	49.07	51.12
2	20	6	14	0.12	6.02	44.18	50.20
3	20	10	10	0.20	9.91	39.65	49.57
4	20	14	6	0.28	13.86	35.64	49.50
5	20	18	2	0.36	17.78	31.61	49.39

class	g	p	q	N	g	p	q	N
1	2	0.98	23.44	24.42	40	2.19	52.52	54.71
2	2	1.44	10.56	12.00	40	7.38	54.13	61.51
3	2	1.65	6.60	8.25	40	13.74	54.98	68.72
4	2	1.86	4.78	6.64	40	21.25	54.65	75.90
5	2	2.13	3.79	5.92	40	29.05	51.64	80.69

class	g	a	b	m	p	q	N
1	20	2	18	0.04	2.06	49.54	51.61
2	20	6	14	0.12	6.29	46.12	52.40
3-5	60	42	18	0.84	41.64	7.93	49.57

This is an exact result and not an approximation [11]. It demonstrates that in a representative case the beta approximation coincides with the exact distribution. It encourages the selection of the beta approximation for the general case. Furthermore, the result is interesting because it demonstrates that in an important case the base rates do not enter the first and second order classification probabilities. The result can easily be generalized to more than two classes and to multivalued symptoms.

5.2. Collapsing Classes

Numerical investigations show that the following principle holds:

Collapsing non-focus classes. The member distribution of the focus class c does not change if any of the non-focus classes in the set are collapsed.

Table 1 shows a numerical example. Consider class 1 in a set of five classes (top). For the base rate weigths $(20, 20, 20, 20, 20)$ its approximate member distribution is the beta $Be(2.04, 49.07)$ with $N = 51.12$. If we collapse the classes $3, 4$, and 5, that is if we consider the same class in a set of three, the distribution is practically unchanged (bottom): $Be(2.05, 49.17)$. The result shows that there are no context effects of partitioning or collapsing non-focus classes. It also justifies the simplification of complicated multi-class calculations by first collapsing all the non-focus classes (as we have done in [8]).

If in a multiway contingency table we observe low cell counts, we often collapse cells to increase the counts and the precision of related inferences. This is not true for member distributions. Inspection of Table 1 shows: If we collapse the focus class with other classes the result is difficult to predict. Often, collapsing does *not* increase the precision.

5.3. Non-Dirichletian weigth of evidence

The member probability $\mu_{c;x}$ gives the probability that an object with symptom x belongs to class c; $1 - \mu_{c;x}$ gives the probability that it belongs to any of the alternative classes. The weigths p and q of the according member distributions give the weigths of the pros and cons for class c. The sum of the weigths expresses the total knowledge about the focus class c and about the non-focus classes taken together. As we change the focus from one class to the next - what happens to the total weigth of evidence? In Tabel 1 we see that the sums of the weigths change. If, in a multiway contingency table, we add up marginals along different dimensions, we always come up with the same total. If the entries of the table are interpreted as the weigths of a Dirichlet distribution then all the marginals are also Dirichlet and their weigths are obtained by simply adding up the according entries. We say that a distribution for which these conditions hold has the Dirichlet property. Not so if we construct a table for the member weigths. The total knowledge depends upon the 'focus' through which we are looking at the system. The joint distribution of the member parameters $f(\mu_{11}, \ldots, \mu_{mk})$ strikingly violates the Dirichlet property. Collapsing categories does not allow adding weigths. There is only one condition, though, in which the system has exactly the Dirichlet property and this is natural sampling.

5.4. Popper effect

Can we give an interpretation to the direction of the deviations from the Dirichlet property? Let $m > 2$. In the case of little base rate knowledge low probability diagnoses are more precise than high probability diagnoses. In the case of substantial base rate knowledge high probability diagnoses are more precise than low probability diagnoses.

In practice, the case of little base rate knowledge is more important than the case of substantial base rate knowledge. We call the result 'Popper effect' because it resembles Popper's claim that in respect to a hypothesis we can learn more from the improbable than from the probable events. If improbable classifications are more precise than probable ones we learn from the same data more about the improbable diagnoses. The effect is illustrated in Figure 4 with data from Table 1 and the rather vague priors with all $g_i = 2$. Affirmative and non-affirmative evidence do not lead to the same precision. We first observed this effect for continuous symptoms [10, 9]. The Popper effect shows that the elimination of improbable alternatives is a safe strategy in the case where little is known about the 'epidemiology' of the diseases.

6. UPDATING AND PROPAGATION

In a system with many symptoms, the member distribution for *independent symptoms* can be obtained in two different ways: simultaneously or sequentially. The simultaneous

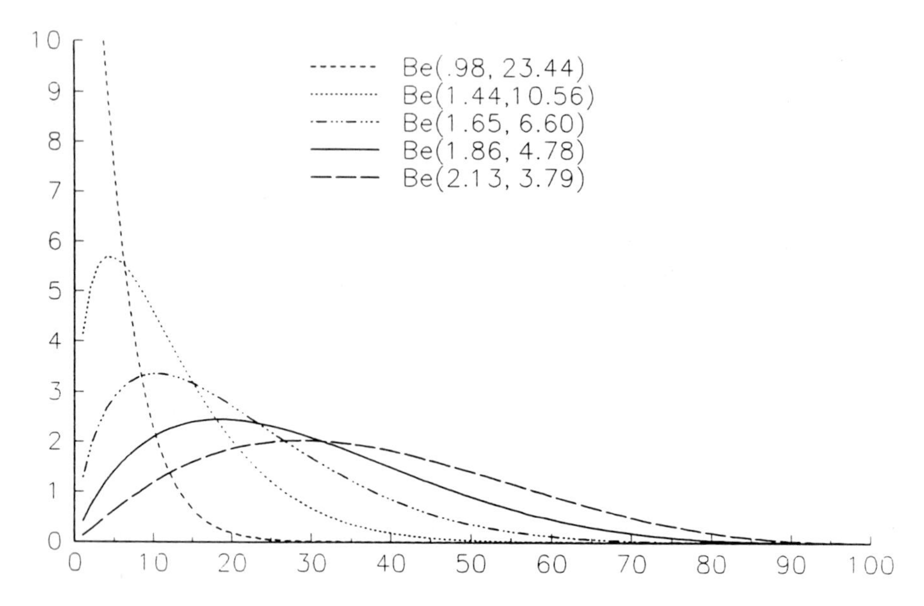

Figure 4. Beta member distributions for the numerical example in Table 1 (middle, on the left). The distributions correspond to the classes 1 to 5 from left to right. Distributions with low mean member probabilities (on the left) are tighter than those for high mean member probabilities (on the right).

method processes all the symptoms in a simultaneous multivariate analysis. The sequential method processes the symptoms one by one. Each stage starts with the member distribution of the previous one, combines it with the current symptom and produces the updated distribution for the next stage. The final result is just the outcome of the last stage when no symptoms are left over. The results of the simultaneous and the sequential method are the same and invariant in respect to the order of the sequential processing.

A necessary condition for the admissibility of the sequential updating scheme is that in each stage the input and output distributions belong to the same family of distributions. The condition for this property to be valid is that the distributions are *natural-conjugate*

Table 2
Two classes with two binary zero-correlated symptoms. The prior base rate distribution is assumed to be $Be(5,5)$.

class 1

	5: 1	2	Σ
1	12	6	18
2	6	3	9
Σ	18	9	27

class 2

	5: 1	2	Σ
1	4	16	20
2	8	32	40
Σ	12	48	60

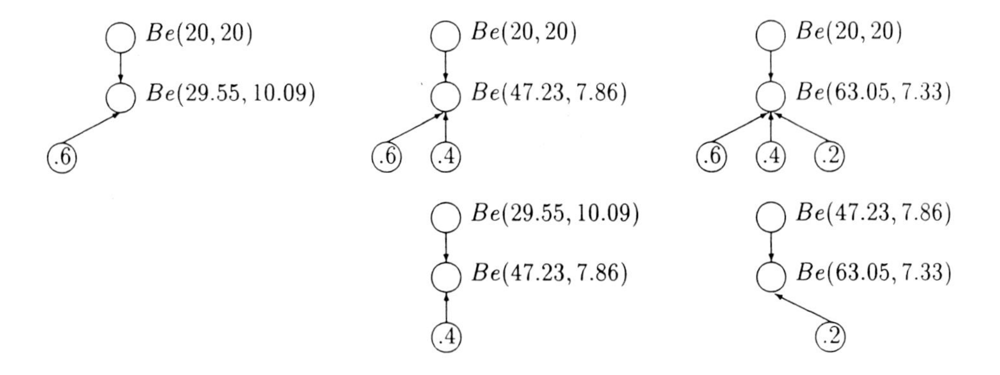

Figure 5. *Left*: Simultaneous revision of the prior $Be(20, 20)$ by one (top), two (middle), and three (bottom) independent normally distributed symptoms. The precision of the member distribution increases from the prior $Be(20, 20)$ to the final member distribution $Be(63.05, 7.33)$. *Right:* Sequential updating. The result of the first analysis $Be(29.55, 10.09$ is used as input for the second updating; the result of the second analysis $Be(47.23, 7.86)$ is used as input for the third updating. The final results of the simultaneous and the sequential methods are equivalent.

[18]. In the present case, both the input and the output distributions of each stage are beta distributions.

Sequential updating. The analysis of a set of symptoms can only be decomposed and performed sequentially if the symptoms are *conditionally independent* given the class node. Two or more subsets of symptom nodes can be updated sequentially if the focus node *separates* the subsets.

Table 2 contains an example with two 2×2 tables of independent binary symptoms. The prior knowledge is expressed by the beta $Be(5, 5)$. Simultaneous updating by the likelihood weigths (12, 15) and (4, 56) leads to the member distribution $Be(11.15, 1.67)$: $Be(5, 5) \rightarrow (12, 15; 4, 56) \rightarrow Be(11.15, 1.67)$. Sequential updating in the order 'prior $\rightarrow$ symptom 1 $\rightarrow$ symptom 2' gives $Be(5, 5) \rightarrow (18, 9; 20, 40) \rightarrow Be(6.57, 3.29) \rightarrow (12, 6; 4, 16) \rightarrow Be(11.27, 1.69)$. Updating in the order 'prior $\rightarrow$ symptom 2 $\rightarrow$ symptom 1' gives $Be(5, 5) \rightarrow (18, 9; 12, 48) \rightarrow Be(8.92, 2.68) \rightarrow (12, 6; 4, 8) \rightarrow Be(11.34, 1.70)$. The three procedures lead to practically identical results.

Figure 5 gives an example for three zero-correlated symptoms for the Normal case. Figure 6 shows the analysis for the same means and symptom values but for correlated symptoms. For the correlated symptoms the sequential updating is not be permissible.

7. PRECISION AS AN OCKAM RAZOR

Monotonically decreasing precision. In a discrete system with natural sampling the precision - measured by the sum of the member distribution weigths $p + q$ - of the inverse Bayes' probability can never exceed the precision of the base rates.

Under natural sampling the member distribution $Be(a_{c,x}, n_x - a_{c,x})$ is determined by

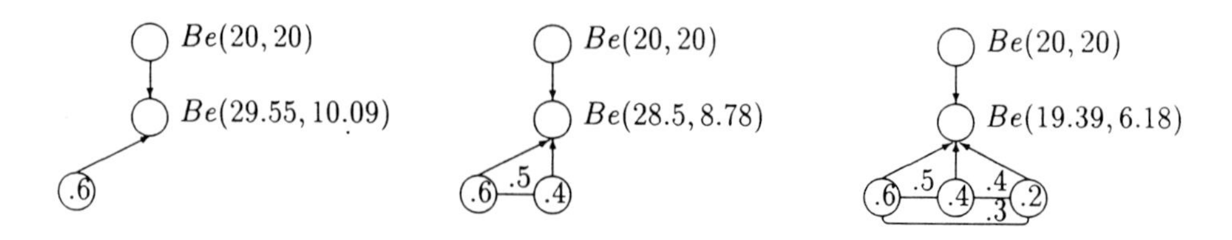

Figure 6. Simultaneous updating of the prior $Be(20,20)$ by one (top), two (middle), and three (bottom) correlated normal distributed symptoms. $n_1 = n_2 = 20$, variance of the Normal distributions in both classes = 1, means of all three symptoms in class one = 1, means of all three symptoms in class two = −1, in both classes $r_{12} = .5$, $r_{13} = .3$, and $r_{23} = .4$. The precision of the member distribution increases from the prior $Be(20,20)$ to $Be(28.5, 8.78)$ after the first symptom but then decreases to $Be(19.39, 6.18)$ after the third one.

the symptom weigths only; natural sampling is defined by $g_i = \sum_{x=0}^{1} a_{i,x}$ for $i = 1, \ldots, m$. Thus, $a_{c,x} \leq g_c$ and $n_x' - a_{c,x} \leq G - g_c$, both symptom weigths, can never exceed their parental base rate weigths. As a consequence, the precision of an inverse Bayes' probability is *monotonically decreasing* as new symptoms are appended to the single case description. While the inclusion of a new symptom may lead to a gain in the first order probability, it always induces a loss in precision. The principle holds for many situations close to natural sampling. Despite the fact that natural sampling is widespread (including all multiway contingency tables with random marginals) the principle of monotonically decreasing precision seems counter-intuitive and deserves a few comments. New information can be added to a knowledge system in different ways. Two frequently occuring possibilities are (i) the increase of the sample size and (ii) the refinement of the symptoms. In the first case we increase the number of observations or the amount of experience and keep the number of variables (dimensions, symptoms) constant; in the second case we increase the number of dimensions and keep the sample size constant. In the first case, we sometimes believe in the following 'principle of monotonically increasing precision': As new observations are added to the database of a statistical model the precision of its inferences always increases. Indeed, this 'principle' is false. It does not hold, e.g., for the investigation of a single proportion by relative frequencies, nor for the investigation of the mean of a normal distribution (as long as the sample variance is not held constant, a condition forbidden by constant probability sampling). In the second case, the consideration of additional variables in a multiway contingency table leads to a progressive subdivision of the table until the smallest cell counts of the symptoms are reached. Under natural sampling the precision of the member distribution then depends upon these cell counts and upon these counts only - the total sample size or any marginals are irrelevant. Note that we add symptoms while keeping the the total experience constant. It may happen that a further symptom is added that imports new information. A typical example is the inclusion of the results of a recent investigation in the field of interest or the inclusion of the knowledge of a new expert. In these cases, though, the precision increases because the knowledge base is changed.

Precision in continuous systems. The proposition of monotonically decreasing precision does not hold for continuous features. In continuous systems the inclusion of additional symptoms may increase or decrease the precision of the member distribution. In a discrete system with little knowledge (small cell counts) the classification of single cases is necessarily imprecise. Not so in continuous cases; the distance information contributes substantially to the precision and makes it possible to have precise inferences with small sample sizes.

The number of symptoms on which a single case diagnosis is based should be small. It does not make sense to include all logically possible symptoms into the analysis because this leads to a low precision of the final diagnosis. A pre-selection should be performed. It can be based on preferences for member distributions. In portfolio selection such criteria have been discussed since the classical work of [15]. Take as an example the following criterion proposed by [3]: Let M and M' be two member distributions with means E and E'; let L and L' be the lower limits of a confidence interval. Baumol proposed to prefer M to M' iff $E > E'$ and $L \geq L'$ (but not if $L < L'$). Keeping the number of variables small prevents the combinatorial explosion in inference systems induced by many symptoms that combine to an exponentially increasing number patterns. The precision may become a criterion for node reduction and pruning in graphical models [2]. The precision takes on the role of an Ockam razor.

It may well be true that there is no reason to complain about the limited capacity for the number of causal factors that can be processed by human cognition. In an environment with highly correlated cues the principles of probabilistic inference imply that inferences about single cases cannot be improved by including more than, say, the 'magical number seven' predictors. On the contrary, too many predictors can easily lead to a deterioration of a diagnostic system. Complex probabilistic inferences are imprecise. Small is not only beautiful but rational!

REFERENCES

1. Aitchison, J. and Dunsmore, I. R. (1975). Statistical Prediction Analysis. Cambridge: Cambridge University Press.
2. Baker, M. and Boult, T. E. (1991). Pruning Bayesian networks for effective computation. In P. P. Bonissone, M. Henrion, and L. N.Kanal and J. F. Lemmer (eds.), Uncertainty in Artificial Intelligence 6, Amsterdam: North-Holland, pp. 225 232.
3. Baumol, W. J. (1963). An expected gain-confidence limit criterion for portfolio selection, Management Science, 10, 174–182.
4. Bishop, Y. M., Fienberg, S. E., and Holland, P. W. (1975). Discrete Multivariate Analysis: Theory and Practice. Cambridge, MA: MIT Press.
5. Coletti, G. and Gilio, A. and Scozzafava, R. (1991). Conditional events with vague information in expert systems. In B. Bouchon-Meunier, R. R. Yager, and L.A. Zadeh (eds.), Uncertainty in Knowledge Bases, Berlin: Springer, pp. 106–114.
6. Fertig, K. W. and Breese, J. S. (1990). Interval influence diagrams. In M. Henrion, R. D. Shachter, L. N. Kanal, and J. F. Lemmer (eds.), Uncertainty in Artificial Intelligence 5, Amsterdam: North-Holland, 149–161.
7. Frydenberg, M. and Lauritzen, S.L. (1989). Decomposition of maximum likelihood in

mixed graphical interaction models, Biometrika, 76, 539–555.

8. Kleiter, G. D. (1992a). Bayesian Diagnosis in Artificial Intelligence. Artificial Intelligence, 54, 1–32.
9. Kleiter, G. D. (1992b). The precision of Bayesian single case classification. In Schader, M. (ed), Analyzing and Modeling Data and Knowledge. Berlin: Springer, pp. 47–54.
10. Kleiter, G. D. (in print a). The precision of Bayesian classification: the multivariate normal case. Journal of General Systems Theory.
11. Kleiter, G. D. (in print b). Natural sampling: rationality without base rates. In G. Fischer and D. Laming (eds.), Contributions to Mathematical Psychology, Psychometric, and Methodology. New York: Springer.
12. Kleiter, G. D. and Kardinal, M. (1993). A Bayesian approach to imprecision in belief nets. Institut für Psychologie, Universität Salzburg.
13. Lauritzen, S. L. (1990). Propagation of probabilities, means and variances in mixed graphical association models. Research Report R–90–18, Inst. Electronic Systems, Aalborg Univ.
14. Lauritzen, S.L. and Wermuth, N. (1989). Graphical models for association between variables, some of which are qualitative and some quantitative. Annals of Statistics, 17, 31–57.
15. Markowitz, H. (1959). Portfolio Selection. New York: Wiley.
16. Neapolitan, R. E. and Kenevan, J. R. (1991). Investigations of variances in belief networks. In B. D. D'Ambrosio, P. Smets, and P. P. Bonissone (eds.), Uncertainty in Artificial Intelligence. San Mateo, CA: Morgan Kaufmann, pp. 232–240.
17. Paaß, G. (1991). Second order probabilities for uncertain and conflicting evidence. P. P. Bonissone, M. Henrion, L. N.Kanal, and J. F. Lemmer (eds.), Uncertainty in Artificial Intelligence 6. San Mateo, CA: Morgan Kaufmann, pp. 447–456.
18. Raiffa, H. and Schlaifer, R. (1961). Applied Statistical Decision Theory. Boston: Harvard University.
19. Shachter, R. D. (1986). Evaluating influence diagrams. Operations Research, 34, 871–882.
20. Spiegelhalter, D. J., Franklin, R. C. G., and Bull, K. (1990) Assessment, criticism and improvement of imprecise subjective probabilities for a medical expert system. In M. Henrion, R. D. Shachter,L. N. Kanal, and J. F. Lemmer (eds.), Uncertainty in Artificial Intelligence 5. Amsterdam: North-Holland, pp. 285–294.
21. Spiegelhalter, D. J. (1991). A unified approach to imprecision and sensitivity of beliefs in expert systems. MRC Biostatistics Unit, Cambridge.

Uncertainty in Intelligent Systems
B. Bouchon-Meunier et al. (Editors)

On some kind of probabilistic relations*

J. Jacas and J. Recasens

Sec. Matemàtiques i Informàtica. E.T.S. d'Arquitectura de Barcelona. Univ. Politècnica de Catalunya. Diagonal,649 - 08028 Barcelona, Spain.

Abstract. Starting from some ideas by T.L. Saaty about consistent matrices a procedure in order to find one-dimensional probabilistic relations "close" to a given one is developed. This method is extended to T-transitive generalized equalities where T is a strict t-norm.

Keywords: T-indistinguishability operator, probabilistic relation, strict t-norm, dimension of a relation, consistent matrix.

1. INTRODUCTION

T-indistinguishability operators introduced by E. Trillas [11] extend the concept of equivalence relation to the fuzzy framework. They include *similarities* [13], *likeness* [8] and *probabilistic relations* [5].

In [12], L. Valverde proved a *Representation Theorem* for T- indistinguishability operators that generalizes a previous result of S. Ovchinikov for probabilistic relations [6].

Roughly speaking, the theorem states that a fuzzy relation defined on a set X is a T-indistinguishability operator if and only if it can be generated in a natural way by a family of fuzzy subsets of X.

Different families can generate the same T-indistinguishability operator E. The minimum of the cardinals of the indices of generating families of E is called its *dimension*.

*Research partially supported by DGICYT p.n. PB91-0334-CO3

If X is a finite set ($\#X = n$) and the dimension of E is "low", it will be more convenient to store a minimal generating family of E rather than the matrix of E. In this sense, it is useful to work with "low dimensional" T-indistinguishability operators.

In [2] it is proved that, if E is a similarity, then its dimension is reasonably low. Nevertheless, if T is an archimedean t-norm, the dimension of E can reach the value $n/2$ [7].

Therefore, instead of storing a high dimensional T-indistinguishability operator E, it will be preferable to select a minimal generating family of a low dimensional T-indistinguishability operator E' *close* to E.

In this paper, using some ideas of T. L. Saaty [9], we introduce a method of generating a one-dimensional probabilistic relation *close* to a given one.

2. PRELIMINARIES

Definition 2.1. (Menger) [5] A *probabilistic relation* E on a set X is a fuzzy relation that satisfies the following properties:

1) *Reflexivity*: $E(x,x) = 1, \forall x \in X$,
2) *Symmetry*: $E(x,y) = E(y,x), \forall x, y \in X$,
3) *Transitivity*: $E(x,y) \cdot E(y,z) \leq E(y,z), \forall x,y,z \in X$.

If $E(x,y) = 1$ implies $x = y$, then E is called a *separating* probabilistic relation.

Representation Theorem 2.2. (Ovchinikov, Valverde) [6,12]. A fuzzy binary relation E on a set X is a probabilistic relation if and only if there exists a family $\{h_i\}_{i \in I}$ of fuzzy subsets of X such that

$$E(x,y) = \operatorname*{Inf}_{i \in I} \left(\operatorname{Min} \left\{ \frac{h_i(x)}{h_i(y)}, \frac{h_i(y)}{h_i(x)} \right\} \right)$$

Definition 2.3. (Jacas, Ovchinikov) [1,6]. The *dimension* of a T indistinguishability operator E is the minimum of the cardinals of generating families of E.

After this definition, Theorem 2 can be reformulated by saying that every probabilistic relation can be generated by one-dimensional ones.

The next theorem gives a characterization of one dimensional probabilistic relations.

Characterization Theorem 2.4. (Jacas, Ovchinikov) [1,6]. Let E be a probabilistic relation on a set X such that $E(x,y) > 0$, $\forall x, y \in X$. E is one-dimensional if and only if there exists a total order $\leq$ in X such that $E(x,y) \cdot E(y,z) = E(x,z)$, with $x \leq y \leq z$.

From this theorem it follows that a one-dimensional separating probabilistic relation E on X defines a *total betweenness relation* B_E on X in the following way:

$(x,y,z) \in B_E$ if and only if $E(x,y) \cdot E(y,z) = E(x,z)$, and $x \neq y \neq z \neq x$.

Theorem 2.4 links one-dimensional probabilistic relations with *consistent reciprocal matrices* introduced by Saaty [9]. For the sake of completeness we recall the definition and the characterization of this kind of matrices.

Definition 2.5. (Saaty) [9]. An $n \times n$ real matrix A with entries $a_{ij} > 0$, $1 \leq i,j \leq n$, is *reciprocal* if and only if $a_{ij} = \dfrac{1}{a_{ji}}$, $\forall i,j = 1,2,\ldots,n$. A reciprocal $n \times n$ matrix is *consistent* if and only if $a_{ik} = a_{ij} \cdot a_{jk}$, $\forall i,j,k = 1,2,\ldots,n$.

Remark: It will be convenient to consider A as a map $A : X \times X \to \mathbb{R}^+$ with $X = \{x_1, x_2, \ldots, x_n\}$ and $a_{ij} = A(x_i, x_j)$.

Consistent reciprocal matrices can be characterized by the following theorem

Theorem 2.6 (Saaty) [9]. An $n \times n$ real matrix is both reciprocal and consistent if and only if there exists a fuzzy subset h of X such that

$$a_{ij} = \frac{h(x_i)}{h(x_j)} \quad \forall i,j = 1,2,\ldots,n.$$

The fuzzy subset h will be called a *generator* of A.

3. CONSISTENT MATRICES AND PROBABILISTIC RELATIONS

Given a reciprocal, consistent matrix A on X with $a_{ij} \neq 1$, $i \neq j$ a betweenness relation B_A on X can be defined as follows

$(x_i, x_j, x_k) \in B_A$ if and only if, for all $i,j,k = 1,2,\ldots,n$ such that $i \neq j \neq k \neq i$

, we have $a_{ik} = a_{ij} \cdot a_{jk}$

Theorem 3.1. The reciprocal consistent matrix A_h and the probabilistic relation E_h on X, both generated by the fuzzy subset h of X, define the same betweenness relation on X (e.g. $B_{A_h} = B_{E_h}$).

Proof. Let us order the elements of X in the following way: $x_i \leq x_j$ if and only if $h(x_i) \leq h(x_j)$. Then,

$$(x_i, x_j, x_k) \in B_E \leftrightarrow E(x_i, x_j)E(x_j, x_k) = E(x_j, x_k),$$

therefore

$$\frac{h(x_i)}{h(x_j)} \cdot \frac{h(x_j)}{h(x_k)} = \frac{h(x_i)}{h(x_k)} \leftrightarrow a_{ij}a_{ik} = a_{ik} \leftrightarrow (x_i, x_j, x_k) \in B_{A'}.$$

On the other hand, for a given reciprocal matrix A Saaty [9] obtains a consistent matrix A' *"close"* to the matrix A. The matrix A' is generated by an eigenvector associated to the greatest eigenvalue of A.

It is worth noting that if A is already consistent, then $A = A'$, and on the other hand, slight modificatíons on the entries of A produce also slight modifications on the entries of A'.

If E is close to a one-dimensional relation, then it "almost fulfils" Theorem 2.4. This suggests the following definition:

Definition 3.2. Given $\epsilon \in [0,1]$, a probabilistic relation $E : X \times X \to [0,1]$ is termed ϵ-one-dimensional if there exists a total order $(\leq)$ in X, such that for any x, y and z of X with $x \leq y \leq z$, the following inequality holds

$$|E(x,y) \cdot E(y,z) - E(x,z)| < \epsilon.$$

If E is an ϵ-one-dimensional probabilistic relation defined on a finite set X ($\#X = n$) and, $x_1 \leq x_2 \ldots \leq x_n$ the total ordering on X determined by E, then the following definition provides a useful relation between reciprocal matrices and probabilistic relations

Definition 3.3. The matrix $A : X \times X \to R^+$ defined by

$$A(x_i, x_j) = \begin{cases} E(x_i, x_j) & \text{if } \ i \leq j, \\ \dfrac{1}{E(x_i, x_j)} & \text{otherwise,} \end{cases}$$

is the ϵ-consistent reciprocal matrix associated to the ϵ-onedimensional probabilistic relation E.

4. THE PROCEDURE

First of all let us observe that, if A is a reciprocal, positive matrix, then the sum of its eigenvalues is n and, if A is consistent, then there exists a unique eigenvalue $\lambda_{\max} = n$ different from zero.

It is also interesting to point out that small variations in the entries of A give also small variations in its eigenvalues. Therefore, let us consider the ϵ-consistent matrix associated to an ϵ-one-dimensional probabilistic relation E. An eigenvector associated to the maximal eigenvalue generates a one-dimensional probabilistic relation E' that is " *ϵ-close*" to the given E.

This procedure is condensed in the following scheme:

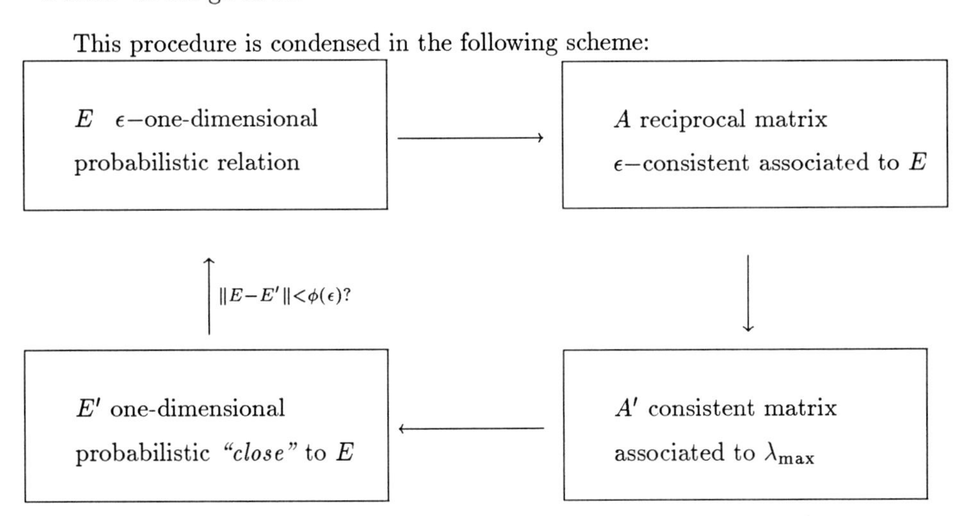

It is an open problem to determine the relation between de value of ϵ and the value $\|E - E'\|$ under a suitable norm.

Example 4.1. Let us consider the probabilistic relation

$$E = \begin{pmatrix} 1.00 & 0.74 & 0.67 & 0.50 & 0.41 \\ 0.74 & 1.00 & 0.87 & 0.65 & 0.53 \\ 0.67 & 0.87 & 1.00 & 0.74 & 0.60 \\ 0.50 & 0.65 & 0.74 & 1.00 & 0.80 \\ 0.41 & 0.53 & 0.60 & 0.80 & 1.00 \end{pmatrix} .$$

It is easy to prove that E is a 0.1-one-dimensional. Its associated 0.1-consistent reciprocal matrix is

$$A = \begin{pmatrix} 1.0000 & 1.3514 & 1.4925 & 2.0000 & 2.4390 \\ 0.7400 & 1.0000 & 1.1494 & 1.5385 & 1.8868 \\ 0.6700 & 0.8700 & 1.0000 & 1.3514 & 1.6667 \\ 0.5000 & 0.6500 & 0.7400 & 1.0000 & 1.2500 \\ 0.4100 & 0.5300 & 0.6000 & 0.8000 & 1.0000 \end{pmatrix},$$

its maximal eigenvalue is 5.0003 and an associated eigenvector

$$h = (1.00, 0.76, 0.67, 0.50, 0.40).$$

This fuzzy set generates the one-dimensional probabilistic relation

$$E' = \begin{pmatrix} 1.00 & 0.76 & 0.67 & 0.50 & 0.40 \\ 0.76 & 1.00 & 0.88 & 0.66 & 0.53 \\ 0.67 & 0.88 & 1.00 & 0.74 & 0.60 \\ 0.50 & 0.66 & 0.74 & 1.00 & 0.81 \\ 0.40 & 0.53 & 0.60 & 0.81 & 1.00 \end{pmatrix}$$

5. STRICT ARCHIMEDEAN T-NORMS

In this section, the preceding results will be generalized to T-indistinguishability operators, where T is a strict archimedean t-norm.

Let us first recall that a t-norm T is *archimedean* if and only if $T(x,x) < x$, $\forall x \in (0,1)$, and *strict* if and only if its set of nilpotent elements is empty [10].

A T-*indistinguishability operator* E on a set X is a reflexive and symmetric fuzzy relation on X that also satisfies the following triangular inequality:

$$T(E(x,y), E(y,z)) \leq E(y,z) \quad \forall x, y, z \in X.$$

It is well known that all strict archimedean t-norms T are isomorphic to the t-norm product Π (e.g.: there exists a continuous bijective map $f : [0,1] \to [0,1]$ such that $f \circ T = \Pi(f \times f)$).

It is also known that if E is a T-indistinguishability operator and T is isomorphic to T' via f, then $f \circ E$ is a T'-indistinguishability operator [1].

On the other hand, a representation theorem for T-indistinguishability operators that generalizes Theorem 2.2 is proved in [12]. Therefore it makes sense to talk about the dimension of a T-indistinguishability operator.

Taking into account all these results we can generalize the procedure defined in section 4 in order to get a one dimensional T-indistinguishability operator (T strict archimedean) close to a given one.

Procedure 5.1 Given a T-indistinguishability operator E on a set X where T is a strict archimedean t-norm, let $f \circ E$ be its isomorphic probabilistic relation. If $f \circ E$ is ϵ-one dimensional, then a one dimensional probabilistic relation E' can be obtained (section 4) close to $f \circ E$. The continuity of f assures us that $f^{-1} \circ E$ is a T-indistinguishability close to E. On the other hand, since the isomorphism preserves the dimension [3], $f^{-1} \circ E$ is one-dimensional. In this case we have the following scheme:

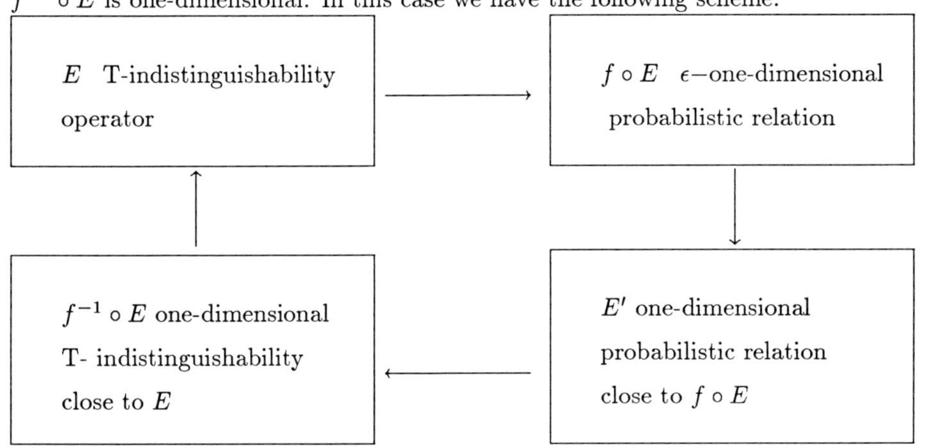

6. CONCLUDING REMARKS

Given a probabilistic relation E "almost one dimensional", a procedure in order to find a one dimensional probabilistic relation E' close to E is found by associating a reciprocal matrix A to E and using Saaty's techniques.

This procedure is easy to implement and can be useful since, when applicable, all the information contained in E' can be stored with fuzzy set (the eigenvector found in the procedure).

This technique is also applicable to T-indistinguishability operators (with T a strict archimedean t-norm).

If E is "far" from being one-dimensional, then it will be useful to find other techniques in order to get a "low-dimensional" probabilistic relation E' close to E.

In this case, E will almost satisfy the betweenness relation determined by E'.

To deal with this problem it seems therefore necessary a suitable definition of fuzzy betweenness in a more general frame.

Total fuzzy betweenness relations were introduced by Katz [4]. The study of general fuzzy betweenness relations will be the subject of a forthcoming paper by the authors.

REFERENCES

[1] Jacas, J., On the generator of T-indistinguishability operator. *Stochastica* XII-1, (1988), 49-63.

[2] Jacas, J., Similarity Relations-The Calculation of Minimal Generating Families. *Fuzzy Sets and Systems* (1990), 151-162.

[3] Jacas, J., Recasens, J., Eigenvectors and generators of fuzzy relations, *Proc. of FUZZ-IEEE'92*, San Diego (1992), 101-112.

[4] Katz, M., Inexact Geometry, *Notre Dame J. of Journal Logica* **21** (1980).

5. Menger, K., Probabilistic theory of relations. *Proc. Nat. Acad. Sci. USA* 37. (1951). 178-180.

[6] Ovchinikov, S., *Representations of transitive fuzzy relations* in H.J. Skala et al. Eds. Aspects of Vagueness (Reidel, Dordrecht (1984), 105-118.

[7] Recasens, J., *Sobre la representació i generació de relacions d'indistingibilitat.* Ph.D. Dissertation. Prog. Matemàtica Aplicada, UPC, Barcelona, 1992.

[8] Ruspini, E., *A theory of Mathematical Classification.* Ph. D. Dissertation. Dept. of System Science. School of Engineering. Univ. of California. Los Angeles (1977).

[9] Saaty, T.L., *The Analytic Hierarchy Process*, McGraw- Hill, (1980).

[10] Schweizer, B., Sklar, A., *Probabilistic Metric Spaces*, North-Holland, Amsterdam (1983).

[11] Trillas, E., Assaig sobre les relacions d'indistingibilitat. *Primer Congrés Català de Lògica Matemàtica.* Barcelona (1982), 51-59.

[12] Valverde, L., On the structure of F-indistinguishability operators.*Fuzzy Sets and Systems* **17** (1985), 313-328.

[13] Zadeh, L.A., Similarity relations and fuzzy orderings. *Inform. Sci.* 3, (1971), 177-200.

Uncertainty in Intelligent Systems
B. Bouchon-Meunier et al. (Editors)
1993 Elsevier Science Publishers B.V.

Description of conditional independence structures by means of imsets: a connection with product formula validity*

Milan Studený
Department of Decision Making Theory, Institute of Information Theory and Automation, Pod vodárenskou věží 4, 182 08 Prague 8, Czech republic.
(studeny@cspgas11.bitnet)

A new approach to mathematical description of structures of stochastic conditional independence, namely by means of so-called imsets, is presented (imset is an abbreviation for **i**nteger-valued **multiset**). It is shown how it is related to the "classical" approach, namely by means of dependency models or semigraphoids. The main result consists in the theorem saying that a probability measure has certain conditional independence structure (CI-structure) if and only if it satisfies the corresponding product formula.

AMS classification : 68T 30 , 62B 10.

INTRODUCTION

The interest in stochastic conditional independence takes its origin from the theory of probabilistic expert systems. To put it shortly, any conditional independence statement can be interpreted as a certain qualitative relationship among symptoms. Therefore there exists a possibility to determine a proper structural model of the probabilistic expert system which is easy to understand. In fact, in the theoretical background of various approaches to qualitative description of probabilistic models (influence diagrams , Markov nets) the concept of *conditional independence* (CI) is hidden. To the best of my knowledge, the importance of CI for probabilistic expert systems was at first highlighted by Pearl [16] but there exist other approaches which more or less explicitly deal with CI [17,18,19,20,26].

The "classical" approach to description of CI-structures (if graphical approaches are omitted) used the concept of dependency model or of semigraphoid [15]. It motivated attempts at "axiomatization" of CI i.e. to characterize relationships among CI-statements in a simple syntactic way. Nevertheless, as proved in [23] there exists no simple dimension-independent deductive system describing relationships among CI-statements. This fact motivated a new approach to description of CI-structures [25], namely by means of imsets. It promises to remove the above mentioned drawbacks. The aim of this article is to give another view on this type of description of CI-stuctures by showing that the complying with such a model of CI-structure is equivalent to the validity of certain product formula. This equivalent definition could make the interpretation of these models more natural.

*This work was made in frame of the project "Explanatory power of probabilistic expert systems: theoretical backgroud" , internal grant of Czechoslovak Academy of Sciences, n.27510.

NOTATION AND BASIC ARRANGEMENTS

Throughout this paper the following situation will be delt with:
A finite set N having at least two elements called the *basic set* is given, i.e. $2 \leq card\, N < \infty$. Having disjoint sets $A, B \subset N$ the juxtaposition AB will stand for their union $A \cup B$ in many examples. The class of all subsets of N will be denoted by $exp\, N$, the class of its *nontrivial subsets* i.e. subsets having at least two elements will be denoted by $\mathcal{U}$: $\quad \mathcal{U} = \{S \subset N,\ card\, S \geq 2\}$.
Having a set $T \subset N$ its *identificator* i.e. the zero-one function on $exp\, N$ (possibly restricted to $\mathcal{U}$) is defined as follows:

$$\delta_T(S) = \begin{cases} 1 & \text{in case } S = T \\ 0 & \text{in case } S \neq T. \end{cases}$$

Having a collection of nonempty finite sets $\{X_i\ ;\ i \in N\}$, an element of the corresponding cartesian product $x \in \prod_{i \in N} X_i$ and a set $\emptyset \neq S \subset N$ the projection of x to $\prod_{i \in S} X_i$ will be denoted by x_S:

$$x_S \in \prod_{i \in S} X_i \quad \text{is specified by} \quad [\ \forall i \in S\ \ (x_S)_i = x_i\ ].$$

Having a pair a, b of real functions on a finite set X their *scalar product* will be denoted by $\langle a, b \rangle$:

$$\langle a, b \rangle = \sum_{x \in X} a(x) \cdot b(x).$$

The set of integers will be denoted by $\mathbb{Z}$, the set of nonnegative integers (including zero) by $\mathbb{Z}^+$, the set of (strictly) positive integer (sometimes called natural numbers) by $\mathbb{N}$ and the set of real numbers by $\mathbb{R}$.
For the sake of brevity, having a probability measure Q on Y (or a function on $exp\, Y$) and $y \in Y$ the symbol $Q(y)$ will be often used instead of $Q(\{y\})$.
Having a function $w : exp\, N \to \mathbb{Z}$ its positive and negative parts will be denoted by w_+ and w_-:

$$w_+(S) = max\, \{w(S), 0\} \qquad \text{for } S \subset N$$
$$w_-(S) = max\, \{-w(S), 0\} \qquad \text{for } S \subset N.$$

1. THE PRODUCT FORMULA GIVEN BY AN IMSET

This section contains basic definitions and the corresponding comments only. Namely, the concepts of imset, its natural extension and finite-domain probability measure are specified. Finally, the product formula given by an imset is introduced and a simple example given.

Def. 1 (imset)
Every integer-valued function on $\mathcal{U}$ is called *imset* (*on* $\mathcal{U}$).
The class of all imsets will be denoted by $Z(\mathcal{U})$.
Nonnegative imsets will be called *multisets*, their class will be denoted by $Z^+(\mathcal{U})$.
Basic operations with imsets like summing, subtracting and multiplying by integers are defined coordinatewisely. An imset $u \in Z(\mathcal{U})$ is called *normalized* iff the collection of numbers $\{u(S); S \in \mathcal{U}\}$ has no common prime divisor.

Trivial examples of imsets are *zero imset* (a function ascribing zero to each set from $\mathcal{U}$) denoted by 0 and identificators for $T \in \mathcal{U}$.

Remark The term multiset is borrowed from Aigner's book about combinatorial theory [1] while the word imset is our abbreviation for **i**nteger-valued **m**ulti**set**.

In some cases it will be convenient to regard imsets on $\mathcal{U}$ as functions on $exp\,N$. The correctness of the definition of the right extension is based on the following lemma; its proof is left to the reader.

Lemma Every imset $u \in Z(\mathcal{U})$ has uniquely determined extension
$\overline{u} : exp\,N \to \mathbb{Z}$ satisfying the following two conditions:
$(N.1)$ $\sum\{\overline{u}(S); S \subset N\} = 0$
$(N.2)$ $\forall r \in N \quad \sum\{\overline{u}(S);\ r \in S \subset N\} = 0$.
This adjudgement defines a one-to-one correspondence between $Z(\mathcal{U})$ and the class of integer-valued functions on $exp\,N$ satisfying $(N.1) - (N.2)$.

Def. 2 (natural extension)
Having an imset u on $\mathcal{U}$ the uniquely determined integer-valued function $\overline{u}$ on $exp\,N$ extending it and satisfying the normalization conditions $(N.1) - (N.2)$ will be called the *natural extension of* u. It will be always denoted by $\overline{u}$ (overline the original symbol).

The focus of our study are CI-structures of finite number of random variables (i.e. random systems). Nevertheless, in this article we limit ourselves to finite-valued random variables. It is a common custom in literature to allude to random variables but in fact deal with probability measures, namely their distributions. In the sequel, we decided both to allude to and deal with probability measures.

Now, the relevant class of probability measures will be specified. As they serve as distributions of random systems indexed by the basic set N their domains are cartesian products indexed by N.

Def. 3 (probability measure over N)
A *probability measure over* N (with finite domain) is specified by a collection of nonempty finite sets $\{X_i\ ;\ i \in N\}$ and by a probability measure on the cartesian product $\prod_{i \in N} X_i$.
Whenever $\emptyset \neq S \subsetneq N$ the *marginal measure of* P is the probability measure P^S on $\prod_{i \in S} X_i$ defined by: $P^S(A) = P(A \times \prod_{i \in N \setminus S} X_i) \qquad$ whenever $A \subset \prod_{i \in S} X_i$.
It will be always denoted by the symbol of the original measure endowed with the upper index identifying the marginal space. Moreover, the marginal measure P on $\prod_{i \in N} X_i$ is introduced as P itself, i.e. $P^N \equiv P$.

As mentioned above every assumption concerning CI-stucture of a probability measure is equivalent to the validity of certain product formula. Now, we are going to explain how these formulas look.

Def. 4 (product formula given by an imset)
Let P be a probability measure over N and u be an imset on $\mathcal{U}$. Say that P *satisfies the product formula given by* u iff it holds:

$$\forall x \in \prod_{i\in N} X_i \quad \prod_{\emptyset\neq S\subset N} (P^S(x_S))^{\overline{u}_+(S)} = \prod_{\emptyset\neq S\subset N} (P^S(x_S))^{\overline{u}_-(S)} \tag{1.1}$$

The condition (1.1) is the above mentioned formula.

Remark Another possible way how to write product formulas is to introduce the conventional symbol $P^{\emptyset}(x_{\emptyset})$ for 1 and write instead of (1.1) :

$$\forall x \in \prod_{i\in N} X_i \quad \prod_{S\subset N} (P^S(x_S))^{\overline{u}_+(S)} = \prod_{S\subset N} (P^S(x_S))^{\overline{u}_-(S)} \tag{1.2}$$

To illustrate this concept a simple example is given.

Example 1
Consider $S, T \in \mathcal{U}$ with $S\cap T = \emptyset$ and $S\cup T = N$. Put $u = \delta_N - \delta_S - \delta_T$. Then $\overline{u} = u + \delta_{\emptyset}$ and the product formula given by u looks:

$$\forall x \in \prod_{i\in N} X_i \quad P(x) = P^S(x_S)\cdot P^T(x_T) \tag{1.3}$$

2. DESCRIPTION OF CI-STRUCTURES

In this section the concept of CI is recalled and the "classical" ways to description of CI-structures are mentioned, especially by means of dependency models. The reason to develop the new approach to description of CI-structures from [25] is explained.

Def. 5 (conditional independence)
Let P be a probability measure over N and $\langle A, B, C\rangle$ is a triplet of pairwise disjoint subsets of N where A and B are nonempty. Say that A is *conditionally independent of* B *given* C *in* P and write $A \perp B|C\ (P)$ iff

$$\forall x \in \prod_{i\in N} X_i \quad P^{ABC}(x_{ABC})\cdot P^C(x_C) = P^{AC}(x_{AC})\cdot P^{BC}(x_{BC}).$$

Another phrase " *P obeys the triplet* $\langle A, B, C\rangle$ " will be often used in the sequel.

There are many equivalent formulations of the statement $A \perp B|C\ (P)$, for example:

- $\forall a, \tilde{a} \in \prod_{i\in A} X_i \;\; b, \tilde{b} \in \prod_{i\in B} X_i \;\; c \in \prod_{i\in C} X_i$
  $P^{ABC}([a,b,c])\cdot P^{ABC}([\tilde{a},\tilde{b},c]) = P^{ABC}([a,\tilde{b},c])\cdot P^{ABC}([\tilde{a},b,c])$
- there exist functions $f : \prod_{i\in AC} X_i \to \mathbb{R}$ and $g : \prod_{i\in BC} X_i \to \mathbb{R}$ such that
  $\forall x \in \prod_{i\in N} X_i \;\; P^{ABC}(x_{ABC}) = f(x_{AC})\cdot g(x_{BC})$
- $\forall c \in \prod_{i\in C} X_i$ with $P^C(c) > 0$ the conditional probability $P_{AB|C}(\cdot|c)$ is a product measure on $(\prod_{i\in A} X_i) \times (\prod_{i\in B} X_i)$.

The last condition leads directly to the common interpretation of $A \perp B|C$: "getting know the values of variables from C the variables from A and B become independent i.e. their probabilistic pieces of information become unrelevant". Thus, the information about CI-structure can be obtained from experts too.

To approach CI to human understanding the CI-structures were usually described by means of graphs in literature. Two trends are distinguishable: by means of undirected graphs (this stems from Markov field theory, the corresponding graph is called **Markov net** [2,6,8,12] and by means of directed acyclic graphs (the long tradition started by geneticist S. Wright [28] led to the concepts of **influence diagram** [5,17,18,19] and **recursive models** [7,27]. Nevertheless both graphical approaches cannot describe all possible probabilistic CI-structures.

Thus, another natural way was proposed: to describe a CI-structure simply by the list of valid CI-statements (i.e. triplets obeyed by the corresponding probability measure). This led to the concept of dependency model introduced by Pearl and Paz [15]; their definition is slightly modified here:

Def. 6 (dependency model, model of CI-structure)
a) Denote by $T(N)$ the set of triplets $\langle A, B, C \rangle$ of pairwise disjoint subsets of N where A and B are nonempty. Every subset of $T(N)$ will be called a *dependency model over* N.
b) Let P be a probability measure over N and I a dependency model over N. Say that I is a *submodel of* CI-*structure of* P iff P obeys every triplet from I.
Further, say that I is the *model of* CI-*structure of* P iff I is exactly the set of triplets obeyed by P.

This terminology emphasizes the presented view on dependency models. Note that authors dealing with dependency models have used also various another phrases:
"I is induced by P" in [26], "P is perfect for I" in [4], "I is conditional independence relation corresponding to P" in [22].

Owing to well-known properties of CI (treated by Dawid [3] resp. Spohn [21] resp. Smith [19]) some of dependency models cannot serve as (complete) models of CI-structures. Therefore Pearl and Paz [15] introduced the concept of **semigraphoid** (= dependency model satisfying the above mentioned properties) to describe CI-structures. As semigraphoids were defined as dependency models closed under 4 inference rules (called axioms by Pearl and Paz) it gives a deductive mechanism to infer valid consequences of input information about CI-structure.

Unfortunately, the original hypothesis from [16] that semigraphoids coincide with the (complete) models of CI-structures appeared untrue [22]. Later, we even found that models of CI-structures cannot be described as dependency models closed under finite number of inference rules [23]. This was strengthened by Geiger and Pearl [4] who showed that "disjunctive" inference rules cannot bring help.

These results led us to an attempt to develop an alternative way to description of CI-structures, namely by means of faces and imsets [25]. The aim of this paper is to give an equivalent view on "imsetal" models of CI-structures which brings an easier interpretation.

3. INFORMATION-THEORETICAL APPROACH

This section recalls the information-theoretical concepts of entropic and multiinformation function and indicates how they enable to describe CI-structures by means of imsets. An initial connection with product formulas is established.

Entropic and multiinformation functions are real functions on $exp\,N$. In fact, the value of entropic function for a set S is the entropy of the marginal measure P^S while the value of multiinformation function is the relative entropy of P^S with respect to the product of its one-dimensional marginals.

Def. 7 (entropic function, multiinformation function)
Let P a probability measure over N.
Its *entropic function* $H : exp\,N \to \mathbb{R}$ is defined as follows:

$$H(\emptyset) = 0$$
$$H(S) = \sum_{x,P(x)>0} P(x) \cdot ln\,(1/P^S(x_S)) \qquad \text{for } \emptyset \neq S \subset N.$$

Its *multiinformation function* $M : exp\,N \to \mathbb{R}$ is defined by:

$$M(\emptyset) = 0$$
$$M(S) = \sum_{x,P(x)>0} P(x) \cdot ln\,(P^S(x_S) / \prod_{i \in S} P^{\{i\}}(x_i)) \qquad \text{for } \emptyset \neq S \subset N.$$

The restriction of the multiinformation function M to $\mathcal{U}$ will be denoted by m.

Remark Multiinformation generalizes the well-known information-theoretical concept of mutual information and thus it serves as a quantitative characteristic of level of stochastic dependence of more than two random variables. This view led us to accept the name "multiinformation" in [22]. Another name "entaxy" was used in [9].

Now, some properties of these functions are mentioned. They explain why these functions are good tools for study of CI. Firstly, a simple computation gives:

$$M(S) = -H(S) + \sum_{i \in S} H(i) \quad \text{whenever } S \subset N \qquad (\text{of course } \sum_{i \in \emptyset} H(i) = 0) \tag{3.1}$$

Moreover, it is shown in [22] §4,5:

$$M(ABC) + M(C) \geq M(AC) + M(BC) \qquad \text{whenever } \langle A, B, C \rangle \in T(N) \tag{3.2}$$

and

$$M(ABC) + M(C) = M(AC) + M(BC) \;\text{ iff }\; A \perp B|C\ (P) \quad \text{for } \langle A, B, C \rangle \in T(N) \tag{3.3}$$

The idea of application of multiinformation function was the main step in the proof of validity of new properties of CI in [22] and [23]. Nevertheless, the connection of entropic and multiinformation functions with CI was recognized earlier - see [9] and [13].

As concerns the ability to describe CI for probability measures with finite domain the entropic and multiinformation functions are equivalent (see below). Nevertheless, their further capacities differ. Entropic function can be also used to describe functional dependencies hidden in a discrete probability measure [11] while multiinformation function can be applied to study of CI for continous or "mixed" probability measures [22].

Lemma 1
Let P be a probability measure over N and u be an imset. Then the following equalities are equivalent:
(a) $\langle m, u \rangle = 0$
(b) $\langle M, \overline{u} \rangle = 0$
(c) $\langle H, \overline{u} \rangle = 0$.

Proof: (a)$\Leftrightarrow$(b) is evident as $M(S) = 0$ for $S \in exp\, N \setminus \mathcal{U}$. To see (b)$\Leftrightarrow$(c) simply write using (3.1) and (N.2): $\langle M, \overline{u} \rangle = \sum_{S \subset N} M(S) \cdot \overline{u}(S) = \sum_{S \subset N} \{-H(S) + \sum_{i \in S} H(i)\} \cdot \overline{u}(S) =$
$= -\sum_{S \subset N} H(S) \cdot \overline{u}(S) + \sum_{S \subset N} \sum_{i \in S} H(i) \cdot \overline{u}(S) = -\langle H, \overline{u} \rangle + \sum_{j \in N} \sum_{T \subset N, j \in T} H(j) \cdot \overline{u}(T) =$
$= -\langle H, \overline{u} \rangle + \sum_{j \in N} H(j) \cdot \{ \sum_{T, j \in T} \overline{u}(T) \} = -\langle H, u \rangle.$ ■

Def. 8 (probability measure complies with imset)
Let P be a probability measure over N and u an imset on $\mathcal{U}$. Say that P *complies with* u iff any of the conditions (a) - (c) from Lemma 1 is fulfilled.

The above defined concept is related to the product formula validity as follows:

Lemma 2
Let P be a probability measure over N and u be an imset on $\mathcal{U}$. Consider the following conditions:
(a) P satisfies the product formula given by u
(b) $\forall\, x \in \prod_{i \in N} X_i$ with $P(x) > 0$ it holds: $\prod_{S \subset N} P^S(x_S)^{\overline{u}(S)} = 1$
(c) P complies with u.
Then (a) $\Rightarrow$ (b) $\Rightarrow$ (c).

Proof: (a)$\Rightarrow$(b) is evident. To show (b)$\Rightarrow$(c) consider some $x \in \prod_{i \in N} X_i$ with $P(x) > 0$ and write using properties of logarithm:
$\sum_{S \subset N} \overline{u}(S) \cdot ln\, P^S(x_S) = \sum_{S \subset N} ln\, P^S(x_S)^{\overline{u}(S)} = ln \prod_{S \subset N} P^S(x_S)^{\overline{u}(S)} = ln\, 1 = 0.$
By multiplying these equalities by $P(x)$ and summing over all such $x \in \prod_{i \in N} X_i$ get :
$0 = \sum_{x, P(x) > 0} P(x) \cdot \sum_{S \subset N} \overline{u}(S) \cdot ln\, P^S(x_S) = \sum_{x, P(x) > 0} \sum_{S \subset N} P(x) \cdot \overline{u}(S) \cdot ln\, P^S(x_S) =$
$= \sum_{S \subset N} \overline{u}(S) \cdot \sum_{x, P(x) > 0} P(x) \cdot ln\, P^S(x_S) = \sum_{S \subset N} -\overline{u}(S) \cdot H(S) = -\langle H, \overline{u} \rangle$
, i.e. the condition (c) from Lemma 1 holds. ■

4. STRUCTURAL IMSETS

The class of structural imsets is introduced in this section and the corresponding dependency model is defined for every such imset. Then it is shown that complying a probability measure with a structural imset introduced in the preceding section can be interpreted as partial description of the CI-structure (namely by means of the corresponding dependency model). Special attention is devoted to the question how to recognize structural imsets. Last result says that any possible model of CI-structure can be completely described in such a way by a structural imset.

The class of all imsets on $\mathcal{U}$ is too wide for our purposes. Certain subclass will be used to describe CI-structures. These imsets, called structural, can be introduced as "combinations" of so-called elementary imsets defined below.

Def. 9 (elementary imset)
An imset $u \in Z(\mathcal{U})$ is called *elementary* iff its natural extension has the form:
$\overline{u} = \delta_{S \cup T} - \delta_S - \delta_T + \delta_{S \cap T}$ where $S, T \subseteq N$ $card\, S \setminus T = card\, T \setminus S = 1$.
The set of elementary imsets will be denoted by E.

The following example illustrates this concept in case $card\, N = 4$.

Example 2
Suppose that $N = \{1,2,3,4\}$. By definition every elementary imset is "produced" by a couple $[S,T]$, necessarily $1 \leq card\, S = card\, T \leq card\, N - 1$.
Thus, elementary imsets can be naturally divided into classes according to the cardinality of "producing" sets. In the considered case three classes can be distinguished:

$E_1 \ldots$ i.e. $card\, S = card\, T = 1$
for instance $S = \{1\}$ and $T = \{2\}$ gives $\overline{u} = \delta_{\{1,2\}} - \delta_{\{1\}} - \delta_{\{2\}} + \delta_{\emptyset}$
and hence $u = \delta_{\{1,2\}}$.
The corresponding list follows :
$\delta_{\{1,2\}}, \delta_{\{1,3\}}, \delta_{\{2,3\}}, \delta_{\{1,4\}}, \delta_{\{2,4\}}, \delta_{\{3,4\}}$.

$E_2 \ldots$ i.e. $card\, S = card\, T = 2$
for instance $S = \{1,2\}$ and $T = \{2,3\}$ gives $\overline{u} = \delta_{\{1,2,3\}} - \delta_{\{1,2\}} - \delta_{\{2,3\}} + \delta_{\{2\}}$
and hence $u = \delta_{\{1,2,3\}} - \delta_{\{1,2\}} - \delta_{\{2,3\}}$.
The corresponding list follows :
$\delta_{\{1,2,3\}} - \delta_{\{1,2\}} - \delta_{\{2,3\}}, \delta_{\{1,2,3\}} - \delta_{\{1,2\}} - \delta_{\{1,3\}}, \delta_{\{1,2,3\}} - \delta_{\{1,3\}} - \delta_{\{2,3\}},$
$\delta_{\{1,2,4\}} - \delta_{\{1,2\}} - \delta_{\{2,4\}}, \delta_{\{1,2,4\}} - \delta_{\{1,2\}} - \delta_{\{1,4\}}, \delta_{\{1,2,4\}} - \delta_{\{1,4\}} - \delta_{\{2,4\}},$
$\delta_{\{1,3,4\}} - \delta_{\{1,3\}} - \delta_{\{1,4\}}, \delta_{\{1,3,4\}} - \delta_{\{1,3\}} - \delta_{\{3,4\}}, \delta_{\{1,3,4\}} - \delta_{\{1,4\}} - \delta_{\{3,4\}},$
$\delta_{\{2,3,4\}} - \delta_{\{2,3\}} - \delta_{\{2,4\}}, \delta_{\{2,3,4\}} - \delta_{\{2,3\}} - \delta_{\{3,4\}}, \delta_{\{2,3,4\}} - \delta_{\{2,4\}} - \delta_{\{3,4\}}$.

$E_3 \ldots$ i.e. $card\, S = card\, T = 3$
for instance $S = \{1,2,3\}$ and $T = \{1,2,4\}$ gives
$u = \overline{u} = \delta_N - \delta_{\{1,2,3\}} - \delta_{\{1,2,4\}} + \delta_{\{1,2\}}$.
The corresponding list follows :
$\delta_N - \delta_{\{1,2,3\}} - \delta_{\{1,2,4\}} + \delta_{\{1,2\}},$
$\delta_N - \delta_{\{1,2,3\}} - \delta_{\{1,3,4\}} + \delta_{\{1,3\}},$
$\delta_N - \delta_{\{1,2,3\}} - \delta_{\{2,3,4\}} + \delta_{\{2,3\}},$
$\delta_N - \delta_{\{1,2,4\}} - \delta_{\{1,3,4\}} + \delta_{\{1,4\}},$
$\delta_N - \delta_{\{1,2,4\}} - \delta_{\{2,3,4\}} + \delta_{\{2,4\}},$
$\delta_N - \delta_{\{1,3,4\}} - \delta_{\{2,3,4\}} + \delta_{\{3,4\}}$.

Thus the total number of elementary imsets is 24 in this case.

It makes no problem to give the formula for total number of elementary imsets: $card\, N \cdot (card\, N - 1) \cdot 2^{card\, N - 3}$.

Hint: the couple S, T can be characterized by the set $(S \setminus T) \cup (T \setminus S)$ of cardinality 2 and by the intersection $S \cap T$ i.e. a subset of the complement.

Def. 10 (structural imset)
An imset $u \in Z(\mathcal{U})$ will be called *structural* iff it holds:

$$\exists n \in \mathbb{N} \quad k_v \in \mathbb{Z}^+ \text{ (for } v \in E) \qquad n \cdot u = \sum_{v \in E} k_v \cdot v \tag{4.1}$$

The following lemma enables to identify CI-statements with structural imsets and to ensure the correctness of further definition.

Lemma Whenever $\langle A, B, C \rangle \in T(N)$ then the imset $u \in Z(\mathcal{U})$ determined by its natural extension $\overline{u} = \delta_{ABC} - \delta_{AC} - \delta_{BC} + \delta_C$ is a structural imset.

Hint: This can be shown by induction according to $card\, AB$. Whenever $card\, AB = 2$ then u is an elementary imset. In case $card\, A \geq 2$ chose $x \in A$ and "extend" $\overline{u}$ by $\pm(\delta_{ABC \setminus \{x\}} - \delta_{AC \setminus \{x\}})$, however in case $card\, A = 1$ take $x \in B$.

Def. 11 (dependency model corresponding to imset)
a) To every triplet $\langle A, B, C \rangle \in T(N)$ assign the structural imset denoted by $i(\langle A, B, C \rangle)$ and specified by its natural extension $\delta_{ABC} - \delta_{AC} - \delta_{BC} + \delta_C$.
b) Let u be a structural imset. The *dependency model corresponding to* u denoted by I_u is defined as follows:
$\langle A, B, C \rangle \in I_u$ iff [$\exists n \in \mathbb{N}$ $n \cdot u - i(\langle A, B, C \rangle)$ is a structural imset].

Remark that dependency models corresponding to structural imsets are called *structural semigraphoids* in [25].

Lemma 3
Let P be a probability measure over N and u a structural imset on $\mathcal{U}$. Then the following two conditions are equivalent:
(a) P complies with u
(b) I_u is a submodel of CI-structure of P.

Proof: Recall that M is the multiinformation function for P and m its restriction to $\mathcal{U}$.
I. $\langle m, v \rangle \geq 0$ whenever v is a structural imset.
By (3.2) the inequality holds for elementary imsets, then use (4.1).
II. $\langle m, u \rangle = 0 \Rightarrow$ (b).
Consider $\langle A, B, C \rangle \in I_u$, take the structural imset $n \cdot u - i(\langle A, B, C \rangle)$ with $n \in \mathbb{N}$ and write: $0 = \langle m, n \cdot u \rangle = \langle m, n \cdot u - i(\langle A, B, C \rangle) \rangle + \langle m, i(\langle A, B, C \rangle) \rangle$.
Owing to I. both terms on the right-hand side are nonnegative and therefore they vanish. Thus $0 = \langle m, i(\langle A, B, C \rangle) \rangle = \langle M, \overline{i(\langle A, B, C \rangle)} \rangle$ gives by (3.3) $A \perp B | C\ (P)$.

III. (b) $\Rightarrow$ $\langle m, u \rangle = 0$.
By Def. 10 write $n \cdot u = \sum_{v \in E} k_v \cdot v$ with $n \in \mathbb{N}$, $k_v \in \mathbb{Z}^+$. Clearly, it suffices to show $\langle m, v \rangle = 0$ for each $v \in E$ with $k_v > 0$. For this purpose find $\langle A, B, C \rangle \in T(N)$ with $v = i(\langle A, B, C \rangle)$. By Def. 11 $\langle A, B, C \rangle \in I_u$ and by (b) get $A \perp B | C \ (P)$; hence by (3.3) derive $\langle m, v \rangle = \langle M, \overline{v} \rangle = 0$. ■

The following question arises in connection with computer implementation of structural imsets: how to recognize whether an imset is structural? The presented definition of structural imset is not suitable for solving this problem. Nevertheless, structural imsets can be characterized in another more appropriate way. To formulate it some concept has to be introduced.

Def. 12 (completely convex set function)
A set function $c : \mathcal{U} \to \mathbb{R}$ is called a *completely convex set function* iff its settled extension $\underline{c}$ (i.e. $\underline{c}(T) = 0$ for $T \in exp\, N \backslash \mathcal{U}$) satisfies the convexity condition: $\underline{c}(K \cup L) + \underline{c}(K \cap L) \geq \underline{c}(K) + \underline{c}(L)$ whenever $K, L \subset N$.

Remark The adjective 'convex' is borrowed from game theory [14] while the adverb 'completely' indicates that the convexity condition concerns the extension.

Assertion 1
a) Let C denotes the class of completely convex set functions. Whenever u is an imset, then it holds: [u is structural] iff [$\forall c \in C \ \ \langle c, u \rangle \geq 0$].
Moreover, C is the largest class satisfying the previous condition.
b) There exists the least finite set of normalized imsets A such that for each imset $u \in Z(\mathcal{U})$ it holds: [u is structural] iff [$\forall a \in A \quad \langle a, u \rangle \geq 0$].
(According to the first part necessarily $A \subset C$.)

Proof: The first part of previous assertion is proved in [25] as Theorem 2.4b. The second part is also mentioned in [25] as Assertion 1.4a, but the essential proof is in [24], Proposition 7b. ■

Thus from the theoretical point of view a clear criterion to recognize a structural imset u is given: simply to check the validity of all inequalities $\langle a, u \rangle \geq 0$ for $a \in A$.
The following result says that structural imsets can describe all possible CI-structures:

Assertion 2
Whenever P is a probability measure over N and I the model of CI-structure of P then there exists a structural imset u such that I corresponds to u.

Proof: see Consequence 2.9 in [25]. ■

Remarks
a) It may happen that different structural imsets have the same corresponding dependency model. Nevertheless, the pertinent equivalence of structural imsets can be grasped

by means of the set A from Assertion 1b, for details see [25].
b) Our original conjecture that the models of CI-structures coincide with the dependency models corresponding to structural imsets appeared unfortunately untrue (see [25]).
c) However, the theory developed in [25] seems to admit modifications which promise to give fitting description of CI-structures for some special "nice" subclasses of probability measures.

5. EQUIVALENCE RESULT

The aim of this paper is to show that a probability measure complies with a structural imset u iff it satisfies the product formula given by u. This result can be shown under certain formal additional assumption on u, called regularity. All structural imsets in case $card\, N \leq 4$ are shown to be regular; we conjecture that every structural imset satisfies this condition. The main theorem contains the desired equivalence result.

Def. 13 (regular structural imset)
Consider a structural imset u and put:
$\mathcal{A}_u = \{S \subset N;\ S \subset T \text{ for some } T \subset N \text{ with } \overline{u}(T) < 0\}$
$\mathcal{B}_u = \{S \subset N;\ S \subset T \text{ for some } T \subset N \text{ with } \overline{u}(T) > 0\}$.
Say that u is *regular* iff only $\mathcal{E} \subset \mathcal{B}_u$ satisfying the following three conditions:
$[a]$ $\mathcal{E}$ is heriditary (i.e. $K \subset L \in \mathcal{E} \Rightarrow K \in \mathcal{E}$)
$[b]$ $\mathcal{A}_u \subset \mathcal{E}$
$[c]$ whenever $K, L \in \mathcal{E}$ with $\langle K\backslash L, L\backslash K, K \cap L\rangle \in I_u$ then $K \cup L \in \mathcal{E}$
is $\mathcal{B}_u$ itself.

Example 3
Every $u \in Z(\mathcal{U})$ of the form $i(\langle A, B, C\rangle)$ for $\langle A, B, C\rangle \in T(N)$ is a regular structural imset. Especially, every elementary imset is regular.
Indeed: u is a structural imset according to the lemma before Def. 11. Clearly $\mathcal{A}_u = \{K \subset N; K \subset AC \text{ or } K \subset BC\}$ and $\mathcal{B}_u = \{K \subset N; K \subset ABC\}$. Supposing $\mathcal{E} \subset \mathcal{B}_u$ satisfies $[a] - [c]$, by $[b]$ get $AC, BC \in \mathcal{E}$. As $\langle A, B, C\rangle \in I_u$ (see Def. 11) by $[c]$ derive $ABC \in \mathcal{E}$ and hence by $[a]$ $\mathcal{B}_u \subset \mathcal{E}$.

Some facts concerning the classes $\mathcal{A}_u$ and $\mathcal{B}_u$ (for a structural imset u) are needed to derive certain sufficient condition for regularity. Firstly, considering $S \in \mathcal{A}_u$ find maximal $T \subset N$ with $[S \subset T\ \&\ \overline{u}(T) \neq 0]$. Denoting $r^T = \sum\{\delta_K; T \subset K\}$ it makes no problem to see $\overline{u}(T) = \langle r^T, \overline{u}\rangle \geq 0$. (for example Assertion 1a resp. $(N.1) - (N.2)$). Hence:

$$\mathcal{A}_u \subset \mathcal{B}_u \tag{5.1}$$

Moreover, evident facts

$$\mathcal{A}_{l\cdot u} = \mathcal{A}_u \quad \mathcal{B}_{l\cdot u} = \mathcal{B}_u \quad I_{l\cdot u} = I_u \text{ whenever } l \in \mathbb{N} \tag{5.2}$$

imply that u is regular iff $l \cdot u$ is regular. To show

$$[u = y + w \quad y, w \text{ structural imsets}] \Rightarrow \mathcal{B}_u = \mathcal{B}_y \cup \mathcal{B}_w \tag{5.3}$$

take $S \in \mathcal{B}_y$ and find a maximal $T \subset N$ with $[S \subset T\ \&\ \overline{y}(T) \neq 0]$. As $\overline{y}(T) > 0$ the hypothesis $S \notin \mathcal{B}_u$ leads to the contradiction: $0 < \langle r^T, \overline{y}\rangle + \langle r^T, \overline{w}\rangle = \langle r^T, \overline{u}\rangle \leq 0$.

Thus $\mathcal{B}_y \subset \mathcal{B}_u$, similarly $\mathcal{B}_w \subset \mathcal{B}_u$ and the inclusion $\mathcal{B}_u \subset \mathcal{B}_y \cup \mathcal{B}_w$ is evident.
You can derive from (5.3) and (5.2) by putting $y = i(\langle K\backslash L, L\backslash K, K \cap L\rangle)$ (see Def. 11):

$$\langle K\backslash L, L\backslash K, K \cap L\rangle \in I_u \Rightarrow K \cup L \in \mathcal{B}_u \qquad (5.4)$$

Lemma 4 Suppose that every structural imset u satisfies:

$$[n \cdot u = \sum_{v \in G} k_v \cdot v \text{ with } n \in \mathbb{N} \;\; \emptyset \neq G \subset E \;\; k_v \in \mathbb{N}] \Rightarrow [\exists w \in G \;\mathcal{A}_w \subset \mathcal{A}_u] \qquad (5.5)$$

Then every structural imset is regular.

Proof: Prove the regularity of a structural imset u by induction according to $r = min\,\{g;\, \exists G \subset E \; card\, G = g$ such that $n \cdot u = \sum_{v \in G} k_v \cdot v$ for $n \in \mathbb{N}$, $k_v \in \mathbb{Z}^+\}$.
In case $r \leq 1$ either $u = 0$ (then $\mathcal{B}_u = \emptyset$) or $u = l \cdot v$ for $l \in \mathbb{N}$ and $v \in E$ (then use the procedure from Example 3 combined with (5.2)). In case $r > 1$ consider a concrete "minimal decomposition" $n \cdot u = \sum_{v \in G} k_v \cdot v$ with $G \subset E$, $card\, G = r$, $n \in \mathbb{N}$, $k_v \in \mathbb{Z}^+$.
Necessarily $k_v > 0$! Using (5.5) find $w \in G$ with $\mathcal{A}_w \subset \mathcal{A}_u$ and put $y = \sum_{v \in G\backslash\{w\}} k_v \cdot v$ i.e. $n \cdot u = y + k_w \cdot w$. To verify the regularity of u consider $\mathcal{E} \subset \mathcal{B}_u$ satisfying $[a] - [c]$ (see Def. 13). Our aim is to show $\mathcal{E} = \mathcal{B}_u$. Owing to (5.3) it suffices to verify $\mathcal{B}_w \subset \mathcal{E}$ and $\mathcal{B}_y \subset \mathcal{E}$.

I. $\mathcal{B}_w \subset \mathcal{E}$

Indeed: Let be $\overline{w} = \delta_{K \cup L} - \delta_K - \delta_L + \delta_{K \cap L}$; as $\mathcal{A}_w \subset \mathcal{A}_u$ by $[b]$ derive $K, L \in \mathcal{E}$ and as $\langle K\backslash L, L\backslash K, K \cap L\rangle \in I_u$ by $[c]$ $K \cup L \in \mathcal{E}$. Hence by $[a]$ $\mathcal{B}_w \subset \mathcal{E}$.

II. $\mathcal{B}_y \subset \mathcal{E}$

Indeed: The desired condition is equivalent to $\mathcal{E} \cap \mathcal{B}_y = \mathcal{B}_y$. Since y is regular by the induction assumption it suffices to verify for $\mathcal{E}' = \mathcal{E} \cap \mathcal{B}_y$ the following three conditions:
$[a']$ $\mathcal{E}'$ is heriditary
$[b']$ $\mathcal{A}_y \subset \mathcal{E}'$
$[c']$ $[K, L \in \mathcal{E}' \;\; \langle K\backslash L, L\backslash K, K \cap L\rangle \in I_y] \Rightarrow K \cup L \in \mathcal{E}'$.
As $\mathcal{B}_y$ is heriditary, $[a']$ follows from $[a]$. It makes no problem to see $\mathcal{A}_y \subset \mathcal{A}_u \cup \mathcal{B}_w$ and hence by $[b]$ and I. $\mathcal{A}_y \subset \mathcal{E}$. Thus $[b']$ follows from (5.1). To show $[c']$ realize that $I_y \subset I_u$ and $[c]$ can be used to derive $K \cup L \in \mathcal{E}$. By (5.4) get $K \cup L \in \mathcal{E}'$. ∎

Consequence 1
In case $card\, N \leq 4$ every structural imset is regular.

Proof: The condition (5.5) will be verified for every structural imset in case $card\, N = 4$ (the same method can be used in simpler cases $card\, N = 3$ and $card\, N = 2$). Consider a concrete "decomposition":
$n \cdot u = \sum_{v \in E} k_v \cdot v$ with $n \in \mathbb{N}$, $k_v \in \mathbb{Z}^+$, $G = \{v \in E, k_v > 0\}$ (define $k_v = 0$ for $v \in E\backslash G$).
Divide E into three classes E_1, E_2, E_3 (see Example 2) and put: $p_i = \sum_{v \in E_i} k_v$ for $i = 1, 2, 3$.
Three basic cases can be distinguished:

I. $p_1 = p_2 = 0$

In this case $n \cdot u = \sum_{v \in E_3} k_v \cdot v$. It suffices to take arbitrary $w \in E_3$ with $k_w > 0$ (note

that $w(S) < 0$ & $card\, S = 3$ implies $n \cdot u(S) < 0$ as $v(S) \leq 0$ for every $v \in E_3$).
II. $p_1 = 0 \;\; p_2 > 0$
In this case $n \cdot u = \sum_{v \in E_2 \cup E_3} k_v \cdot v$ and there exist $v \in E_2$ with $k_v > 0$. Consider the case $\mathcal{A}_v \setminus \mathcal{A}_u \neq \emptyset$ (otherwise put $w = v$) and find $S \subset N$ with $v(S) < 0$ and $[\forall T \supset S \;\; u(T) \geq 0]$. More concretely let $S = \{a, b\}$ $v = \delta_{\{a,b,c\}} - \delta_{\{a,b\}} - \delta_{\{a,c\}}$ where $N = \{a, b, c, d\}$. Moreover consider the following elementary imsets:

$$
\begin{aligned}
w &= \delta_{\{a,b,d\}} - \delta_{\{a,d\}} - \delta_{\{b,d\}} \\
x &= \delta_N - \delta_{\{a,b,d\}} - \delta_{\{a,c,d\}} + \delta_{\{a,d\}} \\
y &= \delta_N - \delta_{\{a,b,d\}} - \delta_{\{b,c,d\}} + \delta_{\{b,d\}} \\
z &= \delta_{\{a,b,c\}} - \delta_{\{a,b\}} - \delta_{\{b,c\}}
\end{aligned}
$$

Our aim is to show that $k_w > 0$ and $\mathcal{A}_w \subset \mathcal{A}_u$. For this purpose write $n \cdot [u(\{a,b\}) + u(\{a,b,d\})] = +k_w - k_v - k_x - k_y - k_z$ (for all remaining $t \in E_2 \cup E_3$ $t(\{a,b\}) + t(\{a,b,d\}) = 0$). Hence $k_w \geq k_v + k_x + k_y$ implies both $k_w \geq k_v > 0$ and $0 > -k_v \geq k_x - k_w \geq n \cdot u(\{a,d\})$ and $0 > -k_v \geq k_y - k_w \geq n \cdot u(\{b,d\})$ (the inequality $n \cdot u(\{a,d\}) \leq k_x - k_w$ follows from the fact $t(\{a,d\}) \leq 0$ for remaining $t \in E_2 \cup E_3$, similarly $n \cdot u(\{b,d\}) \leq k_y - k_w$).
III. $p_1 > 0$
In this case take arbitrary $w \in E_1$ with $k_w > 0$. By (5.1) and (5.3) $\mathcal{A}_w \subset \mathcal{B}_w \subset \mathcal{B}_u$ i.e. for each $S \in \mathcal{A}_w$ $(card\, S = 1)$ get $\sum_{S \subset K} \overline{u}_+(K) > 0$. Nevertheless by (N.2) $\sum_{S \subset K} \overline{u}_-(K) = \sum_{S \subset K} \overline{u}_+(K)$ and hence $S \in \mathcal{A}_u$. ■

So far, we have no example of nonregular structural imset. Our conjecture is that it cannot be found:

Conjecture Every structural imset is regular.

Now, the main result can be proved.

THEOREM
Let P be a probability measure over N and u be a regular structural imset on $\mathcal{U}$. Then the following conditions are equivalent:

(a) P satisfies the product formula given by u (see Def. 4)

(b) $\forall x \in \prod_{i \in N} X_i$ with $P(x) > 0$ it holds $\prod_{S \subset N} P^S(x_S)^{\overline{u}(S)} = 1$

(c) P complies with u (see Def. 8)

(d) I_u is a submodel of CI-structure of P (see Def. 6,11).

Proof: By Lemma 2 (a)⇒(b)⇒(c), by Lemma 3 (c)⇒(d). It remains to show (d)⇒(a). For fixed $x \in \prod_{i \in N} X_i$ two possibilities can occur:

I. $\prod_{S\subset N} P^S(x_S)^{\overline{u}_-(S)} = 0.$

In this case find $K \subset N$ with $[\overline{u}(K) < 0 \;\&\; P^K(x_K) = 0]$. By (5.1) there exists T with $[K \subset T \;\&\; \overline{u}(T) > 0]$. Necessarily $P^T(x_T) = 0$ and hence $\prod_{S\subset N} P^S(x_S)^{\overline{u}_+(S)} = 0$.

II. $\prod_{S\subset N} P^S(x_S)^{\overline{u}_-(S)} > 0.$

Put $\mathcal{E} = \{S \in \mathcal{B}_u; P^S(x_S) > 0\}$. Evidently $\mathcal{E}$ is heriditary and by the assumption $\mathcal{A}_u \subset \mathcal{E}$. Also the condition [c] from Def. 13 is valid, owing to (d):
$\langle K\backslash L, L\backslash K, K\cap L\rangle \in I_u \quad \Rightarrow \quad P^{K\cup L}(x_{K\cup L})\cdot P^{K\cap L}(x_{K\cap L}) = P^K(x_K)\cdot P^L(x_L)$
and hence $P^{K\cup L}(x_{K\cup L}) > 0$. Therefore the regularity of u implies $\mathcal{B}_u = \mathcal{E}$.
By Def. 10 $\;n\cdot u = \sum_{v\in E} k_v \cdot v$ for $n \in \mathbb{N}$, $k_v \in \mathbb{Z}^+$. For every $v \in E$ with $k_v > 0$ consider $\langle A,B,C\rangle \in T(N)$ with $v = i(\langle A,B,C\rangle)$. As $\langle A,B,C\rangle \in I_u$, by (d) $A \perp B|C\,(P)$ and therefore by Def. 5 derive $\prod_{S\subset N} P^S(x_S)^{\overline{v}_+(S)} = \prod_{S\subset N} P^S(x_S)^{\overline{v}_-(S)}$.
All these formulas can be multiplied and therefore it holds:

$$\prod_{S\subset N} P^S(x_S)^{\sum_{v\in E} k_v\cdot\overline{v}_+(S)} = \prod_{S\subset N} P^S(x_S)^{\sum_{v\in E} k_v\cdot\overline{v}_-(S)} \tag{5.6}$$

Put $w = \sum_{v\in E} k_v\cdot\overline{v}_+ - n\cdot\overline{u}_+$, evidently $w = \sum_{v\in E} k_v\cdot\overline{v}_- - n\cdot\overline{u}_-$. Of course $w \geq 0$ and $w(S) > 0$ implies $[\exists v \in E \;\; k_v > 0 \;\; \overline{v}(S) > 0]$ i.e. $S \in \mathcal{B}_v \subset \mathcal{B}_u$ by (5.3) and (5.2). Therefore $S \in \mathcal{E}$ and $P^S(x_S)^{w(S)} > 0$. Together $\prod_{S\subset N} P^S(x_S)^{w(S)} > 0$ and the equality (5.6) can be divided by this number to get

$$\prod_{S\subset N} P^S(x_S)^{n\cdot\overline{u}_+(S)} = \prod_{S\subset N} P^S(x_S)^{n\cdot\overline{u}_-(S)} \tag{5.7}$$

Hence, the desired product formula easily follows. ■

Remark The previous proof can be easily modified to show that for every strictly positive probability measure and every structural imset u the conditions (a), (b), (c), (d) are equivalent.

CONCLUSIONS

Thus, the theorem above relates three approaches to description of CI-structures:
- by means of dependency models
- by means of imsets (information - theoretical definition)
- by means of product formula validity

and shows their equivalence.

Note that the description by means of imsets (and faces which are behind) is systematically treated and illustrated by examples in [25]. It is endowed by a deductive mechanism allowing to infer CI-statements from an input piece of information about CI-structure (finitely-implementable from theoretical point of view).

The description by means of product formula can be understood as a step to interpretation of these CI-structures. It seems to me that the presented models of description of CI-structures have similar reasons (or rights) to be called explicable as hierarchical log-linear models. In fact, a general log-linear model is specified by certain "formula" for the probability measure, namely expressing it as a product of marginal factors. This is close to the presented formulas and in some special cases (decomposable models) even equivalent.

Another interesting analogy of our product formulas can be found in [10] where so-called functional expressions satisfying the unity sum property are delt with. Some of them (for example the expression from Example 4 there) correspond to product formulas representing CI-structure.

Note that as reported in [25] some graphical descriptions of CI-structures can be "translated" to imsets and these can be "forwarded" to product formulas. To inform the reader we give the corresponding imset expressions here (without proof).

Having an influence diagram (= directed acyclic graph) let $\pi(k)$ denotes the set of parents of a node $k \in N$. The corresponding imset can be given by its natural extension: $\overline{u} = \delta_N - \delta_\emptyset + \sum_{k \in N} \{\delta_{\pi(k)} - \delta_{\{k\} \cup \pi(k)}\}$.

Having a decomposable model specified by a triangulated (undirected) graph let $\mathcal{C} \subset exp\ N$ denote the class of its maximal cliques. The corresponding imset is specified by: $\overline{u} = \delta_{\cup \mathcal{C}} + \sum_{\emptyset \neq \mathcal{B} \subset \mathcal{C}} (-1)^{card\, \mathcal{B}} \cdot \delta_{\cap \mathcal{B}}$
($\cup \mathcal{B}$ resp. $\cap \mathcal{B}$ denotes the union resp. intersection of sets from $\mathcal{B}$).

Acknowledgements
I am indebted to Francesco Malvestuto; one of his comments during his visit in Prague last year helped me to overcome certain inconvenience. I am also grateful to my colleague František Matúš for stimulating discussion.

REFERENCES

1. M.Aigner, Combinatiorial Theory, Springer Verlag, Berlin-Heidelberg-New York, 1979.
2. J.N.Darroch and S.L.Lauritzen and T.P.Speed, Markov fields and log-linear interaction models for contingency tables, Annals of Statistics 8 (1980) 522-539.
3. A.P.Dawid, Conditional independence in statistical theory (with discussion), Journal of the Royal Statistical Society ser. B 41 (1979) 1-31.
4. D.Geiger and J.Pearl, Logical and algorithmic properties of conditional independence and graphical models, to appear in Annals of Statistics (1993?).
5. R.A.Howard and J.E.Matheson, Influence diagrams, in Principles and Applications of Decision Analysis 2, Strategic Decisions Group, Menlo Park, California, 1984.
6. V.Isham, An introduction to spatial point processes and Markov random fields, International Statistical Review 49 (1981) 21-43.
7. H.Kiiveri and T.P.Speed and J.B.Carlin, Recursive causal models, Journal of Australian Mathematical Society ser.A 36 (1984) 30-52.
8. S.L.Lauritzen and T.P.Speed and K.Vijayan, Decomposable graphs and hypergraphs, Journal of Australian Mathematical Society 36 (1984) 12-29.
9. F.M.Malvestuto, Theory of random observables in relational data bases, Information Systems 8 (1983) 281-289.
10. F.M.Malvestuto, Existence of extensions and product extensions for discrete probability distributions, Discrete Mathematics 69 (1988) 61-77.

11. F.Matúš, Probabilistic conditional independence structures and matroid theory: backgrounds, in Proceedings of WUPES 1991 (Workshop on Uncertainty Processing in Expert Systems), September 9-12, 1991, Alšovice, Czechoslovakia.

12. J.Moussouris, Gibbs and Markov random systems with constraints, Journal of Statistical Physics 10 (1974) 11-33.

13. K.K.Nambiar, Some analytic tools for design of relational database systems, in Proceedings of Conference on Very Large Data Bases 1980.

14. J.Rosenmüller and H.G.Weidner, Extreme convex set functions with finite carrier: general theory, Discrete Mathematics 10 (1974) 343-382.

15. J.Pearl and A.Paz, Graphoids: a graph-based logic for reasoning about relevance relations, in Advances in Artificial Intelligence II (B. Du. Boulay et al. eds.) 357-363, North-Holland, Amsterdam 1987.

16. J.Pearl, Markov and Bayes networks: a comparison of two graphical representations of probabilistic knowledge, technical report CSD-860024, R-46-1, UCLA, Los Angeles, 1986.

17. J.Pearl and D.Geiger and T.Verma, The logic of influence diagrams, chapter 4 in Influence Diagrams, Belief Nets and Decision Analysis, (R.M. Oliver and J.Q. Smith eds.), John Wiley 1990.

18. R.D.Shachter, Probabilistic inference and influence diagrams, Operations Research 36 (1988) 589-604.

19. J.Q.Smith, Influence diagrams for statistical modelling, Annals of Statistics 17 (1989) 654-672.

20. D.J.Spiegelhalter and S.L.Lauritzen, Sequential updating on conditional probabilities of directed graphical structures, Networks 20 (1990) 579-605.

21. W.Spohn, Stochastic independence, causal independence, and shieldability, Journal of Philosophical Logic 9 (1980) 73-99.

22. M.Studený, Multiinformation and the problem of characterization of conditional independence relations, Problems of Control and Information Theory 18 (1989) 3-16.

23. M.Studený, Conditional independence relations have no finite complete characterization, in Transactions of 11-th Prague Conference on Information Theory, Statistical Decision Functions and Random Processes vol.B 377-396, Academia, Prague, 1992.

24. M.Studený, Convex cones in finite-dimensional real vector spaces, Kybernetika 29 (1993) (to appear).

25. M.Studený, Description of structures of stochastic conditional independence by means of faces and imsets, series of 3 papers submitted to International Journal of General Systems.

26. S.Ur and A.Paz, The representation power of probabilistic knowledge by undirected graphs and directed acyclic graphs: a comparison, in Proceedings of WUPES 1991 (Workshop on Uncertainty Processing in Expert Systems), September 9-12, 1991, Alšovice, Czechoslovakia.

27. N.Wermuth and S.L.Lauritzen, Graphical and recursive models for contingency tables, Biometrika 70 (1983) 537-552.

28. S.Wright, Correlation and causation, Journal of Agricultural Research 20 (1921) 557-585.

Uncertainty in Intelligent Systems
B. Bouchon-Meunier et al. (Editors)

A WEAK COHERENCE CONDITION FOR CONDITIONAL COMPARATIVE PROBABILITIES

G.Coletti

Istituto di Matematica per la Ricerca Operativa, Università di Palermo
viale delle Scienze , 90128 Palermo, Italy

Summary . We give a coherence condition for comparative probabilities, interpretable according to de Finetti's paradigm of coherent bets. Such a coherence condition results necessary and sufficient for the existence of a (de Finetti coherent) numerical probability almost representing the comparative one.

1. INTRODUCTION

A problem that often occurs in Artificial Intelligence is the following : the field expert (a physician, for instance) is not actually able to give a reliable numerical evaluation of the degree of uncertainty of the relevant statements.

However, in some cases it is possible to get from the expert a partial ordinal evaluation of the uncertainty degree only of few uncertain situations, namely the ones strictly related to the problem at hand and known by the expert.

If the framework is a probabilistic one, then one represents uncertainty statments and information by conditional events and introduces, a comparative probability , that is an ordering relation $\leq^*$, expressing the intuitive idea of "*not more probable than*". Moreover one would like that the given ordering relation admits a (not necessarely unique) compatible numerical conditional probability p. That not necessarily with the purpose of making an actual numerical evaluation, but only of looking at the numerical probability model as a pivot.

We must also take into account that the set of interesting events usually constitutes a family without any particular structure (such as Boolean ring, for instance). Moreover, in general, one is not interested in general in the (ordinal or numerical) probability evaluation of the atoms generated by the events. Finally, one may need to enlarge the set of events and relations at any stage. From such considerations it follows that the most suitable probabilistic numerical theory can be the subjective approach of de Finetti , based on the "coherence principles" (see de Finetti, 1930, 1931 or 1970).

In Coletti (1991), and Coletti et al. (1990) and (1991_b), necessary and sufficient conditions (in particular cases and also in the general case) are given for the existence of a de Finetti-coherent conditional probability *p strictly monotone with* (or *representing)* a comparative probability $\leq^*$, that is such that the following conditions hold:

$$E \mid H \leq^* F \mid K \Rightarrow p(E \mid H) \leq p(F \mid K)$$

$$E \mid H <^* F \mid K \Rightarrow p(E \mid H) < p(F \mid K). \qquad (*)$$

All the conditions mentioned above share the good feature ofbeing easily computable: indeed they are equivalent to proving the solvability of a linear system, whose number of unknowns depends on the logical structure of conditional events and whose number of inequalities is equal or proportional to the number of compared pairs of events.

However in some particular context, the existence of a numerical probability, strictly monotone with the comparative one, can be considered a too restrictive condition; indeed it would suffice the existence of a *weakly monotone* (or *almost representing*) numerical probability; that is, such that only the first implication of (*) holds.

The aim of this paper is to characterize comparative probabilities which admitte an almost representing de Finetti-coherent (conditional) probability p.

The proposed condition, characterizing these comparative probabilities, gives rise to a coherence principle in the sense of de Finetti (in fact it is interpretable according to metaphor of the dutch book). Moreover it is linear, as almost all the conditions mentioned above.

Further, when new informations are introduced, the same condition assures the enlargment of the comparative probability in a coherent way, even more, it suggests how to extend the relation, by providing the bounds for them.

The profitable use of coherent probabilities in the menagement of uncertainty in expert systems was fully discussed in Gilio and Scozzafava (1989) and Coletti et al. (1991_a) in the case of numerical evaluations, and in Coletti (1991), Coletti et al. (1990 and 1991_b) in the case of ordinal evaluations.

2. WEAKLY COHERENT COMPARATIVE PROBABILITIES

Given a family $\mathcal{A} = \{A_i\}$ of events, consider the corresponding family $\mathcal{C} = \{C_k\}$ of atoms generated by them, that is all the logical products between the events and their negations (in symbols $C_k = \bigcap_{A_i \in \mathcal{A}} \hat{A}_i$, where $\hat{A}_i = A_i$ or its *contrary* A_i^c).

For every event $A_i \in \mathcal{A}$, consider also the binary vector (for which the same symbol of the relevant event shall be use) $A_i = (x_1^i, \ldots, x_m^i, \ldots)$, where $x_k^i = 1$ when $C_k \subset A_i$ and $x_k^i = 0$ when $C_k \subset A_i^c$.

For a family $\mathcal{E}$ of conditional events $E_i \mid H_i$, we consider, among the atoms generated by the events E_i, H_i, those contained in the logical sum H_0 of the H_i's.

We define *weakly coherent* a (conditional) comparative probability satisfying the following condition :

(WCC) For every $E_i \mid H_i \leq^* F_i \mid K_i$ there exists $\alpha_i > 0$ such that: for every $n \in \mathbb{N}$, and for every $E_i \mid H_i, F_i \mid K_i \in \mathcal{E}$, if $E_i \mid H_i \leq^* F_i \mid K_i \quad (i = 1, \ldots, n)$, then, for any choice of $b_i \in \mathbb{R}$, $a_i, c_i, d_i \geq 0$, with $\Sigma(c_i + d_i) > 0$,one has:

$$\sup \left[\sum_1^n a_i(F_iK_i - \alpha_i E_i H_i) + b_i(\alpha_i H_i - K_i) + c_i H_i + d_i K_i \right] > 0.$$

We notice that, if for some i we have $H_i = K_i$, then it is necessarily $\alpha_i = 1$. In fact, for $\alpha_i \neq 1$, we can have, for an opportune choice of b_i and c_i:
$sup\ [b_i(H_i - H_i) + c_i H_i] < 0$.

Furthermore, if for every i one has $H_i = K_i = \Omega$, (i.e. all the events are unconditioned), then the above condition (WCC) becomes the following (WC), studied in Coletti (1990):

(WC) for every $n \in N$, and $E_i, F_i \in \mathcal{E}$, if $E_i \leq^* F_i \quad (i = 1, \ldots, n)$, then for any choice of $a_i \geq 0$, one has

$\sup \sum_1^n a_i(F_i - E_i) \geq 0$.

To prove the above statment, consider that, if for some $E_i \leq^* F_i$ and $a_i \geq 0$ $(i = 1, \ldots, n)$ we have $\sup \sum_1^n a_i(F_i - E_i) = c < 0$, then it results $\sup \sum_1^n a_i(F_i - E_i) - c\Omega \leq 0$. Vice versa it is obvious.

We remark that, like in the theory of unconditional comparative probabilities it occurs (see Coletti, 1990), condition of weak coherence does not imply any of conditions usually required to a comparative probability (see, for instance Fine, 1970) and *necessary* to the existence of a strictly monotone conditional probability . In fact a comparative probability satisfying (WCC) can be even intransitive or can not satisfy the following axiom of "nonnegativity ":

$\emptyset \leq^* E \mid H$, for every $E \mid H \in \mathcal{E} \quad (\emptyset = \emptyset \mid \Omega$ is the impossible event)

Moreover, if we consider that transitivity or nonnegativity - or any other condition - is a not giving up "*rule of coherence*", then we must *explicitely* require them.

Finally we note that condition (WCC) implies the following property of "positivity" for all conditioning events:

$\emptyset <^* H$, for every $H \in \mathcal{E}$.

To prove such an assertion it is sufficient to consider that, for $E_i \mid H_i \leq^* F_i \mid K_i$ and $H_1 \leq^* \emptyset$, $\quad a_i = b_i = d_i = 0, \quad (i = 1, \ldots, n), c_i = 0, \quad (i = 2, \ldots, n)$ and $c_1 = c \neq 0$, we have

$\sum a_i(F_i K_i - \alpha_i E_i H_i) + b_i(\alpha_i H_i - K_i) + c_i H_i + d_i K_i - cH_1 = 0.$

3. EXTENTION OF WEAK COHERENT COMPARATIVE PROBABILITIES

The following theorem - as the ones in Coletti (1990) related to unconditional comparative probabilities - is inspired by extension theorem of coherent previsions given by de Finetti (see de Finetti 1931 or 1970). In fact his interest is not only on the existence of a (not unique) enlargement of the (ordinal) probabilistic assessment , but also in the fact that the proof actually suggests the way to make the enlargment without introducing incoherences.

THEOREM 3.1. - Let $\leq^*$ be a weakly coherent comparative probability on a set of conditional events $\mathcal{E}$. If $\mathcal{G} \supset \mathcal{E}$, then there exists a (complete) weakly coherent comparative probability $\leq x$ on $\mathcal{G}$, which is an extension of $\leq^*$.

Proof. Let $\Theta = \{(\mathcal{E}_i, \leq_i)$, be such that $\mathcal{E} \subseteq \mathcal{E}_i \subseteq \mathcal{G}$, and $\leq_i$ satisfies condition (WCC) and extends $\leq^*\}$.

It is immediate to check that Θ is an inductive set an so it admits a maximal element $(\mathcal{E}_*, \leq_*)$. We want show that $\mathcal{E}_* = \mathcal{G}$ and also that $\leq_*$ is complete. If either of two statments were false, then there would exists an event $G \mid Z \in \mathcal{G}$, which is not compared with some element of $\mathcal{E}_*$. In this case we can extend $\leq_*$ to $\mathcal{E}_* \cup \{G \mid Z\}$ so that the extention satisfies (WCC).

Define:

$\mathcal{H} = \Big\{E \mid H \in \mathcal{E}_*$: there exists $E_i \mid H_i, F_i \mid K_i \in \mathcal{E}_*$, with $E_i \mid H_i \leq_* F_i \mid K_i$, $(i = 0, \ldots, n)$, such that, for every choice of $\alpha_i > 0$, there exist $b_i \in R$, $a_i, c_i, d_i \geq 0$, $\Sigma(c_i + d_i) \neq 0$, and $(a_0 + b_0) \neq 0$, for which:

$\sup\big[\sum_1^n \big(a_i(F_iK_i - \alpha_i E_iH_i) + b_i(\alpha_i H_i - K_i) + c_iH_i + d_iK_i\big) + a_0(EH - \alpha_0 GZ) + b_0(\alpha_0 Z - H) + c_0Z + d_0H\big] \leq 0\Big\}$;

$\mathcal{K} = \Big\{F \mid K \in \mathcal{E}_*$: there exists $E_i \mid H_i, F_i \mid K_i \in \mathcal{E}_*$, with $E_i \mid H_i \leq_* F_i \mid K_i$, $(i = 0, \ldots, n)$, such that, for every choice of $\beta_i > 0$, there exist $\bar{b}_i \in R$ $\bar{a}_i, \bar{c}_i, \bar{d}_i \geq 0$, $\Sigma(\bar{c}_i + \bar{d}_i) \neq 0$, and $(\bar{a}_0 + \bar{b}_0) \neq 0$, for which:

$\sup\big[\sum_1^n \big(\bar{a}_i(F_iK_i - \beta_i E_iH_i) + \bar{b}_i(\beta_i H_i - K_i) + \bar{c}_iH_i + \bar{d}_iK_i\big) + \bar{a}_0(GZ - \beta_0 FK) + \bar{b}_0(\beta_0 K - Z) + \bar{c}_0Z + \bar{d}_0K\big] \leq 0\Big\}$.

Our aim is now to prove that for every $E \mid H \in \mathcal{H}$ and $F \mid K \in \mathcal{H}$, the relation $F \mid K \leq_* E \mid H$ is incoherent. For such a purpose consider that, if $E \mid H \in \mathcal{H}$, then either of following conditions holds:

1) inequality in $\mathcal{H}$ holds for some choice of $\alpha_i, a_i, b_i, c_i, d_i$ as in $\mathcal{H}$, with $a_0 = 0$.
2) for every choice of $\alpha_i, a_i, b_i, c_i, d_i$ as in $\mathcal{H}$,we have:

$\sup\big[\sum_1^n \big(a_i(F_iK_i - \alpha_i E_iH_i) + b_i(\alpha_i H_i - K_i) + c_iH_i + d_iK_i\big) + b_0(\alpha_0 Z - H) + c_0Z + d_0H\big] > 0$.

Equivalent considerations can be made for elements of $\mathcal{K}$.

Consider now two arbitrary positive number α and $\beta = \alpha/\gamma$. If $E \mid H \in \mathcal{H}$ and $F \mid K \in \mathcal{K}$, then there exist $\hat{a}_i, \hat{b}_i, \hat{c}_i, \hat{d}_i$, ($\hat{a}_i = a_i$ or $\bar{a}_i$), $(i = 0, \ldots, n)$ such that the inequalities in H and in K hold.

By adding up the two inequalities and then analyzing one by one the case (1) and the case (2), one arrives to conclusion that the relation $F \mid K \leq_* E \mid H$ gives incoherence.

Now we can put $F \mid K <_* G \mid Z$, for every $F \mid K \in \mathcal{K}$ and $G \mid Z <_* E \mid H$, for every $E \mid H \in \mathcal{H}$. If $M \mid N \notin \mathcal{H} \cup \mathcal{K}$, any choice of relation between $G \mid Z$ and $M \mid N$ gives a coherent proper extension of comparative probability $\leq_*$.

Such a definition makes sense by previous considerations. The weak coherence of extended comparative probability can be checked by a proof analogous to the previous one.

4. REPRESENTATION OF WEAK COHERENT COMPARATIVE PROBABILITIES

THEOREM 4.1 - Let $\mathcal{F}$ be a finite set of conditional events, containing $\emptyset$ and any event $E \mid H$ together with the event $H = H \mid \Omega$. For a binary relation $\leq^*$ in $\mathcal{F}$, the following statments are equivalent:

i) relation $\leq^*$ satisfies condition (WCC);
ii) there exists a (not unique) dF-coerent conditional probability $p : \mathcal{F} \to \mathbb{R}$, with $p(H_i) > 0$, which almost represents $\leq^*$.

Proof: Let n be the number of events of $\mathcal{F}$ and $r \leq 2^{2n}$ the number of atoms generated by them and contained in H_0 and let m be the number of pairs $E_i \mid H_i \leq^* F_i \mid K_i$.

We introduce the following system $\mathcal{L}$, depending on the positive parameters α_i and with unknown the r-vector $W = (w_1, \ldots, w_r)$ and where the symbol $\times$ denotes the scalar product:

$$\begin{cases} (F_iK_i - \alpha_i E_iH_i) \times W \geq 0 & E_i \mid H_i \leq^* F_i \mid K_i \\ (\alpha_i H_i - K_i) \times W = 0 \\ H_i \times W > 0, K_i \times W > 0 \\ \alpha_i > 0 \\ W \geq 0 \end{cases}$$

We first show that condition (ii) is equivalent to the existence of a solution W for system $\mathcal{L}$.

If there exist $\alpha_1, \ldots, \alpha_m > 0$, for which the system $\mathcal{L}$ admits a solution W, then we put, for any $E \mid H \in \mathcal{F}, p(E \mid H) = (EH \times W)/(H \times W)$. It is easy to prove that the above defined function $p : \mathcal{F} \to [0,1]$ is a (de Finetti coherent) conditional probability, with $p(H) = p(H \mid \Omega) > 0$, and that it represents $\leq^*$.

Viceversa, suppose that there exists a (de Finetti coherent) conditional probability p (with $p(H) > 0$), almost representing $\leq^*$, and let $\{p_1, \ldots, p_r\}$ be a probability extending p on the relevant atoms. If $W = (w_1, \ldots, w_r)$ denotes any vector such that $w_i/\Sigma w_h = p_i$, then it has to be: $H_i \times W > 0$ and $K_i \times W > 0$, for every i, and $F_iK_i \times W/K_i \times W \geq E_iH_i \times W/H_i \times W$, for $E_i \mid H_i \leq^* F_i \mid K_i$.

If $p(E_i \mid H_i)$ is positive - the case equal zero is trivial - the last inequality is verified if and only if the following one holds $K_i \times W/H_i \times W = \alpha_i \leq F_iK_i \times W/E_iH_i \times W$. Moreover the system $\mathcal{L}$ admits a solution.

Now, to prove the theorem, it is sufficient to consider that, for a classical alternative theorem (see for instance Fenchel, 1951), the system $\mathcal{L}$ has a solution if and only if, for every $\alpha_i > 0$, the following system $\mathcal{L}'$, with unknowns a_i, b_i, c_i, d_i, has no solution

$$\begin{cases} \sum_1^m \big(a_i(F_iK_i - \alpha_i E_iH_i) + b_i(\alpha_i H_i - K_i) + c_iH_i + d_iK_i\big) \leq 0 \\ a_i, c_i, d_i \geq 0, \quad \Sigma(c_i + d_i) > 0. \end{cases}$$

Remark . Suppose that a weakly coherent comparative probability $\leq^*$ does not verify some condition necessary for the representability with a (numerical) conditional probability - such as transitivity, for instance -. Then any conditional probability almost representing $\leq^*$ - whose existence is assured by theorems 4.1 and 4.2- must to do equivalent (that is give to them the same probability value) all the elements "*not in order*" (for instance the elements of an intransitive cycle).

THEOREM 4.2 - Let $\mathcal{E}$ be a set of conditional events, containing $\emptyset$ and any event $E \mid H$ together with the event $H = H \mid \Omega$. If a binary relation $\leq^*$ in $\mathcal{E}$ satisfies condition

(WCC), then there exists a (not unique) de Finetti-coherent conditional probability $p : \mathcal{E} \to \mathbb{R}$, which almost represents $\leq^*$.

Proof: The case of $\mathcal{E}$ finite was proved in theorem 2. If $\mathcal{E}$ is infinite, from theorem 1 we still get that for every finite subset $\mathcal{F} \subset \mathcal{E}$, there existis a de Finetti-coherent conditional probability $p_{\mathcal{F}} : \mathcal{F} \to [0, 1]$ wich almost represents the restriction of $\leq^*$ to $\mathcal{F}$.

Now, let Φ be the set of all function from $\mathcal{E}$ to [0,1] and let $\Phi_{\mathcal{F}} \subset \Phi$ the set of elements of Φ whose restriction to $\mathcal{F}$ coincides with $p_{\mathcal{F}}$ and put:

$p = \bigcap\{\Phi_{\mathcal{F}}, \mathcal{F} \subset \mathcal{E}, \mathcal{F}$ finite and containig $\emptyset$ and any event $E \mid H$ together with the event $H\}$.

Mimicking the proof of Lemma 8 in Dubins (1985), one proves that there exists p (that is the intersection is not empty) and also that such a p is a de Finetti-coerent conditional probability (for a more extended proof, see Coletti, 1988).

In order to prove that p almost represents $\leq^*$, it is sufficient to consider any set $\mathcal{F} = \{\emptyset, E \mid H, F \mid K, H, K\}$.

We note that the conditional probability p of theorem 4.2 is not necessarily such that $p(H) > 0$ for every H. In fact the compacteness theorem used in the proof does not grant that the strict inequalities, present in any $p_{\mathcal{F}}$, remain such in p.

REFERENCES.

Coletti,G. (1988); *Conditionally coherent qualitative probabilities.* Statistica, vol.3-4, 235-242.

Coletti,G. (1990); *Coherent qualitative probability*; Journ. Math. Psych., 34, 298-310.

Coletti,G. (1991); *Comparative probabilities ruled by coherence conditions and use in expert systems.* Proceedings of Workshop on Uncertainty Processing in Expert Systems. Alsovice, 1991, Int.J. of General Systems, in press.

Coletti,G.,Gilio,A.,Scozzafava,R. (1990); *Coherent qualitative probability and uncertainty in Artificial Intelligence;* Proc. 8th Inter. Conf. on Cybernetics and Systems (vol. I, ed. C.N. Manicopoulos), N.Y.

Coletti,G.,Gilio,A.,Scozzafava,R. (1991_a); *Conditional events with vague information in expert systems.* In "Uncertainty in knowledg bases" (eds. B.Bouchon-Meunier, R.R. Yager, L.A. Zadeh) Lecture Notes in Computer Sciences, 251, Springer Verlag.

Coletti,G.,Gilio,A.,Scozzafava,R. (1991_b); *Assessment of qualitative judgements for conditional events in expert systems.* in Symbolic and Quantitative Approaches to Uncertainty (eds R.Kruse and P.Siegel). Lecture Notes in Computer Sciences, 548, Springer Verlag.

de Finetti,B. (1930); *Problemi determinati e indeterminati nel calcolo delle probabilità*; Rend. Acc. Naz. Lincei 12 , 367-373.

de Finetti,B. (1931); *Sul significato soggettivo della probabilità*; Fundam. Mathem., 17, 298,329.

de Finetti, B. (1975); *Theory of Probability*, London, Wiley.

Dubins, L.E. (1985); *Finitely additive conditional probabilities, conglomerability and disintegrations*; Ann. Prob.,3 , 89-99.

Fenchel,W. (1951); *Convex cones, sets and functions*; Lectures at Princton University, Spring term.

Fine,T.L. (1970); *Theories of Probability*; Acad. Press N.Y.

Gilio,A.,Scozzafava,R. (1989); *Le probabilità condizionate coerenti nei sistemi esperti;* Atti delle giornnate AIRO su Ricerca Operativa e Intelligenza Artificiale, 317-330.

Uncertainty in Intelligent Systems
B. Bouchon-Meunier et al. (Editors)

Uncertainty and Inference Through Approximate Sets

J. Miró and J. Miró-Julià
Universitat de les Illes Balears
07071 Palma de Mallorca SPAIN

Abstract

Uncertainty is usually displayed through imprecise language. Since inference is a language dependent process, the possibility of inferring in an uncertain environment depends upon the language being used. The concept of approximate set is introduced and a simple language is presented allowing imprecise and approximate statements. The corresponding inference rules are discussed

1. INTRODUCTION

Uncertain statements are known since old times. Indian culture recognized the use of the alternative (OR) as the simplest way of expressing uncertainty. It may be also the oldest form. The concept of probability has provided a significant step toward the computability of measures of uncertainty [1]. More recently, other concepts like credibility, measures of belief [2], plausibility functions, fuzzy theories, and non monotonic logics [3] have extended the treatment of uncertainty in a formal frame. Although it is well known that there is not a unique way to interpret the concept of uncertainty, it is generally agreed [4] that three fundamental problems must be distinguished in its treatment, namely:

i. The representation of uncertain concepts,

ii. The combination of uncertain statements,

iii. Assigning uncertainty measures to inferred statements.

Statements may be used 1) to describe extensionally known realities, or 2) to define intensionally classes of realities not extensionally known. In case 2) the combination of statements is done by simple juxtaposition of statements, and problem iii) is particularly important as it is the only procedure that enables us to gain further knowledge. In every case it is obvious that the inferential process must depend on the structure of the language used in the statements, and therefore that the inference rules are language dependent.

It is a fact that usually the inference rules are postulated. They are not conceptually justified, but simply it is shown that their use leads to results which are consistent with our natural understanding of inference. For example, this has been the case of *Evidence Theory*. Great effort is being made to find a theoretical justification of the rules.

This paper describes the result of our research developed toward gaining a better understanding of the justification behind the inference rules from another point of view, in a particular class of languages of our choice.

The paper consists of the following parts:

- Establishing a basic uncertainty free language,
- Deriving the inference rules in this language,

- Establishing uncertain languages, and
- Deriving the inference rules in the uncertain languages.

2. THE BASIC LANGUAGE AND ITS INFERENCE RULES

Let D be the domain and R the codomain, or range, of a correspondence, or relation, given by $G \subseteq D \times R$. It will be assumed that the elements $d_i \in D, i = 1, 2, \ldots m$ are objects, and $r_i \in R, i = 1, 2, \ldots n$ are binary attributes. Let $D_j \subseteq D$ and $R_k \subseteq R$ be subsets. The structure of a basic statement is of the form

$$D_j - R_k$$

with the following meaning: *All elements* $d_i \in D_j$ *have all the attributes of* $r_h \in R_k$. This basic language is called A-A Language [5, 6], 'A' standing for 'all.'

Therefore, if $D_{a'} \subseteq D_a$ and $R_{a'} \subseteq R_a$ then, because of the meaning (semantics) of the statement it follows:

Lemma 1 *The following statements are correct:*
a) $D_a - R_{a'}$
b) $D_{a'} - R_a$
c) $D_{a'} - R_{a'}$

Since $A \cap B \subseteq A \subseteq A \cup B$ from the above considerations it follows:

Lemma 2 *Let* $D' - R_a$ *and* $D' - R_b$, *then* $D' - (R_a \cup R_b)$

Lemma 3 *Let* $D_a - R''$ *and* $D_b - R''$, *then* $(D_a \cup D_b) - R''$

From the previous lemmas the theorem below follows:

Theorem 4 *Let* $D_a - R_a$ *and* $D_b - R_b$ *be predicated of a certain entity, then:*
a) $D_a \cap D_b - R_a \cup R_b$
b) $D_a \cup D_b - R_a \cap R_b$

Proof: From the above considerations it follows

$$D_a \cap D_b \subseteq D_a - R_a \supseteq R_a \cap R_b$$
$$D_a \cap D_b \subseteq D_b - R_b \supseteq R_a \cap R_b$$

From where it can be written, on the one side

$$D_a \cap D_b - R_a$$
$$D_a \cap D_b - R_b$$

and by Lemma 2

$$D_a \cap D_b - R_a \cup R_b$$

On the other side it can also be written

$D_a \quad — \quad R_a \cap R_b$

$D_b \quad — \quad R_a \cap R_b$

and by Lemma 3

$D_a \cup D_b — R_a \cap R_b$

Q.E.D

Notice that the above result hinges on the validity of Lemma 1 and the existence of a partial order $\subseteq$ and operations $\cup, \cap$ fulfilling the expression $A \cap B \subseteq A \subseteq A \cup B$. From Theorem 4 it follows that, using the representation typical in logic, from premises p1: $D_a — R_a$ and p2: $D_b — R_b$ the conclusions $c1$ and $c2$ may be inferred writing the result of the inference immediately as follows:

p1:	$D_a — R_a$
p2:	$D_b — R_b$
c1:	$D_a \cap D_b — R_a \cup R_b$
c2:	$D_a \cup D_b — R_a \cap R_b$

This result may be generalized. Let $D_i — R_i, i = 1, 2, \ldots n$ and $F(D_1, D_2, \ldots, D_i)$ be a boolean expression with operations $\cup$ and $\cap$. Let F_d be the *dual function*, obtained from D by interchanging the operations as well as the universal set D and the empty set $\emptyset$. It is immediate that

Theorem 5 *Let* $D_i — R_i, i = 1, 2, \ldots n$ *be correct statements, then*

$$F(D_1, D_2, \ldots D_i) — F_d(R_1, R_2, \ldots R_i)$$

is also a correct statement.

The proof is practically immediate.

3. THE UNCERTAIN LANGUAGE

Let $X = \{x_a, x_b, \ldots x_x\}$, be a finite, extensionally defined set, of cardinal x, that is $\mid X \mid = x$ and $x_1, x_2 \in N$. For notation convenience x is used to represent both the generic symbol of the elements of X and the cardinal of X, unless where it is stated otherwise.

> *Definition:* An approximate set $\mathbf{X}$ is a three-tuple given by $\langle x_1, x_2, X \rangle$ where X is a set of cardinal x, and x_1, x_2 are two positive integers, $x_1 \leq x_2 \leq x$.
>
> $\mathbf{X} = \langle x_1, x_2 \rangle X$

3.1. Meaning

The triplet $\langle x_1, x_2 \rangle X$ is the symbol of an approximate statement. $\langle x_1, x_2 \rangle X =_s$ "between x_1 and x_2 elements of the set X," or what is the same:

> $\langle x_1, x_2 \rangle X =_s$ "x' elements of the set X. Where $0 \leq x_1 \leq x' \leq x_2 \leq x$"

This statement describes approximately a subset of X. It is not known neither what elements constitute it, nor exactly how many elements there are; the statement only establishes two bounds: a lower bound x_1 and an upper bound x_2. However, X is supposed to be known.

3.2. Equality

Let $\mathbf{X} = \langle x_1, x_2 \rangle X$ and $\mathbf{Y} = \langle y_1, y_2 \rangle Y$. Let the relation ρ be defined as follows: $X \rho Y$ if and only if:

$$x_1 = y_1, \quad x_2 = y_2, \quad \text{and } X = Y$$

where the two first $=$ signs represent number equality and the last $=$ sign represents set equality.

ρ will inherit all the properties of the $=$ relations, therefore it immediately follows that it will be reflexive, symmetric, and transitive, and consequently since it is an equivalence relation we may simply relace the symbol ρ by $=$, writing $\mathbf{X} = \mathbf{Y}$.

3.3. Order

Let $\mathbf{X} = \langle x_1, x_2 \rangle X$ and $\mathbf{Y} = \langle y_1, y_2 \rangle Y$. Let the relation σ be defined as follows: $X \sigma Y$ if and only if:

$$x_1 \leq y_1, \quad x_2 \leq y_2, \quad \text{and } X \subseteq Y$$

where the two first $\leq$ signs represent number inequality and the last $\subseteq$ sign represents set inclusion.

σ will inherit all the properties of the relations $\subseteq$ and $\leq$, therefore it follows immediately that it will be reflexive, transitive, and antisymmetric, in other words, σ is an order, and consequently, we may simply replace the symbol σ by $\sqsubseteq$, writing $\mathbf{X} \sqsubseteq \mathbf{Y}$. Since two approximate sets are not necessarily related it follows that the order is partial.

3.4. Operations

Let $= \langle y_1, y_2 \rangle Y, |Y| = y$. Two operations $\sqcup$ and $\sqcap$ may be conceived as follows:

- *First operation:*

  Let X and Y be two sets, not necessarily disjoint. Assume that you pick up v elements of $X, x_1 \leq v \leq x_2$, and w elements of $Y, y_1 \leq w \leq y_2$. What can be said about the z elements found in both v elements of X and w elements of Y?

  Let $f = |(X \cup Y)|$. It may happen that there are no common elements to both X and Y, thus there must be at least 0 common elements, however in any case the z elements belong to $X \cap Y$. On the other hand, if

  $$x_1 + y_1 > |(X \cup Y)| = f,$$

  then for sure there will be at least $x_1 + y_1 - f$ common elements. Thus the number of common elements will be at least $\max(0, (x_1 + y_1 - f))$. On the other hand, the highest possible value of the common elements cannot be greater than x_2 or y_2, whichever is smaller. However, if the value of $c = |(X \cap Y)|$ is $c < min(x_2, y_2)$, then the number of common elements cannot exceed c. Thus what can be said about z may be written as follows: $\langle a, b \rangle (X \cap Y)$, where

  $$a = \max(0, (x_1 + y_1 - f)); \quad b = \min(x_2, y_2, c)$$

- *Second operation:*

  Assume now that v elements of $X, x_1 \leq v \leq x_2$, and w elements of $Y, y_1 \leq w \leq y_2$, are put together in a box. What can be said about the elements in the box?

  All the elements belong to $X \cup Y$. The minimum number cannot be smaller than $\max(x_1, y_1)$. On the other hand it cannot exceed neither $f = |(X \cup Y)|$ nor $x_2 + y_2$. What is known about the elements in the box may be written $\langle d, e\rangle(X \cup Y)$, where

$$d = \max(x_1, y_1); \quad e = \min((x_2 + y_2), f)$$

This suggests the possibility of defining the following two closed operations:

Definition:

$$\begin{aligned}\mathbf{X} \sqcap \mathbf{Y} &= \langle a, b\rangle(X \cap Y) \\ \mathbf{X} \sqcup \mathbf{Y} &= \langle d, e\rangle(X \cup Y)\end{aligned}$$

where $\sqcap$ is the symbol for an operation I will call *intersection* of approximate sets, and $\sqcup$ the symbol for the *union* of approximate sets, and

$$\begin{array}{ll} c = |(X \cap Y)| & f = |(X \cup Y)| \\ a = \max(0, (x_1 + y_1 - f)) & b = \min(x_2, y_2, c) \\ d = \max(x_1, y_1) & e = \min((x_2 + y_2), f) \end{array}$$

Let the properties of these operations be investigated.

Commutativity

From the symmetry of the expressions with respect to $\mathbf{X}$ and $\mathbf{Y}$ it is immediate that the two operations are commutative.

Idempotency

Let $\mathbf{X} = \langle x_1, x_2\rangle X$. Since $X \cap X = X \cup X = X$:

$$\mathbf{X} \sqcap \mathbf{X} = \langle a, b\rangle X; \quad \mathbf{X} \sqcup \mathbf{X} = \langle d, e\rangle X$$

where

$$\begin{array}{ll} c = |X| = x & f = |X| = x \\ a = \max(0, (x_1 + x_1 - f)) = max(0, 2x_1 - x) & d = \max(x_1, x_1) = x_1 \\ e = \min((x_2 + x_2), f) = min(2x_2, x) & b = \min(x_2, x_2, c) = min(x_2, x) \end{array}$$

therefore:

Theorem 6 *Neither one of the two operations is idempotent.*

On the other hand, both operations reflect the idempotency of the conventional set operations since if $\mathbf{X} = \langle x_1, x_2\rangle X$, then

$$\mathbf{X} \sqcap \mathbf{X} = \langle a, b\rangle X; \quad \mathbf{X} \sqcup \mathbf{X} = \langle d, e\rangle X$$

where

$$\begin{array}{ll} c = |X| = x & f = |X| = x \\ a = \max(0, 2x_1 - x) = \max(0, 2x - x) = x & d = x_1 = x \\ b = \min(x_2, x) = \min(x, x) = x & e = \min(2x, x) = x \end{array}$$

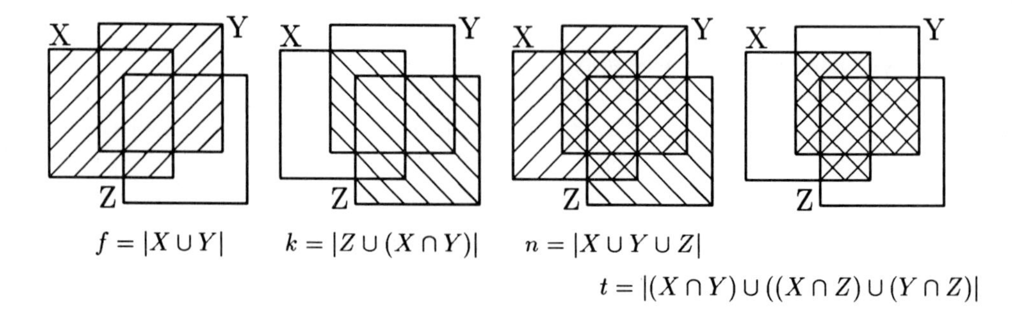

Figure 1. Venn diagrams

3.5. Generalization to Multiple Operands

The classical technique to define an operation for multiple operands consists of two steps: i) defining the operation for two elements and later ii) show that the operation is associative. This is done for the intersection below, assuming that

$$\mathbf{X} = \langle x_1, x_2 \rangle X; \quad \mathbf{Y} = \langle y_1, y_2 \rangle Y; \quad \mathbf{Z} = \langle z_1, z_2 \rangle Z$$

Intersection

The following intermediate results will be used in the proof for the associativity of the intersection.

Lemma 7 *1.-* $\max(a, b) + c = max(a + c, b + c)$
2.- $\max(a, max(b, c)) = max(a, b, c)$
3.- Let

$$f = |(X \cup Y)| \qquad k = |(X \cap Y) \cup Z|$$
$$n = |X \cup Y \cup Z| \qquad t = |(X \cap Y) \cup (X \cap Z) \cup (Y \cap Z)|$$

from the Venn diagrams of Fig.1 it follows that:

$k + f = n + t$

Theorem 8 *The intersection of approximate sets is associative*

Proof:

a. Computing $(\mathbf{X} \sqcap \mathbf{Y}) \sqcap \mathbf{Z}$

$$\mathbf{X} \sqcap \mathbf{Y} = \langle a, b \rangle (X \cap Y)$$

where

$$c = |(X \cap Y)| \qquad f = |(X \cup Y)|$$
$$a = \max(0, (x_1 + y_1 - f)) \qquad b = \min(x_2, y_2, c)$$

therefore

$(\mathbf{X} \sqcap \mathbf{Y}) \sqcap \mathbf{Z} = \langle g, h\rangle(X \cap Y \cap Z)$

where

$$\begin{aligned} g &= \max(0, (a + z_1 - |(X \cap Y) \cup Z)| \\ h &= \min(b, z_2, |(X \cap Y) \cap Z)| \end{aligned}$$

If for convenience we say that $m = |(X \cap Y \cap Z|$, then

$$\begin{aligned} g &= \max(0, (\max(0, (x_1 + y_1 - f)) + z_1 - k) \\ &= \max(0, (\max(z_1 - k, x_1 + y_1 + z_1 - (f + k))) \\ &= \max(0, z_1 - k, x_1 + y_1 + z_1 - (f + k)) \\ h &= \min(\min(x_2, y_2, c), z_2, m) \\ &= \min(x_2, y_2, z_2, c, m) \end{aligned}$$

but

$$f + k = n + t; \quad z_1 - k \leq z - k \leq 0; \quad m \leq c$$

therefore

$$\begin{aligned} g &= \max(0, x_1 + y_1 + z_1 - (n + t)) \\ h &= \min(x_2, y_2, z_2, m) \end{aligned}$$

b. Computing $\mathbf{X} \sqcap (\mathbf{Y} \sqcap \mathbf{Z})$

$\mathbf{Y} \sqcap \mathbf{Z} = \langle a', b'\rangle(Y \cap Z)$

where

$$\begin{aligned} c' &= |(Y \cap Z)| & f' &= |(Z \cup X)| \\ a' &= \max(0, (y_1 + z_1 - f')) & b' &= \min(y_2, z_2, c') \end{aligned}$$

therefore

$(\mathbf{Y} \sqcap \mathbf{Z}) \sqcap \mathbf{X} = \langle g', h'\rangle(Y \cap Z \cap X)$

where

$$\begin{aligned} g' &= \max(0, (a' + x_1 - |((Y \cap Z) \cup X))) \\ h' &= \min(b', x_2, |(Y \cap Z) \cap X)| \end{aligned}$$

if $k' = |((Y \cap Z) \cup X)|$, then

$$\begin{aligned} g' &= \max(0, (\max(0, (y_1 + z_1 - f')) + x_1 - k') \\ &= \max(0, (\max(x_1 - k', y_1 + z_1 + x_1 - (f' + k'))) \\ &= \max(0, x_1 - k', y_1 + z_1 + x_1 - (f' + k')) \\ h' &= \min(\min(y_2, z_2, c), x_2, m)) \\ &= \min(y_2, z_2, x_2, c, m) \end{aligned}$$

but $f' + k' = n + t, x_1 - k' \leq x - k' \leq 0$, and $\leq c$ so

$$g' = \max(0, y_1 + z_1 + x_1 - (n + t)) = g$$
$$h' = \min(y_2, z_2, x_2, m) = h$$

Q.E.D

Therefore the intersection is associative and consequently it can be naturally extended to multiple operands.

Union

Theorem 9 *The union of approximate sets is not associative*

Proof:

i. Computing $(\mathbf{X} \sqcup \mathbf{Y}) \sqcup \mathbf{Z}$

$$\mathbf{X} \sqcup \mathbf{Y} = \langle d, e \rangle (X \cup Y)$$

where

$$f = |(X \cup Y)|; \quad d = \max(x_1, y_1); \quad e = \min((x_2 + y_2), f)$$

$$(\mathbf{X} \sqcup \mathbf{Y}) \sqcup \mathbf{Z} = \langle p, q \rangle (X \cup Y \cup Z)$$

where

$$f = |(X \cup Z)|$$
$$p = \max(\max(x_1, y_1), z_1) = \max(x_1, y_1, z_1)$$
$$q = \min((\min((x_2 + y_2), f) + z_2), n) = \min((x_2 + y_2 + z_2), (f + z_2), n)$$

ii. Computing $\mathbf{X} \sqcup (\mathbf{Y} \sqcup \mathbf{Z})$

It is immediate that

$$\mathbf{X} \sqcup (\mathbf{Y} \sqcup \mathbf{Z}) = \langle p', q' \rangle (X \cup Y \cup Z)$$

where

$$f' = |(X \cup Z)|; \quad p' = \max(x_1, y_1, z_1); \quad q' = \min((x_2 + y_2 + z_2), (f' + x_2), n)$$

and in general $q \neq q'$. Therefore in general union is not associative. *Q.E.D*

This means that the quality of the approximation will depend upon the computation process. Since q and q' are the upper bounds, if one so desires it is legitimate to compute the union in more than one way choosing as an answer the one with the lower value of q.

Corollary 10 *In the particular case of $X = Y = Z$ the union is associative.*

In this case

$$\begin{aligned} f &= |(X \cup Z)| = |X| = x; \quad n = |(X \cup Y \cup Z)| = x \\ p &= \max(\max(x_1, y_1), z_1) = \max(x_1, y_1, z_1) \\ q &= \min((\min((x_2 + y_2), x) + z_2), x) \\ &= \min((x_2 + y_2 + z_2), (x + z_2), x) \\ &= \min((x_2 + y_2 + z_2), x) \end{aligned}$$

therefore $p' = p$ and $q' = q$.

Associativity of the union is not restricted to the particular case above. As a matter of fact the construction of an example showing the lack of associativity may take some effort. Absorption law is not fulfilled either. The two following properties are interesting too.

Theorem 11 $\mathbf{X} \sqcap \mathbf{Y} \sqsubseteq \mathbf{X} \sqsubseteq \mathbf{X} \sqcup \mathbf{Y}$

Proof:

1. $\mathbf{X} \sqcap \mathbf{Y} = \langle a, b\rangle(X \cap Y)$

$$\begin{aligned} f &= |(X \cup Y)| \geq y \geq y_1 \\ a &= \max(0, (x_1 + y_1 - f)) \leq x_1 \\ b &= \min(x_2, y_2, c) \leq x_2 \end{aligned}$$

2. $\mathbf{X} \sqcup \mathbf{Y} = \langle d, e\rangle(X \cup Y)$

$$\begin{aligned} f &= |(X \cup Y)| \geq x \geq x_2 \\ d &= \max(x_1, y_1) \geq x_1 \\ e &= \min((x_2 + y_2), f) \geq x_2 \end{aligned}$$

Q.E.D

The mathematical structure of approximate sets $S = \langle \mathbf{X}, \mathbf{Y}, \ldots \mathbf{Z}, \sqcup, \sqcap \rangle$ deserves further treatment. It is evidently not a lattice on its own, although it is based on the lattice $L = \langle X, Y, \ldots, Z, \bigwedge, \bigvee, \cap, \cup \rangle$. Although further discussion is beyond the scope of this paper, the following simple results are worth mentioning.

Let $s = |\bigvee|$,

Theorem 12 *a)* $\mathbf{0} = \langle 0, 0\rangle \bigwedge$ *is a zero element for* $\sqcap$ *and a unit element for* $\sqcup$.
b) $\mathbf{1} = \langle s, s\rangle \bigvee$ *is a unit element for* $\sqcap$ *and a zero element for* $\sqcup$.

Proof: Let $\mathbf{Y} = \langle 0, 0\rangle \bigwedge$ and $\mathbf{X} = \langle x_1, x_2\rangle X$. In this case:

$(X \cap Y) = \bigwedge$	$(X \cup Y) = X$
$c = \|(X \cap Y)\| = \|\| \leq x_1$	$f = \|(X \cup Y)\| = x \geq x_2$
$a = \max(0, (x_1 + 0 - x)) = 0$	$b = \min(x_2, 0, c) = 0$
$d = \max(x_1, 0) = x_1$	$e = \min((x_2 + 0), x) = x_2$

On the other hand let $\mathbf{Y} = \langle s, s\rangle \mathsf{V}$. In this case:

$$\begin{array}{ll} (X \cap Y) = X & (X \cup Y) = \mathsf{V} \\ c = |(X \cap Y)| = x & f = |(X \cup Y)| = s \geq x \\ a = \max(0, (x_1 + s - s)) = x_1 & b = \min(x_2, s, x) = x_2 \\ d = \max(x_1, s) = s & e = \min((x_2 + s), s) = s \end{array}$$

from where Theorem 12 follows *Q.E.D*

3.6. Concept description

Since "subset of X" and "concept" are two isomorphic concepts, approximate sets may be used to describe concepts approximately. The same concept may be described with more or less approximation, as well as making reference to different sets. The following lemmas establish some simple statements in this concern.

Let $\langle x_1, x_2\rangle X$ be the approximate description of a concept, C.

Lemma 13 *The concept C may be also described by $\langle x_1, x_2\rangle X$ where $Z = X - \{$ all elements of $x \mid \neg C\}$.*

By definition of $Z, |Z| \doteq x_2$ and none of the elements of X included in $\neg C$ is an element of Z.

Lemma 14 *$\langle x_1 - \epsilon, x_2 + \eta\rangle X$ where $x_1 - \epsilon \geq 0, x_2 + \eta \leq |x|$, is also an approximate description of C.*

4. EXTENDING AN IMPRECISE DESCRIPTION TO A SUBSET

Let $\mathbf{X} = \langle x_1, x_2\rangle X$, be an approximate description of a concept C and $\langle z_1, z_2\rangle X$ be an approximate subset of X I will call $\mathbf{Z}$. To extend the description of C to $\mathbf{Z}$ is to describe it by means of an expression $\langle z_1, z_2\rangle Z$.

An example will illustrate the concept Let X be a set of 8 people and it is known that:

1. M of them, $5 \leq m \leq 7$, are married. The concept "married" is approximatetly described by $\langle 5, 7\rangle X$.
2. Between 3 and 6 elements of X are women. $Z = \langle 3, 6\rangle X$.

Question: How many married women are there? That is, what is the approximate description of the concept "married" by means of an expresion of the type $\langle z_1, z_2\rangle Z$?

Finding the answer is a simple matter, as follows. Since the concept of reference is described by $\langle z_1, z_2\rangle X, |X| = x$, by Lemma 14 it is also described by $\langle z_1, z_2\rangle Z$, where $|Z| = x_2$ and the concept "married women" will be described by the intersection

$\langle x_1, x_2\rangle X \cap \langle z_1, z_2\rangle Z$,

In the case of our example the answer will be

$\langle 5, 7\rangle X \cap \langle 3, 6\rangle Z = \langle a, b\rangle Z$

where

$$\begin{array}{ll} c = |(X \cap Z)| = |Z| = 6 & f = |(X \cup Z)| = |X| = 8 \\ a = \max(0, (x_1 + y_1 - f)) = 0 & b = \min(x_2, y_2, c) = 6 \end{array}$$

5. EXTENDING THEOREM 4 TO APPROXIMATE SETS

Let us check whether expressions similar to those of Theorem 4 hold for approximate sets.

Let $\mathbf{D_a} - \mathbf{R_a}$ and $\mathbf{D_b} - \mathbf{R_b}$ be approximate statements predicated of a certain entity. $\mathbf{D_a} \sqcap \mathbf{D_b}$ represents "what can be said about the z elements found in both v elements of D_a and w elements of D_b." Therefore, the following statements will also be correct

$$\mathbf{D_a} \sqcap \mathbf{D_b} \quad - \quad \mathbf{R_a}$$
$$\mathbf{D_a} \sqcap \mathbf{D_b} \quad - \quad \mathbf{R_b}$$

On the other hand after having placed $\mathbf{R_a}$ and $\mathbf{R_b}$ in a box $\mathbf{R_a} \sqcup \mathbf{R_b}$ represents "what can be said about the elements in the box." If $\mathbf{R_a}$ can be said about the entity $\mathbf{D_a} \sqcap \mathbf{D_b}$ and $\mathbf{R_b}$ can be said about the same entity, $\mathbf{R_a} \sqcup \mathbf{R_b}$ can be said too. Therefore the statement

$$\mathbf{D_a} \sqcap \mathbf{D_b} - \mathbf{R_a} \sqcup \mathbf{R_b}$$

is also correct.

Similarly, from

$$\mathbf{D_a} \quad - \quad \mathbf{R_a}$$
$$\mathbf{D_b} \quad - \quad \mathbf{R_b}$$

it follows that

$$\mathbf{D_a} \quad - \quad \mathbf{R_a} \sqcap \mathbf{R_b}$$
$$\mathbf{D_b} \quad - \quad \mathbf{R_a} \sqcap \mathbf{R_b}$$

are also correct statements and therefore

$$\mathbf{D_a} \sqcup \mathbf{D_b} - \mathbf{R_a} \sqcap \mathbf{R_b}$$

is also a correct statement. We may therefore state:

Theorem 15 *Let* $\mathbf{D_a} - \mathbf{R_a}$ *and* $\mathbf{D_b} - \mathbf{R_b}$ *be approximate statements predicated of a certain entity, then*
a) $(\mathbf{D_a} \sqcap \mathbf{D_b}) - (\mathbf{R_a} \sqcup \mathbf{R_b})$
b) $(\mathbf{D_a} \sqcup \mathbf{D_b}) - (\mathbf{R_a} \sqcap \mathbf{R_b})$
are also correct statements.

Since the union of approximate sets is not associative, when extending Theorem to multiple premises the final result will depend upon the sequence of operations. This means the the final statement will be more or less approximate depending upon the sequence, although it will always be correct.

Similarly Theorem 15 can also be generalized. Let $\mathbf{D}_i - \mathbf{R}_i, i = 1, 2, \ldots n$, and let $E(\mathbf{D}_1, \mathbf{D}_2, \ldots \mathbf{D}_i)$ be a syntactically well formed expression made up of $\mathbf{D}_1, \mathbf{D}_2, \ldots \mathbf{D}_i$ with the connectives $\sqcap$ and $\sqcup$, and adequate parenthesis. Let now $E_d(\mathbf{R}_1, \mathbf{R}_2, \ldots \mathbf{R}_i)$ be the expression obtained from E by leaving the elements in the same ordered sequence but interchanging $\sqcap$ and $\sqcup$ on the one hand and **0** and **1** by the other hand. The following theorem is an extension of Theorem 15, of almost immediate proof.

Theorem 16 *Let* $\mathbf{D}_i - \mathbf{R}_i, i = 1, 2, \ldots n$ *be correct statements,*

$$E(\mathbf{D}_1, \mathbf{D}_2, \ldots \mathbf{D}_i) - Ed(\mathbf{R}_1, \mathbf{R}_2, \ldots \mathbf{R}_i)$$

is also a correct statement.

6. CONCLUSION

Although the concepts presented here may have far reaching consequences the main conclusion we were aiming at when undertaking this research was to illustrate that inference rules, usually intuitively established on evidence grounds, are not necessarily primitive ideas, but formal conclusions obtained through correct reasoning. In this specific example, dealing with an approximate language, this has been the case.

REFERENCES

1. D. Dubois and H. Prade. Fuzzy sets and Systems. Academic Press. New York, 1980
2. G. Shafer. A Mathematical Theory of Evidence. Princeton University Press. 1976
3. D. McDermot and J. Doile. Non Monotonic Logic. Artificial Intelligence . 13(1,2):41-72. 1980
4. R.K.Bhatanagar and L.N.Kanal. Handling Uncertain Information: A Review of Numeric and Non-Numeric Methods. In L.N. Kanal and J.F.Lemner, editors, Uncertainty in Artificial Intelligence. North Holand, Amsterdam, 1986[
5. J.Miró-Juliá. Contribución al Estudio de la Demostración Automática. Tesis Doctoral. Universitat de les Illes Balears, 1988
6. J. Miró. Lectures on Direct Inference. Presented at the University of Warsaw, Oct. 1989 Universitat de les Illes Balears. August 1989

Uncertainty in Intelligent Systems
B. Bouchon-Meunier et al. (Editors)

An Efficient Tool for Reasoning with Belief Functions

Hong Xu

IRIDIA - Université libre de Bruxelles, 50 av. F. Roosevelt, CP 194/6,
B-1050, Brussels, Belgium

In this paper, we present an efficient implementation of Dempster-Shafer theory of belief functions. The system is called TRESBEL (a Tool for REaSoning with BELief functions). By using some optimization techniques focusing on the computational efficiency, TRESBEL has its distinct advantage of performing fast computations and incremental changes. A particular attention has also been devoted to the effectiveness of the user-interface. Some comparisons of the features with those of other similar existing systems are given.

1. INTRODUCTION

Dempster-Shafer theory [6, 11] is generally regarded as a useful tool for the task of representing and manipulating uncertain knowledge. It provides flexible input requirements and an effective method for combining information obtained from multiple knowledge sources.

Using Dempster-Shafer theory, knowledge can be represented in a so-called belief function network. A belief function network consists of a set of variables, a set of subsets of variables, and a set of belief functions defined on the variables or the subsets of variables. The belief functions represent the available evidence. The set of variables, denoted by $\mathbf{X}$, represents the universe of discourse of the problem. For each variable X, we use $\mathcal{W}_X$ to denote the set of its possible values, and call $\mathcal{W}_X$ *the frame of X*. Any non-empty subset h of $\mathbf{X}$ represents the existence of a relationship among the variables in it, and its frame $\mathcal{W}_h$ is the Cartesian product of all $\mathcal{W}_X$ for X in h. we call the elements of $\mathcal{W}_h$ as *configurations of h*. A *basic probability assignment (bpa)* m on X is a function which assigns a value in [0, 1] to every subset $\boldsymbol{a}$ of $\mathcal{W}_X$ and satisfies the following axioms: (1) $m(\emptyset) = 0$; and (2) $\Sigma\{m(\boldsymbol{a}) \mid \boldsymbol{a} \subseteq \mathcal{W}_X\} = 1$. A *belief function* Bel on X can be stored as a set of bpa's $m(\boldsymbol{a})$ where $\boldsymbol{a} \subseteq \mathcal{W}_X$. The subsets $\boldsymbol{a}$ for which $m(\boldsymbol{a}) > 0$ are called *focal elements* of Bel. The simplest belief function is the one with $m(\mathcal{W}_X) = 1$, called *vacuous belief function*..

To evaluate a belief function network, we have to (1) extend all belief functions to the global frame and combine them, the resulting belief function is called the *global belief function* ; (2) project the global belief function to the frame of each variable, the results are called *the marginals for each variables.*

However, the computational complexity of Dempster's rule of combination, the pivotal mechanism of this theory, has been a major obstacle to its applications. To solve this problem, Shafer and Shenoy proposed a local computation technique for computing belief functions [7,

8, 9]. Based on this technique and the observation that an arbitrary belief function network can be represented as a hypergraph [2], which can be in turn be embedded in what is called a Markov tree [16], several implementations of the Dempster-Shafer theory have been developed [1, 15].

Using local computations, the marginal for each variables can be computed by propagating belief functions through the Markov tree in a message-passing scheme. After the marginals for all the nodes have been computed, one may want to change one or more of the prior belief functions. Then the impact of the changes will be re-propagated to all the other nodes. To this respect, there may be in general repeated computations during both the propagation and the re-propagation processes, which greatly affects the efficiency of the computation. Moreover, the large number of set-intersections and set-comparisons needed in Dempster's rule of combination may heavily impact the speed of computation.

This paper attempts to present a more efficient implementation of Dempster-Shafer theory to ameliorate the above problems. Two optimization techniques are presented: the first one avoids all the repeated computations in the propagation process, the second one uses bit-arrays in the storage and computation of the belief functions. The system TRESBEL is developed using these two techniques, thus performing fast computations and incremental changes is one of its distinct features. TRESBEL also has a user-friendly interface, which makes itself an easy-to-use flexible system.

The rest of the paper is organized as follows: In section 2, we briefly present the optimization techniques used by TRESBEL to speed up the inference process. In section 3, we demonstrate TRESBEL through an example, focusing on the features' comparison with other similar existing systems. Finally in section 4, we give some conclusions.

2. OPTIMIZATION TECHNIQUES

In this section, we briefly describe the optimization techniques used for an efficient implementation in design of the system TRESBEL. Details of these techniques can be found in [12, 14].

2.1. Efficient Propagation Scheme

It has been mentioned [6] that if the belief function network can be represented as a so-called Markov tree, the belief functions can be "propagated" in the Markov tree, producing a result in the marginals of the global belief function for each of the nodes. In a nutshell, the belief function propagation can be described as a message-passing scheme: each node in the Markov tree sends its message to one of its neighbours after it has received the messages from all of its other neighbours, and the result of the propagation on each node is computed by combining its own belief function (prior belief function) and the messages from all of its neighbours. The propagation process can be illustrated as in Figure 1.

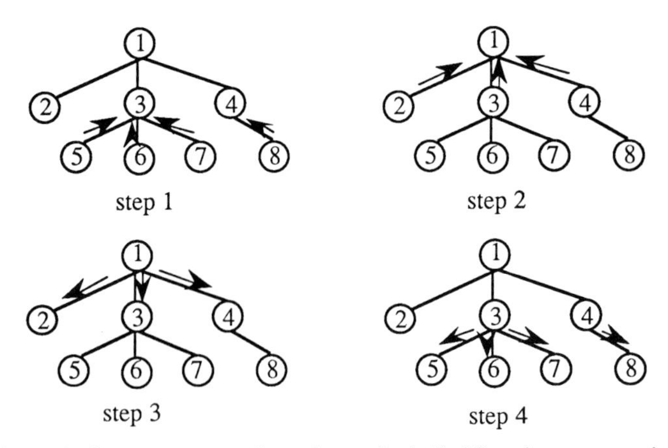

Figure 1. the message-passing scheme for belief function propagation

In TRESBEL, some intermediate results are stored and reused during propagation, which avoids all the repeated computations. For example, at step4 of Figure 1, node 3 is sending the messages to node 5, 6, and 7. According to the message-passing scheme, message to node 5 is computed by combining messages from node 1, 7, and 6 to node 3 and the belief function of node 3; message to node 6 is computed by combining messages from node 1, 7, and 5 to node 3 and the belief function of node 3. Obviously, there are some repeated computations. In TRESBEL, all these repeated computations are avoided. Table 1 compares the number of combinations performed at each node of the Markov tree during propagation between using and not using the intermediate variables. Given a rooted Markov tree G, for each node v, $\mathcal{P}_v$ is the father node of v, and $\mathcal{Ch}_v$ is the set of children nodes of v.

Table 1
Comparison of computation between using and not using intermediate variables

Number of neighbors	Number of combinations at each node	
	using intermediate variables	not using intermediate variables
$\lvert\{\mathcal{P}_v\}\rvert = 1, \lvert\mathcal{Ch}_v\rvert = 0$	1	1
$\lvert\{\mathcal{P}_v\}\rvert = 1, \lvert\mathcal{Ch}_v\rvert = n\ (>0)$	3n	$(n + 1)^2$
$\lvert\{\mathcal{P}_v\}\rvert = 0, \lvert\mathcal{Ch}_v\rvert = n\ (>0)$	3n - 3	$n^2 - n + 1$

Moreover, when some prior belief functions are changed after propagation, the corresponding messagess passed between the nodes and intermediate variables will be changed, and the marginals for the nodes will be recomputed. Making full use of the unchanged messages and of intermediate variables, we can re-propagate only the changed values, thus improving the efficiency of propagation.

2.2. Using Bit-arrays for Belief Function Storage and Computation

Another technique of TRESBEL is using bit-arrays for storing and computing belief functions. In general, a belief functions is stored as a set of pairs ($\mathbf{a}$, v) where the subset $\mathbf{a}$ is a focal element and v = m($\mathbf{a}$) is the value assigned on $\mathbf{a}$. On the average, less space is required for storing subsets by using bit-array representations than using list representations. For example, suppose we have a belief of 0.9 that the true value of a variable A with frame {red, green, yellow} is in {red}. Using bit-arrays, the belief function is stored as: ((#*100 0.9) (#*111 0.1)), instead of (((red) 0.9) ((red green yellow) 0.1)).

The big number of set-intersection needed in Dempster's combination is a factor strongly influencing the complexity of computation. Using bit-arrays to represent the subsets, the intersection of two focal elements just need "AND" operation on the two bit-array representations. Moreover, when two belief functions defined over different frames are to be combined, they have to be "projected" or "extended" to the same frame first. The operations of projection and extension require a lot of set-manipulations as well. In TRESBEL, algorithms are defined on "projection" and "extension" of the bit-array representations of the subsets, where all those set-manipulations are substituted by the bit-array operations. For example, for two variables A and B in a belief function network, $\mathcal{W}_A$ = (a b)×(s t), $\mathcal{W}_B$ = (a b)×(x y)×(s t), and $\mathbf{a}$ = ((a x s) (a y s) (b x s) (b y s) (b y t)) is a focal element of the belief function of B. Suppose we need to project this belief function to the frame of A, we need to project $\mathbf{a}$ to $\mathcal{W}_A$ first. This can be done as follows:

((a x s) (a y s) (b x s) (b y s) (b y t))
↓ ↓ ↓ ↓ ↓
(a s) (a s) (b s) (b s) (b t)

Thus, the result is ((a s) (b s) (b t)), a subset of $\mathcal{W}_A$. If we use bit-array representation, $\mathbf{a}$ can be represented as #*10101011, and the projection of $\mathbf{a}$ to $\mathcal{W}_A$ can be implemented as follows:

$\underline{10}\ \underline{10}\ \underline{10}\ \underline{11}$
or or
$\underline{10}$ $\underline{11}$

Because local computation mainly consists of applying projection, extension and combination, improving the efficiency of set-manipulation is essential to the overall speed of the propagation process.

3. TRESBEL: A TOOL FOR REASONING WITH BELIEF FUNCTIONS

3.1. Overview of TRESBEL

TRESBEL is a new implementation of Dempster-Shafer theory using the above techniques, embedded in a graphical environment to develop Dempster-Shafer models. It is implemented in Allegro Common Lisp with Common Windows (by Franz Inc), running on a SUN-3/60 Workstation under SUN OS 4.0.3.[13]. The environment allows both creation (structural and quantitative) and evaluation of belief function networks. A typical working session with TRESBEL is illustrated in Figure 2.

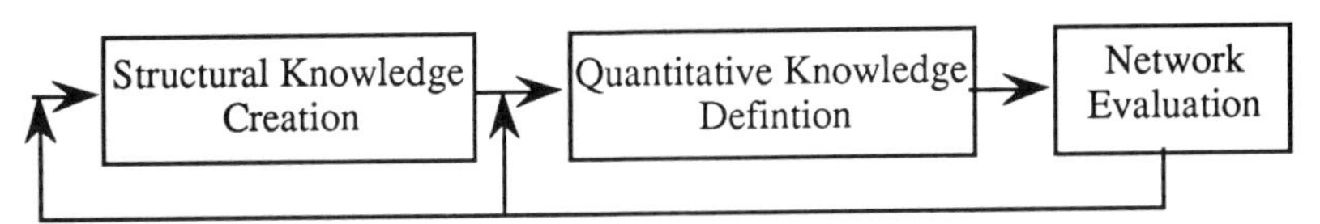

Figure 2. working session with TRESBEL

One of the most distinct features of TRESBEL is the response time. TRESBEL has a speed advantage compared with similar previous systems such as Delief [15] and MacEvidence [1]. Table 2 illustrates the speed comparison between these systems where $|\mathcal{X}|$, $|\mathcal{H}|$ and $|\mathcal{W}_{\mathcal{X}}|$ denote the number of the nodes, the number of the relations and the size of the global frame of a belief function network, respectively. The user can query the computation result of a particular variable once a time or of all variables simultaneously. Furthermore, when one or more belief functions are changed after the initial evaluation of the belief function network, the system can provide a quicker response (T_2 in Table 2).

Table 2
Comparison of propagation time (T_1) and re-propagation time (T_2) among three systems

Size of the network			Propagation Time (seconds) MacEvidence		Delief*		TRESBEL	
$\|\mathcal{X}\|$	$\|\mathcal{H}\|$	$\|\mathcal{W}_{\mathcal{X}}\|$	T_1	T_2	T_1	T_2	T_1	T_2
3	2	12	5	5	1.22	1.22	0.24	0.14
7	5	2^7	51	51	11.52	11.52	1.36	0.92
15	18	2^{15}	1168	1168	202.36	202.36	6.82	4.94
12	8	3^7*2^9	1351	1351	222.27	222.27	9.42	6.98

As the nucleus of TRESBEL is based on Delief, TRESBEL has its genesis in Delief: it constitutes a graphical specification environment, into which the inference mechanism (calculus of belief function) is incorporated. Users can easily interact with the system to solve problems through the interface. Besides these features, TRESBEL provides both a graphical and a functional interface for creating and evaluating knowledge schema as belief function networks. Users can create and evaluate the network by using graphical facilities such as mouse buttons, menu facilities etc., or by calling the Lisp functions alternatively. The graphical representation is available according to the users' requirement. Thus, TRESBEL may be used both as a Lisp module implementing uncertain reasoning with belief functions inside a more complex system, or as a stand-alone tool.

* Here is the previous version of TRESBEL using the same propagation scheme as that of Delief.

3.2. Using TRESBEL

In this section, we will demonstrate the system through the three steps in Figure 2 by considering the following example, abstracted from [3].

> Shortness-of-breath (dyspnoea) may be due to tuberculosis, lung cancer or bronchitis, or none of them. A recent visit to Asia increases the chances of tuberculosis, while smoking is known to be a risk factor for both lung cancer and bronchitis. The results of a single chest X-ray do not discriminate between lung cancer and tuberculosis, as neither does the presence or absence of dyspnoea.

3.2.1. Structural Knowledge Creation

The framework of a belief function network is created as a graphical network consisting of variable nodes and joint variable nodes (also called relation nodes). The creation of a structural belief function network is essentially a process of creating variables, defining their frames, and defining relations among the variables (creating joint variables).

Variables are created by clicking the left mouse button over any blank area of the main window. In response, the system records the click position and pops up a definition window for the user to specify the name and the frame of the variable. Variables can also be created by calling Lisp functions. For example, calling function:

```
(new-variable "Tuberculosis" '(yes no))
```

will create a variable named "Tuberculosis" with frame {yes, no}.

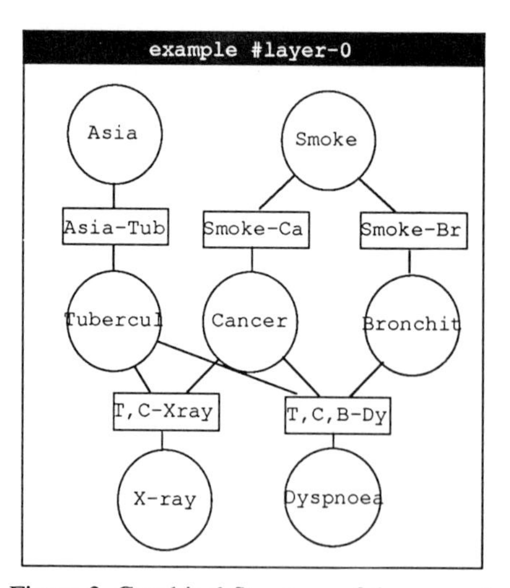

Figure 3. Graphical Structure of the example

Joint variables are created by selecting at least two variable nodes to be included as members of the joint variables. Selecting multiple variable nodes is achieved by pressing the shift-key and clicking the right (or middle) mouse button over the area of the variable nodes. When all the member variable nodes have been selected, the user can choose a suitable location and click the left mouse button to specify the position where the joint variable node should be. A pop-up window appears for the user to enter the name of the new node. The joint variable can also be created by calling Lisp functions. Calling function:

```
(new-relation "Asia-Tub" '("Asia" "Tuberculosis"))
```

yields the creation of the node "Asia-Tub". Figure 3 illustrates the graphical network of the example.

3.2.2. Quantitative Knowledge Definition

Quantitative knowledge definition is to provide evidence for each node in the network. Clicking the right mouse button on the node's area will open the operation menu of the node. Selecting the "Evidence" item will invoke another window which is used as belief function definition facility, as shown in Figure 4. It comprises two sub-windows: the left one displays the belief function defined so far, and the right one contains a list of the node's frame elements and the operation choices buttons.

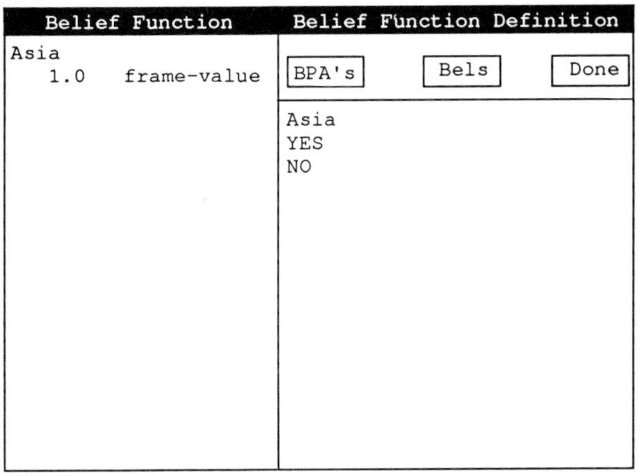

Belief Function | Belief Function Definition
Asia
1.0 frame-value
BPA's | Bels | Done
Asia
YES
NO

Figure 4 the evidence definition facilities

Suppose we have a belief of 0.8 that the patient in the question has visited Asia recently. This means we need to define a focal element, the set {yes}, and assign it a value of 0.8. To do so, we select the frame element "yes" with the mouse button, the system highlights the selected element. Then, we click the "BPA's" button, a pop-up window appears and prompts for a value to assign.

After the user entering the number, the system returns the new focal element added to its belief function and the value of the entire frame approciately adjusted. Clicking the button "Done", the system returns to the main window.

Belief functions for joint variable nodes are defined in exactly the same way. Alternatively, the belief function for a node can be defined by calling functions. For example, if we want to express a belief of 0.6 on the relation that a recent visit to Asia implied the presence of Tuberculosis, we can define a belief function for "Asia-Tub" by calling:

```
(set-bpa "Asia-Tub" '((((yes yes) (no yes) (no no)) 0.6))).
```

Figure 5 shows the screen dump of the system including the inputs of the example.

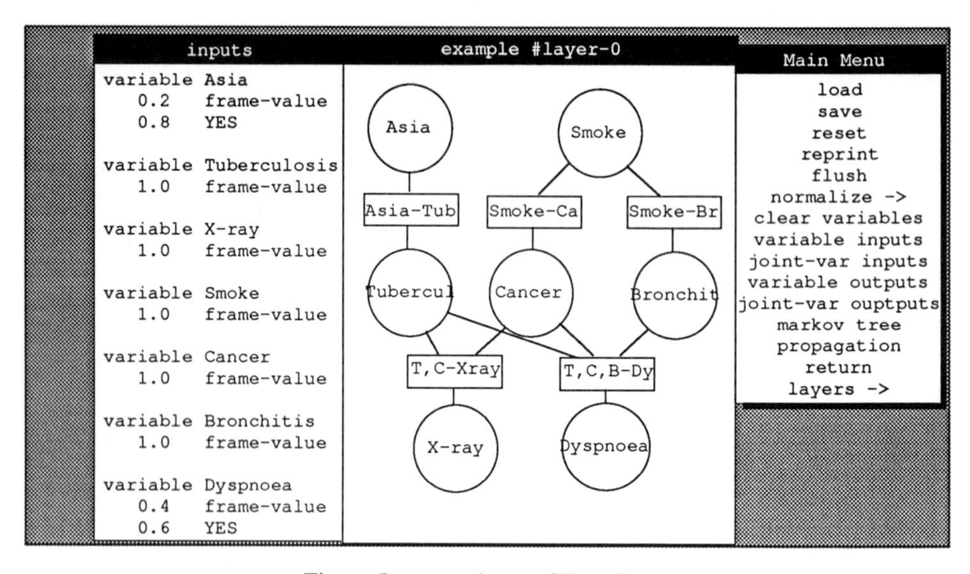

Figure 5. screen dump of TRESBEL

3.2.3. Network Evaluation

To propagate evidence in the belief function network, the user can simply select the main menu item "Propagation" or call the Lisp function `(propagation)`. The system will do the simultaneous propagation, and display the Markov tree used for propagation and the resulting values of the variables. The user may also view the propagation results for joint variables by selecting the appropriate menu item.

If the user is only interested in the result of one or two of the variables, he/she can select the "output" item on the menu for that variable. The system can perform partial propagation in order to calculate the result of a specific variable, thus the result is computed quicker. This can also be achieved by calling Lisp functions. For example, by calling the Lisp function,

```
(get-bpa "Dyspnoea")
```

the user can get the result for the variable "Dyspnoea".

Moreover, it is also important that the environment be able to display the Markov tree employed for the belief function propagation and examine the intermediate results of the propagation with the Markov tree. TRESBEL provides this facility for the users, helping them to better understand why certain results are obtained. Figure 6 shows an example for displaying the message passing through the nodes in the Markov tree.

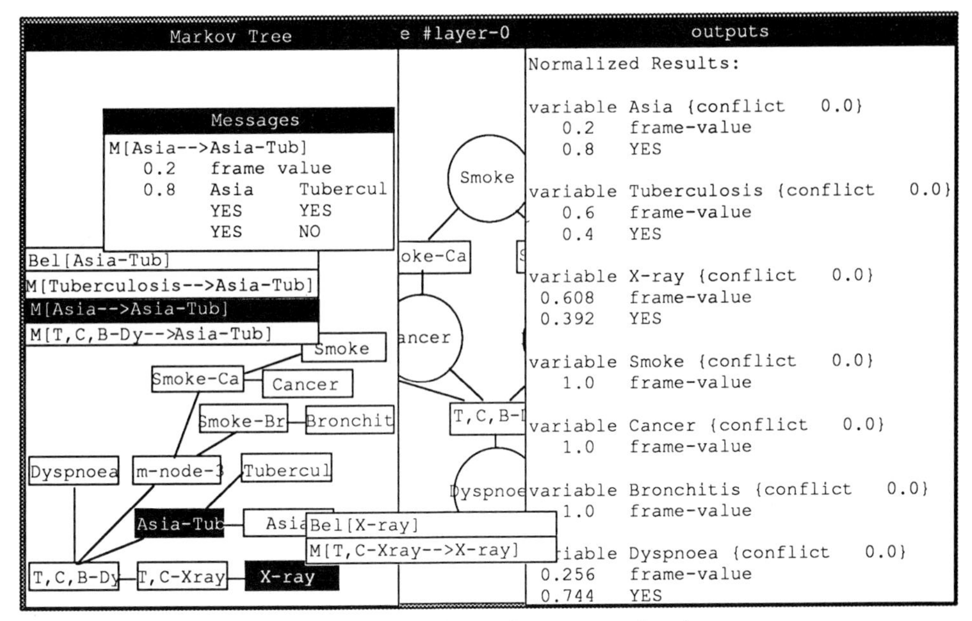

Figure 6. A typical case for message showing

CONCLUSIONS

This paper has presented a new efficient tool for belief function propagation, including some optimization techniques. Besides addressing issues of computation efficiency, we have tried to make TRESBEL an easy-to-use flexible system. For instance, a full graphical interface has been built for it. As a result, TRESBEL appears as a self-contained tool used for developing Dempster-Shafer models. Moreover, it is being used as an independent module for creating and evaluating belief function network in a family of experimental system for integrating knowledge representation and uncertainty management [4].

Though it has been first developed for belief functions, the local computation technique may be generalized to other uncertainty calculi [10]. Correspondingly, the implementation techniques discussed here may be applied to any calculus to which local computation applies. Abstract form TRESBEL, a general uncertainty propagation implementation [5] for propagating belief functions, probabilities and possibilities has been implemented.

ACKNOWLEDGMENTS

The author greatly acknowledge Philippe Smets, Alessandro Saffiotti and Yen-Teh Hsia for the useful discussion and worthy comments on the drafts. The author would also like to thank Robert Kennes and Elisabeth Umkehrer for their help in this work. This work has been supported by a grand of IRIDIA, Universié libre de Bruxelles.

References

1. Hsia Y. and Shenoy P. P. (1989) "MacEvidence: A Visual Environment for Constructing & Evaluating Evidential Systems",*Working Paper* No. 211, School of Business, University of Kansas, Lawrence, KS.
2. Kong A. (1986) "Multivariate Belief Functions & Graphical Models", Ph.D dissertation, Department of Statistics, Harvard University, Cambridge, MA.
3. Lauritzen S. L. and Spielgelhalter D. J. (1988) "Local Computation with Probabilities on Graphical Structures and Their Application to Expert Systems", *Journal of the Royal Statistical society*, Series B, 50, No. 2, pp. 157-224.
4. Saffiotti A. (1991) "Integrating Uncertainty and Knowledge Representaion: the Mikic+TresBel Experiment", ARCHON Task 710 Public Delivable, IRIDIA, Univerisé libre de Bruxelles.
5. Saffiotti A. and Umkehrer E. (1991) "Pulcinella: A General Tool for Propagating Uncertainty in Valuation Networks", *Proc. 7th Uncertainty in Artificial Intelligence* D'Ambrosio B. D., Smets Ph. and Bonissone P. P. (eds.) San Mateo, Calif.: Morgan Kaufmann. pp. 323-331.
6. Shafer G. (1976) *A Mathematical Theory of Evidence*, Princeton University Press.
7. Shenoy P. P. and Shafer G. (1986) "Propagating Belief Functions with Local Computations", *IEEE Expert*, 1(3), pp.43-52.
8. Shafer G.and Shenoy P. P. (1988) "Local Computation in Hypertrees",*Working Paper* No. 201, School of Business, University of Kansas, Lawrence, KS.
9. Shafer G., Shenoy P. P. and Mellouli K. (1987) "Propagating Belief Functions in Qualitative Markov Trees", *International Journal of Approximate Reasoning*, 1:349-400.
10. Shenoy P. P. (1989) "A Valuation-Based Language for Expert Systems", *International Journal of Approximate Reasoning*, 3:383-411.
11. Smets Ph. (1988) "Belief Functions", *Non Standard Logics for Automated Reasoning* Smets Ph., Mamdani A., Dubois D. and Prade H. (eds.), Academic Press, London, pp. 253-286.
12. Xu H. (1991) "An Efficient Implementation of Belief Function Propagation", *Proc. 7th Uncertainty in Artificial Intelligence* D'Ambrosio B. D., Smets Ph. and Bonissone P. P. (eds.) San Mateo, Calif.: Morgan Kaufmann, pp. 425-432.
13. Xu H. (1991) "TresBel User's Manual", *Technical Report* TR/IRIDIA/91-11, IRIDIA, Université Libre de Bruxelles.
14. Xu H. and Kennes R. (1992) "Steps Towards an Efficient Implementation of Dempster-Shafer Theory" *Advances in the Dempster-Shafer Theory of Evidence* Fedrizzi M., Kacprzyk J., and Yager R. R. (eds.), To appear.
15. Zarley D., Hsia Y. and Shafer G. (1988) "Evidential Reasoning Using DELIEF", *Proc. of the 7th National Conference on Artificial Intelligence* Paul St.(eds.), MN, 1, pp. 205-209.
16. Zhang L. (1988) "Studies on Finding Hypertree Covers of Hypergraphs" *Working Paper* No. 198, School of Business, University of Kansas, Lawrence, KS.

CHAPTER 3:

FUZZY SET METHODS

A Fuzzy Expert System For On-Line Diagnosis

Madjid Fathi-Torbaghan
Dirk Danebrock
Joachim Peter Stöck

Universität Dortmund, FB Informatik LS 1, Postfach 500500,
W-4600 Dortmund 50, Germany

The development of expert systems for monitoring and diagnosis is significant for safety management of machines and plants. Development tasks for such systems are the reduction of interrupt situations and the improvement of reliability and lifespan of workpieces and plant components. The task of technical diagnostic is to detect breakdowns by fault type, fault position, fault reason and the time of occurrence[1.2].
Presently the interest in systems with online-relation to the plant is increasing. Such online systems diagnose defects alone from data that they get via the sensors from the plant.
This paper describes an architecture for a real time application in fault diagnosis. The architecture will use the blackboard model and the theory of fuzzy sets for technical diagnostic.

1. Motivation

Expert systems have modeled uncertainty and imprecision in various ways. Most of the methods of dealing with uncertainty and imprecision in expert systems have been ad hoc, in the sense that there is no underlying theory to support them. We applied the well-established theory of fuzzy sets and fuzzy logic to the problem of modeling imprecision and uncertainty in expert systems under real-time constraints.
Hard real-time systems are those systems in which the correctness of the system depends not only on the logical result of computation, but also on the time at which the results are produced. Examples of this type of real-time system are command and control systems, process control systems and flight control systems.
In summary, real-time systems differ from traditional systems in that deadlines or other explicit timing constraints are attached to tasks, the systems are in a position to make compromises, and faults, including timing faults, may cause catastrophic consequences. Thus, real-time systems solve the problem of missing deadlines in ways specific to the requirements of the target application. However, it should be said that the sooner a system determines that a deadline is going to be missed, the more flexibility it will have in dealing with the exception.

2. The System Structure

This section describes the construction of an online-diagnosis system. The scheme of the system is shown in Figure 1.

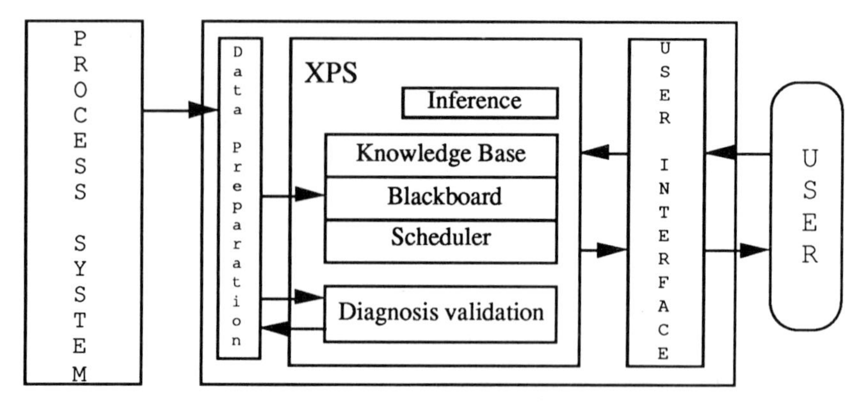

Figure 1: Scheme of an Online expert system

The task of the process system is to monitor the plant. Therefore a process computer reads significant data from the process by scanning sensors. With this data the process computer decide whether to transmit the data to the expert system computer or, if the data represent an alarm situation, to interrupt the plant. Data that represent the normal state of the plant will not be transmitted.
Because of the fact that determine a diagnosis takes much more time than scanning the sensors and transmitting their values to the expert system computer, the amount of data to diagnose must be reduced, buffered and ordered by their priority [4, 7, 8]. This is done by the data preparation module, which is a parallel process to the expert system process. The reduction of data is done by classifying numerical data in several linguistic values (e.g. small, normal, high etc.) [9, 10, 11]. The data set with the highest priority is transmitted to the expert system, which has the task to determine defects. It is possible to interrupt the expert system if data with higher priority occur and to restart it with new values.
The expert system takes the values of the data preparation module and tries to make a diagnosis. The architecture of the expert system is based on the blackboard model, where different knowledge sources communicate by a common database (the blackboard). Knowledge sources model single functional groups (e.g. the cooling system of a plant). The moderator has to determine how much a knowledge source can contribute to solve the problem and activates the knowledge source with the highest priority. The knowledge is represented by rules and frames, where rules have a fuzzy part in the premise and the conclusion [1].
In the time needed for a diagnosis, the state of diagnosis can change in such a way, that the actual diagnosis is faulty. Therefore it is important to check the diagnosis with the help of actual data. By this means it is possible for example to recognize the faults based on transfer faults.

3. The data preparation module

Let us now look to the structure of the data preparation module. As said above this module reduces the amount of data that has to be diagnosed and orders them by their priority. To do this, each value from the sensors is converted to linguistic variables. The scope of each variable consists of values which give the expert system a point for a possible diagnosis. Reduction is done by considering similar data sets only once, because they points to the same diagnosis. If we have two data sets S an T and sensor values s_i,t_i with $S=s_1,...,s_n$ and $T=t_1,...,t_n$ this data sets are similar if $\forall i \in \{1,...,n\}$: inRange($s_i$)=inRange($t_i$), where inRange is

a function that returns the name of the linguistic variable with the highest membership of the sensor value.
To order the data sets we use a number we have called dangerous level. The dangerous level describes the possibility that a damage at the machine will occur. For each linguistic variable we compute this dangerous level and combine this numbers to a weight for the whole data set. Typically a sensor value consists of five ranges: the high danger range, the high range, the normal range, the low range and the low danger range. The normal range consists of values that are near the normal state of the system. These values aren´t transmitted to the XPS and will only be considered for building trends. The high/low range consist of values above/under the normal expected values of the sensor. If they are considered in time, a breakdown of the machine could be prevented. These ranges are often subdivided into smaller ranges, where each range points to a diagnosis. The danger ranges consist of values near the alarm states. Values in these ranges must be considered immediately to prevent more damage.
Each range of a sensor is associated with a number. Higher numbers means that the values in this range are more dangerous than values in ranges with lower numbers. By executing the inRange function we get one range for each sensor and therefore we get one number. This number is called weight of the sensor (ws).

Example: Imagine two flow sensors. Each of them is divided into the five ranges explained above. The ranges are associated with the following numbers: high and low danger range $\Rightarrow$ 2, high and low range $\Rightarrow$ 1 and normal range $\Rightarrow$ 0. Now if the inRange function returns for sensor 1 the high range, then the weight for this sensor is 1.

As said above this single weights must be combined to one number that expresses the dangerous level of the whole data set. This number is called the *weight of the data set (wds)*. This combination is done with an array initialized with zero (the array starts with index 0). For each sensor we increment the field in the array that has the weight of the sensor as index. The resulting array is converted to a number. The first digit of the number is the array field with the highest index and the last digit is the field with index 0. This number is used as the sort key for our data sets.

Example: We have the above explained two flow sensors with their associated values. Flow sensor 1 has the weight 2 and flow sensor 2 has the weight 1. At the start the array has only zero values (0,0,0). For flow sensor 1 we have to increment the array field with index 1. We get 0,1,0. For flow sensor 2 we must increment the second field and we get 0,1,1. We convert it to an integer and get as he weight for our data set the number 110. This is our sort key for this data set.

The sort key corresponds to our requirements that

- a data set with only one sensor in a range nearest to the danger range is more dangerous than a data set with many sensors that are not so near.

- a data set with many sensors in a range nearest to the alarm range is more dangerous than a data set with only one sensor in this range and no sensor in more dangerous ranges.

The data sets are sorted so that the first buffered data set has always the highest weight. The diagnoses expert system always receives this, first, data set.

4. The diagnosis expert system

The data preparation transmits the linguistic values to the diagnosis expert system which tries to make the diagnosis. The architecture of the expert system is based on the blackboard model. The knowledge source are conducive to model single function groups of the plant (e.g. the cooling system). Several knowledge sources communicate over a common database (the blackboard). The scheduler evaluates the contribution of the knowledge source to the problem solving and activates the one with the highest priority. The knowledge in the knowledge sources will be represented with frames and rules. For the representation of fuzzy knowledge the theory of fuzzy sets will be applied.

Due to the structure of rules in different knowledge sources the number of rules to examine is small in comparison to the total number of rules contained in the system. In addition the expressiveness of each rule can be intensified by using fuzzy rules. The application of fuzzy rules, the blackboard model and inference strategies, which constraints the diagnosis space, leads to an efficient diagnosis.

4.1 The blackboard model

The blackboard model consists of several knowledge sources, the scheduler and the blackboard. The scheme of the blackboard system is shown in Figure 2. The three elements of the blackboard model will be described in the following sections.

4.1.1 The blackboard

The blackboard is devided into several abstraction levels and contains the data and the computed solutions of the knowledge sources. The abstraction levels represent the problem solving progress. The concept introduced in figure 2 contains seven abstraction levels.

Level 1,"sensors data" This level embodies only the current linguistic value of the data preparation.

Level 2,"tendency values" This level contains the tendency (in the data preparation evaluated value) of every sensor data as a linguistic value. The sensor data are used for a short term diagnosis whereas the tendency values are used for a long term diagnosis.

Level 3,"KS valuation" Level 3 contains the valuation of each knowledge source. These values are used to decide which knowledge source can be triggered.

Level 4,"KS priority" Level 4 embodies the priority of each knowledge sources. These values are important for the decision which knowledge source must be triggered first.

Level 5,"practicable KS" This level contains a list with all practicable knowledge sources.

Level 6,"diagnosis hypothesis" This level contains the diagnosis hypotheses from the actual triggered knowledge source. This level also include the degree of belief of the diagnosis hypotheses.

Level 7,"diagnosis valuation" Finally level 7 includes the solution of the diagnosis expert system.

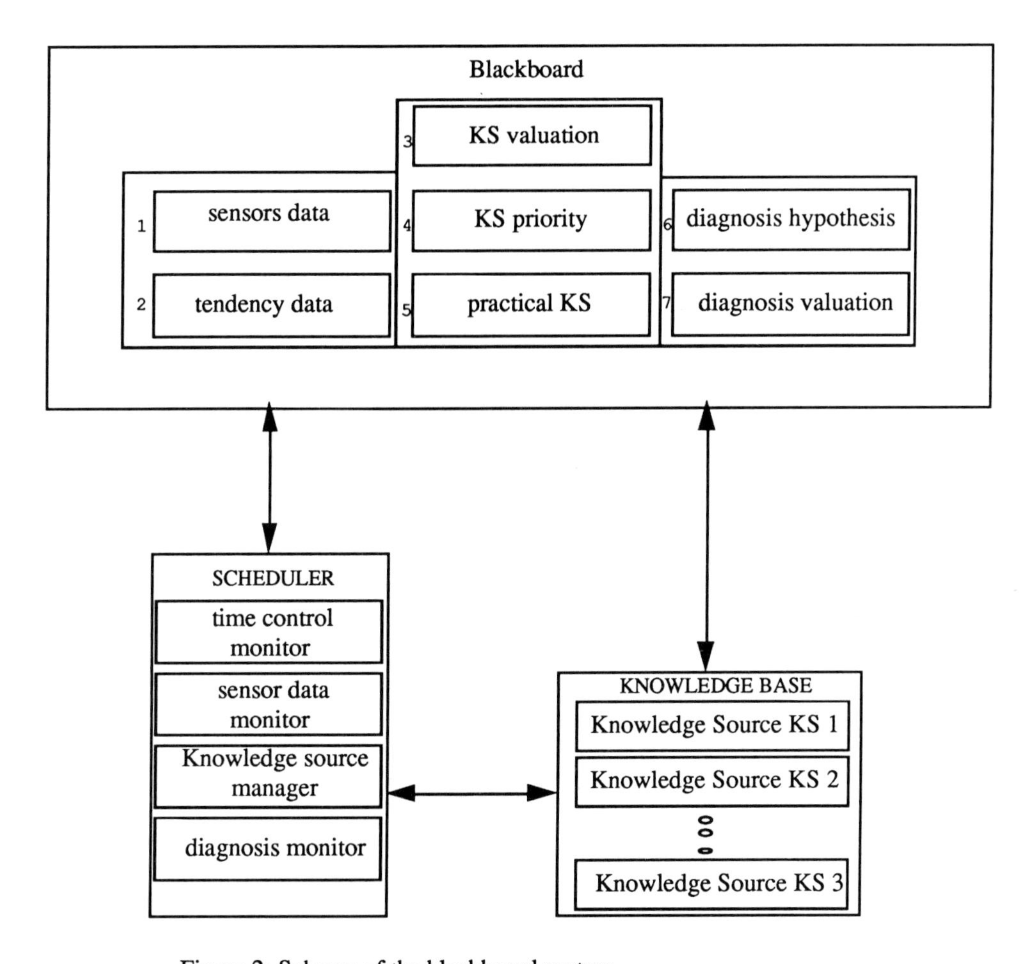

Figure 2: Scheme of the blackboard system

In summary these seven abstraction levels can be combined to three abstraction levels:

- State of plant, existing of sensor and tendency data,
- Executability of knowledge sources, existing of knowledge source valuation, knowledge source priority and executable knowledge sources,
- Diagnosis state, existing of diagnosis hypothesis and diagnosis valuation

The advantage of the balckboard with regard to the real time requirements are discussed in the next part. The blackboard is a data structure, that represents centrally for all knowledge sources the state of the plant (machine). This avoids that the different knowledge sources to a specified point of time t_a derive the diagnosis from different knowledge sources. Due to the ability of the knowledge sources to build solutions on the blackboard it is possible to consider the reasoning process and estimate the solution progress. The analysis of the solution progress is important for real time applications because that is the way bad solutions can be detected and time can be spared.

4.1.2 The knowledge base

The knowledge base is subdivided into different knowledge sources. The knowledge source includes the knowledge of the plant components. The advantage of this graduation is considerable because the knowledge source compared to the knowledge base has only few rules and methods. The knowledge source contains only knowledge of plant components (e.g. the cooling system). The selection of a knowledge source generate a rough diagnosis hypothesis. This inference strategy has the ability to focus on several problems by invoking the knowledge source needed to deal with a particular set of circumstances. Other advantages are the expansion abilities because the architecture of the system is modular. For example, it is possible to take in consideration new or extended control systems without modifying large parts of the implementation. New knowledge sources for other control systems can be added only by modifying scheduler and blackboard. Another important demand on real time expert systems is reasoning about events. Reasoning about events is typically for blackboard systems because knowledge sources are triggered by events.

All knowledge sources have their own interfaces to the blackboard and so it is possible to use one processor for each knowledge source.

4.1.3 The scheduler

The scheduler is responsible for the selection of knowlegde sources in blackboard systems. The scheduler selects the knowlegde source with the highest priority for the plant state. The priority classifies faults by quality and so it is possible to find out faults which are important for the process. The early recognization of faults is important for the reduction of faults in the plant. The scheduler has a facility to interrupt a current diagnosis. A diagnosis will be interrupted if the data preparation transmits data with higher priority. The data preparation has two different kinds of signals. The first kind of signal gives the scheduler the information to break off the current diagnosis. The second kind of signal cause the scheduler to prove the validity of current diagnosis. The validity of the diagnosis is dependent on the used time which is important for guaranteeing time intervals. The structure of the scheduler is shown in figure 2.

The scheduler is subdivided into four parts:

Sensor data monitor The sensor data monitor has the task of controlling the plant state. If sensor data with higher priority occur the sensor data monitor has to decide if a diagnosis is to be broken off or continued. This decision is not necessary if the data preparation signals to break off the diagnosis. In this case the sensor data monitor and the time monitor decide the following action. If time is not sufficient for a new diagnosis the scheduler informs the process computing system to execute and appropriate action to prevent further damages of the machine. If time is sufficient for a diagnosis the scheduler will start the diagnosis expert system with new values. In the case that the data preparation only signals to prove the current diagnosis and the time monitor decides that there is sufficient time for a new diagnosis the knowledge source monitor is triggered.

Time monitor The time monitor controls the used time for diagnosis.

Knowledge source monitor The knowledge source monitor finds out which knowledge source can be triggered. Finally the scheduler selects the knowledge source with the highest priority.

Diagnosis monitor The diagnosis monitor controls the diagnosis hypotheses which was built in the knowledge source.

5. The diagnosis validation

After the expert system has made a diagnosis it must prove if the actual state of the machine can be explained by this diagnosis, or if the state has changed once more so that this diagnosis doesn´t describe the state any more. Therefore the diagnosis validation gets the latest data set from the data preparation module, that was transmitted by the process computer. The validation is done by backward chaining through the diagnostic graph to get to a point, which tells how the data set must be constituted so that the diagnosis is true. If the actual data set is constituted in the same manner, then the diagnosis is still true, or else the diagnosis doesn´t represent the state of the machine and is worthless.

6. Conclusion

We tried to demonstrate with this paper that AI based systems using fuzzy logic are feasible for industrial application. The first application that uses this concept was implemented on a SUN SPARC station and has the task of monitoring a welding machine. Experts who participated in its development have already tested its results and acknowledge their validity.

References

[1] Dirk Danebrock and Joachim P. Stöck.: Fuzzy-Logik in Diagnoseexpertensystemen unter Echtzeitbedingungen. Masterthesis, University Dortmund, March 1992 (in german).

[2] Madjid Fathi and Dirk Danebrock and Joachim Peter Stöck. Ein Konzept für Fuzzy-Logik in On-Line-Diagnosesystemen. *Technische Anwendungen von Fuzzy-Systemen, VDE-Fachtagung,* 1992.

[3] H.Kern and M Fathi ,An On-Line Expert System for Diagnosis and Prognosis in Industrial Application. VDI Verlag 1990, pp. 173-180.

[4] Madjid Fathi and H. Kern.: A Concept for Combining Features of Real-time Systems and Expert Systems for On-Line Plant Diagnosis, *Proceedings of the IPCCC*, 1990.

[5] H.Kern and M.Fathi. An Expert System for Diagnosis of Low Pressure Plasma Spraying Devices. *National Thermal Spraying Conference*, Cincinnati, USA, 1988.

[6] Madjid Fathi. A Real Time Expert System for an Industrial Plant. *Conference of Real-Time Operating System*, Pittsburgh, PA 1989.

[7] Madjid Fathi et al. An Expert System for Machine Diagnosis of Plants based on Tool S.1. *IEEE ,IPCCC'91 , 24-26 of March 1991, AZ, USA,pp825-831*

[8] Madjid Fathi. A fuzzy concept for real-time monitoring, on 12th World congeress Intr. Federation of Automatic Control, Sydney, Australia, 18-23 July. 1993

[9] Madjid Fathi .et al. An Expert System Approach. *AEING-90*, Boston USA, 1990.

[10] Gorge J. Klir and Tina A. Folger. *Fuzzy Sets, Uncertainty, and Information*. Prentice Hall, London, 1988.

[11] Lotfi A. Zadeh. The concept of a linguistic variable and its application to approximate reasoning-I. *Information Science*, Vol. 8:199-250, 1975.

[12] Lotfi A. Zadeh. The role of fuzzy logic in the management of uncertainty in expert systems. *Fuzzy Sets ans Systems*, Nr. 11:199-227, 1983.

Uncertainty in Intelligent Systems
B. Bouchon-Meunier et al. (Editors)

Fuzzy Cellular Automata - a Practical Approach to Fuzzy Differential Equations

J.F.Baldwin* T.P.Martin & Y.Zhou+

A. I. Group, Advanced Computing Research Centre,
University of Bristol, Bristol BS8 1TR
United Kingdom

ABSTRACT

Many physical systems can be modelled using differential equations, although there are usually a number of approximations in this process. These approximations can be classified as (i) *conceptual model uncertainty*, ie does a set of equations model the physical process accurately enough for our purposes, and (ii) *parameter uncertainty*, ie given a particular model, how accurately do we know the values to feed into it. The latter point is addressed in this paper. Typically in the past, uncertain parameters have been treated using probabilistic methods irrespective of the underlying cause of uncertainty. More recently there have been attempts to use fuzzy techniques to model parameter uncertainty, and these have been particularly successful in cases where analytic solutions to the differential equations are available. We propose here a different approach, in which the system is modelled using a cellular automaton rather than differential equations, and we show how cellular automata may be fuzzified. The method is illustrated by an example of diffusion in which parameters are known fuzzily.

KEYWORDS : Fuzzy, Differential Equations, Cellular Automata, Uncertainty

1. INTRODUCTION

The use of differential equations to model the behaviour of evolving physical systems is very common in engineering and science. In many cases, such models may be approximate in a number of ways - terms may be neglected, infinite series may be truncated, parameters may not be known accurately, initial values and boundary conditions may be estimated roughly, experimental quantities may be subject to error, etc. These approximations result in uncertainty in the relation of the mathematical model to the physical system under consideration, and in uncertainty within the mathematical model. The widespread use of differential equations in modelling is partly due to their success in representing actual behaviour, but is also influenced by the historical background of the subject [1]. Early engineers were forced to perform all calculations manually and could only solve a limited class of problems exactly using analytical methods. In order to solve a real problem it was first ap-

---

* Professor Baldwin is a Senior SERC Research Fellow.

+ This work was partially funded by the CEC (DG XII) under contract FI2W-0091

proximated by an idealized model, this model was solved exactly, and from this a solution to the real problem could be obtained. More recently, the advent of digital computers has enabled the use of numerical methods to solve problems which are not amenable to analytic solution, but the underlying modelling approach using differential equations is unchanged. When obtaining a numerical solution, there are three levels of approximation :
(a) the physics of the system is modelled using a set of partial differential equations
(b) the parameters of the model are given values by measurement, estimation, etc
(c) the partial differential equations are approximated using a discrete space and time model (possibly involving terms expressed as infinite series which are truncated) and this is implemented on a digital computer which truncates real values to a fixed precision.

As Feynman observed [2]

> "It always bothers me that, according to the laws as we understand them today, it takes a computing machine an infinite number of logical operations to figure out what goes on in no matter how tiny a region of space, and no matter how tiny a region of time. How can all that be going on in that tiny space? Why should it take an infinite amount of logic to figure out what a tiny piece of space/time is going to do? So I have often made the hypothesis that ultimately physics will not require a mathematical statement, that in the end the machinery will be revealed, and the laws will turn out to be simple, like the checker board with all its apparent complexities"

Cellular automata [3; 4] can be used to simulate calculations normally performed using partial differential equations. They are simple systems, which are inherently massively parallel and are a promising alternative to traditional numerical methods. In essence, a cellular automaton consists of a large array of cells each containing a value drawn from a very restricted range, typically {0, 1}. Computation proceeds in discrete time steps, and at each step the contents of each cell is updated according to a simple rule involving the contents of the cell and some of its immediate neighbours. Solutions to many standard partial differential equations can be obtained using cellular automata, and in addition, more realistic models may be treated, thus removing some or all of the approximations in the modelling process. In this paper, we are concerned with investigating calculations normally performed using partial differential equation models, and therefore we assume an idealized system which matches the mathematical model very closely, ie we assume that approximation (a) above is exact.

In such a system, there may still be uncertainties in the values of various parameters (eg values taken from experimental data), and in initial values and boundary conditions. Usually this uncertainty is treated using probability distributions and Monte Carlo simulation, ie the calculation is repeated with many combinations of values from the probability distributions.

If the underlying source of uncertainty is not random, it may be better to use fuzzy set theory and possibility theory to model some or all of the uncertain values. Several authors have proposed methods of incorporating fuzzy numbers into differential equations (see for example [5] and references therein), although few have actually been applied to real problems. Shaw and Grindrod [6] examined various methods proposed in the literature and found that none were satisfactory within the fuzzy framework, as additional assumptions were needed about the evolution of uncertainty. In particular, they found that for a simple first order differential equation

$$\frac{\partial f(t)}{\partial t} = -\lambda f(t)$$

where only the initial value was fuzzy, it was not possible to obtain a unique solution, and that many valid solutions had rapidly diverging uncertainties. As an alternative, they proposed an approach in which the differential equation is treated as a mapping from input values to output, and this mapping is fuzzified in the usual way using the extension principle. A similar method was used by Protopescu *et al* [7]. An illustration of this approach in a practical system is given in [6] by examining fuzzy diffusion, in which both the initial amount of material and the diffusion coefficient are known only fuzzily. The method is particularly suited to exactly solvable systems where the inverse of the mapping may be determined; in other cases, it is necessary to use a sampling approach which detracts somewhat from the utility of the method.

In this paper, we examine methods of incorporating fuzzy values into the cellular automaton model, concentrating on a system that models the diffusion equation.

2. Cellular Automaton for Diffusion

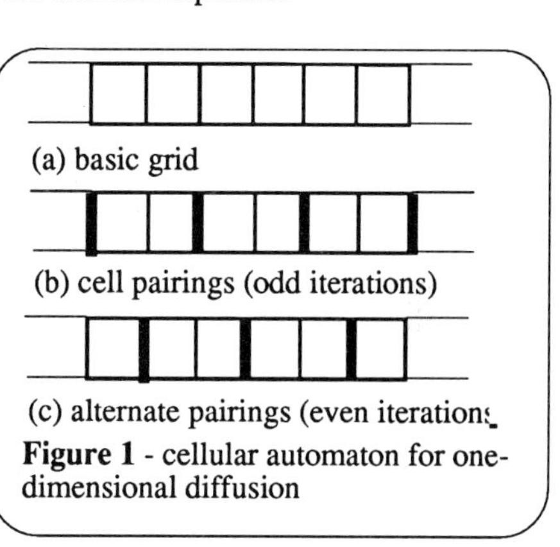

Figure 1 - cellular automaton for one-dimensional diffusion

It is straightforward to model diffusion using a cellular automaton. In one dimension, an array of cells is partitioned into pairs (Figure 1) and a random bit is generated for each pair. If the random bit is 1, the contents of the two cells are swapped; otherwise the pair is left unchanged. This operation is performed simultaneously in one time step of the automaton for all pairs. In the next time step, the alternate grid is used and the updating is repeated. It is not immediately clear how this process is related to diffusion. In any partial differential equation model, it is assumed that some measurable quantity in a physical system can be represented by a continuous function - in the case of diffusion, the quantity measured is the concentration of the diffusing material, which is taken to be a function of position and time. The modelled quantity (concentration) is a mathematical abstraction which is assumed to have a value at every point in the system - in fact, there is no such thing as the concentration *at a point*, since concentration is defined to be an average amount in some finite volume. Physically, we choose an arbitrary volume and define the concentration at a point as the amount of substance in the volume centred at that point; this gives a smoothly varying quantity. Clearly this is sensitive to the volume chosen. In the mathematical model of the system, the concentration at a point is defined in the limiting case as the volume goes to zero, although this has no physical meaning. In the cellular automaton, we have an array of cells each containing 0 or 1. It is not reasonable to think of the cell value as an approximation to the concentration at a point, but we can define the concentration as the average of the cell contents over a region centred on the cell in question. The size of the region determines the accuracy of the calculation, and the concentration emerges as a smoothly varying quantity which behaves in the same way as the concentration modelled by the differential equation model. The automaton is formally equivalent to a finite difference solution of the one-dimensional diffusion equation.

In order to show this, we consider the evolution of the concentration at an arbitrary point in the automaton. If $S(j,t)$ is the contents of cell j at time step t and the concentration is defined as the average occupation over a range of 2n+1 cells, then the concentration $\rho_i(t)$ at any cell i and time step t is defined by

$$\rho_i(t) = \sum_{j=i-n}^{i+n} S(j, t)$$

as illustrated in Figure 2. The difference in concentration between two adjacent cells is dependent on the occupations of the cells at the edges of the range (see Figure 2). Thus we can define the derivative with respect to x as

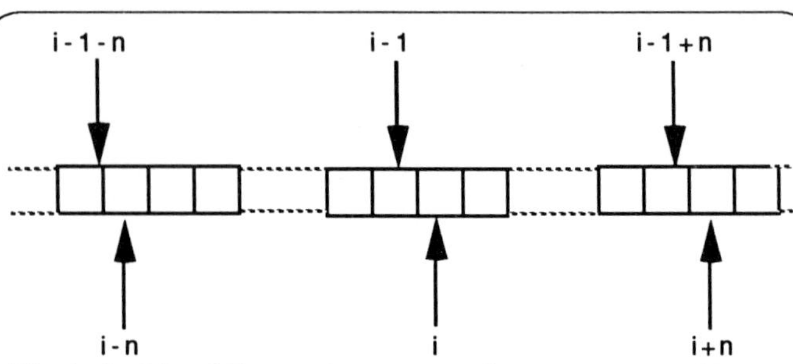

Figure 2 - The difference in concentration between an arbitrary point i in the cellular automaton and its neighbour $i-1$ is dependent on the occupation of cells at the extremes of the averaging region ie at ±n cells from the point in question

$$\rho_i(t) - \rho_{i-1}(t) = \sum_{j=i-n}^{i+n} S(j, t) - \sum_{j=i-n-1}^{i+n-1} S(j, t)$$
$$= S(i+n, t) - S(i-n-1, t)$$

where most terms are common to both summations and hence cancel. Similarly, we define the second derivative as

$$(\rho_{i+1}(t) - \rho_i(t)) - (\rho_i(t) - \rho_{i-1}(t)) = \rho_{i+1}(t) - 2\rho_i(t) + \rho_{i-1}(t)$$
$$= S(i+n+1, t) - S(i+n, t) + S(i-n-1, t) - S(i-n, t)$$

A derivative with respect to time at any point i is defined as the change in concentration between two consecutive time steps, t and $t+1$. Again, the change depends only on the cells at the extremes of the averaging range, since any swaps in other pairs of cells will not affect the total number of 1's in the range. Taking the cells at one end of the range, say $i-n$ and $i-n-1$, the contribution will be unchanged if a 0 is generated and $S(i-n-1, t) - S(i-n, t)$ if a 1 is generated. Since 0 and 1 are equally likely, the expected change is

$$\rho_i(t+1) - \rho_i(t) = \sum_{j=i-n}^{i+n} S(j, t) - \sum_{j=i-n}^{i+n} S(j, t+1)$$
$$= \frac{1}{2}S(i-n-1, t) - \frac{1}{2}S(i-n, t) + \frac{1}{2}S(i+n+1, t) - \frac{1}{2}S(i+n, t)$$

Thus we have

$$\frac{\partial \rho}{\partial t} = \frac{1}{2} \frac{\partial^2 \rho}{\partial x^2}$$

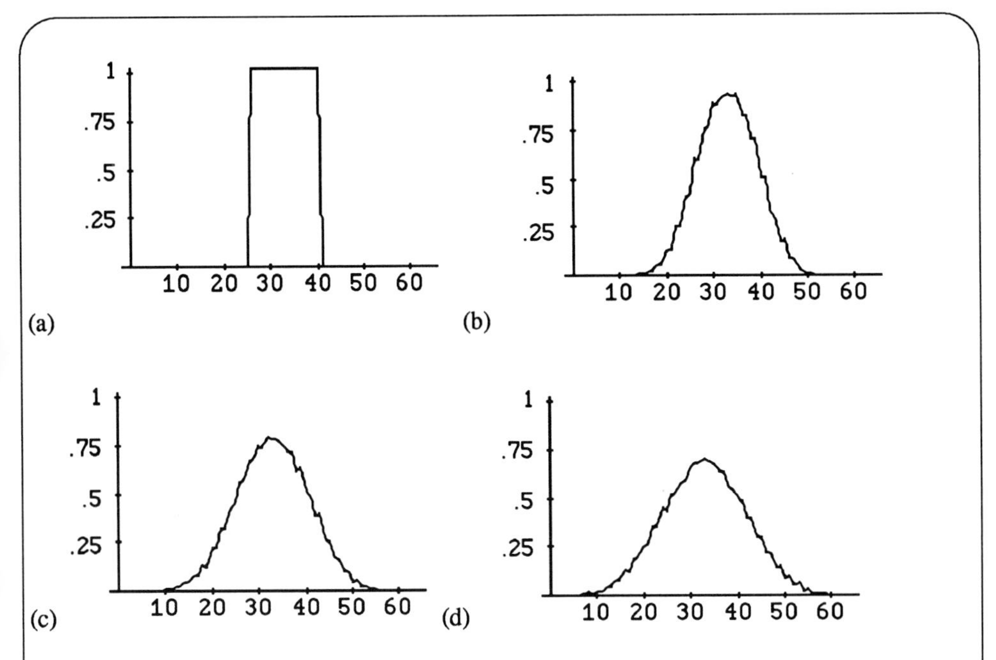

Figure 3 - Evolution of cellular automaton for diffusion after (a) 0 iterations (b) 10000 iterations (c) 20000 iterations (d) 30000 iterations. The vertical scale is concentration and the horizontal scale represents distance in arbitrary units.

More correctly, we should write

$$\frac{\partial \rho}{\partial t} = \frac{1}{2} D \frac{\partial^2 \rho}{\partial x^2} \qquad \text{where D is a diffusion coefficient given by} \qquad D = \frac{\Delta x^2}{\Delta t}$$

and Δx and Δt are respectively the cell spacing and time step used in the automaton. Since we choose unit spacing for the lattice and unit time intervals, $D = 1$ in this example. Sample plots of density against time for a series of time points are shown in Figure 3.

3. Fuzzification of the Diffusion Automaton

We consider three methods of fuzzifying the one-dimensional diffusion automaton. The arguments are extendible to the two or three dimensional cases, but one dimension is taken for simplicity. We note first some important features of the diffusion automaton:

(i) at any time *t*, the average number of bits in a predefined range centred on a cell gives the density at that cell

(ii) the total number of bits is preserved at all times during the computation (ie the mass of diffusing material is conserved),

(iii) when a partition of the automaton is used in a time step, the total number of bits in each element of the partition remains constant. In the one dimensional case, the number of bits in two cells paired by the grid partition is the same at the beginning and end of a time step. Because computation is purely local within each element of the partition, this ensures that condition (ii) holds automatically.

3.1. Cell Occupation as a Characteristic Function

Each cell of the automaton can be interpreted as corresponding to a small element of the physical system being modelled, and a 1 or 0 corresponds (respectively) to the presence or absence of diffusing material in that cell. If we consider the set of full cells in an automaton, then the cell contents S(i, t) can be interpreted as a characteristic function, ie any cell with S(i, t) = 1 is a member of the set of full cells, and any cell with S(i, t) = 0 is not a member. An obvious fuzzification is to allow intermediate degrees of membership in the set of full cells - each cell is filled *to some degree* rather than being definitely full or definitely empty. If the contents of the cell *i* at time step *t* is denoted by S(i, t), then instead of $S(i, t) \in \{0, 1\}$ we have $S(i, t) \in [0, 1]$. A further fuzzification is to allow intermediate degrees of exchange between two cells, rather than the 'all-or-nothing' swap used in the crisp automaton.

The total number of bits in the crisp system is the cardinality of the set of full cells. In the fuzzy case, we can take either the scalar cardinality or the fuzzy cardinality. The random bit for each pair of cells can be taken either as a crisp value, or it can be generalised in the same way as the cell contents. In the former case, we simply swap or leave unchanged the contents of two cells, according to the valuer of the random bit. In the latter case, the obvious interpretation is to treat the random value as a *proportion* of the cell contents to be swapped - thus if R(i,t) is the random value generated for cells *i* and *i+1*, at time *t*, the occupations of these cells at *t+1* are

$$S(i, t+1) = R(i, t) * S(i+1, t) + (1 - R(i, t)) * S(i, t)$$
$$S(i+1, t+1) = R(i, t) * S(i, t) + (1 - R(i, t)) * S(i+1, t)$$

This ensures that condition (iii) above is valid.

Since we are interested in the average behaviour of the automaton, and the expectation value in both cases will be 0.5, it makes no difference whether a random value is taken from {0, 1} or [0,1]. However, the latter method is considerably quicker in a simulation on a sequential computer.

The density at a cell *i* is the cardinality of the subset of cells centred on *i*, divided by the number of cells. This gives an "average" membership in the region, and again gives a smoothly varying quantity, but it is not clear what it represents. By taking an average of the membership levels, we are implying that the membership level of a particular cell simply represents the proportional occupation of that cell. This suggests that we could devise a finer scale (ie smaller cells) which contain only 0 or 1, and that the fuzzy automaton is a summary of this finer scale, with each cell in the fuzzy automaton corresponding to several in the crisp automaton.

An alternative approach is to use fuzzy cardinality, ie consider how many cells have membership greater than 0.9, 0.8, etc. The density at any cell is then a possibility distribution over a range of values between 0 and 1. Such possibility distributions can be non-normalised (if there is no cell with a membership of 1 in the range), and are discrete, not continuous. If fuzzy cardinality is used to determine the density at a point, we should also take the fuzzy cardinality of the whole set to determine the total number of "bits" in the automaton. This gives a possibility distribution, implying that there is uncertainty as to the total amount of material in the system being modelled. If the random value is taken from {0, 1} and a straightforward swap is used, the possibility distribution representing the total will not change; on the other hand, using a random value from [0, 1], it is not clear that conditions (ii) and (iii) are satisfied as the fuzzy cardinality of a pair of cells can change at each iteration, and hence the fuzzy cardinality of the whole automaton is not necessarily constant. For example, consider two adjacent cells containing 0.8 and 0.6 at time t, and a random value of 0.4. Initially, the cardinality is {1:0.8, 2:0.6} where the notation x : $\mu(x)$ indicates that element x has a membership of $\mu(x)$ in the fuzzy set. At time $t+1$, the cells contain 0.72 and 0.68, giving a fuzzy cardinality of {1:0.72, 2:0.68}. Clearly although scalar cardinality is preserved, fuzzy cardinality is not. We can however make sense of this by interpreting each cell membership as the *proportion* of diffusing material in the corresponding volume element of the physical system. It is possible to devise a crisp cellular automaton which exhibits identical behaviour to the fuzzy cases discussed here. The crisp automaton is two-dimensional* instead of one dimensional, but can be treated as a series of one-dimensional automata. Each of the one-dimensional automata uses the updating rule for diffusion described in Section 2. We consider the one dimensional automata as rows, which are stacked on top of each other (Figure 4). Each cell in the fuzzy automaton is represented by a column of cells in the crisp automaton, and the proportion of filled cells in a column determines the membership level of the corresponding fuzzy cell. Clearly the exact number of rows (ie copies of the one-dimensional automaton) depends on the degree to which we wish to approximate the continuous set of membership levels with a discrete set of possible values. In principle this can be made as accurate as necessary. At each time step, adjacent bits in two columns are exchanged (or not) according to the usual rules - if a single random bit is used for each pair of columns this will correspond to the wholesale swapping ie random values from {0, 1}, and if a random bit is generated for each row-pair in the two columns, this will correspond to proportional swapping. It is clear that the update steps are the same in the fuzzy and crisp cases, and hence the global behaviour of the two automata will be the same.

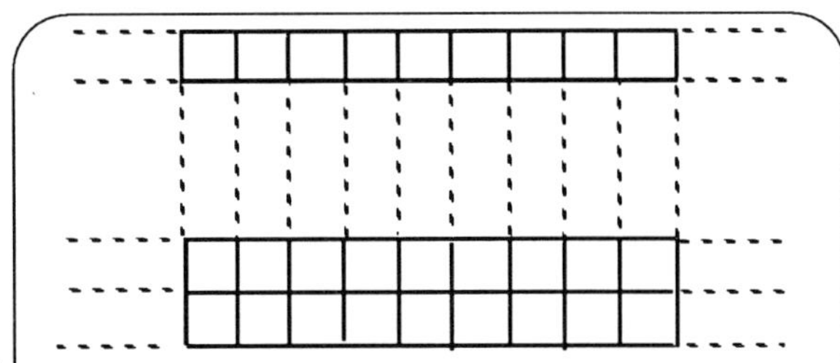

Figure 4 - a crisp automaton which can simulate the fuzzified automaton described in Section 3.1. Each row is a one-dimensional diffusion automaton, as described in Section 2; each column corresponds to one cell of the fuzzified automaton, with a membership equal to the proportion of filled cells in the column.

Thus generalising the cell memberships to the interval [0, 1] instead of the set {0, 1} does not generate any new features, and cannot properly be described as a fuzzification. There may however be efficiency advantages depending on the hardware used.

3.2. Possibility Distributions as Cell Contents

If the value in a cell is interpreted physically as an estimate of the proportion of diffusing material in the corresponding volume element, it is difficult to see how this could be judged exactly. A more reasonable approach would be to define a set of linguistic categories, such as *completely occupied, completely empty, about half, about 20%* etc. These correspond to possibility distributions on [0, 1], so that we could use fuzzy numbers in place of cell occupations. The total number of "bits" in the automaton is a fuzzy number, obtained using extended addition. If we assume that piecewise linear possibility distributions are used, and consider a pair of cells, it is clear that under either crisp or proportional swapping, the sum

* It is possible that a one-dimensional automaton could be designed to perform this calculation, but this has not been investigated.

of the contents will be the same before and after a swap, ie

$$\widetilde{S}(i, t) \oplus \widetilde{S}(i+1, t) = \widetilde{S}(i, t+1) \oplus \widetilde{S}(i+1, t+1)$$

where $\widetilde{S}(i, t)$ is a fuzzy number representing the contents of cell i at time step t, and $\oplus$ is the extended addition operator.

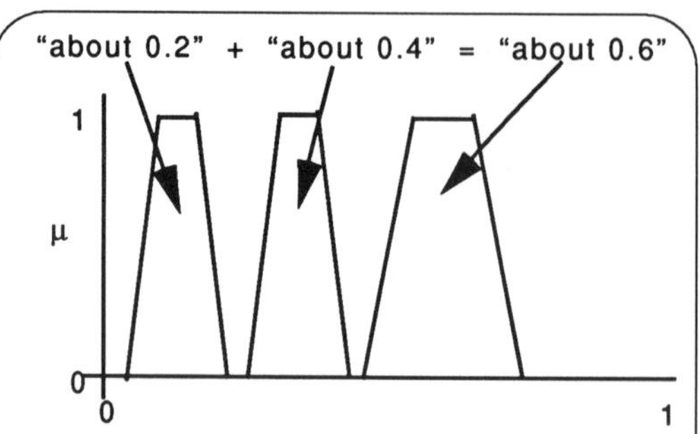

Figure 5 - fuzzy numbers and extended addition - each vertex in the sum, *about 0.6* is computed from the corresponding vertices in the fuzzy numbers *about 0.2* and *about 0.4*

Calculating the density at a point also requires a number of fuzzy additions, yielding a possibility distribution. If all fuzzy numbers are assumed to have the same shape eg triangular, trapezoidal, or a more general LR-form then the result of any swap (crisp or proportional) will also have the same shape. Each vertex of a fuzzy number only enters into the calculation in association with the corresponding vertex in other fuzzy numbers. Thus we could consider four separate crisp automata of the type described in the previous section. For trapezoidal fuzzy numbers, the first automaton would represent all lower left points, the second all upper left points, etc. Taking average values from each of these will yield the four points for the average possibility density; thus this fuzzification is also equivalent to a slightly larger crisp automaton.

We note that a third option is available for generating random values in this case, namely to generate a fuzzy random value. Then we have to use extended multiplication to swap the contents of each cell. This leads to a number of complications. In the first place, extended multiplication does not preserve the linear nature of the multiplicands - ie given two piecewise linear fuzzy numbers, their product is *not* piecewise linear. More seriously, the uncertainty of each fuzzy number increases rapidly - for example, if two adjacent cells contain
[0.1:0 0.2:1 0.3:1 0.4:0] and [0.6:0 0.7: 1 0.8:1 0.9:0],
and a random value of
[0.4:0 0.5:1 0.6:1 0.7:0] is generated, the swapping process (approximated by piecewise linear fuzzy numbers) will yield
[0.22:0 0.38:1 0.58:1 0.82:0] and
[0.27:0 0.43: 1 0.63:1 0.87:0]). Clearly, after a number of iterations the cell contents will tend towards [0:0 0+δ : 1 1-δ : 1 1:0], where δ is a small number.

3.3. Fuzzy Space and Time in a Cellular Automaton

The final approach to fuzzification considered here is to retain a crisp automaton, but to fuzzify the correspondence between the automaton and the physical system. This is inspired by the fact that in order to extract a diffusion model from the automaton, we need to interpret the cell contents in a certain way to obtain the quantities of interest.. We can incorporate fuzziness into this interpretation process, rather than in the automaton model. Instead of defining the distance between two cells to correspond to some precise distance in the physical world, we take a fuzzy distance. In a similar fashion, the time step could be defined as a fuzzy interval.

The density at any point d and time t is defined to be the number of bits in a fixed length centred on i at the appropriate time step. However, using fuzzy time and space we do not know which cell corresponds to the point d, nor how many cells correspond to the fixed length, nor exactly how many time steps correspond to t. If the distance between cells is

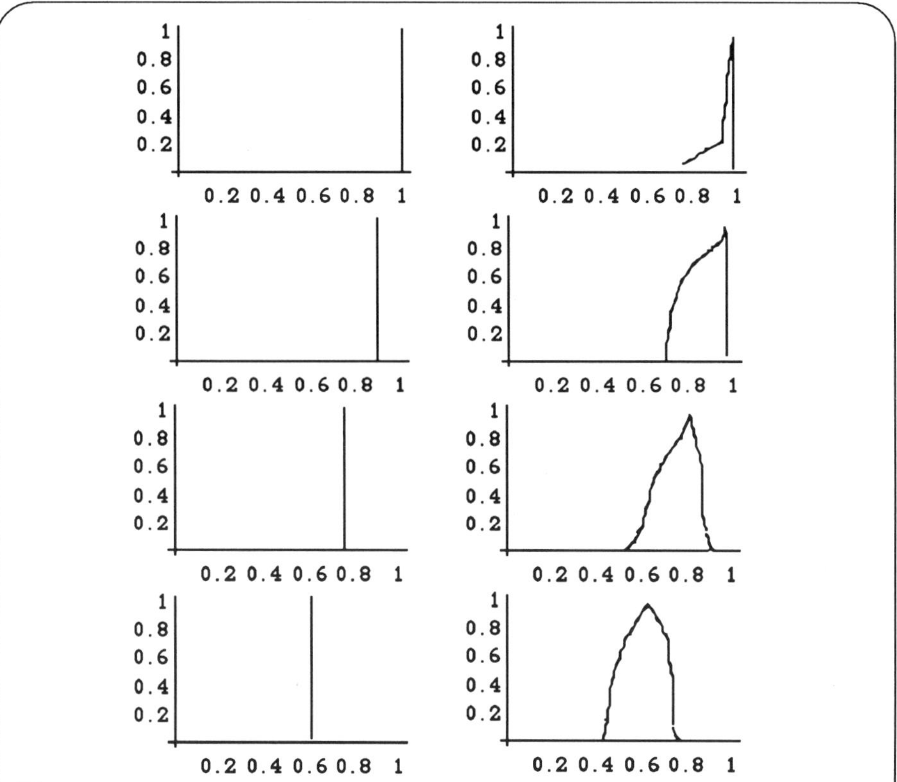

Figure 6 - evolution of the concentration at a single point. The vertical axis is membership, and the horizontal axis represents the concentration of diffusing material at the point. In the crisp case (left), the concentration is a crisp value and gradually decreases from its starting value of 1. In the fuzzy case (right), the concentration is a possibility distribution which starts at about 1, and decreases.

given by a fuzzy number $\widetilde{\Delta x}$ then the cell at distance *d* from the origin is

$$\tilde{i} = \left\{ j \mid j \in \mathbb{Z} \wedge j \in \frac{d}{\widetilde{\Delta x}} \right\}$$

For example, assume $\widetilde{\Delta x} = \{0.95{:}0\ 0.98{:}1\ 1.02{:}1,\ 1.05{:}0\ \}$, (in arbitrary units) and consider the point 100 units from the origin at some time *t* (we treat *t* as a crisp parameter for simplicity). This corresponds to the discrete possibility distribution over the *i*th cell from the origin, $\tilde{i} = \{96{:}0.3,\ 97{:}0.6, 98{:}1, 99{:}1, 100{:}1, 101{:}1, 102{:}1, 103{:}0.6, 104{:}0.3\ \}$ where the memberships have been rounded to 1 significant figure for clarity. Similarly, for any cell *i*, the density is defined as the total number of bits in a range $\pm\, n\Delta x$ from *i*. In the fuzzy case,

$$\tilde{\rho}_i(t) = \left\{ \sum_{j=i-k}^{i+k} S(j,t) \mid k \in \mathbb{Z} \wedge k \in n\widetilde{\Delta x} \right\}$$

If n is (say) 50, and the fuzzy cell spacing is as defined previously, then the density will be given by the possibility distribution

$$\left\{\left(\sum_{j=i-48}^{i+48} S(j,t)\right):.3, \left(\sum_{j=i-49}^{i+49} S(j,t)\right):1, \left(\sum_{j=i-50}^{i+50} S(j,t)\right):1, \left(\sum_{j=i-51}^{i+51} S(j,t)\right):1, \left(\sum_{j=i-52}^{i+52} S(j,t)\right):.3\right\}$$

Thus we can derive an expression for the fuzzy density at a distance d from the origin,

$$\tilde{\rho}_d(t) = \{r \mid r \in \tilde{\rho}_i(t) \wedge i \in \tilde{i}\}$$

where $\mu_{\tilde{\rho}_d}(y) = \max_{y \in \tilde{\rho}_j(t)} \min(\mu_{\tilde{i}}(j), \mu_{\tilde{\rho}_j}(y))$

The relation between automaton time steps and physical time can be similarly fuzzified. Here, we have investigated the result of fuzzifying the correspondence between automaton cells and physical space. The results are best understood by considering the diffusion shown in Figure 3, and focussing on a point just to the left of centre at approximately x=30. The concentration at this point is initially 1, and decreases as the material diffuses. This is shown in the left hand column of Figure 6, with the corresponding fuzzy case is shown in the right hand column. It can be seen that the concentration in the fuzzy case is no longer a crisp point value, but is a possibility distribution, roughly following the trajectory of the crisp case.

4. SUMMARY

The fuzzification of the interpretation process yields a practical approach to handling fuzzy uncertainty in modelling diffusion. Uncertainties in the diffusion constant and the amount of material present can be handled without the need to repeat the calculation many times or to modify the underlying automaton. The cellular automaton has been linked to code written in Fril [8], so that the fuzzy parameters may be easily controlled from a high-level AI-based front end[9]. As Fril includes many facilities for handling fuzzy sets, it is an ideal environment in which to embed the fuzzy cellular automaton.

Further work is needed to extend this approach in order to handle more complex partial differential equations. In principle, it is possible to fuzzify any cellular automaton, and hence derive fuzzy solutions to the corresponding partial differential equation. In practice, given an arbitrary partial differential equation, it is not always clear how to design an automaton that performs the same computation, and this may be a drawback to the method. However, it is known that a simple cellular automaton (the lattice gas cellular automaton) can be used to model fluid flow in one, two, or three dimensions. This can be fuzzified in the same way as the diffusion model discussed here, enabling fluid flow problems to be solved in cases where fuzzy uncertainty is present. Future work will investigate this case.

5. REFERENCES

[1] **Toffoli T**, (1984) "Cellular automata as an alternative to (rather than an approximation of) differential equations in modelling physics", Physica 10D 117-127

[2] **Feynman R.P**, (1967) "The Character of Physical Law", pp 57-58, MIT Press

Cambridge MA.

[3] **Vichniac G.Y**, (1984)"Simulating Physics with Cellular Automata", Physica 10D 96-116

[4] **Doolen G.D.**(Ed) (1990) "Lattice Gas Methods for Partial Differential Equations", Addison-Wesley

[5] **Dubois D. and Prade H**, (1987) "On Several Definitions of the Differential of a Fuzzy Mapping", Fuzzy Sets and Systems 24 ,117-120

[6] **Shaw W. and Grindrod P**, (1989) "Investigation of the potential of fuzzy sets and related approaches for treating uncertainties in radionuclide transfer predictions", CEC Report EUR 12499EN (DG XII)

[7] **Protopescu V, Yager R, and Dockery J**, (1992) "Combat Modeling with Imprecise Data", Int.J.Intelligent Systems 7 277-291

[8] **Baldwin J.F, Martin T.P, and Pilsworth B.W,**(1988)"FRIL Manual, version 4.0", FRIL Systems Ltd, Bristol ITeC, St. Anne's House, Bristol BS4 4AB, UK .

[9] **Baldwin J.F, Martin T.P, and Zhou Y**, "Uncertainty in Safety Assessment - the PSACOIN Level-E Model", CEC Workshop on Modelling with Uncertainty and Variability, Riso, Denmark 1991; University of Bristol Report ITRC166

Uncertainty in Intelligent Systems
B. Bouchon-Meunier et al. (Editors)
1993 Elsevier Science Publishers B.V.

Analogical Reasoning and Fuzzy Resemblance

Bernadette Bouchon-Meunier[a] and Llorenç Valverde[b] [1]

[a]LAFORIA, Université Paris VI, Boite 169. 4 place Jussieu. 75252 Paris cedex 05. France

[b]Dept. de Matemàtiques i Informàtica. Universitat de les Illes Balears Carretera de Valldemossa, 07071 Palma. Spain

Abstract

We propose a general approach to analogical reasoning based on the definition of a so-called resemblance between imprecise and/or uncertain facts. As it is known, this kind of facts is directly related to the use of a fuzzy logic to model the reasoning processes that are based on the matching -even partial- between facts. The resemblance relations that we propose can be related to generalizations of similarity relations: in fact, they are simply fuzzy binary relations that are reflexive, transitive and monotone with respect to the set inclusion. We give their properties and their characterization using the representation theorems for fuzzy binary transitive relations. We also provide some examples of resemblance relations

1. INTRODUCTION

Natural reasoning is based on the utilization of both facts: knowledge about the domain concerned with the problem to solve, and information about previously solved problems that could be related to the given problem in some sense. Aspects of analogical reasoning are parts of natural reasoning in many cases, which means that a kind of resemblance is recognized between the situation we have to cope with and situations that have been observed in the framework of problems with an already known solution. Obviously, the details of the resemblance are not listed in the natural reasoning process and they are taken into account in a global approach.

For instance, how can we determine that the price of an object *Ob* we want to buy is either reasonable or excessive? We have in mind the approximate prices of objects of the same kind and we compare the various attributes of *Ob* with those of the other objects. If we find a great resemblance for most of the attributes, and the price of *Ob* is analogous to the prices of the other objects, even smaller, we conclude that *Ob* is reasonable.

Approximate reasoning has been presented by Lofti A. Zadeh ([13]) as underlying the "remarkable human ability to ... make rational decisions in complex and/or uncertain environments." It is then natural to establish links between approximate reasoning and

[1]Research partially supported by the DGICYT project nr PB91-0334

some aspects of analogical reasoning. This approach has already been chosen by several authors ([7], [8], [9]), they have proposed measures of the proximity between the observed situation and already known situations and they have presented methods of deriving a conclusion from the solution of the already solved problem, which depends on the produced proximity.

We propose a more general approach by defining so-called resemblances between imprecise and/or uncertain facts that are directly related to the use of fuzzy logic for the reasoning that follows the identification of a resemblance between facts. These resemblances can be regarded as generalizations of similarity relations i.e., they are fuzzy binary relations that are reflexive, transitive and monotone with respect to set inclusion. We give their properties and we provide examples of resemblances, as well as their characterization using the representation theorems for fuzzy relations ([10]).

The general framework for our approach may be described as follows: suppose that some experiment provides a piece of knowledge concerning a variable V represented, for instance, by an imprecise characterization A', and we look for a characterization B' of a second variable W related to V by means of a link β. We know another experiment in which characterizations A of V and B of W have been obtained. Are we able to construct B' from the knowledge of A, B, A' and β? It will be possible if we point out a resemblance R between A and A'.

More precisely, for every set Ω, we denote by $F(\Omega)$ the set of its fuzzy subsets. We consider two variables V, defined on a universe X, and W, defined on a universe X. The fuzzy characterizations A and A' of V belonging to $F(X)$, B and B' belonging to $F(Y)$, are associated with membership functions μ_A, μ'_A and μ_B, μ'_B, respectively defined on X and Y and lying on $[0,1]$. We consider the relationship β between fuzzy propositions "V is A" and "W is B" described by means of the fuzzy rule "If V is A then W is B ". We note $A\beta B$ this relationship between A and B. The following scheme summarizes the various elements of the analogical reasoning we consider:

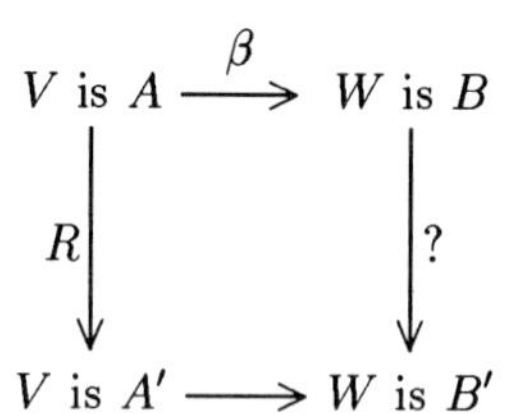

We notice that, if we associate the fuzzy rule with the values of a fuzzy implication r_{AB}, usually defined through residuation of a given t-norm T, i.e.

$$r_{AB}(u,v) = T^*(\mu_A(u) \mid \mu_B(v)) \tag{1}$$

where T^* stands for the quasi-inverse of the triangular t-norm T, i.e.

$$T^*(x \mid y) = \sup\{\alpha \in [0,1]; T(\alpha, x) \leq y\} \tag{2}$$

then, obviously, the Compositional Rule of Inference (CRI) i.e., the generalized Modus Ponens procedure, allows to obtain a characterization B' of W deduced from the fact that $A\beta B$, that is to say that $B' = CRI(A', A\beta B)$, where

$$\mu_{B'}(y) = sup_{x \in X} T(\mu_{A'}(x), r_{AB}(x, y)) \tag{3}$$

for every $y \in Y$, T being a triangular t-norm used to define r_{AB}.

To model a kind of natural analogical reasoning, the question we have to answer is the following: how we can directly construct B' from B, knowing that B is related to A by means of β, and knowing that A' resembles A. For instance, a solution could be the following: B' is a fuzzy characterization of W that resembles B in the same way and with the same degree as A' resembles A.

Our work proves that such a simply way of characterizing W is possible and lies on strong mathematical reasons, if we choose the definition of a resemblance R and a fuzzy implication β in a proper way. In addition, we also show that the fuzzy inference processes which use the CRI fulfills, with respect to a very special kind of resemblance relations, our requirements for an analogical reasoning process.

2. RESEMBLANCE RELATIONS

Following the intuitive properties of a resemblance between facts which arise in an analogical reasoning, we define an analogy on $F(\Omega)$ by means of a function

$$R : F(\Omega) \times F(\Omega) \longrightarrow [0, 1] \tag{4}$$

satisfying the following properties for every A, A', A'' in $F(\Omega)$:

1. $R(A; A) = 1$ (reflexivity).
2. $T(R(A; A'), R(A'; A'')) \leq R(A; A'')$ (T-transitivity), where T stands for a continuous triangular norm.
3. If $A \supseteq A'$, then $R(A; A') = 1$, and
4. If $A \cap A' = \emptyset$, then $R(A; A') = 0$

As it is usual with fuzzy binary relations, a resemblance relation can be regarded as the membership function of a fuzzy relation from $F(\Omega)$ or as a fuzzy subset of $F(\Omega) \times F(\Omega)$. Notice that the idea of using transitive fuzzy binary relations to measure resemblance is not new in the literature. Thus, for instance, if $T = Min$, a resemblance is simply a reflexive fuzzy ordering ([12]).If, for instance $T(u, v) = u.v$ or $T(u, v) = max(u + v - 1, 0)$, then we have a weaker form of transitivity (since $T \leq Min$, for any t-norm T). If R is symmetric i.e., if $(R(A; A') = R(A'; A)$, then it is associated with a similarity relation ([12]). For any t-norm T, a resemblance is a preorder and a symmetric resemblance is a T-indistinguishability operator ([10]).

In all these well known cases, a resemblance further satisfies additional properties (3) and (4) that are generally not satisfied since the concerned fuzzy relations are defined on a general universe $\Omega \times \Omega$ on which no ordering and intersection operation are considered and/or defined.

An example of resemblance relation is the following function defined on $F(\Omega) \times F(\Omega)$ and lying in $[0, 1]$:

$$R(A; A') = inf_{x \in \Omega} T^*(\mu_{A'}(x) \mid \mu_A(x)) \quad (5)$$

The fact that R is a resemblance is a consequence of the following properties satisfied by the quasi-inverse T^* of any continuous t-norm T([10]):

- $T^*(u \mid u) = 1$ for any $u \in [0, 1]$,
- $T(T^*(u \mid v), T^*(v \mid w)) \leq T^*(u \mid w)$ for any u, v and w, and
- $T^*(u \mid v) = 1$ if, and only if, $u \leq v$.

If φ is an additive generator of T, then $\varphi(R(A; A'))$ evaluates the proximity of A' with respect to A.

In the particular case of being $T(u, v) = max(u + v - 1, 0)$, it is $\varphi(u) = 1 - u$ and $T^*(u \mid v) = min(1 - u + v, 1)$. The function

$$R(A; A') = inf_{x \in \Omega} min(1 - \mu_{A'}(x) + \mu_A(x), 1) \quad (6)$$

defines a resemblance in the set $F(\Omega)$, which has associated the following pseudodistance (non symmetric):

$$D(A; A') = sup_{x \in \Omega} max(\mu_{A'}(x) - \mu_A(x), 0) \quad (7)$$

that is related to the distance on $F(\Omega)$ defined by

$$d(A, A') = sup_{x \in \Omega} |\mu_{A'}(x) - \mu_A(x)| \quad (8)$$

On the other hand, it is worth noticing that distances on a given set are associated, in a rather natural way, with fuzzy T-transitive relations on the same set. Thus, as it has been shown in ([10]), if d is a distance on a set X and if f is a continuous and strictly decreasing bijection from $[0, \infty]$ into $[0, 1]$, then

$$R(x, y) = f(d(x, y)) \quad (9)$$

is a transitive relation with respect to the t-norm $T(u, v) = f(f^{-1}(u) + f^{-1}(v))$. Although resemblance relations do not need to be symmetric, the above result gives a method to construct such kind of relations, starting from any non symmetric pseudodistance in $F(\Omega)$, as the one defined by (7). Thus, for instance,

$$R(A; A') = \exp -D(A; A') \quad (10)$$

is a resemblance relation, that is transitive with respect to the t-norm $T(u, v) = u.v.$

In fact, all resemblance relations have the same structure of those defined by formula (5). This is a straightforward corollary of the representation theorems for fuzzy transitive relations, given in ([10]):

Theorem 1. [Representation Theorem] Let R be a reflexive fuzzy binary relation on a given set X. Then for a given t-norm T, R is T-transitive if, and only if, there exists a family $\{h_j\}_{j \in J}$ of fuzzy subsets of X for which

$$R(x;y) = \inf_{j \in J} T^*(h_j(y) \mid h_j(x)) \tag{11}$$

From this standpoint, it is easy to check the following corollary:

Corollary 1. Let R be a fuzzy binary relation on $F(\Omega)$ and let T be a t-norm . Then R is a resemblance relation if, and only if, there exists a family $\{h_j\}_{j \in J}$ of functions from $F(\Omega)$ into $[0,1]$ such that

1. $R(A; A') = \inf_{j \in J} T^*(h_j(A') \mid h_j(A))$
2. For any $j \in J$, h_j is an increasing function, i.e. $h_j(A') \leq h_j(A)$ whenever $A' \subseteq A$
3. For any pair of fuzzy subsets (A, A') of Ω with disjoint support there exists $j \in J$ for which $h_j(A') \neq 0$ and $h_j(A) = 0$.

Thus, for instance, the resemblance relations (5) are generated by the functions $h_x(A) = \mu_A(x)$, where $J = \Omega$.

It is worth noting that other kinds of analogy measures have been used in the literature about analogical reasoning. Thus, for instance, in ([8]) the following family of relations is considered

$$R(A; A') = \frac{\sigma(A \cap A')}{\sigma(A')} \tag{12}$$

where σ is an arbitrary function from $F(\Omega)$ into $[0,1]$. It is easy to check that, such a family satisfy the following form of weak-transitivity

$$R(A; A' \cap A'') \cdot R(A'; A'') \leq R(A \cap A'; A'') \tag{13}$$

as conditional probabilities do. It is easy to check that, as it happens with resemblance relations, this kind of relations can also be completely characterized. Nevertheless, by now, we restrict ourselves to the resemblance relations given by the above corollary.

3. RESEMBLANCE RELATIONS AND FUZZY LOGIC

As it has been mentioned before, the fuzzy reasoning procedure which uses the Compositional Rule of Inference may be viewed from an analogical reasoning point of view. To this end, let us remind that the inverse-truth functional qualification process as introduced by Baldwin ([1]) i.e.

$$\tau_{A'A}(x) = \begin{cases} Sup\{A'(\alpha); \alpha \in A^{-1}(\{x\})\} & \text{if } A^{-1}(\{x\}) \neq \emptyset \\ 0 & \text{otherwise} \end{cases} \tag{14}$$

is simply a special kind of resemblance relation in $F(\Omega)$ with values in $F([0,1])$. In fact, $\tau_{A'A}$ is, in some sense, the best solution to the inequality

$$\tau \circ A' \geq A \tag{15}$$

Therefore, the procedure to get these fuzzy truth-labels τ may be represented by means of a function

$$\tau : F([0,1]) \times F([0,1]) \longrightarrow F([0,1]) \tag{16}$$

which maps $(A, A') \longmapsto \tau_{A'A}$, and satisfies the following properties:

1. $\tau_{AA} = j$ (j being the identity in $[0,1]$).

2. $\tau_{A'A} \circ \tau_{A''A'} \geq \tau_{A''A}$.

3. If $A' \leq A$ then $\tau_{A'A} \leq j$.

4. If $supp(A) \cap supp(A') = \emptyset$ then $\tau_{AA'} = \tau_o$. i.e. $\tau_{AA'}(x) = 0$, if $x \neq 0$; $\tau_{AA'}(0) = 1$.

As it is known, functions τ measure to what extend A and A' match and it turns out that if $B' = CRI(A', A\beta B)$ then

- If $\tau_{A'A} \leq j$ then $\tau_{B'B} = j$, and

- If $\tau_{A'A} \geq j$ then $\tau_{B'B} \geq j$.

the function j acts as a resemblance threshold, i.e. if the resemblance of two possibility distributions is less than j then the thesis B itself is the only suitable output. On the other hand, if the resemblance measured through τ- is bigger than j, then the inferred possibility distribution has a resemblance with the thesis which is also bigger than j.

Therefore, we can say that, for a given threshold ($j \in F([0,1])$, an observation A' analogous to the reference description A with regard to the variable V for the threshold j (i.e. $R(A; A') \geq j$), leads to a characterization B' of W analogous to the reference characterization B for the same threshold ($R(B; B') \geq j$). Nevertheless, from the analogical reasoning point of view, such measure of resemblance is extremely limited, because only less restrictive distributions can be inferred, i.e. the inference procedure is sensitive only for possibility distributions which are resemblant to the thesis but less restrictive.

Therefore, a first approach to analogical reasoning process may consists of a refinement using some aditional resemblance relations- of the above inference procedure, i.e. given R and R^* resemblance relations in the hypothesis and thesis spaces, respectively, under what conditions it can be assured that for a given $s*$ chosen as a threshold level of analogy in the thesis space, there exists a threshold level s in the hypothesis space such that whenever $\min(R(A; A'), R(A'; A)) \geq s$ we have $R^*(CRI(A'; A\beta B); B) \geq s*$. This is, in fact, a continuity problem: the $R - R^*$- continuity of the function CRI. Moreover, as it will be shown in the next section, analogical reasoning procedures can be defined using only this feature i.e., as continuous transformations.

4. A RESEMBLANCE-BASED DEFINITION OF ANALOGICAL REASONING FUNCTIONS

According to the above considerations, the analogical reasoning process may be represented by means of continuous -with respect to some given resemblance relations- functions from $F(X) \times F(X \times Y)$ into $F(Y)$. In other words:

Definition 1. Let R and R^* be resemblance relations in $F(X)$ and $F(Y)$, respectively, and let $A\beta B$ represent some causal links between the elements A and B of $F(X)$ and $F(Y)$. A function

$$\mathcal{A} : F(X) \times F(X \times Y) \longrightarrow F(Y) \tag{17}$$

is termed $R - R^*$-analogical reasoning function if it satisfies

1. $\mathcal{A}(A, A\beta B) = B$, and
2. If B' stands for $\mathcal{A}(A', A\beta B)$, then for any $s*$ there exists s such that if $\min(R(A; A'), R(A'; A)) \geq s$, then $\min(R^*(B'; B), R^*(B; B')) \geq s*$

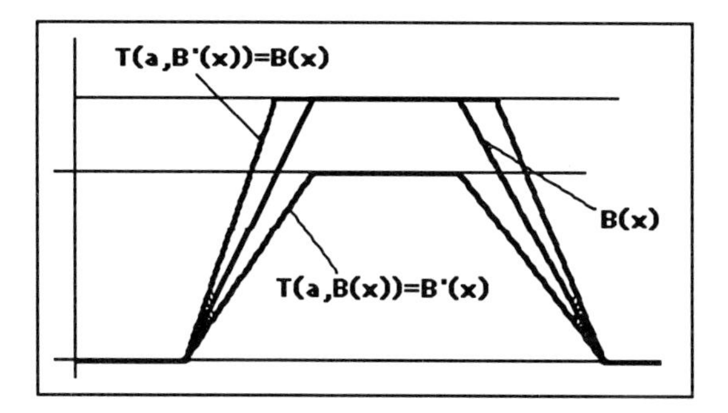

Fig. 1. Upper and lower resemblant distributions for a strict t-norm T

The characterization of resemblance relations given in Corollary 1, allows to give upper and lower bounds for the inferred analogical possibility distributions, once the threshold level $s*$ is given: Since $R^*(B; B') = \inf_{j \in J} T^*(h_j(B') \mid h_j(B))$, it turns out that, if $A\beta B$ and $\min(R(A; A'), R(A'; A)) \geq s$, then $R(B; B') \geq s*$ if, and only if, $T^*((h_j(B') \mid h_j(B)) \geq s*$ for any $j \in J$, which is equivalent to assert that $T(s*, h_j(B')) \leq h_j(B)$ and, since functions h_j are increasing, the above inequality gives a lower bound for the inferred resemblant possibility distribution. Similarly, if we consider that also $R(B'; B) \geq s*$, it turns out that $T(s*, h_j(B)) \leq h_j(B')$, which gives an upper bound for the value of B'.

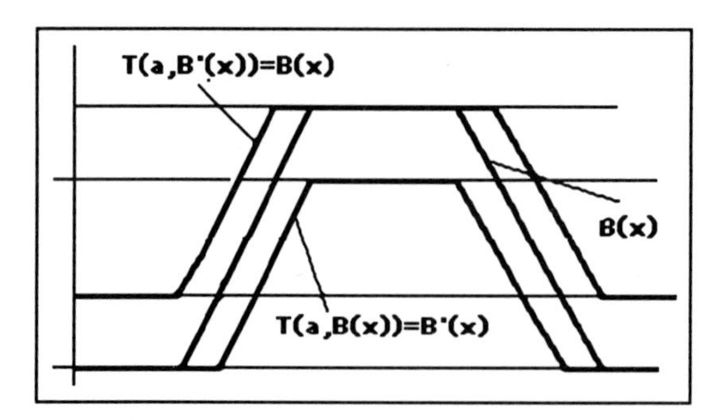

Fig. 2. Upper and lower resemblant distributions for a non-strict t-norm T

Thus, if R^* is the resemblance relation (5) i.e., $J = X$ and $h_x(B) = \mu_B(x)$, then these lower and upper bounds are given by the fuzzy sets defined by $\mu_{B'}(x) = T(\mu_B(x), s*)$ and $\mu_{B'}(x) = T^*(s * | \mu_B(x))$, respectively.(See Fig. 1 and Fig. 2).

In other words, if we take $s = s*$, then the following functions:

$$\mathcal{A}_1(A', A\beta B)(x) = \mu_{B'_1}(x) = T(\mu_B(x), \min(R(A; A'), R(A'; A))) \tag{18}$$

and

$$\mathcal{A}_2(A', A\beta B)(x) = \mu_{B'_2}(x) = T^*(\min(R(A; A'), R(A'; A)) | \mu_B(x)) \tag{19}$$

are examples of analogical reasoning functions, according to the above definition. Notice that functions $\mathcal{A}_1$ and $\mathcal{A}_2$ only depend on the particular form of the resemblance relation R^* taken in the thesis space.

5. CONCLUDING REMARKS

Through out this article we have been concerned with the definition of the analogical reasoning processes using fuzzy resemblance relations. Those relations have been introduced as a kind of fuzzy binary transitive relations and we have given a complete characterization of them.

We have also shown that the Compositional Rule of Inference of Fuzzy Logic can be regarded from an analogical reasoning point of view, and we have given a general definition of analogical reasoning transformations as well as some examples. Further research will be carried out to clarify the links between the analogical reasoning functions and the Compositional Rule of Inference

REFERENCES

1. Baldwin, J.F.: A New Approach to Approximate Reasoning Using Fuzzy Logic. Fuzzy Sets and Systems, 2, 1979. pp. 309-325
2. Bouchon, B.: How to replace computations by simple rules in the framework of fuzzy logic. Proc. COGNITIVA, AFCET. Madrid, 1990.

3. Bouchon, B.: Logique floue et analyse des similitudes. Actes des Journées Pôle-A-Pôle E du PRC Intelligence Artificialle, CNRS, Plestin-les-Grèves (1991). Rapport LAFORIA 92/10, 1992
4. Bouchon, B. and Valverde, L.: Fuzzy relations and analogy. Proceedings International Conference IPMU'92, Palma, 1992
5. Bouchon, B. and Valverde, L.: Analogy Relations and Inference. Proceedings Second IEEE International Conference on Fuzzy Systems (FUZZ-IEEE'93), San Francisco, 1993. pp. 1140-1144.
6. Godo, L.; Jacas, J. and Valverde, L.: Fuzzy Values in Fuzzy Logic. Int. J. of Intelligent Systems, vol. 6.1991.pp. 199-212
7. Mukaidono, M.; Ding, L. and Shen, Z.: Approximate reasoning based on revision principle. Proc. NAFIPS'90. Toronto, 1990.
8. Turksen, I.B. and Lucas, C.: A pattern matching inference method and its comparison with known inference methods. Proc. IFSA meeting. Brussels, 1991.
9. Turksen, I.B. and Zhao Zhong: An approximate analogical reasoning approach based on similarity measures. IEEE Trans. on Systems, Man and Cybernetics 18(1988), 1049-1056
10. Valverde,L.: On the structure of F-indistinguishability operators. Fuzzy Sets and Systems, 17(1985). 313-328
11. Zadeh, L.A.: The concept of linguistic variable and its application to approximate reasoning I, II, III Information Sciences (1975) 8 pp. 199-249; pp. 301-357; 9 pp 43-80
12. Zadeh, L.A.: Similarity relations and fuzzy orderings. Information Sciences 3(1971). 117-200
13. Zadeh, L.A.: A theory of approximate reasoning. In J. Hayes, D. Michie and L.I. Mikulich (Eds.): Machine Intelligence, vol 9. Halstead Press, New York, 1979. 149-194.
14. Zwick, R.; Carlstein, E. and Budescu, D.V.: Measures of similarity among fuzzy concepts: a comparative analysis. Int. J. Appr. Reasoning 1 (1987), 221-242.

Uncertainty in Intelligent Systems
B. Bouchon-Meunier et al. (Editors)

Triangular fuzzy relational compositions revisited

B. De Baets [1] and E. E. Kerre[a]

[a] Department of Applied Mathematics and Computer Science, University of Gent, Krijgslaan 281 (S9), 9000 Gent, Belgium

Abstract

This paper discusses the fuzzy relational compositions introduced by Bandler and Kohout. It is shown that these compositions are subject to some improvement. Alternative definitions are suggested.

1. INTRODUCTION

Fuzzy relational compositions and corresponding fuzzy relational equations and inequalities are undeniably one of the key issues in fuzzy set theory. The most popular fuzzy relational composition, due to L. Zadeh [7], is the sup-min composition. In the late seventies W. Bandler and L. Kohout [1] introduced several new ways of composing fuzzy relations, the so-called triangular compositions. These compositions are very promising in view of applications. Recent research on this topic [3] brought to light that Bandler and Kohout's definitions show some deficiencies. This paper addresses these shortcomings and provides ways to get around them.

2. HARSH TRIANGULAR FUZZY RELATIONAL COMPOSITIONS

2.1. Triangular compositions of crisp relations

2.1.1. Preliminary definitions

A (crisp) relation R from a universe X to a universe Y is a subset of $X \times Y$, i.e., $R \subseteq X \times Y$. The formula $(x, y) \in R$ is abbreviated as xRy, and one says that x is in relation R with y. The R-afterset xR of $x \in X$ is the set in Y defined as $xR = \{y \mid xRy\}$ and the R-foreset Ry of $y \in Y$ is the set in X defined as $Ry = \{x \mid xRy\}$. The domain $\text{dom}(R)$ of R is the set in X defined as $\text{dom}(R) = \{x \mid xR \neq \emptyset\}$ and the range $\text{rng}(R)$ of R is the set in Y defined as $\text{rng}(R) = \{y \mid Ry \neq \emptyset\}$. The converse relation R^T of R is the relation from Y to X defined as $R^T = \{(y, x) \mid xRy\}$.

2.1.2. Triangular compositions

Consider a relation R from X to Y and a relation S from Y to Z. The classical definition of the composition of the relations R and S is given by

$$R \circ S = \{(x, z) \mid (\exists y \in Y)(xRy \text{ and } ySz)\}.$$

[1]Research Assistant of the National Fund for Scientific Research (Belgium)

The composition $R \circ S$ is a relation from X to Z and consists of those couples (x, z) for which there exists *at least one* element of Y that is in relation R^T with x and that is in relation S with z. The relation $R \circ S$ is read as R *before* S or R *followed by* S. This definition can be written in terms of after- and foresets in the following way

$$R \circ S = \{(x, z) \mid xR \cap Sz \neq \emptyset\}.$$

Inspired by this style of notation, Bandler and Kohout have introduced several new compositions based on the notions of aftersets and foresets [1].

Definition 2.1 (Bandler-Kohout)

$$\begin{aligned} R \triangleleft_{bk} S &= \{(x, z) \mid xR \subseteq Sz\} \\ R \triangleright_{bk} S &= \{(x, z) \mid Sz \subseteq xR\} \\ R \diamond_{bk} S &= \{(x, z) \mid xR = Sz\} \end{aligned}$$

These compositions are called *products* by Bandler and Kohout, more specifically *round product* $\circ$, *subproduct* $\triangleleft_{bk}$, *superproduct* $\triangleright_{bk}$ and *square product* $\diamond_{bk}$. The subproduct and superproduct are also called *triangular products*. The subcomposition consists of those couples (x, z) for which *all* elements of Y that are in relation R^T with x are also in relation S with z. A similar interpretation holds for the supercomposition.

These definitions do not require any non-emptiness condition of the aftersets and foresets. This is a regrettable shortcoming. One easily verifies that

$$\begin{aligned} \text{co}(\text{dom}(R)) \times Z &\subseteq R \triangleleft_{bk} S \\ X \times \text{co}(\text{rng}(S)) &\subseteq R \triangleright_{bk} S. \end{aligned}$$

The first expression means that x is in relation $R \triangleleft_{bk} S$ with all elements of Z even if there is no element of Y that is in relation R^T with x. A similar remark holds for the second expression. In this way, the compositions $R \triangleleft_{bk} S$, $R \triangleright_{bk} S$ and $R \diamond_{bk} S$ can contain a lot of unwanted couples. It is clear that only those couples can be accepted for which both components are involved in the relations. An apparent solution would be to consider only relations for which the domain and the range coincide with the universes. However, this becomes too big a restriction when one wants to consider several relations between the same universes. It is unrealistic that all of these relations would have the same domain and range. Moreover, such a restriction could not solve similar problems that arise in the corresponding fuzzy relational compositions.

In order to overcome the foregoing difficulties, improved definitions are suggested.

Definition 2.2

$$\begin{aligned} R \triangleleft S &= \{(x, z) \mid \emptyset \subset xR \subseteq Sz\} \\ R \triangleright S &= \{(x, z) \mid \emptyset \subset Sz \subseteq xR\} \\ R \diamond S &= \{(x, z) \mid \emptyset \subset xR = Sz\} \end{aligned}$$

An extensive overview of the relationships between these compositions and of their properties, such as monotonicity, interaction with union and intersection, and associativity, can be found in [4]. There it is shown that several relationships and properties valid for the Bandler-Kohout compositions are no longer valid or exist only in weakened versions for the improved definitions.

Example 2.1 [2]
Consider a set of patients P, a set of symptoms S and a set of illnesses I. Let F be the relation from P to S defined by

$$pFs \Leftrightarrow \textit{ patient } p \textit{ shows symptom } s$$

and D the relation from S to I defined by

$$sDi \Leftrightarrow \textit{ s is a symptom of illness } i.$$

We assume that each of the illnesses has at least one symptom. The compositions of F and D are given by

- $p(F \circ D)i \Leftrightarrow$ *patient p shows at least one symptom of illness i,*
- $p(F \triangleleft D)i \Leftrightarrow$ *all symptoms shown by patient p are symptoms of illness i and patient p shows at least one symptom,*
- $p(F \triangleright D)i \Leftrightarrow$ *patient p shows all symptoms of illness i,*
- $p(F \diamond D)i \Leftrightarrow$ *the symptoms shown by patient p are exactly those of illness i.*

Two sets of equivalent expressions for definitions 2.2 are particularly interesting, since they will lead to two alternative ways of fuzzifying these relational compositions. The first set expresses that the improved compositions can be seen as the intersection of the Bandler-Kohout compositions and the cartesian product of the domain of the first relation and the range of the second relation. The second set expresses that they can be seen as the intersection of the Bandler-Kohout compositions and the classical composition.

Properties 2.1

$$\begin{aligned} R \triangleleft S &= (R \triangleleft_{bk} S) \cap (\mathrm{dom}(R) \times \mathrm{rng}(S)) \\ R \triangleright S &= (R \triangleright_{bk} S) \cap (\mathrm{dom}(R) \times \mathrm{rng}(S)) \\ R \diamond S &= (R \diamond_{bk} S) \cap (\mathrm{dom}(R) \times \mathrm{rng}(S)) \end{aligned}$$

Properties 2.2

$$\begin{aligned} R \triangleleft S &= (R \triangleleft_{bk} S) \cap (R \circ S) \\ R \triangleright S &= (R \triangleright_{bk} S) \cap (R \circ S) \\ R \diamond S &= (R \diamond_{bk} S) \cap (R \circ S) \end{aligned}$$

To conclude this paragraph, it is mentioned that the square composition is nothing else but the intersection of the triangular compositions.

Properties 2.3

$$\begin{aligned} R \diamond_{bk} S &= (R \triangleleft_{bk} S) \cap (R \triangleright_{bk} S) \\ R \diamond S &= (R \triangleleft S) \cap (R \triangleright S) \end{aligned}$$

2.1.3. Characteristic mappings

It is well-known that a relation R from X to Y can be identified with its characteristic mapping as follows

$$\begin{aligned} R : X \times Y &\to \{0,1\} \\ (x,y) &\mapsto 1 \quad \text{if } xRy \\ (x,y) &\mapsto 0 \quad \text{if } \neg(xRy). \end{aligned}$$

The characteristic mapping of the round composition is given by

$$R \circ S(x,z) = \sup_{y \in Y} R(x,y) \wedge_B S(y,z)$$

where $\wedge_B$ stands for the boolean conjunction operator.

Bandler and Kohout have shown [1] that the characteristic mappings of their compositions can be expressed in the following way.

Properties 2.4

$$\begin{aligned} R \triangleleft_{bk} S(x,z) &= \inf_{y \in Y} R(x,y) \Rightarrow_B S(y,z) \\ R \triangleright_{bk} S(x,z) &= \inf_{y \in Y} R(x,y) \Leftarrow_B S(y,z) \\ R \diamond_{bk} S(x,z) &= \inf_{y \in Y} R(x,y) \Leftrightarrow_B S(y,z) \end{aligned}$$

In these expressions $\Rightarrow_B$ en $\Leftrightarrow_B$ stand for the boolean implication operator and the boolean equivalence operator, while $a \Leftarrow_B b$ has to be understood as $b \Rightarrow_B a$. Taking the boolean conjunction of these expressions with an extra term that takes into account the non-emptiness conditions, the characteristic mappings of the improved definitions can be written in the following way.

Properties 2.5

$$\begin{aligned} R \triangleleft S(x,z) &= \left(\inf_{y \in Y} R(x,y) \Rightarrow_B S(y,z)\right) \wedge_B \left(\sup_{y \in Y} R(x,y)\right) \\ R \triangleright S(x,z) &= \left(\inf_{y \in Y} R(x,y) \Leftarrow_B S(y,z)\right) \wedge_B \left(\sup_{y \in Y} S(y,z)\right) \\ R \diamond S(x,z) &= \left(\inf_{y \in Y} R(x,y) \Leftrightarrow_B S(y,z)\right) \wedge_B \left(\sup_{y \in Y} R(x,y)\right) \\ &= \left(\inf_{y \in Y} R(x,y) \Leftrightarrow_B S(y,z)\right) \wedge_B \left(\sup_{y \in Y} S(y,z)\right) \end{aligned}$$

2.2. Triangular compositions of fuzzy relations

2.2.1. Preliminary definitions

It is well-known that existential and universal quantification can be represented by means of supremum and infimum. To ease the notations, the height and plinth operators are introduced. A study of the properties of these operators can be found in [3, 6].

The height $\text{Hgt}(A)$ of a fuzzy set A in X is defined as $\text{Hgt}(A) = \sup_{x \in X} A(x)$. The plinth $\text{Plt}(A)$ of a fuzzy set A in X is defined as $\text{Plt}(A) = \inf_{x \in X} A(x)$.

A fuzzy relation R from a universe X to a universe Y is a fuzzy set in $X \times Y$, i.e., a mapping from $X \times Y$ to $[0,1]$. $R(x,y)$ is called the degree of relationship between x and y in R. The R-afterset xR of $x \in X$ is the fuzzy set in Y defined by $xR(y) = R(x,y)$. The R-foreset Ry of $y \in Y$ is the fuzzy set in X defined by $Ry(x) = R(x,y)$. The domain $\mathrm{dom}(R)$ of R is the fuzzy set in X defined by $\mathrm{dom}(R)(x) = \mathrm{Hgt}(xR)$. The range $\mathrm{rng}(R)$ of R is the fuzzy set in Y defined by $\mathrm{rng}(R)(y) = \mathrm{Hgt}(Ry)$.

To simplify the notations in the sequel, a $[0,1]^2 - [0,1]$ mapping $\mathcal{M}$ is extended to a $\mathcal{F}(X)^2 - \mathcal{F}(X)$ mapping in the following way. Let A and B be fuzzy sets in X then $\mathcal{M}(A,B)$ is the fuzzy set in X defined by $\mathcal{M}(A,B)(x) = \mathcal{M}(A(x),B(x))$.

2.2.2. Triangular compositions

The classical composition of relations has been extended to fuzzy relations by L. Zadeh in his very first paper on fuzzy sets [7]. Consider a fuzzy relation R from X to Y and a fuzzy relation S from Y to Z. The sup-min composition $R \circ S$ of R and S is the fuzzy relation from X to Z defined by

$$R \circ S(x,z) = \sup_{y \in Y} \min(R(x,y), S(y,z)).$$

Other authors have introduced the sup-$\mathcal{T}$ composition by replacing the fuzzy intersection operator minimum by a general triangular norm $\mathcal{T}$, i.e.,

$$R \circ S(x,z) = \sup_{y \in Y} \mathcal{T}(R(x,y), S(y,z))$$

which can be written as

$$R \circ S(x,z) = \mathrm{Hgt}(\mathcal{T}(xR, Sz)).$$

This more general definition is adopted here. Note that the degree of relationship between x and z in $R \circ S$ is determined by the strongest of the connections between x and z via an element y of Y, where the strength of such a connection is given by $\mathcal{T}(R(x,y), S(y,z))$. Since $\mathcal{T}(R(x,y), S(y,z)) \leq \min(R(x,y), S(y,z))$ it follows that the strength of such a connection is not greater than the strength of the connection between x and y and the strength of the connection between y and z. This is a mathematical interpretation of the expression "a chain is as strong as the weakest of its links".

Bandler and Kohout have extended their triangular compositions to fuzzy relations by replacing the boolean implication operator $\Rightarrow_B$ in the characteristic mappings of properties 2.4 by a fuzzy implication operator $\mathcal{I}$, i.e., a $[0,1]^2 \to [0,1]$ mapping that satisfies the boundary conditions $\mathcal{I}(0,0) = \mathcal{I}(0,1) = \mathcal{I}(1,1) = 1$ and $\mathcal{I}(1,0) = 0$. As far as we know, they have never mentioned how they have extended the square composition. One possibility is that they have defined the square composition as the intersection of the triangular compositions similar to properties 2.3. A second possibility is that they have replaced the boolean equivalence operator $\Leftrightarrow_B$ in the characteristic mapping by a fuzzy equivalence operator $\mathcal{E}$, i.e., $[0,1]^2 \to [0,1]$ mapping that satisfies the boundary conditions $\mathcal{E}(0,0) = \mathcal{E}(1,1) = 1$ and $\mathcal{E}(0,1) = \mathcal{E}(1,0) = 0$. In this paper we adopt the first possibility.

Definition 2.3 (Bandler-Kohout)

$$\begin{aligned} R \triangleleft_{bk} S(x,z) &= \inf_{y \in Y} \mathcal{I}(R(x,y), S(y,z)) \\ R \triangleright_{bk} S(x,z) &= \inf_{y \in Y} \mathcal{I}(S(y,z), R(x,y)) \\ R \diamond_{bk} S(x,z) &= \min(R \triangleleft_{bk} S(x,z), R \triangleright_{bk} S(x,z)) \end{aligned}$$

It has already been argued that in the crisp case these compositions show some shortcomings when the aftersets or foresets involved are empty. Additional problems arise in the fuzzy case, as is shown in the following example.

Example 2.2
Consider a set of patients P, a set of symptoms S and a set of illnesses I. Let F be the fuzzy relation from P to S defined by

$F(p,s)$ is the degree to which patient p shows symptom s

and D the fuzzy relation from S to I defined by

$D(s,i)$ is the degree to which s is a symptom of illness i.

Let $S = \{s_1, \ldots, s_6\}$ and i_k be an illness with D-foreset given by

$Di_k = \{(s_1, 0.3), (s_2, 0.7), (s_3, 1), (s_4, 0.8), (s_5, 0.2), (s_6, 0.9)\}$.

Let p_i and p_j be two patients with F-aftersets given by

$$\begin{aligned} p_i F &= \{(s_1, 0), (s_2, 0.2), (s_3, 0), (s_4, 0), (s_5, 0), (s_6, 0)\}, \\ p_j F &= \{(s_1, 0.2), (s_2, 0.6), (s_3, 1), (s_4, 1), (s_5, 0), (s_6, 0.8)\}. \end{aligned}$$

Consider the Luckasiewicz fuzzy implication operator $\mathcal{I}(x,y) = \min(1, 1-x+y)$, then $F \triangleleft_{bk} D(p_i, i_k) = 1$ and $F \triangleleft_{bk} D(p_j, i_k) = 0.8$.
This means that the degree to which all symptoms shown by patient p_i are symptoms of illness i_k is equal to 1, although patient p_i is only showing symptom s_2 to degree 0.2. Such surprising results stem from the fact that definitions 2.3 do not take into account the degree of emptiness or non-emptiness of the aftersets and foresets involved.

Two sets of improved definitions are suggested here, based on properties 2.1, properties 2.2 and properties 2.3. The reason why we have not based our definitions on the characteristic mappings of properties 2.5 is quite obvious. The characteristic mapping of $R \triangleleft S$ for instance has been derived based on the fact that the condition $\emptyset \subset xR \subseteq Sz$ is equivalent with $xR \subseteq Sz$ and $xR \neq \emptyset$ (and immediately implies $Sz \neq \emptyset$). In the fuzzy case no such observation holds. Therefore, other ways are used to take into account the degree of non-emptiness of both xR and Sz.

Definition 2.4

$$\begin{aligned} R \triangleleft_b S(x,z) &= \min(\mathrm{Plt}(\mathcal{I}(xR, Sz)), \mathrm{Hgt}(xR), \mathrm{Hgt}(Sz)) \\ R \triangleright_b S(x,z) &= \min(\mathrm{Plt}(\mathcal{I}(Sz, xR)), \mathrm{Hgt}(xR), \mathrm{Hgt}(Sz)) \\ R \diamond_b S(x,z) &= \min(R \triangleleft_b S(x,z), R \triangleright_b S(x,z)) \end{aligned}$$

Definition 2.5

$$
\begin{aligned}
R \triangleleft_k S(x,z) &= \min(\mathrm{Plt}(\mathcal{I}(xR, Sz)), \mathrm{Hgt}(\mathcal{T}(xR, Sz))) \\
R \triangleright_k S(x,z) &= \min(\mathrm{Plt}(\mathcal{I}(Sz, xR)), \mathrm{Hgt}(\mathcal{T}(xR, Sz))) \\
R \diamond_k S(x,z) &= \min(R \triangleleft_k S(x,z), R \triangleright_k S(x,z))
\end{aligned}
$$

The explicit expressions of these compositions are given by

$$
\begin{aligned}
R \triangleleft_b S(x,z) &= \min(\inf_{y \in Y} \mathcal{I}(R(x,y), S(y,z)), \sup_{y \in Y} R(x,y), \sup_{y \in Y} S(y,z)) \\
R \triangleright_b S(x,z) &= \min(\inf_{y \in Y} \mathcal{I}(S(y,z), R(x,y)), \sup_{y \in Y} R(x,y), \sup_{y \in Y} S(y,z)) \\
R \diamond_b S(x,z) &= \min(R \triangleleft_b S(x,z), R \triangleright_b S(x,z))
\end{aligned}
$$

and

$$
\begin{aligned}
R \triangleleft_k S(x,z) &= \min(\inf_{y \in Y} \mathcal{I}(R(x,y), S(y,z)), \sup_{y \in Y} \mathcal{T}(R(x,y), S(y,z))) \\
R \triangleright_k S(x,z) &= \min(\inf_{y \in Y} \mathcal{I}(S(y,z), R(x,y)), \sup_{y \in Y} \mathcal{T}(R(x,y), S(y,z))) \\
R \diamond_k S(x,z) &= \min(R \triangleleft_k S(x,z), R \triangleright_k S(x,z)).
\end{aligned}
$$

Remark 2.1
As already indicated, an alternative way of defining the square composition is by introducing a fuzzy equivalence operator $\mathcal{E}$, i.e.,

$$
\begin{aligned}
R \diamond_b^{\mathcal{E}} S(x,z) &= \min(\mathrm{Plt}(\mathcal{E}(xR, Sz)), \mathrm{Hgt}(xR), \mathrm{Hgt}(Sz)) \\
R \diamond_k^{\mathcal{E}} S(x,z) &= \min(\mathrm{Plt}(\mathcal{E}(xR, Sz)), \mathrm{Hgt}(\mathcal{T}(xR, Sz))).
\end{aligned}
$$

It is easily verified that definitions 2.5 are more restrictive, i.e., they yield lower degrees of relationship, than definitions 2.4, and that these in turn are more restrictive than definitions 2.3.

Properties 2.6

$$
\begin{array}{ccccc}
R \triangleleft_k S & \subseteq & R \triangleleft_b S & \subseteq & R \triangleleft_{bk} S \\
R \triangleright_k S & \subseteq & R \triangleright_b S & \subseteq & R \triangleright_{bk} S \\
R \diamond_k S & \subseteq & R \diamond_b S & \subseteq & R \diamond_{bk} S
\end{array}
$$

Remark 2.2
It is easily verified that for definitions 2.4 and definitions 2.5 the height of a composition of two fuzzy relations cannot be greater than one of the heights of these fuzzy relations, i.e.,

$$\mathrm{Hgt}(R * S) \leq \min(\mathrm{Hgt}(R), \mathrm{Hgt}(S))$$

where $* \in \{\circ, \triangleleft_b, \triangleright_b, \diamond_b, \triangleleft_k, \triangleright_k, \diamond_k\}$. In fact, it is the lack of this property for definitions 2.3 that leads to the problems of the Bandler-Kohout compositions.

An extensive overview of the relationships between these new compositions and of their properties can be found in [5].

3. MODERATE TRIANGULAR FUZZY RELATIONAL COMPOSITIONS

3.1. Introduction

Bandler and Kohout consider definitions 2.3 as a special case of a more general relational product. Consider a fuzzy relation R from X to Y and a fuzzy relation S from Y to Z, then Bandler and Kohout define an abstract fuzzy relational product $R \star S$, analogous to the matrix product, in the following way

$$R \star S(x,z) = \bigoplus_{y \in Y} R(x,y) \odot S(y,z)$$

where, as Bandler and Kohout write [1], "$\odot$ is likely to be something other than multiplication and $\oplus$ to refer to something other than summation".

In the discussion of the characteristic mappings of the triangular compositions of crisp relations we have already seen that for $\star \in \{\circ, \triangleleft_{bk}, \triangleright_{bk}, \diamond_{bk}\}$ the operator $\oplus$ is an element of the set $\{\sup, \inf\}$ and the operator $\odot$ is an element of the set $\{\wedge_B, \Rightarrow_B, \Leftarrow_B, \Leftrightarrow_B\}$. These compositions have been extended to fuzzy relations using triangular norms and fuzzy implication operators.

3.2. Mean compositions of crisp relations

Other instances of this general relational product are the *mean* compositions. In case of finite universes Bandler and Kohout suggest another choice for the operator $\oplus$, namely the (arithmetic) mean or averaging operator. Consider a relation R from X to Y and a relation S from Y to Z and assume that the universe Y has a finite cardinality $\#Y$. The mean compositions of R and S are the *fuzzy* relations from X to Z defined next.

Definition 3.1 (Bandler-Kohout)

$$\begin{aligned}
R \triangleleft_m S(x,z) &= \frac{1}{\#Y} \sum_{y \in Y} R(x,y) \Rightarrow_B S(y,z) \\
R \triangleright_m S(x,z) &= \frac{1}{\#Y} \sum_{y \in Y} R(x,y) \Leftarrow_B S(y,z) \\
R \diamond_m S(x,z) &= \frac{1}{\#Y} \sum_{y \in Y} R(x,y) \Leftrightarrow_B S(y,z)
\end{aligned}$$

These compositions are called *mean subproduct*, *mean superproduct* and *mean square product* by Bandler and Kohout. Notice that in general a mean composition of crisp relations already yields a fuzzy relation. Bandler and Kohout's motivation for the introduction of these mean compositions is interesting. They argue that in some situations the infimum operator in properties 2.4 and definitions 2.3 is too strict (*harsh* membership criterion). For instance, both of the following cases

1. $\#xR = 10$ and $\#(xR \cap Sz) = 9$
2. $\#xR = 10$ and $\#(xR \cap Sz) = 0$

yield $R \triangleleft_{bk} S(x,z) = 0$ even when in the first case 90% of the elements of xR belong to Sz.

Replacing the infimum operator by the mean operator, they have chosen a less strict operator (*moderate* membership criterion). Indeed, one easily verifies the following properties.

Properties 3.1

$$\begin{array}{lllll} R \triangleleft S & \subseteq & R \triangleleft_{bk} S & \subseteq & R \triangleleft_m S \\ R \triangleright S & \subseteq & R \triangleright_{bk} S & \subseteq & R \triangleright_m S \\ R \diamond S & \subseteq & R \diamond_{bk} S & \subseteq & R \diamond_m S \end{array}$$

In this way $R \triangleleft_m S(x,z)$ is interpreted as the mean degree to which xR is a subset of Sz. Comparing definitions 3.1 and properties 2.5 raises the question whether there is a need for extra terms, analogous to the non-emptiness conditions. Indeed, although the idea behind the new compositions is appealing, they suffer from similar problems as definitions 2.1. Consider for instance the mean subcomposition $\triangleleft_m$. One easily verifies that

$$R \triangleleft_m S(x,z) = 1 - \frac{\#(xR \cap \mathrm{co}\, Sz)}{\#Y} = \frac{\#(\mathrm{co}\, xR \cup Sz)}{\#Y}.$$

When $xR = \emptyset$ it immediately follows that $R \triangleleft_m S(x,z) = 1$. This means that the degree of relationship between x and z in $R \triangleleft_m S$ is equal to 1 even if there is no element of Y that is in relation R^T with x.

Compared to the infimum operator the mean operator *seems* to offer a better gradation: $R \triangleleft_m S$ is the sum of the percentage of elements of Y not belonging to xR and the percentage of elements of Y belonging to $xR \cap Sz$. Apart from the boundary problems, this gradation still does not completely fulfil the purpose. Assume that $\#Y = 20$ and consider the following four cases

1. $\#xR = 3$ and $\#(xR \cap Sz) = 1$,
2. $\#xR = 4$ and $\#(xR \cap Sz) = 2$,
3. $\#xR = 6$ and $\#(xR \cap Sz) = 4$,
4. $\#xR = 1$ and $\#(xR \cap Sz) = 0$.

In the first three cases $R \triangleleft_m S(x,z) = 0.9$, although respectively $\frac{1}{3}$, $\frac{1}{2}$ and $\frac{2}{3}$ of the elements of xR belong to Sz. The fourth case yields $R \triangleleft_m S(x,z) = 0.95$, although no element of xR belongs to Sz. These unexpected results are due to the fact that $R \triangleleft_m S$ is always equal to or greater than the percentage of elements of Y not belonging to xR.

Therefore it is clear that definitions 3.1 cannot be improved be trying to add extra terms. Alternative definitions for the mean triangular compositions can be defined as follows.

Definition 3.2

$$R \triangleleft_p S(x,z) = \begin{cases} \dfrac{\#(xR \cap Sz)}{\#xR} & , \textit{ if } xR \neq \emptyset \\ 0 & , \textit{ else} \end{cases}$$

$$R \triangleright_p S(x,z) = \begin{cases} \dfrac{\#(xR \cap Sz)}{\#Sz} & , \textit{ if } Sz \neq \emptyset \\ 0 & , \textit{ else} \end{cases}$$

Of course, these definitions are no longer instances of the general relational product. In this way $R \triangleleft_p S(x,z)$ is interpreted as the percentage of elements of xR belonging to Sz. In the foregoing four cases the value of $R \triangleleft_p S(x,z)$ is respectively $\frac{1}{3}$, $\frac{1}{2}$, $\frac{2}{3}$ and 0.

3.3. Mean compositions of fuzzy relations

The mean subcomposition and supercomposition have been extended to fuzzy relations by replacing the boolean implication operator $\Rightarrow_B$ by a fuzzy implication operator $\mathcal{I}$. Consider a fuzzy relation R from X to Y and a fuzzy relation S from Y to Z and assume that the universe Y has a finite cardinality $\#Y$.

Definition 3.3 (Bandler-Kohout)

$$R \triangleleft_m S(x,z) = \frac{1}{\#Y} \sum_{y \in Y} \mathcal{I}(R(x,y), S(y,z))$$

$$R \triangleright_m S(x,z) = \frac{1}{\#Y} \sum_{y \in Y} \mathcal{I}(S(y,z), R(x,y))$$

Having in mind the problems with the mean compositions of crisp relations it is obvious that these definitions will not capture the desired interpretation. Alternative definitions can be obtained by extending definitions 3.2 to fuzzy relations, in the following way.

Definition 3.4

$$R \triangleleft_p S(x,z) = \begin{cases} \dfrac{\sum_{y \in Y} \mathcal{T}(R(x,y), S(y,z))}{\sum_{y \in Y} R(x,y)} & , \textit{if } xR \neq \emptyset \\ 0 & , \textit{else} \end{cases}$$

$$R \triangleright_p S(x,z) = \begin{cases} \dfrac{\sum_{y \in Y} \mathcal{T}(R(x,y), S(y,z))}{\sum_{y \in Y} S(y,z)} & , \textit{if } Sz \neq \emptyset \\ 0 & , \textit{else} \end{cases}$$

4. CONCLUSION

Fuzzy relational compositions have been approached from a critical point of view. The triangular products of Bandler and Kohout have been improved. The shortcomings of their mean products have been demonstrated and alternative definitions have been suggested.

REFERENCES

1. W. Bandler and L. J. Kohout, *Fuzzy relational products as a tool for analysis and synthesis of the behaviour of complex natural and artificial systems*, in Fuzzy Sets: Theory and Application to Policy Analysis and Information Systems (S. K. Wang and P. P. Chang, eds.), Plenum Press, New York and London, 1980, pp. 341–367.

2. W. Bandler and L. J. Kohout, *A survey of fuzzy relational products in their applicability to medicine and clinical psychology*, in Knowledge Representation in Medicine and Clinical Behavioural Science, (W. Bandler and L. J. Kohout, eds.), Abacus Press, Cambridge and Turnbridge Wells, 1986, pp. 107–118.
3. B. De Baets, *Theoretical basis for some modules from an environment for fuzzy relational calculus*, Postgraduate Thesis in Knowledge Technology, University of Gent, Belgium, 1991. (Dutch)
4. B. De Baets and E. Kerre, *A revision of Bandler-Kohout compositions of relations*, submitted to Mathematica Pannonica.
5. B. De Baets and E. Kerre, *Fuzzy relational compositions*, submitted to Fuzzy Sets and Systems (Pap. No. 92073).
6. E. Kerre, *Introduction to the Basic Principles of Fuzzy Set Theory and some of its Applications* (E. Kerre, ed.), Communication & Cognition, Gent, Belgium, 1991.
7. L. Zadeh, *Fuzzy sets*, Information and Control **8** (1965), 338–353.

Uncertainty in Intelligent Systems
B. Bouchon-Meunier et al. (Editors)

Modelling support system for ecological application based on fuzzy logic

A. Salski[a] and P. Kandzia[b]

[a]Ecosystem Research Center, University of Kiel
Schauenburgerstr.112, D-2300 KIEL 1
E-Mail: arek@pz-oekosys.uni-kiel.dbp.de
[b]Institut of Computer Science, University of Kiel
Olshausestr.40, D-2300 KIEL 1

Abstract:
The paper deals with some uncertainty problems in ecological modelling, namely uncertainty of ecological data and uncertainty of expert knowledge. A fuzzy logic approach to solve these problems is proposed and the modelling support system FLECO developed in Ecosystem Research Center at the University of Kiel is presented.
Keywords: knowledge-based systems,modelling, fuzzy set theory,ecosystems.

1. UNCERTAINTY PROBLEMS IN ECOLOGICAL RESEARCH

The problem of uncertainty in ecological research seems to be essential. It is a problem of uncertainty of ecological data and uncertainty of expert knowledge.

Ecological data bear a large inherent uncertainty not only as a consequence of inaccuracy of data, inaccuracy of interpolation methods and unreliability of measurement tools (statistics are dealing with these issues). Another problem is that some ecological parameters or variables are not even measurable (for example the number of fishes in a lake or the number of birds in a forest). The values of such parameters can be only approximated. One can say, for example, that there are "a lot of fishes" or "only a few fishes" in the particular lake. The question is, how to use and to process such unprecise information.

Sometimes the relations between the components of an ecosystem are not exactly known, sometimes we don't have any analytical model for these relations, or we don't have enough data for statistical analysis. In such a case the idea is to build a model based on the expert knowledge which is also often not precise. Ecologists often have to use vague, ill-defined natural language to describe their knowledge.

A fuzzy logic approach to deal with these uncertainty problems was proposed at IPMU'90 - Conference in Paris (Salski and Sperlbaum 1991).

2. FUZZY LOGIC APPROACH TO ECOLOGICAL MODELLING

Compared to the conventional expert system the application of the fuzzy set theory offers better applicability of unprecise data and vague knowledge, i.e.:

- the representation and processing of unprecise data in the form of fuzzy sets,
- the representation and processing of vague knowledge in the form of linguistic rules with unprecise statements defined as fuzzy sets.

A fuzzy logic approach to ecological modelling allows to work with unprecise knowledge of relations between ecosystem components and to build a model based only on qualitative information.

Since 1989 two fuzzy knowledge-based models have been created at Ecosystem Resarch Center at the University of Kiel (Sperlbaum 1991):

- LARKS- a model of the breeding success of larks,
- SOIL - a model of soil qualifiers.

Because of promising results (the difference between the simulation results and first results of field research is acceptable) we continue this research and provide further models using the modelling support system presented below.

3. ARCHITECTURE OF FLECO - THE MODELLING SUPPORT SYSTEM BASED ON FUZZY LOGIC

In order to facilitate the construction of fuzzy knowledge-based models of complex ecological processes the Modelling Support System based on Fuzzy Logic (FLECO) was developed (Schepers 1991, Bui 1991, Salski and Kandzia 1991). FLECO employs fuzzy logic to handle inexact reasoning and fuzzy sets to handle uncertainty of ecological data. The main tasks of this system are:

- to facilitate the creation of the fuzzy knowledge base,
- to simplify the changes within the fuzzy knowledge base,
- to create and facilitate the connections between sub-models,
- to offer a set of inference methods,
- to facilitate the simulation process.

The architecture of FLECO seems to be similar to an architecture of fuzzy expert systems, but it is oriented more to a modelling support. It integrates properties and requirements of the modelling technique with properties of the qualitative reasoning systems. FLECO enables the user to create and to connect separate sub-models of complex ecological systems. Each sub-model contains a set of linguistic rules which can be interpreted as the "fuzzy approximation" (or "fuzzy interpolation") of the relation to be modelled and forms a separate module of the knowledge base.

FLECO consists of four main subsystems (see Fig.1):
- the modelling subsystem /knowledge acquisition subsystem,
- the fuzzy knowledge base,
- the simulation driver,
- the system management module.

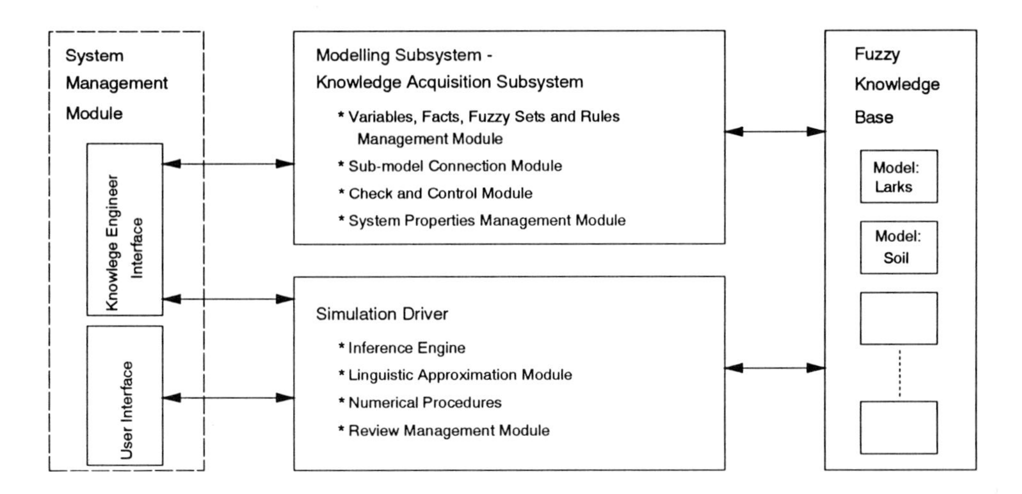

Fig.1 Global architecture of FLECO.

Modelling Subsystem /Knowledge Acquisition Subsystem

The knowledge acquisition subsystem contains management modules for variables, facts, fuzzy sets and rules. A variable is a basic entity in the model (for example a variable "population size" of an animal type) and describes a certain component of an ecosystem. Variable management module defines, modifies or deletes variables in the system.

By a "fact" we understand a proposition in the form:

<variable> is <value>.

Variables in FLECO can take fuzzy or crisp values. Fuzzy values are defined as fuzzy sets. The rules have the form of implication ("IF-THEN") and describe the relationship between the ecosystem components.

A very important point in the modelling process is to make possible the connections between sub-models. In FLECO this problem is solved by using the rules. The sub-model connection module of FLECO automatically generates the rules describing these connections.

Fuzzy Knowledge Base

Each model created by FLECO has its own knowledge base which includes all knowledge entities, such as the objects, facts, fuzzy set definitions and rules, determinate by using the knowledge acquisition subsystem. All knowledge bases are under common management of system (system-properties management module).

Simulation Driver

The main tasks of this subsystem are to process the model knowledge base for certain fuzzy or crisp input values, to facilitate the input procedure and to store the output values for each simulation process. The output values can be formulated in the form of fuzzy sets or transformed into crisp (numerical) values (so-called defuzzification process) or approximated to one of the linguistic terms we have defined for the output variable (using linguistic approximation module). The main part of the simulation driver is an inference engine, which contains a set of inference methods (actually only two methods: the first proposed by Mamdani (Mamdani and Assilian 1975) and the second proposed by Mizumoto, Fukami and Tanaka (Mizumoto et al. 1979) used also by Leung and Lam (Leung and Lam 1988)). The set of inference methods will be extended and should contain particularly suitable methods for the modelling process. They should have the following properties:

- to process both crisp and fuzzy input values in one simulation run,
- to calculate the output values with the lowest possible degree of fuzziness (particulary important in the case of the series connections between sub-models),
- to simplify the matching of the model to the system to be modelled.

System Management Module

FLECO can be used by the expert (knowledge engineer) to build a fuzzy knowledge- based model and also by the ordinary user to realize the simulation. The user trustee rights are determinated by system management module.

4. MODELLING PROCESS

To create a fuzzy knowledge-based model one should go through the following stages (partially supported by FLECO):

a) determination of model structure,
b) formulation and construction of the fuzzy knowledge base,
c) choice of fuzzy knowledge processing methods,
d) calibration,
e) validation.

First, one has to determine the model structure, i.e. input and output variables, the number of sub-models, connections between submodels etc.

In the process of the formulation of the knowledge representation in the form of the knowledge base the following steps must be taken:

- determination of linguistic rules,
- defining the fuzzy sets which are used to describe the terms in linguistic rules ("low", "high",etc.-values of the model variables).
- construction of the fuzzy knowledge base using, in our case, the knowledge acquisition subsystem of FLECO.

The major problem of "fuzzy" modelling is finding an appropriate set of linguistic rules describing the system to be modelled. They may be taken directly from the expert's experience. Nevertheless it is a very difficult process to define these rules on this basis, because the knowledge of the expert may be too complex to be written as a limited set of rules.
The set of linguistic rules should be complete and provide a correct answer for every possible input value. Thus the obvious condition should be satisfied that the sum of all input values (union of fuzzy sets) should 'cover' the value space of the input variable. If this condition is not satisfied the number of linguistic rules should be increased. If this is impossible one can try to increase the so-called degree of fuzziness of the fuzzy sets.

To process the created fuzzy knowledge base the following methods should be chosen from the set of methods offered by FLECO:
- inference method,
- defuzzification method,
- linguistic approximation method (based on the distance measure between fuzzy sets).

In order to improve the model's performance one can adjust the parameters of the model (for example parameters of fuzzy sets) and find the best combination of parameter values by means of simulation tests (calibration process). The relation between these parameters and the performance of the fuzzy model is not yet completely known theoretically. In a fixed application one can check it by means of experiments.
The validation process (further simulation tests for a wide range of independent data) should give the answer to the question: how well the model fits the real system.

5. SIMULATION PROCESS

The set of linguistic rules, definitions of fuzzy sets and the facts (data) compose a main part of the fuzzy model: the fuzzy knowledge base (see fig.2). Using a fuzzy inference method one can

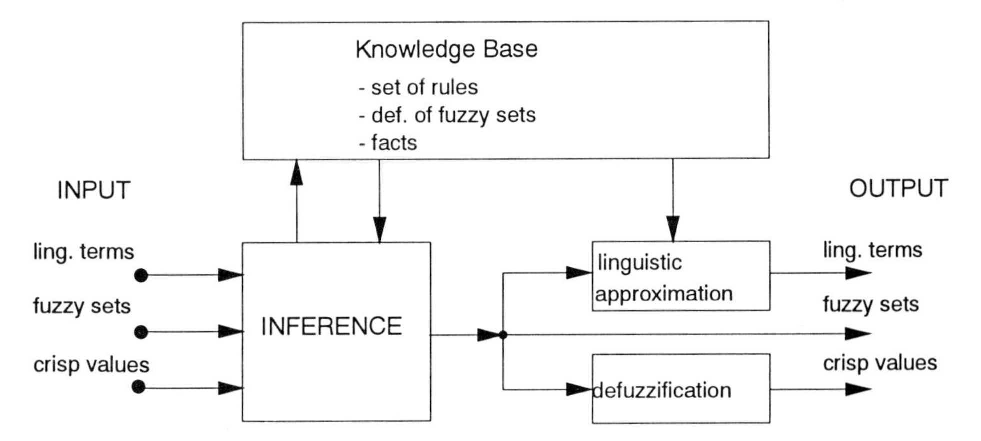

Fig.2. Information flow in the fuzzy knowlege-based model.

process this knowledge and compute output values for certain input values. The input values can take crisp (numerical) or fuzzy set form. Linguistic terms (represented also by fuzzy sets) are also allowed as input. The output values have the form of fuzzy sets too. These fuzzy sets can be transformed into a numerical value (defuzzification process) or approximated to one of the linguistic terms we have defined for the output variable. This so-called linguistic approximation can be accomplished by means of distance calculation between fuzzy sets.
An example of a fuzzy knowledge-based model in ecological research (model LARKS) with simulation results can be found in (Salski & Sperlbaum 1991).

6. FINAL REMARKS

FLECO is developed for ecological modelling, but the use of this system for fuzzy rule-based modelling in other fields is possible. FLECO has been written in Pascal (because of access to ORACLE- data base using the Pascal- Precompiler for ORACLE) and implemented on VAX 3400.
Actually, we are using FLECO for creating some further models, namely:
- model of dry deposition,
- model of population dynamics.

At the same time other inference methods are being investigated with respect to their suitability to modelling processes.

LITERATURE

Bui M. T.: Weiterentwicklung des Unterstützungssystemes zur wissensbasierten Modellierung unter dem Einsatz der Fuzzy-Set-Theorie. Diplomarbeit, Inst. f. Informatik und Praktische Mathematik, Universität Kiel (1993), pp. 96.

Leung K.S. and Lam W.: Fuzzy concepts in expert systems. IEEE Computer MC-21 (1988), pp. 43-56.

Mamdani E.H. and Assilian S.: An experiment in linguistic synthesis with a fuzzy logic controller. IJMMS 7 (1975), pp.1-13.

Mizumoto M., Fukami S. and Tanaka K.: Some methods of fuzzy reasoning. In Gupta M.M., Radage R.K., and Yager R.R. (Eds.): Advances in fuzzy set and applications. North-Holland (1979) pp.117-136.

Salski A. and Kandzia P.: Einsatz der Fuzzy-Set-Theorie in der Ökosystemforschung. In M. Hälker, A. Jaeschke (Eds.): Informatik für den Umweltschutz. Springer (1991), pp.303-310.

Salski A. and Sperlbaum C.: Fuzzy logic approach to modelling in ecosystem research. In Bouchon-Meunier B., Yager R. R., Zadeh L.A. (Eds.): Uncertainty in Knowledge Bases. Springer LNCS 521 (1991), pp.520-527.

Schepers M.: Ein Unterstützungssystem zur wissensbasierten Modellierung und Simulation von ökologischen Prozessen. Diplomarbeit, Inst. f. Informatik und Praktische Mathematik, Universität Kiel (1991), pp.115.

Sperlbaum C.: Einsatz der Fuzzy-Set-Theorie in der Modellierung ökosystemspezifischer Prozesse. Diplomarbeit, Inst. f. Informatik und Praktische Mathematik, Universität Kiel (1991), pp.120.

Uncertainty in Intelligent Systems
B. Bouchon-Meunier et al. (Editors)

On the least models of fuzzy DATALOG programs *

Attila Kiss
Department of General Computer Sciences, Loránd Eötvös University
P.O. Box 157, Budapest 112, H-1502, Hungary

In the paper we define fuzzy DATALOG programs as sets of Horn-formulae with degrees and we give their meaning by showing that when the least model exists then it can be computed by an algorithm which iterate certain expressions of the extended fuzzy relational algebra.

1. Introduction

In the knowledge-base systems there are given some facts which represent certain knowledge and some rules which in general mean that some kind of information implies another kind of information. In classical relational database theory [11] the DATALOG data model is widely spread in which the facts form a so-called extensional database and the rules are in a restricted (safe and rectified) set of Horn-formulae. The meaning of a DATALOG program is the least (if exists) or a minimal model which contains the facts and satisfies the rules.

There are known algorithms which use expressions of relational algebra to compute the meaning of a DATALOG program [11]. The aim of the paper is to give a possible extension of DATALOG programs to fuzzy relational databases [1-7],[15] using lower bounds of degrees of uncertainty in facts and rules. Allowing to apply different kinds of implication operators we prove the existence of the least model of fuzzy DATALOG programs (no matter they are recursive or not).

The least model, which is a fuzzy relational database in fact, can be given by using a modified version of the algorithms of classical case. This modification depends on the used implication operators. As a special case we get that adding some modifiers to the set of operators of fuzzy relational algebra [9],[10],[12], in non-recursive case the least fuzzy database is an expression of this extended fuzzy relational algebra using the fact fuzzy relations. As a consequence the least fuzzy database is finite (the set of tuples with non-zero degrees is finite). The algorithm for recursive case gives the solutions of interesting and important problems e.g. to find the transitive closure of a binary fuzzy relation. Our results make it possible to consider many problems formulated with

* This research was made at Liège University in the framework of a Hungarian-Belgian research scholarship, 1991.

the tools of logic [8] in the framework of fuzzy relational databases, besides to find the solutions of these problems we use only the expressions of extended fuzzy relational algebra.

2. Fuzzy DATALOG programs (fDATALOG)

To introduce the new concept of fDATALOG we need a few basic definitions.
$\mathcal{C} = \{C_1, C_2, \ldots\}$ is the set of constants and $\mathcal{X} = \{X, Y, \ldots\}$ is the set of variables. Let $p_1, \ldots, p_m$ be ordinary predicate and $b_1, \ldots, b_n$ be built-in predicates (such as APPROX-EQUAL (X,Y), CLOSE (X,Y,Z) etc [14]). There is only semantic difference between the two kinds of predicates. (The fuzzy relations correspond to built-in predicates are the same in all interpretations, which will mean that the truth value of $b(C_1, \ldots, C_k)$ is always the same known value, and this value can be computed by a known algorithm (subroutine).)

A Horn-formula is of form : $q_1(\underline{A}_1) \wedge q_2(\underline{A}_2) \wedge \ldots \wedge q_k(\underline{A}_k) \rightarrow q(\underline{X})$, where
- $\underline{A}_i$ is a tuple of variables and constans with arity that corresponds to q_i,
- $\underline{X}$ is a tuple of variables with arity that corresponds to q,
- q_i is ordinary or build-in predicate,
- q is ordinary predicate.

The condition part of a Horn-formula is called the body and the consequence part is called the head of the formula. A Horn-formula is safe:
- if $q_i(\underline{A}_i) = b_i(\underline{A}_i)$ then each variable in $\underline{A}_i$ appears in some $q_j(\underline{A}_j) = p_j(\underline{A}_j)$, (the variables are limited),
- the variables of $\underline{X}$ are in the set of variables of $\underline{A}_1, \ldots, \underline{A}_k$.

For example:
$loves(X, "Ann") \wedge friends(Y, X) \wedge similar("Ann", Z) \wedge friends("Ann", Z) \rightarrow loves(Y, Z)$
is safe, but
$loves(X, "Ann") \wedge loves(Y, X) \wedge similar("Ann", Z) \rightarrow loves(Y, Z)$
is not safe (where *similar* is the only built-in predicate).

This condition is needed because we intend to use only finite fuzzy relations for evaluating the truth values correspond to ordinary predicates.

The set of Horn-formulae is rectified iff
- when two formulae have the heads $q(\underline{X})$ and $q(\underline{Y})$ respectively then $\underline{X} = \underline{Y}$.

For example:
$young(X) \rightarrow beautiful(X)$
$friends(X, Y) \rightarrow beautiful(Y)$
is not a rectified set, but
$young(X) \rightarrow beautiful(X)$

$friends(Z,X) \rightarrow beautiful(X)$
is already a rectified set. It is only a technical assumption, because we can rename the variables to get a rectified set from a not rectified one.

Associate a dependency graph with a finite set of safe, rectified Horn-formulae:
- the nodes are the ordinary predicates,
- there is an arc $p_i \rightarrow p_j$ iff there is a formula with the head p_j and with the body containing p_i.

We call EDB (Extensional Database) predicates the nodes to where there are no arcs and IDB (Intensional Database) predicates all the others [11]. Thus input data are stored in EDB's and applying the rules we can get the data in IDB's.

For example:
$edge(X,Z) \rightarrow path(X,Z)$
$edge(X,Y) \wedge path(Y,Z) \rightarrow path(X,Z)$
$path(X,X) \rightarrow on\_cycle(X)$
where the only EDB is *edge*.

When we want to give the meaning for our formulae we have to use implication operators corresponding to $\rightarrow$ in our formulae. We allow to use for each formula any different implication operators from a given set. Thus we fix a set of implication operators $\mathcal{I}$. For example if we want to apply always the same implication operator I then $\mathcal{I} = \{I\}$.

Now we give what we mean fuzzy datalog programs. First we define the syntax where we associate some degrees for the truth of the facts and the rules.

Definition 1. An fDATALOG program $\mathcal{P} = (\mathcal{F}, \mathcal{R})$ is a finite set of facts and rules, where
- a fact $f \in \mathcal{F}$ is a pair $(p_f(C_1,\ldots,C_N), \alpha_f)$,
- a rule $r \in \mathcal{R}$ is a triplet (h_r, I_r, α_r), where p_f is an EDB predicate, $C_1,\ldots,C_N \in \mathcal{C}$, $\alpha_f \in (0,1], h_r$'s form a rectified set of safe Horn-formulae, $I_r \in \mathcal{I}, \alpha_r \in (0,1]$.

For example: the facts are
$(fruit(\text{"}Apple\text{"}), 1)$
$(fruit(\text{"}Peach\text{"}), 1)$
$(loves(\text{"}John\text{"}, \text{"}Apple\text{"}), 0.85)$
$(needs\_similar\_climate(\text{"}Apple\text{"}, \text{"}Peach\text{"}), 0.7)$
$(friends(\text{"}John\text{"}, \text{"}Tom\text{"}), 0.78)$
and the rules are
$(loves(Z,T) \rightarrow likes(Z,T)), I_1, 1)$
$(likes(X,Y) \wedge fruit(Y) \wedge friends(X,Z) \wedge needs\_similar\_climate(Y,T) \rightarrow likes(Z,T)), I_1,\ 0.7)$
where $\mathcal{I} = \{I_1\}$ and $I_1(x,y) = \begin{cases} 1, & \text{if } x \leq y, \\ 0, & \text{otherwise} \end{cases}$.

Hence we assign some (maybe infinite) fuzzy relations $B_1, \ldots, B_n$ to $b_1, \ldots, b_n$. A valuation v is a mapping $\mathcal{X} \cup \mathcal{C} \to \mathcal{C}$, where $\mathcal{X}$ is a set of variables, and v is the identity on $\mathcal{C}$. For a tuple $t = (A_1, \ldots, A_l)$ let $v(t) = (v(A_1), \ldots, v(A_l))$.

Let $\mathcal{D} = \{P_1, \ldots, P_n\}$ be a fuzzy relational database ($P_i : D^{n_i} \to [0,1]$), where D is a domain of possible values and P_i corresponds to the predicate p_i. Let $v(p_i(\underline{A}_i)) = P_i(v(\underline{A}_i))$ and $v(b_j(\underline{A}_j)) = B_j(v(\underline{A}_j))$.

For example: if $v(X) = a$, $v(Y) = b$ and $P(X, c, Y) = 0.78$ where a, b, c are constans then $v(p(X, c, Y)) = 0.78$

Definition 2. $\mathcal{D}$ is a model of $\mathcal{P}$ iff
- $(p_f(C_1, \ldots, C_N), \alpha_f) \in \mathcal{F}$ then $P_f(C_1, \ldots, C_N) \geq \alpha_f$
- $(q_1(\underline{A}_1) \wedge \ldots \wedge q_k(\underline{A}_k) \to q(\underline{X}), I_r, \alpha_r) \in \mathcal{R}$ then for all valuation v:

$$I_r(\bigwedge_{i=1}^{k} v(q_i(\underline{A}_i)), v(q(\underline{X})) \geq \alpha_r,$$

Where $\wedge$ is *minimum*. To be short we denote $\bigwedge_{i=1}^{k} v(q_i(\underline{A}_i))$ by v(body) and $v(q(\underline{X}))$ by v(head) if the corresponding formula is clear from the context. So the condition can be written in a simpler way $I_r(v(body), v(head)) \geq \alpha_r$.

For example: the $fDATALOG$ program and the implication operator
$I_2(x, y) = \begin{cases} 1, & \text{if } x \leq y, \\ y, & \text{otherwise} \end{cases}$,
$(p(X, Y) \to q(X), I_2, 0.5)$
$(q(Y) \wedge r(Y, X) \to q(X), I_2, 0.6)$
$(p(a, b), 0.7)$
$(r(a, b), 0.8)$
$(r(a, c), 0.4)$
has the model $\mathcal{D}_1 = \{P, Q, R\}$ where
$P(a, b) = 0.7$, $R(a, b) = 0.8$, $R(a, c) = 0.4$, $Q(a) = 0.5$, $Q(b) = 0.5$, $Q(c) = 0.4$,
and the next fuzzy database is also a model $\mathcal{D}_2 = \{P, Q, R\}$ where
$P(a, b) = 0.8$, $R(a, b) = 0.9$, $R(a, c) = 0.6$, $Q(a) = 1$, $Q(b) = 0.6$, $Q(c) = 0.8$.

3. The least model of a fDATALOG program

When fuzzy relational databases have components of the same number and the same arity then inclusion, intersection, union, difference can be defined by components [9],[12]. $\mathcal{D}_1 = \{P_1, Q_1, \ldots\} \subseteq \mathcal{D}_2 = \{P_2, Q_2, \ldots\}$ iff $P_1 \subseteq P_2$, $Q_1 \subseteq Q_2$, ..., where $P_1 \subseteq P_2$ iff for all t $P_1(t) \leq P_2(t)$. Thus we have a partial ordering on models of $\mathcal{P}$. In the last example $\mathcal{D}_1 \subseteq \mathcal{D}_2$.
$\mathcal{D}_1 \cap \mathcal{D}_2 = \{P_1 \cap P_2, Q_1 \cap Q_2, \ldots\}$ where for all t $P_1 \cap P_2(t) = min(P_1(t) \cap P_2(t))$.
$\mathcal{D}_1 \cup \mathcal{D}_2 = \{P_1 \cup P_2, Q_1 \cup Q_2, \ldots\}$ where for all t $P_1 \cup P_2(t) = max(P_1(t) \cup P_2(t))$.

Hence we denote 'non-increasing' by $\searrow$, 'non-decreasing' by $\nearrow$ and minimum(x,y), maximum(x,y) by $\wedge$, $\vee$ respectively.

Proposition 1. If for all $I \in \mathcal{I}$, $I(\searrow, \nearrow)$ holds and $\mathcal{D}_1, \mathcal{D}_2$ are models then $\mathcal{D}_1 \cap \mathcal{D}_2$ is a model, too.

Proof: If $I(a,b) \geq \alpha$ and $I(c,d) \geq \alpha$ then $I(min(a,c),b) \geq \alpha$ and $I(min(a,c),d) \geq \alpha$ and this follows $I(min(a,c), min(b,d)) \geq \alpha$. $\square$

We note that this property is not valid if we allow negations in our formulae. It causes that in case of using negation the least model does not always exist.

Proposition 2. If for all $I \in \mathcal{I}$, $I(x,y)$ is continuous from right in y, then $\mathcal{D}^0 := \bigcap_{\mathcal{D} \text{ model}} \mathcal{D}$ is the least model.

Proof: If $I(a,b) \geq \alpha$ where $a \in A$ and $b \in B$ then for all $b \in B$ $I(inf(a|a \in A), b) \geq \alpha$ thus $I(inf(a|a \in A), inf(b|b \in B)) \geq \alpha$. $\square$

Proposition 3. If the dependency graph is acyclic, then there is a sequence $p_1, \ldots, p_m$ that if $p_i \to p_j$ then $i < j$ and there is not any IDB predicate before EDB predicates.

Proof: This is the so-called topological ordering well-known in graph-theory. See [11]. $\square$

We will examine the special set $\mathcal{I}_0 = \{I_1, I_2, I_3\}$, where

$$I_1(x,y) = \begin{cases} 1, & \text{if } x \leq y, \\ 0, & \text{otherwise} \end{cases}, \quad I_2(x,y) = \begin{cases} 1, & \text{if } x \leq y, \\ y, & \text{otherwise} \end{cases},$$

$$I_3(x,y) = \begin{cases} 1, & \text{if } x \leq y, \\ 1-(x-y), & \text{otherwise} \end{cases}.$$

Proposition 4. Let $I_{i,\alpha}(x) = \min_{I_i(x,y) \geq \alpha} y$. Then
i) $I_{i,\alpha}(x)$ is well defined and $I_{i,\alpha}(\nearrow), i = 1,2,3$,
ii) $I_{1,\alpha}(x) = x$,
iii) $I_{2,\alpha}(x) = x \wedge \alpha$,
iv) $I_{3,\alpha}(x) = 0 \vee (x - (1-\alpha))$.

Proof: It is a simple calculation. $\square$

Proposition 5. Define a partial ordering $(x,y) \leq (x',y')$ by $x \leq x'$ and $y \leq y'$. Then for all $(0,0) \leq (x_0, y_0) \leq (1,1)$ there exists the least point of the set

$$\{(x,y) | I_i(x,y) \geq \alpha, (0,0) \leq (x,y) \leq (1,1), (x_0,y_0) \leq (x,y)\},$$

and this point is $(x_0, I_{i,\alpha}(x_0) \vee y_0)$ where $i = 1,2,3$.

Proof: It is clear if we draw this set. □

Proposition 6. (Proposition 5. for k+1 dimensional case) Define a partial ordering $(\underline{x}, y) \leq (\underline{x}', y')$ by $x_i \leq x_i'$ for the components of $\underline{x}, \underline{x}'$ and $y \leq y'$. Then for all $\underline{x}_0 = (x_1^0, \ldots, x_k^0)$ and $(\underline{0}, 0) \leq (\underline{x}_0, y_0) \leq (\underline{1}, 1)$ there exists the least point of the set $\{(\underline{x}, y) | I_i(x_1 \wedge \ldots \wedge x_k, y) \geq \alpha, \underline{x} = (x_1, \ldots, x_k), (\underline{0}, 0) \leq (\underline{x}, y) \leq (\underline{1}, 1), (\underline{x}_0, y_0) \leq (\underline{x}, y)\}$ and this point is $(\underline{x}_0, I_{i,\alpha}(x_1^0 \wedge \ldots \wedge x_k^0, y_0))$ where $i = 1, 2, 3$.

Proof: It is easy to verify using Proposition 5. □

We remind the fuzzy relational algebra given by Umano in 1983 [12]. The fuzzy relational algebra contains five operators and expressions can be given by finite compositions of these operators. Let U be the set of the possible names of the columns in fuzzy relations (U is the set of attributes). The operators are the following.

1) Projection
If $X \subseteq U$ then $\Pi_X(R)(s) := \bigvee_{t:t[X]=s} R(t)$.

2) Union
$R \cup S(t) := R(t) \vee S(t)$.

3) Selection
If $\forall t:\ F(t) \in [0,1]$ then $\sigma_F(R)(t) := R(t) \wedge F(t)$.

4) Cartesian product
$R \times S(uv) := R(u) \wedge S(v)$.

5) Subtraction
$R - S(t) := 0 \vee (R(t) - S(t))$.

The natural join $\bowtie$ is expressible in the relational algebra. Instead of giving the exact definition [11] we illustrate it with an example. If we have $p(X,Y)$ and $q(X,Z)$ and $P : dom(X) \times dom(Y) \to [0,1]$, $Q : dom(X) \times dom(Z) \to [0,1]$ then $P \bowtie Q : dom(X) \times dom(Y) \times dom(Z) \to [0,1]$ such that $P \bowtie Q(t) = P(t[X], t[Y]) \wedge Q(t[Y], t[Z])$.

We call extended fuzzy relational algebra the case when we fix a finite number of functions $W_1, W_2, \ldots$ where $W_i : [0,1] \to [0,1]$ for all i, and introduce the operations

6i) Aggregation W_i
$W_i(R)(t) = W_i(R(t))$.

For example λ-cuts, negation are aggregations. [6]

Now we assign an expression of fuzzy relational algebra to the body of a given Horn-formula. Consider the Horn-formula $h_r = q_1(\underline{A}_1) \wedge \ldots \wedge q_k(\underline{A}_k) \to q(\underline{X})$ and suppose that we have a fuzzy relation P for each ordinary predicates p.

If $q_i(\underline{A}_i) = p_{l_i}(A_1, \ldots, A_k)$ then let F_i be the conjunction of $\$j = C$ if $A_j = C$, and $\$j = \$n$ if $A_j = A_n$ and denote the set of variables in $q_i(\underline{A}_i)$ by V_i.

For example: if $p(X, Y, a, X)$ where a is a constans then the corresponding expression is $\sigma_{\$1=\$4 \wedge \$3=a} P$

Definition 3. EVAL-RULE $(h_r, P_1, \ldots, P_m) := \sigma_F(\bowtie_i \prod_{V_i}(\sigma_{F_i}(P_{l_i})))$, where the natural join is taken for all i, where $q_i(\underline{A}_i)$ is an ordinary predicate and the condition F in the selection is the conjunction of $b_j(\underline{A}_j)$ which are the built-in predicates in the body [11].

For example: if
$p(X, Y, a) \wedge q(X, Z) \wedge b(Z, a) \rightarrow q(Z, Y)$
where b is a built-in predicate then the corresponding expression is
$\sigma_{b(Z,a)}(\Pi_{X,Y}(\sigma_{\$3=a} P) \bowtie (\Pi_{X,Z} Q))$.

Proposition 7. If the variables in the body are $t = (X_1, \ldots, X_n)$ and V is a valuation, then v(body)=EVAL-RULE $(h_r, P_1, \ldots, P_n)(v(t))$.

Proof: The steps of the two evaluations can be assigned to each other. □

Proposition 8. i) EVAL-RULE $(h_r, P_1, \ldots, P_m)$ is finite if $P_1, \ldots, P_n$ finite,
ii) EVAL-RULE $(h_r, \nearrow, \ldots, \nearrow)$.

Proof: EVAL-RULE uses only Cartesian Products, Selections, Projections and all of them are non-decreasing and preserve finiteness. □

Proposition 9. Let $\mathcal{D} = \{P_1, \ldots, P_n\}$ be a model and $r_j, j = 1, \ldots, K$ be the rules with the head $p(x_1, \ldots, x_k)$. Define EVAL $(p, \mathcal{D})$ by the formula

$$\bigcup_{j=1}^{K} I_{r_j, \alpha_{r_j}}(\prod_{X_1, \ldots, X_k}(EVAL-RULE(h_{r_j}, P_1, \ldots, P_n))).$$

Then i) EVAL$(p, \nearrow)$ and it is an expression of the extended fuzzy relational algebra,
ii) EVAL $(p, \mathcal{D}) \subseteq P$.

Proof: The aggregator $I_{i,\alpha}$ is non-decreasing for $i = 1, 2, 3$. ii) holds because of the definition of $I_{i,\alpha}$. □

Algorithm 1. (non-recursive case) Let $p_1, \ldots, p_k, q_1, \ldots, q_{n-k}$ be a sequence in Proposition 3., where p_i's are EDB, q_j's are IDB predicates. Let $\mathcal{D}_0$ be $(P_1, \ldots, P_k, \emptyset, \ldots, \emptyset)$ where the fuzzy relations P_i have a tuple $(C_1, \ldots, C_L)$ with the degree α iff there exists a fact $f \in \mathcal{F}$, that $f = (p_i(C_1, \ldots, C_L), \alpha)$.
For the j-th IDB predicate q_j define $Q_j = EVAL(q_j, \mathcal{D}_{j-1})$ and $\mathcal{D}_j = \mathcal{D}_{j-1} \cup (\emptyset, \ldots \emptyset, Q_j, \emptyset, \ldots, \emptyset)$.

Theorem 1. i) $\mathcal{D}_0 \subseteq \mathcal{D}_1 \subseteq \ldots \subseteq \mathcal{D}_{n-k}$,
ii) $\mathcal{D}_{n-k}$ is the least (finite) model.

Proof: We do not give the details, just give some hints. $\mathcal{D}$ is defined by adding new tuples for the previous database which implies the increaing sequence. The last database is a model because in each step we add as few tuples as possible to our database, so ii) can be proved by induction on the number of steps. □

To give a similar algorithm for the recursive case we need a technical statement which says that the possible degrees form a finite set when $\mathcal{I} = \mathcal{I}_0$.

Proposition 10. If $G \subseteq [0,1]$ is finite and $\alpha \in \mathcal{A}$ where $\mathcal{A}$ is a finite set of degrees, then there exists a finite $\overline{G}, G \subseteq \overline{G}$ which is closed for $\wedge, \vee, I_{i,\alpha}, i = 1,2,3,$.

Proof: $I_{1,\alpha}$ and $I_{2,\alpha}$ add new values only from $\mathcal{A}$.
The set $\{I_{3,\alpha}(x),\ I_{3,\alpha}(I_{3,\alpha}(x)),\ I_{3,\alpha}(I_{3,\alpha}(I_{3,\alpha}(x))), \ldots\}$ is finite for all x and α. □

Algorithm 2. (recursive case) Let $p_1, \ldots, p_n; q_1, \ldots, q_{n-k}$ be a sequence where p_i's are EDB, q_j's are IDB predicates. Let $\mathcal{D}_0$ be as in Algorithm 1. Define $\mathcal{D}_i = (P_1, \ldots, P_k, EVAL(q_1, \mathcal{D}_{i-1}), \ldots, EVAL(q_{n-k}, \mathcal{D}_{i-1}))$.

Theorem 2. i) $\mathcal{D}_0 \subseteq \mathcal{D}_1 \subseteq \ldots$,
ii) Algorithm 2. halts in finite time : $\mathcal{D}_0 \subseteq \ldots \subseteq \mathcal{D}_L = \mathcal{D}_{L+1} = \mathcal{D}_{L+2}$,
iii) $\mathcal{D}_L$ is the least (finite) model.

Proof: The algorithm halts in finite time, because the set of degrees and constants in the program and in the EDB databases is finite, so the number of possible $\mathcal{D}$'s is finite. To prove that the last database is the least model is based on induction similarly to Theorem 1. □

4. Summary

As a special case of $\mathcal{I} = \{I_1\}$ and all degrees $\alpha = 1$, we get the classical case (DATALOG). If $\mathcal{P}$ has no variables then we have the least model of a set of formulae of fuzzy propositional logic [4],[13]. So this method can be used in approximate reasoning and we use only the operators of an extended fuzzy relational algebra and iteration to get the least model. We can also compute the transitive closure of a binary fuzzy relation by Algorithm 2. and it can be shown that the transitive closere is not equivalent to any expression of extended fuzzy relational algebra. We note that our algorithms are not suitable for the case of negated predicates, but we plan for the future to give a restriction of programs with negated predicates, when natural meaning of the program can be defined, as a special minimal fixpoint which is to be calculated by some similar algorithm to Algorithm 2.

References

1. G.Chen, J. Vardenberghe, E.E. Kerre : A step towards the theory of fuzzy relational database design. Proceedings of IFSA'91 Brussels, Computer, Management & Systems Science, pp 44-47.
2. S. Dutta: Approximate Reasoning by Analogy to Answer Null Queries. IJ. AR, 1991, Vol. 5. pp 373-398.
3. Special Issues on Fuzzy Information and Database Systems Edited by J. Kacprzyk, B.P. Buckles, F.E. Petry; Fuzzy Sets and Systems, 1990, Vol. 38. Number 2.
4. A. Kiss : The truth values of dependencies of fuzzy databases. Proceedings of First Finnish-Hungarian Workshop on Computer Science, Szeged, Hungary, Aug. 8-11, 1989, pp 177-186.
5. A. Kiss: On fuzzy relational databases. Proceedings of Mathematical Sciences Past and Present, Hamburg, March 18-25, 1990, Vol. 3. pp 1183-1193.
6. A. Kiss : λ-decomposition of fuzzy relational databases. Proceedings of Third Joint IFSA-EC, EURO-WG Workshop on Fuzzy Sets Visegrád, Hungary, Dec 11-13, 1990. pp 133-142.
7. W. Jr. Lipski: On databases with incomplete information. J. ACM, 1981, Vol. 28. pp 41-47.
8. T.P. Martin, J.F. Baldwin, B.W. Pilsworth : The implementation of fPROLOG- a fuzzy PROLOG interpreter. Fuzzy Sets and Systems, 1987, Vol. 23. pp 119-129.
9. H. Prade, C. Testimale : Generalizing database relational algebra for the treatment of incomplete information and vague queries; Inf. Sci. 1984, Vol. 34. pp 115-143.
10. K.V.S.V.N. Raju, A.K. Majumdar : The study of joins in fuzzy relational databases. Fuzzy Sets and Systems, 1987, Vol. 21. pp 19-34.
11. J.D. Ullman; Principles of database and knowledge-base systems. Computer Science Press, Rockville, 1988.
12. M. Umano : Retrieval from fuzzy database by fuzzy relational algebra. IFAC Fuzzy Information, Knowledge Information Systems, Marseille, 1983, pp 1-6.
13. R. Vardenberghe, A. Van Schooten, R. De Caluwe, E.E. Kerre: Some practical aspects of fuzzy database techniques. Information System 1989, Vol. 14. pp 465-472.
14. L.A. Zadeh : Fuzzy sets. Inf. control, 1965, Vol. 8. pp 338-353.
15. M. Zemankova-Leech, A. Kandel : Fuzzy relational databases - A Key to expert systems. Verlag TUV, Cologne, Germany, 1984.

Uncertainty in Intelligent Systems
B. Bouchon-Meunier et al. (Editors)

RESOLUTION CRITERIA FOR ALGEBRAIC EQUATIONS WITH FUZZY PARAMETERS

Saliha TEGHBIT and Noël MALVACHE

Laboratoire d'Automatique Industrielle et Humaine, UA CNRS 1118.

Université de Valenciennes et du Hainaut Cambrésis. Le Mont-Houy, 59326 Valenciennes Cedex - France.

ABSTRACT

This paper deals with the problem of solving algebraic equations containing fuzzy parameters. Criteria of solvability of the algebraic equations A±X=C, A÷X=C and A∗X=C, with A and C defined by continuous triangular membership functions, are determined. An approach, using classical methods for solving algebraic equations and truncation with a level α, is proposed. It yields a single solution $[w_{inf}, w_m, w_{sup}]$. This solution is defined by its support $[w_{inf}, w_{sup}]$, an interval within which the membership function is nonzero, and a mean value w_m for which the membership function is equal to one (1). Depending on the used operator (+, -, ∗, ÷), criteria for solving the various equations are deduced.

Key words : algebraic equation, α cut, α truncation, support, mean value, extension principle, criteria of solution.

I.INTRODUCTION

Automatics is one of the fields where the fuzzy sets theory, proposed by ZADEH [1], has emerged as one of the most active areas for research. Used for fuzzy models conception and fuzzy for purposes of control, regulation and computer-aided control, it becomes an essential tool in processing subjective information.

To design, identify or regulate the models of processes containing ill-defined information, it is imperative to know how to solve the algebraic equations describing these models. There are two types of equations containing fuzzy parameters, those with imprecise arithmetic operators, met when the relationship among the parameters of the described process is imprecise, and those with precise arithmetic operators, met when the action to be performed is specified (addition, subtraction, multiplication, division). For example, to maintain a temperature T at a value approximately equal to 25 degrees (C), it is possible, taking $A+X=C$ and $B-X=C$, to calculate the quantity (X) to be added, if the temperature is below 25 degrees (A), or subtracted, if it is above 25 degrees (B).

YAGER, [2], studied the first case. The proposed solution is based on intuition. If it satisfies the equation, it is considered valid; else, another solution is suggested. Considerable interest has been devoted to the resolution of the second type equations. There is no general method for solving these equations and classical ones cannot be applied. Several approaches have been studied. SANCHEZ, [3], proposed a non-standard inverse operation $\mathfrak{R}^{-1}$, which associates the solution $X=C\mathfrak{R}^{-1}A$ to the relation $A\mathfrak{R}X=C$. Other approaches, [7], are based on the definition of fuzzy numbers, [4], through the use of the extension principle, [5], proposed by ZADEH, [6], and the cut at a level α.

The classical mathematical method for solving algebraic equations $A\pm X=C$ is to add the term $\pm A$ to each side of the equation, yielding the solution $X=C\pm A$. This method is not applicable to algebraic equations containing fuzzy parameters because the variable X cannot be isolated from the equation $A\pm X=C$ by adding $\pm A$, insofar as $A+(-A)\neq 0$. Likewise, the classical solution for the equation $A\div X=C$ and similarly for $A*X=C$ is found by dividing either terms of the equation respectively by C with $C\neq 0$ and by A with $A\neq 0$, to yield respectively $X=A\div C$ and $X=C\div A$. This is not valid for equations containing fuzzy parameters, insofar as $C*(1\div C)\neq 1$ and $A*(1\div A)\neq 1$, respectively.

The study proposed by ZHAO, [7], examines the case in which the support of C is higher than that of A; in the reverse case, the equation is considered insolvable, limiting, thereby, the possibilities for solving algebraic equations to a particular case. To overcome that inconvenient, an approach whereby algebraic equations can be solved, whatever the value of the supports of A and C, is proposed. Depending on the operator used (+, -, ∗, ÷), solvability criteria for each equation are deduced. The representation of the membership functions of the fuzzy sets studied is triangular and continuous. The same representation is used in this study.

II. BACKGROUND

II.1. Definition of fuzzy set

A fuzzy set, with a triangular and normalized membership function, is defined by a triplet $[w_{inf}, w_m, w_{sup}]$. w_m is the mean value for which the membership function reaches its maximal value, one because normalized. The interval $W_s=[w_{inf}, w_{sup}]$ - inf and sup indicating lower and upper boundaries - called support, is the values set $w\in W$ for which the membership function is nonzero. The membership function of such a number is defined by

$$\mu_X(w) = \begin{cases} \dfrac{w - w_{inf}}{w_m - w_{inf}} & \text{if } w_{inf} \leq w \leq w_m \\ \dfrac{w - w_{sup}}{w_m - w_{sup}} & \text{if } w_m \leq w \leq w_{sup} \\ 0 & \text{else} \end{cases} \quad \text{(II.1)}$$

Each value of $\mu_X(w)$ is obtained for two values, w_L and w_R - L and R indicating the left and right boundaries of W_s - such as :

$$w_L = \mu_X(w)\,(w_m - w_{inf}) + w_{inf} \quad \text{(II.2)}$$

$$w_R = \mu_X(w)\,(w_m - w_{sup}) + w_{sup} \quad \text{(II.3)}$$

II.2. Cut at a level α (α cut)

A cut at a level α, $\alpha \in [0, 1]$, of the membership function $\mu_X(w)$, defines a closed interval, W_α, on which $\mu_X(w)$ is at less equal to α, fig II.1.

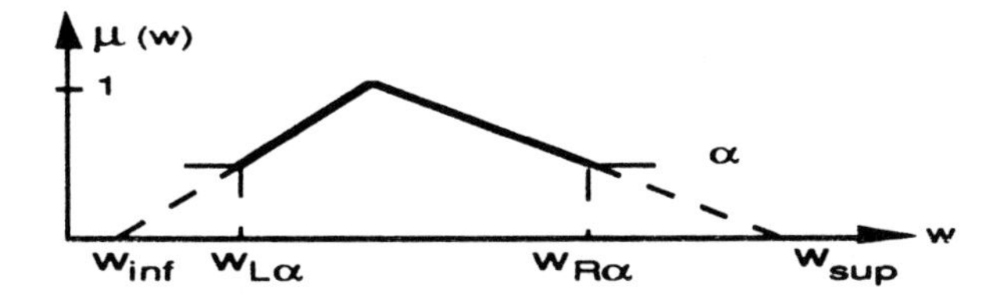

Figure II.1 : α level cut ; $W_\alpha = [w_{L\alpha}, w_{R\alpha}] = \{w \,/\, \mu_{X\alpha}(w) \geq \alpha\}$

II.3. Truncation with a level α.

A truncation with a level α, of the membership function $\mu_X(w)$ defines a closed interval, on which $\mu_X(w)$ is at most equal to α, fig II.2.

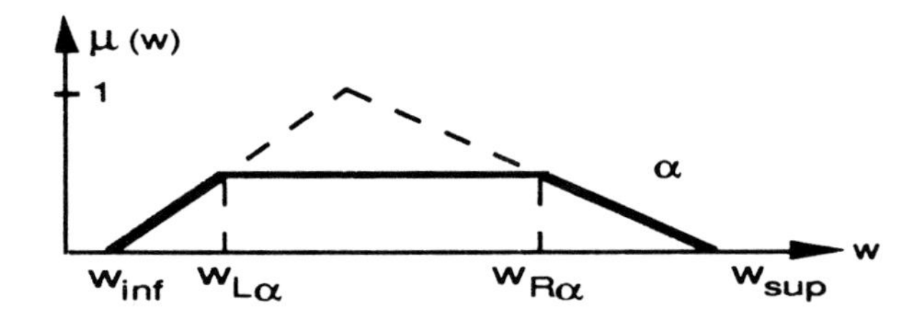

Figure II.2 : truncation with a level α; $\mu_{X\alpha}(w) = \min(\alpha, \mu_X(w))$

and $W_\alpha = [w_{L\alpha}, w_{R\alpha}] = \{w \,/\, \mu_{X\alpha}(w) = \alpha\}$ stands for the interval on which the membership function is maximal, so equal to α.

II.4. Extension principle

The extension principle, proposed by ZADEH, [5], is applied to each value w of the fuzzy interval W_α. This principle has a very significant importance in evolving the fuzzy arithmetic, [2]. Let be A and C two fuzzy sets defined, respectively, by the triplets $[u_{inf}, u_m, u_{sup}]$ on U and $[v_{inf}, v_m, v_{sup}]$ on V and a relation o such as AoX=C; the application of extension principle at a level α gives :

$$(AoX)_\alpha = A_\alpha o X_\alpha = C_\alpha \tag{II.6}$$

III. SOLUTION BY FUZZY INTERVÁLS

Solving the equation AoX=C means the calculation, applying II.6, of the triplet $[w_{inf}, w_m, w_{sup}]$, defining the solution X, using the fuzzy intervals one obtain :

$$[u_{L\alpha}, u_{R\alpha}] \text{ o } [w_{L\alpha}, w_{R\alpha}] = [v_{L\alpha}, v_{R\alpha}] = [u_{L\alpha} \text{ o } w_{R\alpha}, u_{L\alpha} \text{ o } w_{R\alpha}]$$

- at the level $\alpha = 0$, $\quad [v_{inf}, v_{sup}] = [u_{inf} \text{ o } w_{inf}, u_{sup} \text{ o } w_{sup}]$ (III.1)

- at the level $\alpha = 1$, $\quad v_m = u_m \text{ o } w_m$ (III.2)

Knowing $[u_{inf}, u_m, u_{sup}]$ and $[v_{inf}, v_m, v_{sup}]$, one can deduce, applying III.1 and III.2, the triplet $[w_{inf}, w_m, w_{sup}]$ defining the fuzzy set X, solution of the equation AoX=C. Note that the equality III.1 is verified if and only if $u_{inf} \text{ o } w_{inf} \leq u_{sup} \text{ o } w_{sup}$, ZHAO studied this particular case out of which the equation is considered without solution. In fact, the solution of such equations depends on the used operator and the support of the parameter A by comparison with the support of the parameter C.Let use the operator '+', then II.1 becomes :

$[v_{inf}, v_{sup}] = [u_{inf} + w_{inf}, u_{sup} + w_{sup}]$, which gives

$$winf = vinf - uinf \text{ and } wsup = vsup - usup \tag{III.3}$$

which are verified if and only if $v_{inf} - u_{inf} < v_{sup} - u_{sup}$

Therefore, if and only if the support of A is lower than that of C, else the equation is considered without solution. As it will be shown, in the following, That is not always true.

This solution presents the drawback of limiting resolution possibilities of algebraic equations to a single particular case. To provide against this inconvenient, an approach, of solving these equations independently of the fuzzy sets supports, is proposed. Depending of the used operator (+, -, ∗, ÷), criteria for solving each equation are deduced.

IV SOLUTION BY THE α TRUNCATION.

IV.1. Proposed approach

let be $A=[u_{inf}, u_m, u_{sup}]$ on U and $C=[v_{inf}, v_m, v_{sup}]$ on V, two fuzzy sets. The equation AoC=X is defined by :

$$\forall\ v \in V \quad \mu_C(v) = \sup_{\{(u,\, w)\ :\ u o w = v\}}[\min(\mu_A(u), \mu_X(w))] \tag{IV.1}$$

at the α level truncation, $\mu_{A\alpha}(u) = \alpha$ on U_α and $\mu_{C\alpha}(v) = \alpha$ on W_α, so IV.1 becomes :

$$\mu_{C\alpha}(v) = \sup_{\{(u,\, w)\ :\ u o w = v\}}[\min(\alpha, \mu_{X\alpha}(w))] = \alpha \tag{IV.2}$$

which involves $\mu_{X\alpha}(w) \geq \alpha$, as, at the level α, $\mu_{X\alpha}(w) \leq \alpha$, then $\mu_{X\alpha}(w) = \alpha$. What means that if $v \in V_\alpha$ and $u \in U_\alpha$ then $w \in W_\alpha$, one has only to calculate boundaries of the interval $W_\alpha=\{w / \mu_{X\alpha}(w) = \alpha\}$. So, the boundaries are given by :

$$w_1 o\ u_{L\alpha} = v_{L\alpha} \text{ and } w_2 o\ u_{R\alpha} = v_{R\alpha} \tag{IV.3}$$

and the interval $[w_{L\alpha}, w_{R\alpha}]$ is defined by :

$$w_{L\alpha}, w_{R\alpha}] = [\min(w_1, w_2), \max(w_1, w_2)] \tag{IV.4}$$

IV.2. Support calculation

Applying II.2 and II.2', at the level α=0, to the fuzzy sets A and C, one obtain :

$$u_{L0} = u_{inf} \text{ and } v_{L0} = v_{inf} \tag{IV.5}$$

$$u_{R0} = u_{sup} \text{ and } v_{R0} = v_{sup} \tag{IV.6}$$

Then, the équalities in IV.3 become :

$$w_1\ o\ u_{inf} = v_{inf} \text{ and } w_2\ o\ u_{sup} = v_{sup} \tag{IV.7}$$

so, the support is defined by the interval $[w_{inf}, w_{sup}]$ such as :

$$w_{inf} = \min(w_1, w_2) \tag{IV.8}$$

$$w_{sup} = \max(w_1, w_2) \tag{IV.8'}$$

IV.3. Mean value calculation

The application of II.2 and II.2', at the level α=1, gives :

$$w_{L1} = w_{R1} = w_m \text{ then } w_m\ o\ u_m = v_m \tag{IV.9}$$

which means that the membership function is maximal for a single value of w.

V. CRITERION FOR SOLVING EQUATIONS AoX=C

V.1. Criterion for solving the equation A+X=C

a) Support Calculation

The support of the solution X is defined, using the operator (+) in IV.7 , by :

$w_1 + u_{inf} = v_{inf}$ and $w_2 + u_{sup} = v_{sup}$

then $w_1 = v_{inf} - u_{inf}$ and $w_2 = v_{sup} - u_{sup}$

therefore

$$[w_{inf}, w_{sup}] = [\min(v_{inf} - u_{inf}, v_{sup} - u_{sup}), \max(v_{inf} - u_{inf}, v_{sup} - u_{sup})] \quad \text{(V1.1)}$$

b) Mean value calculation

The operator (+) used in IV.10 gives $w_m+u_m=v_m$ which involves:

$$w_m = v_m - u_m \quad \text{(V1.2)}$$

The triplet $[w_{inf}, w_m, w_{sup}]$, calculated above, form a solution for the equation A+X=C if and only if the criterion defined in the following paragraph is satisfied.

c) Resolution criterion

In order that w_{inf}, w_m, w_{sup} may form a solution for the equation A+X=C, they may satisfy the definition conditions of the membership function described by II.1; such as :

$w_{inf} \leq w_m \leq w_{sup}$ as $w_1 = v_{inf} - u_{inf}$, $w_2 = v_{sup} - u_{sup}$ and $w_m = v_m - u_m$

and, therefore, $[w_{inf}, w_{sup}] = [\min(w_1, w_2), \max(w_1, w_2)]$

So, the necessary and sufficient criterion for solving the equation A+X=C is given by :

$$\min(v_{inf} - u_{inf}, v_{sup} - u_{sup}) \leq v_m - u_m \leq \max(v_{inf} - u_{inf}, v_{sup} - u_{sup}) \quad \text{(V1.3)}$$

If this condition is satisfied, the triplet $[w_{inf}, w_m\ w_{sup}]$, above, is the single solution.

d) Examples

Example 0 : Let be two fuzzy sets such as A=[1, 2, 6] and C=[3, 6, 9]. Applying (V1.1) and (V1.2), one obtain w_1=2 and w_2=3 then W_s=[2, 3] but w_m=4 do not belong to the support W_s, so the criterion (V1.3) is not satisfied, the equation has no solution.

Example 1 : let be A=[2, 4, 6] and C=[4, 7, 10], solving the equation A+X=C with these parameters gives w_1=2 and w_2=4 so $W_s=[w_1, w_2]$=[2, 4] contains w_m= 3 then the triplet [2, 3, 4] satisfies the criterion and form the unique solution of the equation, figure V.1.1.

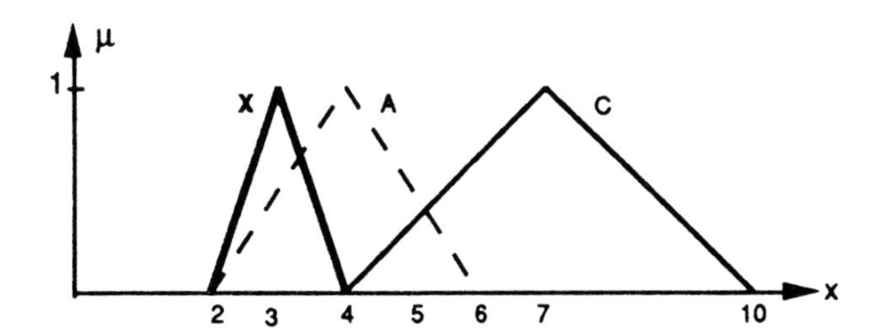

Figure V.1.1 : result of the example 1

Example 2 : Let be A=[2, 5, 11] and C=[9, 10, 12] two fuzzy sets, used in the equation A+X=C they give w_1=7, w_2=1 and w_m=5, W_s=[w_2, w_1]=[1, 7] contains the mean value w_m, then the criterion is satisfied and the solution is given by the triplet [1, 5, 7], figure V.1.2.

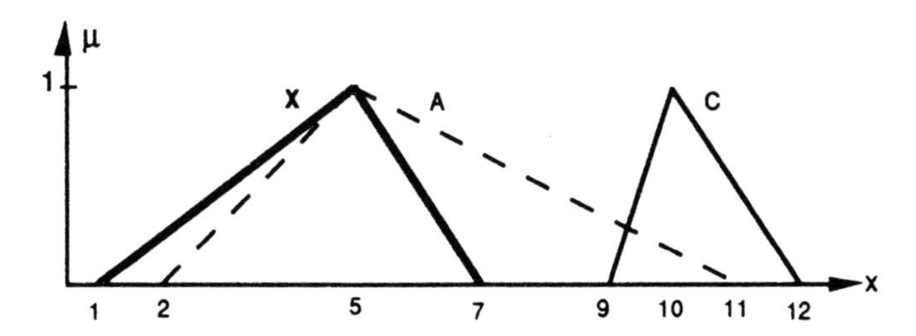

Figure V.1.2 : result of the example 2

Whatever the equation studied may be, the reasoning is the same, so, in the following , for the operators (-, *, +), one just give the results and the deduced criteria of each equation.

V.2 Resolution criterion for the equation A-X=C

a) Support calculation

For the operator "+", the support of the solution X, of the equation A-X=C, is given by :

$w_1 = u_{inf} - v_{inf}$ and $w_2 = u_{sup} - v_{sup}$

then

$$[w_{inf}, w_{sup}] = [\min (w_1, w_2), \max (w_1, w_2)] \qquad (V2.1)$$

b) Mean value calculation

The mean value of the fuzzy set X, solution of the equation A-X=C, is given by :

$$w_m = u_m - v_m \qquad (V2.2)$$

c) Resolution criterion

The solving criterion for the equation A-X=C is given by :

$$\min(u_{inf} - v_{inf}, u_{inf} - v_{inf}) \leq u_m - v_m \leq \max(u_{inf} - v_{inf}, u_{sup} - v_{sup}) \qquad \text{(V2.3)}$$

which is identical with the criterion for solving the equation A+X=C.

proof:

if $\quad v_{inf} - u_{inf} \geq v_{sup} - u_{sup} \Rightarrow u_{inf} - v_{inf} \leq u_{sup} - v_{sup}$

then $\quad \max(v_{inf} - u_{inf}, v_{sup} - u_{sup}) = -\min(u_{inf} - v_{inf}, u_{sup} - v_{sup})$

and $\quad \min(v_{inf} - u_{inf}, v_{sup} - u_{sup}) = -\max(u_{inf} - v_{inf}, u_{sup} - v_{sup})$

replacing, both of the last equalities in V.2 gives :

$$-\max(u_{inf} - v_{inf}, u_{inf} - v_{inf}) \leq -(u_m - v_m) \leq -\min(u_{inf} - v_{inf}, u_{sup} - v_{sup})$$

multiplying by -1 each side of the double inequality gives :

$$\min(u_{inf} - v_{inf}, u_{sup} - v_{sup}) \leq u_m - v_m \leq \max(u_{inf} - v_{inf}, u_{sup} - v_{sup})$$

then $\quad x_{inf} \leq x_m \leq x_{sup}$

If this criterion is satisfied, the triplet calculated above is the single solution of the equation A-X=C, this criterion is identical with the criterion for solving the equation A+X=C.

d) Examples

Example 1 : Let be two fuzzy sets A=[7, 8, 10] and C=[3, 5 , 9]. Solving the equation, applying (V2.1) and (V2.2), A-X=C gives w_1=4, w_m=3 and w_2=1, so W_s=[w_2, w_1]=[1, 4]. w_m belong to W_s then the criterion(V2.3) for solving the equation A-X=C is satisfied, the solution is X=[1, 3, 4], figure V.2.1.

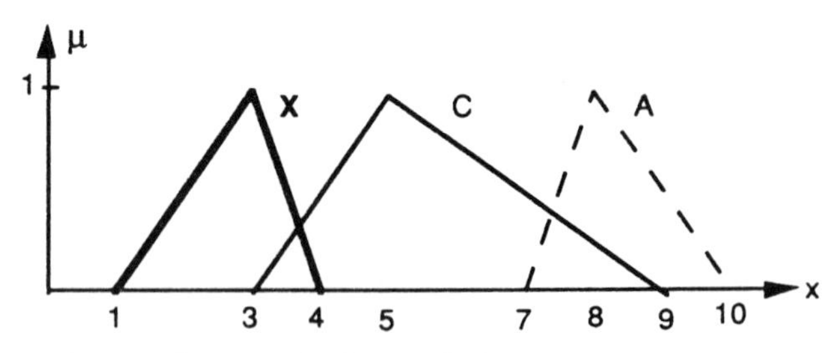

Figure V.2.1 : Result of example 1.

Example 2 : Let be two fuzzy sets A=[3, 7, 11] and C=[2, 3, 5], solving the equation A-X=C gives w_1=1, w_m=4 and w_2=6, then $W_s=[w_1, w_2]$=[1, 6]. w_m belong to W_s then the criterion is satisfied and the solution is given by X=[1, 4, 6], figure V.2.2.

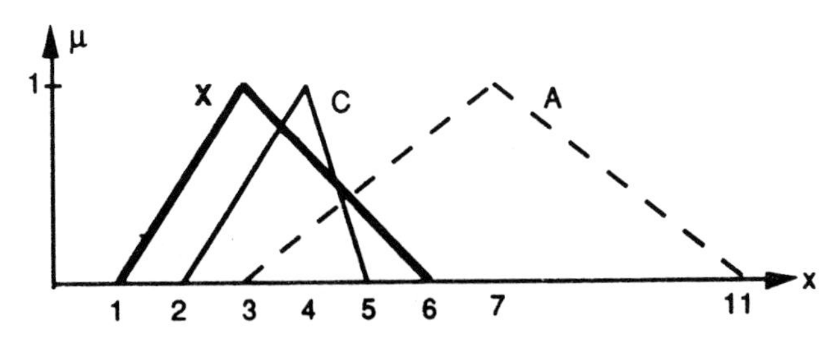

Figure V.2.2 : Result of example 2.

V.3. Criterion for solving the equation A∗X=C

a) support determination

Solving the equation A∗X=C, with $u_{inf} \neq 0$, $u_{sup} \neq 0$ and $u_m \neq 0$, gives :

$$w_1 = \frac{v_{inf}}{u_{inf}} \text{ and } w_2 = \frac{v_{sup}}{u_{sup}}$$

$$[w_{inf}, w_{sup}] = [\min(\frac{v_{inf}}{u_{inf}}, \frac{v_{sup}}{u_{sup}}), \max(\frac{v_{inf}}{u_{inf}}, \frac{v_{sup}}{u_{sup}})] \quad \text{(V3.1)}$$

b) mean value determination

The mean value of the fuzzy set X, solution of the equation A∗X=C, is given by :

$$wm = \frac{v_m}{u_m} \quad \text{(V3.2)}$$

c) Resolution criterion

The solving criterion of the equation A∗X = C is given by :

$$\min(\frac{v_{inf}}{u_{inf}}, \frac{v_{sup}}{u_{sup}}) \leq \frac{v_m}{u_m} \leq \max(\frac{v_{inf}}{u_{inf}}, \frac{v_{sup}}{u_{sup}}) \quad \text{(V3.3)}$$

If this condition is satisfied, $[w_{inf}, w_m, w_{sup}]$ is the single solution of the equation

d) Examples

Example 1 : Let be two fuzzy sets A=[3, 7, 20] and C=[15, 21, 25]. Solving A∗X=C, using (V3.1) and (V3.2), gives w_1=5, w_2=1.2 and w_m=3 belongs to W_s=[1.2, 5], then the criterion (V.3) is satisfied, The solution is given by the triplet [1.2, 3, 5], figure IV.3.1.

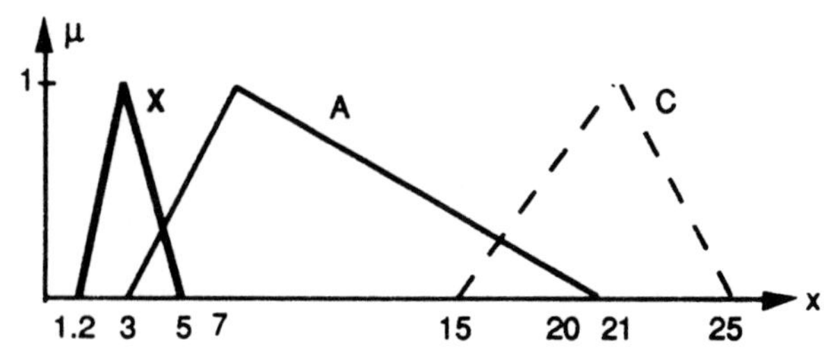

Figure V.3.1 : Result of example 1.

Example 2 : Let be two fuzzy sets A=[4, 7, 10] and C=[6, 14, 40], Solving the equation A∗X=Cgives w_1=1.5, w_2=4 and w_m=2 so W_s=[w_1, w_2]=[1.5, 4] contains w_m, the criterion is satisfied, The triplet [1.5, 2, 4], figure V.3.2, is solution.

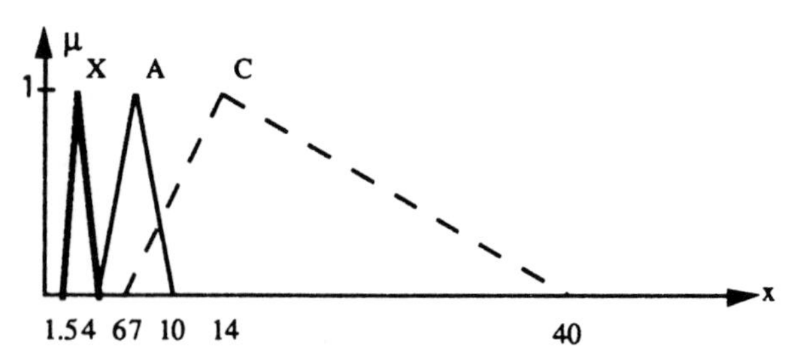

Figure V.3.2 : Result of example 2.

V.4. Criterion for solving the equation A÷X=C

a) Support calculation

Solving the equation A÷X=C, with $v_{inf} \neq 0$, $v_{sup} \neq 0$ and $v_m \neq 0$ gives :

$$w_1 = \frac{u_{inf}}{v_{inf}} \quad \text{and } w_2 = \frac{u_{sup}}{v_{sup}}$$

$$\text{then, } [w_{inf}, w_{sup}] = [\min(\frac{u_{inf}}{v_{inf}}, \frac{u_{sup}}{v_{sup}}), \max(\frac{u_{inf}}{v_{inf}}, \frac{u_{sup}}{v_{sup}})] \quad \text{(V4.1)}$$

b) Mean value calculation

The mean value calculation of the solution X, for the equation A÷X=C, is given by :

$$w_m = \frac{u_m}{v_m} \quad \text{(V4.2)}$$

c) Resolution criterion

The deduced criterion for solving the equation A+X=C is given by :

$$\min\left(\frac{u_{inf}}{v_{inf}}, \frac{u_{sup}}{v_{sup}}\right) \leq \frac{u_m}{v_m} \leq \max\left(\frac{u_{inf}}{v_{inf}}, \frac{v_{sup}}{u_{sup}}\right) \qquad \text{(V4.3)}$$

which is identical with the criterion for solving the equation A∗X=C.

proof : It is obvious that if $\frac{v_{inf}}{u_{inf}} \leq \frac{v_{sup}}{u_{sup}}$ then $\frac{u_{inf}}{v_{inf}} \geq \frac{u_{sup}}{v_{sup}}$

which implies $\max\left(\frac{v_{inf}}{u_{inf}}, \frac{v_{sup}}{u_{sup}}\right) = \min\left(\frac{u_{inf}}{v_{inf}}, \frac{u_{sup}}{v_{sup}}\right)$

and $\min\left(\frac{v_{inf}}{u_{inf}}, \frac{v_{sup}}{u_{sup}}\right) = \max\left(\frac{u_{inf}}{v_{inf}}, \frac{u_{sup}}{v_{sup}}\right)$

replacing the expressions above gives:

$$\min\left(\frac{u_{inf}}{v_{inf}}; \frac{u_{sup}}{v_{sup}}\right) \leq \frac{u_m}{v_m} \leq \max\left(\frac{u_{inf}}{v_{inf}}, \frac{u_{sup}}{v_{sup}}\right)$$

d) Examples

Example 1 : Let be two fuzzy sets A=[15, 21, 25] and C=[3, 7, 20]. Solving A+X=C, using (V4.1) and (V4.2), gives $w_1=5$, $w_2=1.2$ and $w_m=3$ belongs to $W_s=[w_2, w_1]=[1.2, 5]$, then the criterion (V4.3) is satisfied, the triplet [1.2, 3, 5], figure IV.4.1, is solution.

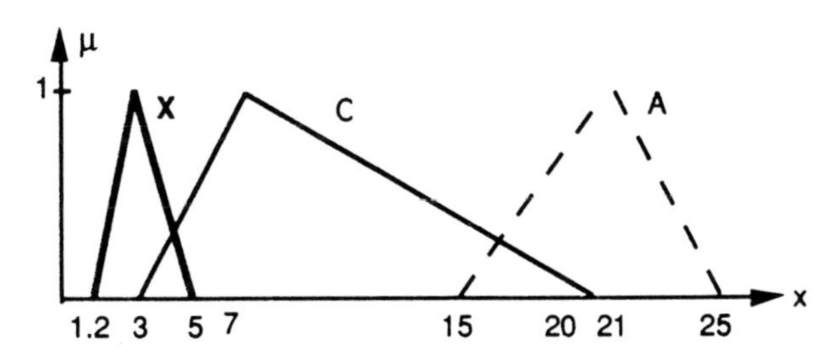

Figure V.4.1 : Result of example 1.

Example 2 : Let be two fuzzy sets A=[9, 20, 30] and C=[6, 10, 12]. Solving the equation A+X=C gives $w_1=1.5$, $w_2=2.5$ and $w_m=2$, $W_s=[w_1, w_2]=[1.5, 2.5]$. w_m belongs to W_s, so the criterion for solving the equation is satisfied and the solution is given by the triplet [1.5, 2, 2.5], figure V.4.2.

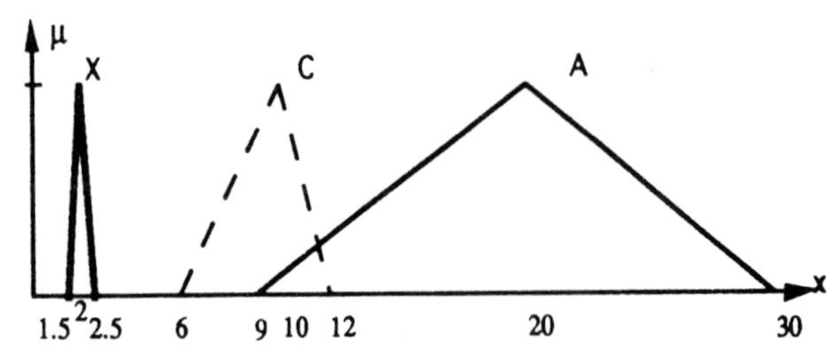

Figure V.4.2 : Result of example 2.

VI. CONCLUSION

The approach presented in this article proposes a method for solving algebraic equations containing fuzzy parameters, that is independent of the values for the supports of A and C, namely the support of A "less than or equal" or "greater than or equal" to that of C. For each operator studied, two examples are given to illustrate each of the two cases. The single, necessary and sufficient condition for solving algebraic equations is to verify that the mean value of the resulting set is a member of its support. If that criterion is satisfied, then the solution is considered valid. In this article, the approach is applied to fuzzy sets with a triangular representation, that most commonly used, but it may also be applied to other representations verifying, at least partially, the property of continuity. For the trapezoidal representation, for example, the criteria for solving algebraic equations are the same, but instead of the mean value, the mean interval (kernel) of the resulting fuzzy set has to be contained in its support.

BIBLIOGRAPHY

1. L.A.ZADEH, Fuzzy sets Information control 8, 338-353, 1965.

2. R.R.YAGER, A characterization of the extension principle, Fuzzy sets & systems 18, 205-217, 1986.

3. E.SANCHEZ, Solutions of fuzzy equations with extended operators, Fuzzy sets & systems 12, 237-248, 1984.

4. D.DUBOIS & H.PRADE, Fuzzy real algebra : some results, Fuzzy sets & systems 2, 327-383, 1979.

5. R.R.YAGER, On solving fuzzy mathematical relationships, Information & control, n°41, 29-55, 1979.

6. L.A.ZADEH, The concept of linguistic variable and its application to approximate reasoning, Information Sciences 8, 199-251, 1975.

7. Z.ZHAO & R.GOVIND, Solutions of algebraic equations involving generalized fuzzy numbers Information Sciences 56, 199-243, 1991.

Uncertainty in Intelligent Systems
B. Bouchon-Meunier et al. (Editors)

A Petri net based fuzzy PLC for linear interpolation between control steps

JC. Pascal and R. Valette

LAAS-CNRS, 7 Av. du Colonel Roche
31077 TOULOUSE CEDEX - France

This work is an attempt to develop fuzzy PLCs with a similar approach as that of fuzzy controllers. The formal model used to specify control sequences, concurrency and synchronization is Petri nets. Their combination with possibilistic theory is directly derived from the work by J. Cardoso (Petri nets with fuzzy markings)[1-2] .

In a first section, Petri nets with fuzzy markings are shortly presented. It is shown how the concepts developped for Petri nets with objects can be adapted to the ordinary safe Petri nets which are typically used for PLCs. *"Fuzzy"* events (non instantaneous evolutions from one state to another) are attached to transitions in such way that an interpretated Petri net with fuzzy markings can be used as an input language for a fuzzy PLC. In the following section, some hints will be given in order to adapt the typical token player (specific inference engine for firing the Petri net transitions in real-time) to the case of ordinary safe Petri nets with fuzzy markings.

1. INTRODUCTION

PLCs are widely used for real-time control of industrial processes. They are well suited for discrete control because they are able to directly implement the control sequences specified by means of standard languages such as Grafcet or formal models such as Petri nets. However they are inadequate even in the case of simple regulation problems. Only boolean inputs and outputs are handled and it is impossible to generate outputs which are continuously varying in relation to the controlled system state.

Moreover, it is impossible to specify the control in a qualitative way such as *"if the automated guided vehicle is near the contact, then stop it gently"* or *"if the value of sensor C is changing then slow down"* . The development of fuzzy controllers has shown the practical interest of qualitative specification of control strategies in order to obtain a friendly human interface (security is increased) and a compact implementation based on the notion of compromise and linear interpolation between typical situations represented by rules.

2. PETRI NETS WITH FUZZY MARKINGS

2.1. Ordinary safe Petri nets for PLCs

Places are seen as conditions (or local partial internal states) which can be associated with actions (commands to the actuators). Transitions correspond to instantaneous events to which instantaneous actions (pulses) can be attached. A place can be empty (the corresponding condition is false) or contain at most one token (the condition is true). When all the input places of a transition contain a token, the transition is said to be enabled which means that the corresponding event (typically described by a boolean expression involving sensor values) may be taken into account. The occurrence of an event results instantaneously in the firing of the corresponding enabled transition. The net marking (a list of places containing a token) is a representation of the current controller state.

Let us just remember that the major advantage of using Petri nets for the specification of discrete control is that they allow both formal analysis for validation and direct implementation by means of a token player.

2.2. Fuzzy markings

Fuzzy markings have been defined by J. Cardoso in the context of Petri nets with objects. In such a model tokens are individuals (object instances) and it is possible to use the token locations as an alternative representation of the marking. With each token, the place containing it is associated (ordinary marking) or the set of places likely to contain it (imprecise marking) or the fuzzy set of places likely to contain it (fuzzy marking). These sets of places can be interpreted as a representation of the possible token locations.

Here, we are interested in ordinary Petri nets for which tokens cannot be differentiated. However in the context of PLCs, the Petri nets used for specifying the desired control sequences are safe, i.e. they are such that a place cannot contain more than one token. Moreover, it is typically required that this property should result from the structure of the net. This means that the net is covered by a set of linear P-invariants (subset of places) whose global token load is invariant and equal to one. Consequently, for a structurally safe Petri net, it is possible to define the marking by associating to each elementary invariant the name of its place whose token load is different from zero. This application has to be consistent (a place cannot be simultaneously marked for an invariant and empty for another one). Doing so, it is possible to extend this definition to imprecise or fuzzy markings.

A fuzzy marking of a structurally safe ordinary Petri net is a consistent mapping associating with each elementary linear P invariant a fuzzy sub-set of places. This sub-set S_k represents the possible location of the token in a set of places P within the invariant k. The location of the token is defined by the possibility distribution $\pi_{]k}(p)$ of the token location k in the set of places P. For a place p belonging to P, the numeric estimation of the possibility for the token location to be the place p, which is marked for P-invariant k, is defined by:

$$\pi_{]k}(p): P \longrightarrow [0,1]$$

Fuzzy marking, for an invariant k is: $M_k = S_k$

Let I be a set of P-invariant covering the net (with one token associated to each P-invariant) and $\mathcal{P}(P)$ the set of fuzzy sets over P, fuzzy marking on the net is defined by:

$$M \text{ fuzzy}: I \longrightarrow \mathcal{P}(P)$$

The application is consistent when the membership functions of each fuzzy sub-sets has the same value for a given place:

$\pi_{lk1}(p) = \pi_{lk2}(p)$

A fuzzy marking represents a graduated current state of a controller. Actually, it is a join description of two (or more) states. In the case of PLCs, it is a way of describing a fine grained state within a continuous evolution from one coarse grained state (an ordinary marking) to another one. As for ordinary PLCs, the marking trajectories have to be synchronous with the controlled system ones (the role of the token player is to ensure that). Therefore, a fuzzy marking is a function of time.

2.3. Evolution rules

2.3.1. Enabling transition

In ordinary Petri nets, a transition t is enabled by a marking M, if " $p \in P$, $M(p) \geq Pre(p,t)$. If we consider a transition t with only one input place p, t is enabled in the fuzzy Petri net as soon as p is a possible location for the token of the corresponding p-invariant i.e. $\pi_{lk}(p) > 0$.

When t has more than an input place and that the conditions described by them are independent, the possibility of firing t is:

$\Pi_t = \min \{\pi_{lk}(p1),....\pi_{lk}(pi)\} > 0$

This means that in a fuzzy Petri net, a transition t is enabled as soon as, for each P-invariant k passing by t, the token location possibility of the input place of t is positive.

2.3.2. Firing a transition

In ordinary Petri nets, a transition is firable when it is enabled by the marking and its associated condition is true. Similarly, we have to define external conditions in fuzzy Petri nets. Let us consider a sensor whose value is x. We compare it to a fuzzy set E defining the possible truth values of the condition at a given time. This condition is defined by a possibility distribution π_E, attached to the transition t, delimited by the fuzzy set of possible sensor values x for the firing of the transition (figure 1).

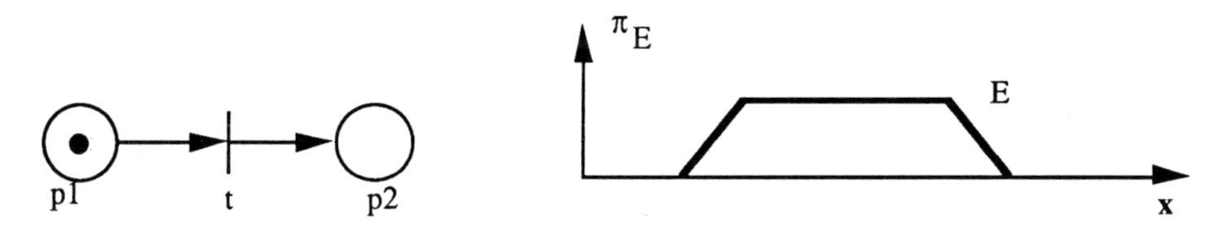

Figure 1: Possibility distribution π_E attached to transition t

Consequently, in a fuzzy Petri net, in order to fire a transition, the two following conditions must be satisfied:

- the transition must be enabled, i.e. $P_t > 0$
- the attached condition must be possibly true, that is $p_E(x)>0$ (x belongs to the support of E).

Remark: Till now, time is not explicit in our model. It is only taken into account implicitely by the fact that the sensor values are varying in function of time. Consequently, this model differs from the fuzzy time Petri net defined by Cardoso.

2.4. Marking computation

Let us consider a transition t and its attached condition defined by the fuzzy set E and the possibility distribution π_E. Firing this transition is no longer an instantaneous event in a fuzzy Petri net. Two steps have to be considered: the beginning of the firing and the termination of the firing. Knowing the token location possibility $\pi_{lk}(p)$ and the possibility π_X at time point $\tau-\delta\tau$, we calculate their value at time point t.

2.4.1. Beginning of firing

The token location possibility for the output place of t is increasing from zero to one according to the increasing values of the token location possibility in the input place and to the truth value of the condition. This step terminates when all these possibilities are 1, i.e. the condition is totally possibly true as well as the token location is totally possibly the input place as well as the output place of t.

The marking of the input place p1, at current time point τ, is:

$$\pi_{lk}(p1,\tau) = \pi_{lk}(p1,\tau-\delta\tau) \tag{1}$$

The marking of the output place p2, at current time point τ, is:

$$\pi_{lk}(p2,\tau) = \max [\ \pi_{lk}(p2,\tau-\delta\tau),\ \min \{\ \pi_{lk}(p1,\tau),\ \pi_E(x,\tau)\ \}] \tag{2}$$

2.4.2. Termination of firing

The token location possibility for the input place of t is decreasing from one to zero according to the decreasing possibility π_E. This step terminates when the condition attached to the transition is false (π_E=0) and the transition t is no longer enabled ($\pi_{lk}(p)$=0 for input place p).

The marking of the input place p1, at current time point τ, is:

$$\pi_{lk}(p1,\tau) = \min \{\ \pi_{lk}(p1,\tau-\delta\tau),\ \pi_E(x,\tau)\ \} \tag{3}$$

The marking of the output place p2, at current time point τ, is:

$$\pi_{lk}(p2,\tau) = \pi_{lk}(p2,\tau-\delta\tau) \tag{4}$$

Remark: The reason of the introduction of the maximum in the expression (2) and the minimum in the expression (3) derives from the irreversibility of the marking evolution (see § 2.5.2).

2.4.3. Marking definition with the notion of "possibly before or possibly after the event"

It is possible to define, in an other way, the token location possibility into places by considering sensor values that are "possibly before the event" and "possibly after the event", at given time point.

The firing of a transition corresponds to a fuzzy event, similar to a fuzzy date as defined by Dubois and Prade [3]. The token location possibility of the input place, at current time point, describes the proposition "possibly before the event" and the token location possibility of the output place corresponds to "possibly after the event".

Let us consider a transition t (see figure 1) on which is attached the fuzzy event defined by the possibility distribution π_E (x) delimited by a fuzzy set E.

Let say that this event signifies "the sensor value x is about a". The fuzzy set $(-\infty,E]$ describes sensor values that are possibly before "x is about a". This defines the membership function of the value x to this fuzzy interval, at a given time point t (xt is the value of x at time point τ) :

$$\tau \in \mathcal{T},\ \mu_{(-\infty,E]}(x) = \sup_{x \geq x\tau} \pi_E(x)$$

The fuzzy set $[E,+\infty)$ describes sensor values that are possibly after "x is about a". This defines the membership function of the value x to this fuzzy interval, at a given time point τ :

$$\tau \in \mathcal{T},\ \mu_{[E,+\infty)}(x) = \sup_{x \leq x\tau} \pi_E(x)$$

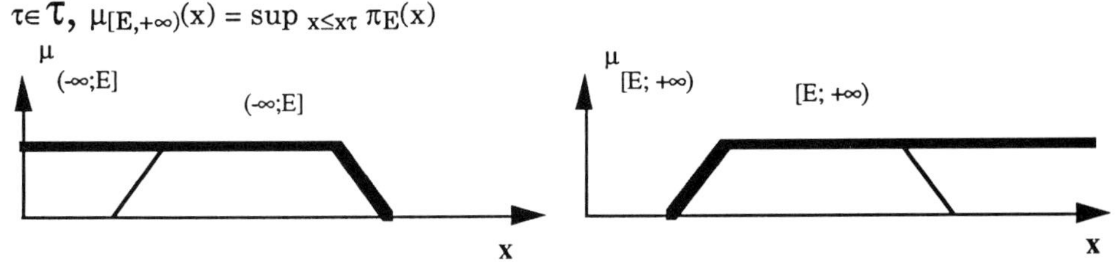

Figure 2: Membership functions of sensor values possibly before and after the event

If we suppose that the variation of the sensor values during time is monotonous (we will see later, in § 2.5.2, how to take into account a non monotonous variation to respect the constraint of irreversibility of the marking), the token location possibility into place p2, at a given time point τ, can be defined by:

$$\pi_{|k}(p2,\tau) = \sup_{x \leq x\tau} \min \{\pi_{|k}(p1,\tau),\pi_E(x)] \}$$

2.5. Constraints

Some constraints must be imperatively respected concerning the using of the fuzzy theory in Petri nets.

2.5.1. Sequential constraint

In an ordinary Petri net, when the attached condition of a transition t which is not enabled becomes true and, after a short duration turns false again, the corresponding event is lost. The same pathological situation may occur in fuzzy Petri nets when the condition varies faster than the token location possibility in the input place. The situation where simultaneously the token location possibility is one as well as the condition is true may never occur and therefore the beginning of the firing of the transition never terminates. In a way we can say that the Petri net does not succeed synchronizing with the event described by the condition attached to the transition. Such a situation is only possible if the input place p of transition t is the output of another transition t' which is being fired simultaneously. This case signifies that the sequential constraint described by Petri net is not respected.

Let suppose the sequence shown figure 3. For the constraint to be respected, the event e1 must occur before the event e2. Indeed, to take into account the event e2 associated to the transition t2, the place p2 must be marked. In order to place p2 be marked, the event e1 associated to the transition t1 has to have occurred. Therefore, the beginning of the event e1 must be before the beginning of the event e2. The end of the transition firing necessarily imposes the location of the token into the place p3, thus the transition t1 must be necessarily fired before t2. Therefore, the end of the event e1 must be before the end of the event e2.

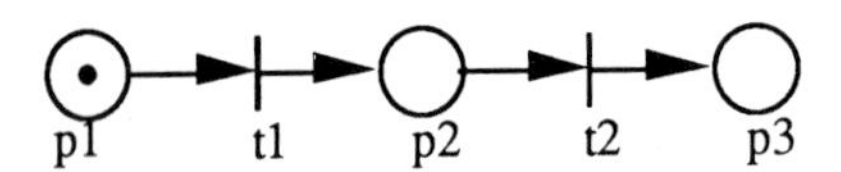

Figure 3: Sequence

2.5.2. Irreversible evolution of the marking

This constraint concerns the irreversible evolution of the marking. This signifies that the marking increases from zero to one before decreasing from one to zero. A non monotonous variation would mean that the token "go back" in the sequence. That is why the token location possibility into places are made monotonous, as it is shown in figure 4. This figure represents a non monotonous variation in function of time of the value x of the sensor s. The evolution of the token location possibility into places during time must be "smoothed down" in order to impose a monotonous behavior.

This is done considering the maximum between the token location possibilities into place at time points τ and $\tau-\delta\tau$, for the increase, i.e.

$$\max (\pi_{lk}(p,\tau),\pi_{lk}(\tau-\delta\tau))$$

and the minimum for the decrease, i.e.

$$\min (\pi_{lk}(p,\tau),\pi_{lk}(\tau-\delta\tau)).$$

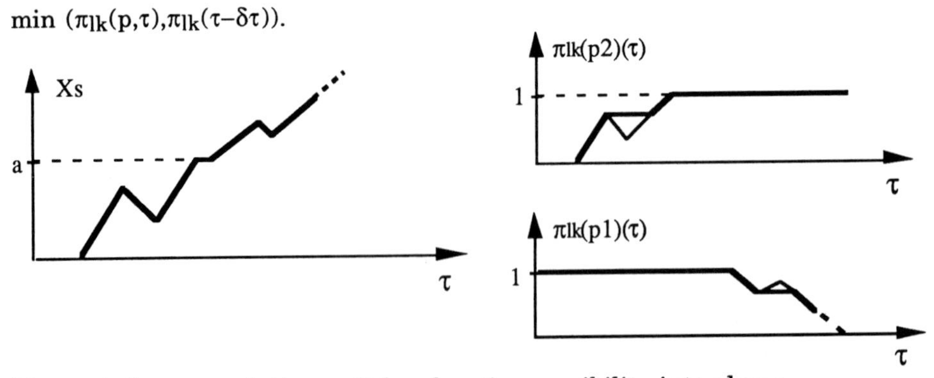

Figure 4: Sensor evolution - Token location possibility into places

2.5.3. Consistent marking

A place, belonging to two or more P-invariants, can't be marked in one and empty in others. The values of $\pi lk(p)$ has to be the same for all the k.

2.6. Firing a sequence

Let us note that ei is the event attached to a transition ti and is defined by a possibility distribution π_{Ei}, delimited by a fuzzy set Ei of possible sensor values xi.

The sequential constraint must be respected. Event e1 must occur before event e2. Fuzzy sets E1 and E2 defining the conditions are in figure 5. The evaluation of the token location possibility into places is done directly in relation to these variables.

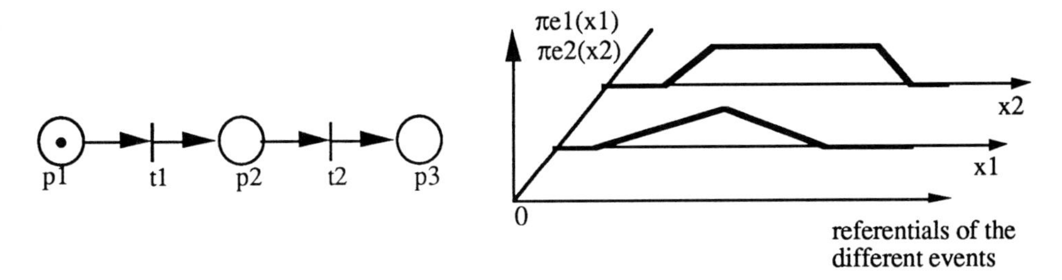

Figure 5: Possibility distributions associated to events ei

The marking of place p2 depends of the firing of transition t1 which increases its marking and to the firing of transition t2 which decreases its marking. The token location possibility in the place p2, at current time point τ, is given by the minimum between its location possibility in the place p1 and the possibility distribution attached to the transition t2. This, because the token location possibility in p2 can't upper than the firing possibility of the transition t2. We obtain:

$$\pi_{lk}(p2,\tau) = \sup_{x1 \leq x1\tau,\ x2\tau \leq x2} \min \{ \min[\pi_{lk}(p1,\tau),\pi_{e1}(x1)],\ \pi_{e2}(x2) \}$$

Let note that, to take non monotonous evolution of the sensor values with time into account, we must apply the "smoothing" as is explain in § 2.5.2.

3. FUZZY PROGRAMMABLE LOGIC CONTROLLER (PLC)

The general architecture of fuzzy PLCs is exactly the same as that of fuzzy controllers [4-5-6] as shown in figure 6.

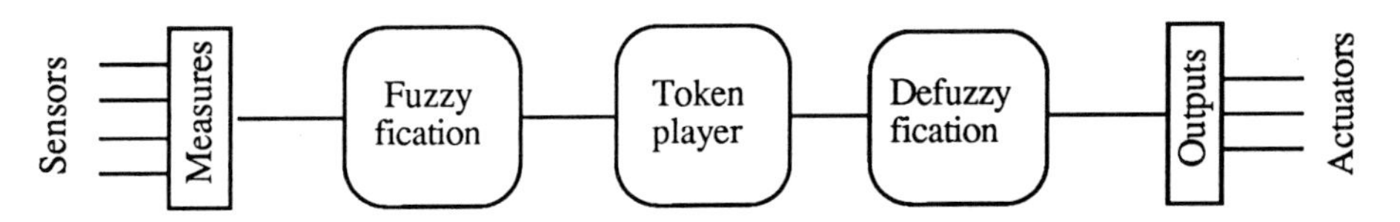

Figure 6: Fuzzy PLC general architecture

The different steps are developped in the following paragraphs

3.1. Measurement

Information from the sensors may be precise, imprecise or fuzzy. In the scope of this work, we shall only consider precise sensors.

3.2. Fuzzyfication

The boolean expressions involving sensor values and attached to the transitions define the corresponding events. This events are represented by possibility distributions in the case of fuzzy events. Fuzzyfication consists in attaching a possibility measure to the expression. This can result from comparing a precise sensor value to the support of this possibility distribution, the evolution will do in relation to the process state, or by generating a fuzzy temporization from a precise event, the evolution will do in relation to time.

The first case corresponds to expressions such as *"fire the transition when the value of the sensor is about 10"* . The second case corresponds to *"when the value of the sensor is equal to 10 fire softly the transition"* . Softly is defined by a temporization, delimiting the firing date and during which it can be said that it is possible that the transition has not yet been fired as well as that the transition has already been fired.

3.3. Token player

It is a token player working on a safe Petri net with fuzzy markings. From the marking of the places, the token player determines the enabled transitions. It searchs for the enabled transitions that are firable. For each of them, it verifies if the possibility measure (resulting to the fuzzyfication) attached to their expression is different from zero. If it is the case, the transition is placed into the list of the transitions in process to firing (it will remain until it would be necessarily fired).

So, the marking of the input and output places is updated. These become the location more or less possible for the tokens. It a transition becomes firable during the same cycle, it is examined. The cycle terminates when there are no more firable transitions. This is named the stable state.

Thus, the marking of the places depending to a fuzzy event will changed in the following way. When the possibility measure increases from zero to one, the token location possibility into the output places take the same value; they becomes possible location for the token just like the input places. When the possibility measure decreases from one to zero, the possibility distribution of the input places take the same value. This means that the input places are progressively no longer possible locations for the tokens.

So that a transition, which a non fuzzy event is associated, would be firable, he'll have to its input places would be necessarily marked. If the event is true, it is instantaneously fired, the input places are put to zero and the output places to one.

In the case of transitions in conflict, afetr the updating of the marking, the marking of the input places, shared with the transitions in conflict, is masked for those that are not fired (the token is reserved by the fired transition).

3.4. Defuzzification

Each output corresponds to a sub-set of places of a P-invariant. The defuzzyfication consists in the determination of the value to apply to each output according to the token location possibility into the places with this output is associated. So that, we calculate the center of gravity of the output according to the desired value this output for each place, weighted by the marking of each place.

REFERENCES

1. J. Cardoso, R. Valette, D. Dubois, Petri nets with uncertain markings, 10th International Conference on Applications of Petri nets, Bonn, Germany, June 1989.
2. R. Valette, J. Cardoso, D. Dubois, Monitoring manufacturing systems by means of Petri nets with imprecise marquings, IEEE International Symposium on Intelligent Control 1989, Albany, New York, September 1989.
3. D. Dubois, H. Prade, Processing fuzzy temporal knowledge, IEEE Transactions on Systems, Man and Cybernetics, n° 4 (1989).
4. M. SUGENO, *An introduction survey of fuzzy control*, Information Sciences, 36, 1985.
5. Y. Y. Chen, T. C. Tsao, A description of the dynamical behavior of fuzzy systems, IEE Transactions on systems, Man, and Cybernetics, vol. 19, no 4, July/August 1989.
6. S. Boverie, B. Demaya, A. Titli, Fuzzy logic control compared with other automatic control approaches, 30th Conference on Decision and Control, Brighton, England, December 1991.

CHAPTER 4:

ANALYSIS OF UNCERTAIN DATA

Uncertainty in Intelligent Systems
B. Bouchon-Meunier et al. (Editors)

A fuzzy logic based qualitative modeling of image data

Makishi NAKAYAMA*, Toshio NORITA# and Anca RALESCU+ ++

Laboratory for International Fuzzy Engineering Research
Siber Hegner Bldg., 4Fl. 89-1 Yamashita-cho Naka-ku Yokohama 231, JAPAN

This paper is concerned with the management of uncertainty aspect of qualitative modeling of image data. By qualitative modeling we mean a description of the image by means of a collection of statements "X is F", where X is an object, a component, or a characteristic thereof, and F is a label of a fuzzy set such as "big", "small", "medium", etc. Of special interest are complex images for which such descriptions are necessarily summaries, approximations, and hence require measures of uncertainty. The approach presented integrates (i) image processing techniques, (ii) statistical arguments, (iii) fuzzy logic and (iv) artificial intelligence techniques.

1. INTRODUCTION

Modeling is an important aspect of information processing. It consists of producing a description of a system, process, phenomenon, from observing its behavior. According to the language used for the description the model may be mathematical (equations), in terms of rules, or both. Recently, substantial progress has been made in producing linguistic/qualitative models of systems based on fuzzy logic concepts, and integrating artificial intelligence techniques [5], [6].

In this paper we consider the task of producing a linguistic model of a still image (photograph). For the purpose of illustrating the approach we chose the task of modeling the image of a human face. The reasons for choosing this task are as follows: (a) face photographs come in a large variety, (b) a face photograph is the image of a natural concept (as such this concept cannot be guaranteed to have a unique, precise description), (c) people are very good at describing photographs, in terms of a few elements of the photograph, thus it is interesting to see to what extent this task can be implemented in a computer program. Producing linguistic descriptions of face photographs provides the basis for defining higher level concepts such as "standard(average) man's face", "a happy face", "family resemblance", etc. These are expected to be of use in a comprehensive information processing system, which can interact with the environment, and as such can 'understand' such descriptions, and relate them to physical features of photographs.

The approach presented in this paper is an integrated approach, making use of
(i) image processing techniques, and artificial intelligence techniques for processing physical features of a photograph,
(ii) fuzzy logic, statistical techniques, and fuzzy reasoning for management of uncertainty, in the step of modeling of image data.

---

* Currently with the Electronics Research Laboratory of Kobe Steel, Co. Ltd.
#Currently with with Minolta Camera Co.
+ On leave from the Computer Science Department, University of Cincinnati, USA.
++ This work was partially supported by the NSF Grant INT-9108632.

This approach provides a good approximation to the way humans produce such models, offers tuning opportunities, and is open for inspection/explanation.

2. LINGUISTIC MODELING OF A FACE IMAGE

By a linguistic model of a photograph we mean a collection of statements "X is F", where X denotes the object in the photograph, components, or attributes of it, and F denotes a label associated to a fuzzy set defined in the universe of discourse of X . The meaning of "X is F", refers to the *size* of X. In some cases this simply means a characteristic which can be measured directly, such as length, width, area, but in the most interesting cases, this characteristic is complex, a combination of other features.

In this paper F can take the values "small", and "big" only. The approach can, however be extended to more values, assumed available in a dictionary of fuzzy sets. We assume that

$$1 - \mu_{small}(x) \geq \mu_{big}(x) \text{ i.e. not(small)} \supseteq \text{big} \quad (1)$$

$$\max(\mu_{small}(x), \mu_{big}(x)\} > 0 \quad (2)$$

for every x in the universe of discourse. That is, "small", and "big" form a fuzzy partition of the universe of discourse. Throughout this paper $\mu_F(\_)$ denotes the membership function of the fuzzy set F. We limit ourselves to linguistic modeling methods which produce descriptions of the type *"The man with big eyes, short full hair"*. For instance, we do not consider expressions involving hedges.

We view our problem as a system modeling task . In this system we have the following:
input: a face photograph. More precisely, it is the result of the image processing applied to this photograph. To extract the data we use a top-down, task aided, model-based image processing of the photograph, as developed in [3].
output: the response given by a user to statements of the form "X is F" concerning various parts of the photograph. This response, which we call index of impression, and denote by I, takes values in a prespecified range. It conveys the degree to which the user agrees with the statement "X is F".

To model the component X, means to produce membership functions for fuzzy sets associated to physical characteristics of X, such that the result of the modeling agrees, in a sense to be made clear later, with the index of impression associated to X. We make the following assumptions:
(A1) For a complex (natural) image, such as that of face, the degree (that is the index of impression) to which a statement "X is F" holds, appears different from user to user.
(A2) In assessing the index of impression a user will take into account local evidence, obtained from intrinsic characteristics of the component X, as well as global evidence, obtained from the interaction between different components of the photograph. This interaction is expressed qualitatively (heuristics).
(A3) There is a psychological dimension to the evaluation of the index of impression, which may not be possible to express explicitly.
Remarks:
1. It follows from (A1) that to obtain robustness of the model we must take into consideration several users.
2. From (A2) it follows that we must consider combining local and global evidence in order to account for the index of impression.
3. Finally, (A3) may mean that the result of modeling using the assumptions (A1) and (A2) is not necessarily identical to I. In fact, the discrepancy between the final result of the modeling and I can be viewed as a complexity measure of the image under consideration.

In the remainder of this paper, for purpose of illustration, we consider the task of producing the description "eyes are big"/"eyes are small", from a collection of photographs. In our study 23 photographs were considered. For each of these photographs, a sample of ten users were required to assess the eye size, using the index of impression I. In this paper I takes values in [1, 3], with 1 corresponding to "small", 2 to "medium", and 3 to "big". A more refined representation of I would have it take values in [1, 5], or, [1, 7]. For each photograph, and each component, the average index of impression is assigned. Since from now on we refer to this average as the index of impression we will denote it by I as well.

2.1. Extracting physical features

At this stage, the image processing operations are guided by (i) a model of the eye, and (ii) image processing expert knowledge. The model is hierarchical, describing the eye in terms of its components (such as iris, pupil, etc.) and their attributes (such as location, color, shape, length, width, height). For a detailed description of this procedure the reader is referred to [3]. Attributes or combinations of attributes are used to generate the descriptions "eye are big"/"eyes are small". Table 1 contains the result of image processing of the eye area and values of attribute combinations. In addition, the table shows the values of I for each of the photographs, as well the correlations between I and the attribute combinations used. The photographs are recorded in increasing order of the values of I. Omitted from the table are those photographs for which the image processing were not satisfactory. For our example we considered the following attribute combinations: eye_height/face_width, pupil_area/eye_area, eye_area/(face_width)^2, eye_height/eye_width. In what follows we will refer to attributes, both in the case of attributes and combinations of the attributes.

The modeling will consist in producing the membership functions associated to some of the attributes of the eye, such that the average I is well approximated by the model. The use of the average index of impression, lends our method the usual properties associated to using a sample mean, that is robustness. More precisely, if we think of the index of impression as a random variable then, we know from statistical considerations, that the sample average value is an unbiased estimator of the expected index of impression. In addition, under assumptions on the distribution of the index of impression, of finite first and second moments, we know that the probability of deviations between the index of impression and its expected value becomes smaller as these deviations become larger (Chebyshev inequality), that is large deviations have small probabilities. The robustness follows from these two facts.

Table 1
Attributes and average index of impression for modeling the "eye":
(1): eye_size/(face_width)^2, (2): dark area/eye area, (3): eye_witdth/face_width,
(4): eye_height/face_width (5): average index of impression

file#	(1)	(2)	(3)	(4)	(5)
1	1.22	86.24	22.47	6.74	1.6
2	1.07	95.44	23.00	5.50	1.7
10	1.23	93.86	24.10	7.69	1.7
4	1.50	73.69	26.52	8.29	1.8
18	1.08	76.73	19.05	7.41	2.0
8	1.37	75.55	22.34	9.64	2.1
16	1.37	84.10	24.18	7.69	2.1
20	1.52	81.64	22.04	9.68	2.1
11	1.43	88.68	22.96	8.67	2.2
3	1.39	84.75	24.51	8.33	2.3
23	1.41	81..10	19.79	9.63	2.3
13	1.66	75.31	23.86	10.80	2.4
7	2.17	86.15	25.61	11.73	2.8
correlation	0.75	-0.52	0.10	0.79	

2.2 Identification (Choosing important attributes)

In general, we have no indication which of the attributes detected at the image processing step are to be included in the model of "eye is big". In order to identify these attributes we make use of I. The correlation values are used to choose the attributes of interest: if the correlation is above a preassigned threshold then the attribute in question is used to describe the component eye. In our example we choose the threshold to be 0.7. With this threshold the attributes selected are: eye_size/(face_width)^2, and (eye_height/face_width). Next, we start with some membership function for the fuzzy sets "small", and "large" which will be associated to each of these attributes. During the modeling process these fuzzy sets will, eventually, be altered, such that the index of impression I is approximated as well as possible. In addition we define the fuzzy sets "small", "medium", and "big" on [1, 3], for the index of impression and we assume that these are the same for "size of eyes".

2.3. Local support

The relationship between the selected attributes and "eye_size" is conveyed via heuristics, expressed as fuzzy if-then rules. They may be obtained directly from the data, by linguistic modeling methods [5], [6], or can be user defined. Such heuristics could be:

"If (eye_area/face_width^2) is **big** then eyes are **big**" (3)
"If (eye_height/face_width) is **small** then eyes are **small**" (4)

2.4. Fitting the index of impression

Fuzzy inference as proposed in [1] is used. The sets "big", and "small" are defined for: "eye_height/face_width", "eye_area/face_width^2". Their max/min values as recorded in Table 1 are used to determine their initial universe of discourse. We assume that "small", "medium", and "big" for "eye_size" and I are the same. The fuzzy sets just mentioned are shown below:

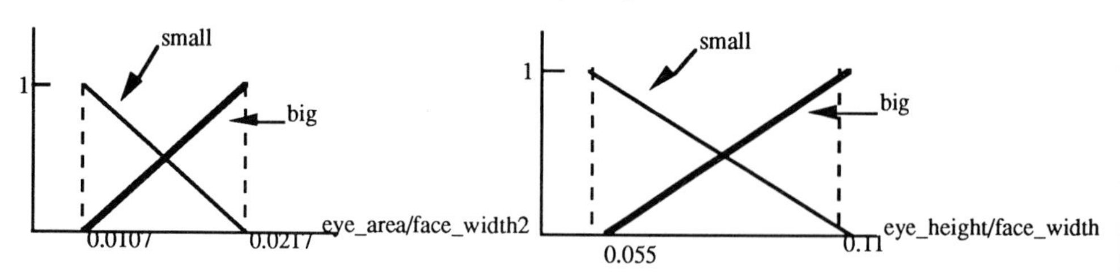

Figure 1. Initial fuzzy sets associated to attribute combinations

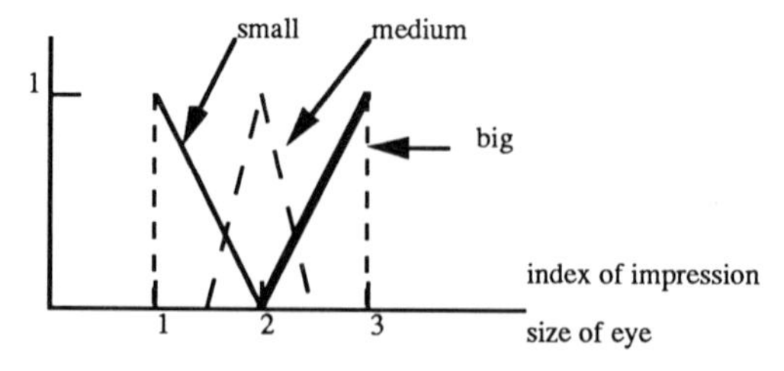

Figure 2. Fuzzy sets associated to the index of impression on the eye size.

Suppose now that for a photograph the measurements for the eye component are such that (eye_area/(face_width^2)=x1 and (eye_height/eye_width)=x2 Then from (3) and (4) we have:

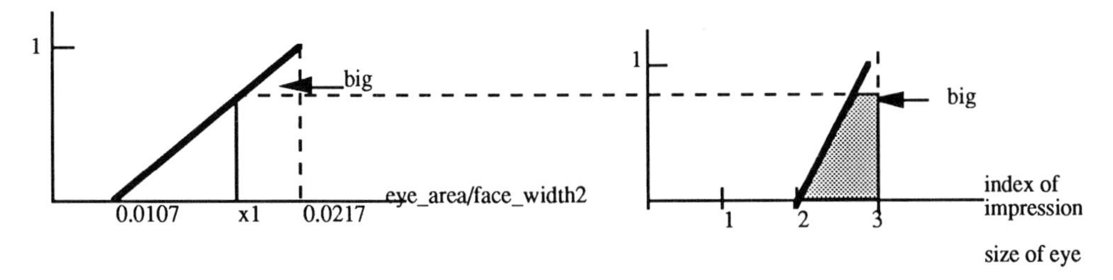

Figure 3. Applying rule (3.1) at a value, x1, of eye_area/face_width^2

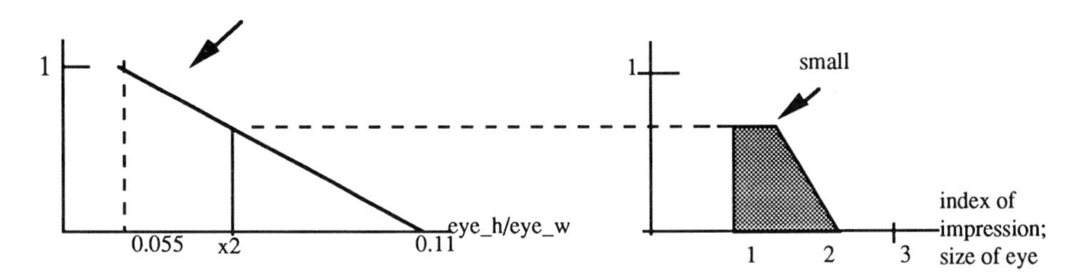

Figure 4. Applying rule (3.2) at a value x2 of eye_height/face_width

By taking the union of the conclusion parts (i.e. maximum) we obtain the result of applying botl (3) and (4):

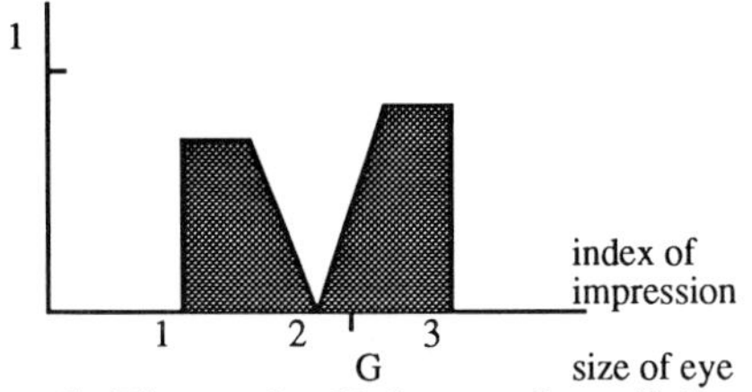

Figure 5. The result of inference from (3.1) and (3.2)

G denotes the center of gravity of the shaded region. Based on G, (> 2, < 2, =2), we model "eyes are big", "eyes are small", "eyes are medium". In our example, since G>2, the modeling will produce the description "eyes are big". We define

$$g\_local(\text{"eyes are big"}) = \mu_{big}(G) \quad (5)$$

where μ_{big} refers to the fuzzy "big" for I.
g_local("eyes are big/small") is computed in parallel for all the photographs in the data base of photographs.

This inference procedure yields Gj, j=1,..., N, where N is the number of images; here N=23. Next Gj and Ij, (Ij, is the corresponding index of impression) are compared. Typically, we want $|Gj-Ij,| \leq \delta$ for some $\delta > 0$ If this condition is satisfied then the modeling ends: It means that the definitions of the fuzzy sets for the attributes selected and the index of impression, are consistent under the heuristics and inference mechanism considered, and within an error δ. In general, it cannot be guaranteed. that Ij and Gj will be within the desired error. We improve the fit between Gj and Ij. This is done in two steps:

(i) by modifying the supports of the fuzzy sets associated to attribute values.This is done in parallel, for each of the fuzzy sets appearing in the heuristics in order to improve the fit simultaneously for all photographs. The criterion for these modifications is the *minimization of a total error function based on $|Gj - Ij|$, $j = 1,..., 23$, such that (1.1), and (1.2) continue to hold.*

(ii) by modifying g_local, such as to bring it within $\nu \geq 0$ of $\mu_{big}(I)$: When the result of (i) fails to provide the desired approximation of the index of impression we improve the approximation indirectly, at the level of the membership function values. In effect this means altering the shape of the membership function of the fuzzy sets associated with the index of impression. Under our assumption that these fuzzy sets are describing the size of eyes, it means that we produce a model of the linguistic description associated with this size.

We illustrate this process of updating by using the lines of the Table 1 corresponding to photographs 20, 23, and 13, and for the heuristic (3):

The fuzzy set "big" associated to eye_area/(face_width^2) is $\mu_1(x) = 90.9x - 0.973$.
Then, using the data and the inference we obtain, Table 2:

Table 2
Results of using (3) for the photographs 20, 23, 13

#	x1	$\mu_1(x1)$	G	I	$\mu_{big}(G)$
20	0.0152	0.40868	2.6	2.1	0.6
23	0.0141	0.30869	2.5	2.3	0.5
13	0.0166	0.53594	2.6	2.4	0.6

Since $G_j > 2$, j=13, 20, 23, it means that for each photograph the description "eyes are big" will be produced. The total error associated with the result of the inference is equal to: error = $|G_{20}-I_{20}| + |G_{23}-I_{23}| + |G_{13}-I_{13}| = 0.9$. If $\delta \geq 0.9$ the modeling is finished.

We try to modify the fuzzy set big associated to (eye_area/(face_width^2), by decreasing its support, from [0.0107, 0.0217], to [0.0107+ε, 0.0217], with $0<\varepsilon<0.011$. Thus $\mu_1(x)$=\F((x-ε-0.0107),(0.011-ε)). The computations of Table 2 will now result in G depending on ε, more precisely, we have:

#	x1	$\mu_1(x1)$	G	I
20	0.0152	$h_{\varepsilon,20}$	$G_{\varepsilon,20}$	2.1
23	0.0141	$h_{\varepsilon,23}$	$G_{\varepsilon,23}$	2.3
13	0.0166	$h_{\varepsilon,13}$	$G_{\varepsilon,13}$	2.4

where

$h_{\varepsilon,20}=\dfrac{(0.0045-\varepsilon)}{(0.011-\varepsilon)}$, $h_{\varepsilon,23}=\dfrac{(0.0034-\varepsilon)}{(0.011-\varepsilon)}$, $h_{\varepsilon,13}=\dfrac{(0.0059-\varepsilon)}{(0.011-\varepsilon)}$, and from the condition that $h_{\varepsilon,20} > 0$, $h_{\varepsilon,23} > 0$, $h_{\varepsilon,13} > 0$ we have $0<\varepsilon<0.0034$. For $0<\varepsilon<0.0034$ we have $0<h_{\varepsilon,20}$, $h_{\varepsilon,23}$,$h_{\varepsilon,13} \leq \frac{2}{3}$ in which case the corresponding centers of gravity are given by

$G_{\varepsilon,i} = 2.5 + 0.25 h_{\varepsilon,i}$ i=20, 23, 13

The error is now a function of ε and we must minimize it subject to the constraints on ε:

$\text{Error}(\varepsilon) = |G_{20,\varepsilon}-I_{20}| + |G_{23,\varepsilon}-I_{23}| + |G_{13,\varepsilon}-I_{13}|$

However, it can be shown that when using the rule (3) the center of gravity in the conclusion part, G will always fall in the interval $[2.5, 2+\frac{1}{\sqrt{2}}]$. Thus no modification of the fuzzy set in the antecedent part can result in a center of gravity outside of this interval. In the example which we considered here, the best we can do is to approximate 2.5. That is, the error function to minimize is

$$\text{Error}(\varepsilon) = |G_{20,\varepsilon} - 2.5| + |G_{23,\varepsilon} - 2.5| + |G_{13,\varepsilon} - 2.5| = \frac{0.5(0.0138-3\varepsilon)}{0.011-\varepsilon} \qquad 0<\varepsilon<0.0034$$

The minimum is obtained for $\varepsilon=0.0034$, and it is equal to $\text{Error}_{min} = \text{Error}(0.0034)=0.23$. This means that in effect the error between the centers of gravity and the true indices of impression is 0.8. Again if $\delta \geq 0.8$ the modeling is finished. If $\delta < 0.8$ we proceed to the second step of the modeling, that is of approximating $\mu_{big}(I)$ for each photograph. In any case the support for the fuzzy set "big" associated to (eye_area/(face_width^2) is be modified, from [0.0107, 0.0217] to [0.0141, 0.0217].

2.5. Global support

We now proceed with the second step of the modification process: that means that the fuzzy sets for attributes have been modified as much as possible, and that Gj and Ij, j=1,..., 23 are not within acceptable errors. At this moment we consider the interaction effect between components and we compute a global support, g_global, based on these interactions: g_global has two components t_global, support for "eyes are big" obtained directly, 1-f_global, obtained from negating the support "eyes are small". In general, the interaction between components is expressed qualitatively by the following heuristics:

(R1) *the degree for "eyes are big" is decreased by "C is big", and it is increased by "C is small", for any component C≠"eyes"*, and
(R2) *the degree against "eyes are big" is increased by "C is big" and decreased by "C is small", for any component C≠"eyes".*

The quantitative relation between components is conveyed by a symmetric, zero diagonal weight matrix, $\mathbf{W}=(w_{C,C'})$, $0 \leq w_{C,C'} \leq 1$. **W** is used to tune the fit between the final support (which takes into account g_local and g_global) and the index of impression, more precisely, the value of the membership function evaluated at the index of impression.. Thus, taking these into account we define: t_global and f_global for "eyes are big", given "C is F",as follows:

t_global("eyes are big"/ "C is F") = g_local("C is F") $(1- \delta_{big,F})(w_{eyes,C})$ (6a)

f_global("eyes are big"/"C is F") = g_local("C is F")$(\delta_{big,F})(w_{eyes,C})$ (6b)

where C is any other component of the face, and $\delta_{x,y} = 1$ if x=y and $\delta_{x,y} = 0$ otherwise.
W can be viewed as filling the gap between measurements of physical characteristics and the psychological dimension of the perception of these characteristics.

The support for "eyes are big" given a component C is obtained by combining the above support over all possible assignments to the set F (in this paper we considered F to be "small", or "big"). Let us denote these components by t_global("eyes are big"/C) and f_global("eyes are big"/C). From (1) it follows that

t_global("eyes are big"/C) + f_global("eyes are big"/C) $\leq$ 1 (7)

Thus we can define the global support as

g_global("eyes are big") = [t_global("eyes are big"), 1-f_global("eyes are big")] (8)
where

t_global("eyes are big")= ***F1***(t_global("eyes are big"/C); C face component, C≠F }
f_global ("eyes are big") = ***F2***(f_global("eyes are big"); C face component, C≠F}

and where ***F1*** and ***F2*** denote some rules of combination, such that (5) holds for t_global("eyes are big") and f_global("eyes are big") as well. Several rules of combination, max, min, and aggregations, or using a fuzzy integral are possible. In the current implementation ***F1***= max, while ***F2*** = min.

2.6. Total support

Finally, g_local("eyes are big") and g_global("eyes are big") are combined to obtain g_total("eyes are big"), the final support. Several rules of combinations, ***F***, have been experimented with. In the current implementation of the system a particular case of the Dempster-Shafer rule has been used. That is:

g_total ("eyes are big") = ***F***(g_local("eyes are big"), g_global ("eyes are big")) (9)

The weights W are adjusted so that the difference between g_total("eyes are big") and $\mu_{big}(I)$ is minimized, or equal to the preassigned value ν.

Figure 6 summarizes the system we just described:

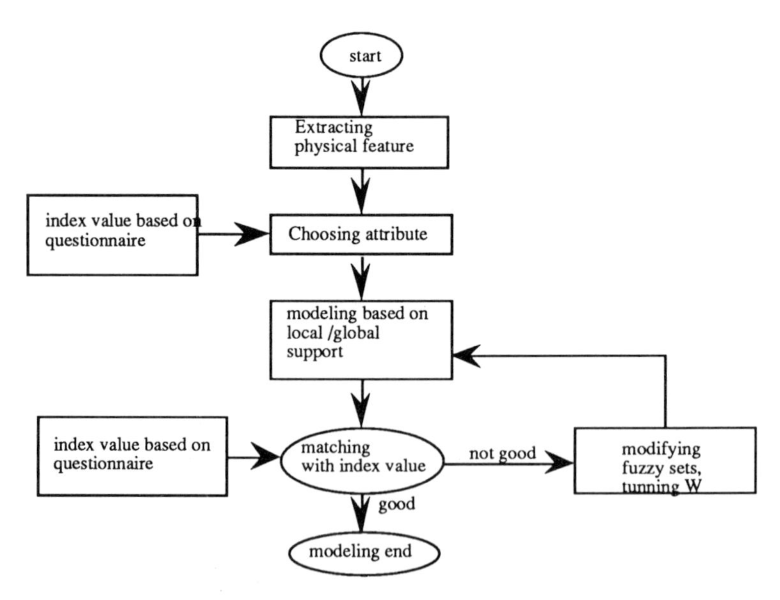

Figure 6. Summary of the linguistic modeling of image data. For simplicity the steps (i) and (ii) of the approximation of the index of impression are depicted in one loop only.

We established a correspondence between the index of impression and the model, and thus the probability arguments concerning the suitability of the index of impression can be transferred to the model itself. The results of using this modeling technique for retrieval purposes depend on the sample of users who provided the index of impression, and the errors, δ and ν when the index of impression was approximated.

3. IMPLEMENTATION

The modeling method described in this paper has been implemented in an image understanding and retrieval system based on linguistic descriptions. We have implemented this system on a sun4 sparc workstation with some peripheral interface systems shown in Figure 7. The program module of this system is written in C language using the X_view tool kit on UNIX operating system. This system has two image processing units. The system is divided into two blocks. The first is a top down image processing block, which has an image processing unit and which can automatically generate the linguistic descriptions of photographs as indicated in this article. The second block is the retrieval and estimation block which uses the other image processing unit as the display for the estimation.

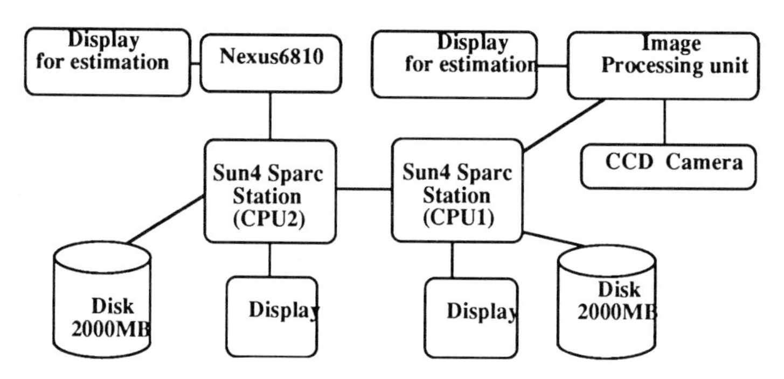

Figure 7. System Diagram

4. CONCLUSIONS

We have presented an approach to a fuzzy logic based linguistic modeling of the image data of complex object/concepts. Important points of this approach are as follows: the modeling method attempts to fit the definitions of the linguistic variables to the data, that is to the index of impression. Due to the fact that the model is an approximation/summary of the data, issues on management of uncertainty arise. In connection with these we have assumed that for any evaluation of a statement "X is F" evidence comes from two sources, **local,** from the structure of X, and **global**, from the interaction between X and other object/components. Both these kinds of evidence provide **direct** supports for "X is F" and **indirect** supports through the lack of support for "X is F' ", F'≠F. The model is biased by the results of the questionnaire used to assess the index of impression. It is therefore possible that when a atypical user will queries the system, in the retrieval mode, the results of the retrieval may not be satisfactory. In connection with this problem our next goal is to provide an adjustment facility.

REFERENCES AND RELATED BIBLIOGRAPHY

1. Mamdani, E.H. "Applications of fuzzy algorithms for control of simple dynamic plants" Proc. IEEE 121(12):1585-1588(1974).

2. Marr D., Hildredth E. "Theory of edge detection", Proc. Royal Society of London, B-207, pp. 187-217,(1980).
3. Miyajima K., Nakayama M., Iwamoto H., and Norita T. "Top-down image processing using fuzzy reasoning", Proceedings of IFES'91, Yokohama Nov. 13-15, 1991.
4. Provan M. "Model-based object recognition: A truth maintenance approach", Proc. of the 4th Conf. on Artificial Intelligence applications, Mar. 14-18, 1988.
5. Ralescu A., and Narazaki H. "Integrating artificial intelligence techniques in linguistic modeling from numerical data" Proceedings of IFES'91, Yokohama Nov. 13-15, 1991.
6. Sugeno M., and Yasukawa T. "Linguistic modeling based on numerical data" IFSA-91 Belgium, July 7-12, 1991.

Uncertainty in Intelligent Systems
B. Bouchon-Meunier et al. (Editors)

Detection of significant points in 2-D outlines using statistical criteria [1]

J. Fdez-Valdivia, J. A. García and N. Pérez de la Blanca

Departamento de Ciencias de la Computación e I.A. Universidad de Granada. 18071. Granada. España. E-mail: jfv@robinson.ugr.es

Abstract

Finding the most outstanding perceptual point set on a planar closed outline is the first step in the shape characterization of such curves. In this paper we present an approach to this problem from the joint information provided by a set of outstanding point and an interpolation procedure defining the shape between them. The main feature of the paper is the optimization criterion for determining the class of outstanding points. The algorithm firstly calculates the graph of curvature and determines the local extremes on it, secondly it identifies the landmark points according to a criterion of importance and thirdly it calculates the interpolated curve from the landmark points and measures the fitting error from the interpolated curve to the observed outline.

1. INTRODUCTION

In planar outlines representation, significant or dominant (high curvature, in general) points are an important attribute of the shape. The locations of the detected dominant points must be accurate and the number of points must provide a good representation of the shape without excessive redundancy.

There are different possible ways of approaching the problem of characterizing planar outlines using interpolation between significant points. One is to use criteria of curvature versus arc length to detect the local extremes of curvature and to use polygonal interpolation,[1] is a survey of some of these methods). The main drawback of all these approaches is the attempt of characterizing planar outlines using only one set of distinguishing points without paying attention to the curve behaviour between such as points.

Our approach to the above problems have been to consider as landmark points (knots) the local extremes of curvature, together with those other points necessary to be able to use spline in tension as the family of approximation functions . As we later show, these others landmark points we consider are some of the zero crossing points of the graph of curvature, which is in agreement with the scale-space approach to locate significant points (e.g, [2, 3]).

In this paper we present an algorithm to fit planar curves from the information provided by a set of estimated landmark points and an interpolation procedure. Since we work from

[1]This work has been supported by the Spanish Direction General de Ciencia y Tecnología (DGCYT) under grant PM-0093-c02-02

the curvature values, in section 2 we briefly discuss the approach we have used to estimate the curvature values and the perceptual criterion used for measuring errors. Section 3 describes the algorithm used in the approach. In this section we also propose criteria to estimate the optimum number of landmark and the tension values associated to the splines. Section 4 shows some test examples.

2. DOMINANT POINT SETS, CURVATURE ESTIMATION AND PERCEPTUAL CRITERION

We suppose that all the information necessary to characterize a 2-D shape is on its outline, so we get a boundary estimation without gaps and one pixel width. The landmark concept such as it was introduced by [4], and supported by psychological experiments ([5]), is associated to those points being local maximum or minimum of the graph of curvature, however the concept is dependent on the interpolation procedure used to join them ([6]). Here we consider a set of points as a dominant set if it is possible to reconstruct the curve from these points using some interpolation method (polygonal, in the simplest case).

We use the information provided by the graph of curvature for estimating errors and we calculate the gaussian curvature at each point of the curve using the method proposed in [7]. This method has been analyzed on some biological images in [8]. It estimates the curvature value at a point P from the angle formed by the orthogonal regression tangent lines fitted in a sample points (N) at equal distances to the left and right of the point P. This procces guarantees the necessary degree of smoothing in the graph of curvature for eliminating errors without loosing of significant points.

Once estimated the local extreme of curvature we want to determine which points of these are really important to define the shape of the original outline. The procedure starts calculating on each point what we call its *importance value*. Let be the sequence of N integer-coordinate points describe a closed curve C,

$$C = \{(x_i, y_i), \quad i = 1, ..., N\}$$

where (x_{i+1}, y_{i+1}) is a neighbour of (x_i, y_i) (modulo N), and let be C^k

$$C^k = \{p_i = (i, k_i) \mid i = 1, ..., N, \quad -\pi \leq k_i \leq \pi\}$$

the sequence of points on the graph of curvature, with k_i the value of curvature in (x_i, y_i).

Given a set of points S in C^k,

$$S = \{p_{i_j} \in C^k \mid j = 1, \cdots, s, s \leq N\}$$

we can calibrate the importance of each point in S:

Let $p_{i_{j-1}} = \left(i_{j-1}, k_{i_{j-1}}\right)$, $p_{i_j} = \left(i_j, k_{i_j}\right)$ and $p_{i_{j+1}} = \left(i_{j+1}, k_{i_{j+1}}\right)$ three successive points in S, with p_{i_j} the point in which we want to measure the error.

If we denote $arc(p_{i_{j-1}}, p_{i_{j+1}})$ the set of points in the graph of curvature values between $p_{i_{j-1}}$ and $p_{i_{j+1}}$, the error in p_{i_j} is represented by

$$r_{i_j} = \sum_{p_l \in arc(p_{i_{j-1}}, p_{i_{j+1}})} \frac{\mid d_l \mid}{L}$$

with

$$d_l = l(k_{i_{j-1}} - k_{i_{j+1}}) + k_l(i_{j+1} - i_{j-1}) \\ +(k_{i_{j+1}} i_{j-1} - k_{i_{j-1}} i_{j+1})$$

and

$$L = \sqrt{(k_{i_{j-1}} - k_{i_{j+1}})^2 + (i_{j+1} - i_{j-1})^2}$$

d_l being the perpendicular distance from p_l to the segment linking the succesive points $p_{i_{j-1}}$ and $p_{i_{j+1}}$, and L the segment length. The quantity r_{i_j} is called the importance value of p_{i_j}. We can think in the importance value in the following sense:

Let S be a set of s points of C^k and let note $E(S)$ the global error calculated on the polygonal approximation with s points. If we eliminate of S a point p_k we have the importance in p_k defined by $E(S - \{p_k\})$. In other words, the global error calculated on the resultant polygonal with $s-1$ points once p_k has been eliminated.

3. LANDMARK POINT ESTIMATION AND INTERPOLATION PROCEDURE

3.1. Landmark point estimation

By landmark point on a curve we mean a point with high importance value helping to characterize its shape. First, we choose the set of local extremes of curvature S according to the simple rule of selecting a point if it is the curvature extreme in a neighbourhood centered in it. Our problem is then, to calculate the set of landmark points $D \subset S$ in the graph of curvature C^k. We use the following algorithm:

Algorithm CSD;

A.1 Using the importance of each point, divide S in K classes $C_1, ..., C_k$ in such way that the probability of dominant point will be larger in C_p that in C_{p+1} $\quad p = 1, ..k-1$.

A.2 From $\{C_1, ..., C_k\}$ and using a statistical criterion, obtain the set of dominant points D.

End.

There are a lot of different ways of making the step 1 in the algorithm. However, similar results have been obtained from all methods:

Method 1:

The cluster procedure start calculating the median of the differences of each two successive values $\lambda = Median(r_j^* - r_{j+1}^*, j = 1, \cdots, n-1)$ from the ordered importance value set. Using λ as a central value of the distribution of the importance values we cluster the values according to the following rule:

A point p belongs to a class C iff there exists a point $p^* \epsilon C$ such that $|p - p^*| \leq k_1\lambda$, $k_1 \leq 1$, and for all the points $p^{**} \in C$ $\max_{p^{**}} |p^{**} - p| \leq k_2\lambda$, $k_2 \geq 1$.

The k_1 and k_2 value determine the partition granularity, $k_1\lambda$ being the maximum distance between a point and its nearest neighbour inside the class and $k_2\lambda$ being the minimum distance between classes.

Method 2:

Let be $R = \{r_1, ..., r_M\}$ the sequence of *importance values* of points $\{P_j \in S\ j = 1, ..., M\}$ ordered in decreasing order ($M = card(S)$), in such a way that r_1 represent the *importance value* of the point P_k such that its elimination produces the largest error, and similarly r_M the lowest. Split R in k classes $C_1, ..., C_k$ using the confidence intervals $I_i = [u_i, v_i]$ calculated on the points $\{r_M, r_{M-1}, ..., r_i\}$ following the next steps:

split-K ();

```
{
nc=K=0; /*K=number of classes*/
do {
    nc++;
    if (r_nc ∈ I_nc+1) r_nc is in the same class that r_nc+1;
    else {
        K++;
        assign r_nc to the class C_k
        }
} while (nc < M);
}
```

With respect to step 2 in the algorithm some different procedures have been designed:

Clustering methods:

1.-Group $\{C_1, ..., C_k\}$ in two initial clusters D^* and $S - D^*$ using hierarchical clustering. We choose the criterion function as:

$$F_T = \sum_k f_k$$

with

$$f_k = \sum_{r \in C_k} (r - m_k)^2$$

and

$$m_k = \frac{1}{card(C_k)} \times \sum_{r \in C_k} r$$

(sum of squared error criterion, with r the corresponding *importance value*). With this function we merge C_s and C_t if the following distance is minimum:

$$d^*(C_s, C_t) = \sqrt{\frac{n_s \times n_t}{n_s + n_t}} \times \mid m_s - m_t \mid$$

with $n_* = \text{card}\ (C_*)$.

2.-From D^* and $S - D^*$ like an initial solution and using an iterative optimization procedure produce the final partition of S in two clusters D and S-D, first of which contains the dominant points. In this iterative procedure, we transfer r from the class D_i to the class $D_j \quad j = 1, 2$ if

$$\frac{n_i}{n_i - 1}(r - m_i)^2 > \frac{n_j}{n_j + 1}(r - m_j)^2.$$

so the procedure is as follow:

Optimization();

(1) Let be D_1^* and D_2^* the initial partition;
Calculate F, m_1 and m_2.

(2) Select a value r. Let suppose that $r \in D_i$.

(3) If $n_i = 1$ go to (6). In other case calculate

$$\rho_j = \begin{cases} \frac{n_j}{n_j+1}(r - m_j)^2 & \text{if } j \neq i \\ \frac{n_i}{n_i-1}(r - m_i)^2 & \text{if } j = i \end{cases}$$

(4) Transfer r to D_k if $\rho_k \leq \rho_j, \ \forall \ \mathrm{j}\ , j = 1, 2$

(5) Update F, m_i and m_k.

(6) If F don't change in $card(S)$ attemps, stop. In other case go to (2).

Optimality criteria:

Let us denote by $\epsilon(C^k, n)$ the cuadratic fitting error to C^k from the polygonal constructed using $n \geq 2$ points T_j. Because of we want to get a compromise between the estimated number of landmark points and the fitting error we use the following criterion:

$$\arg \max_{y \in [0, n-2]} \{f(y) = y \cdot (g(y))\} \tag{1}$$

with

$$g(y) = (\epsilon(C^k, y + 2) - \epsilon(C^k, n) \quad and \quad \epsilon(C^k, y + 2) = \sum_{j=1}^{y-1} \sum_{k_s \in (T_j, T_{j+1})} (k_s - \tilde{k}_s)^2 \ \forall \ y = 0, 1, \cdots$$

$\epsilon(C^k, y + 2)$ being the sum of the interpolation error in the y segments and $\tilde{k}_s$ being the interpolated value. It is clear that $g(y)$ is a decreasing monotone and that $f(y)$ has at least one maximum because of $f(0) = f(n - 2) = 0$ and $f(y) \geq 0$ for all y.

To better understand the meaning of the above criterion let us consider the smallest y verifying $f(y) - f(y + 1) \leq 0$. This condition in terms of $g(y)$ is,

$$g(y + 1) < y \cdot |g(y) - g(y + 1)| \tag{2}$$

Eq.2 shows we obtain maximum in all the points y in which the residual error fitting after adding a new point is lower that y times the decreasing by the new point ([9] for details). Since our interest is to consider first those point with higher importance value we sort such as set of values, $\{r_i^*\}$, in decreasing order, calculating the values of $\epsilon(C^k, m)$ from the polygonal interpolation constructed with the m highests values of the sorted set.

A way of calculating the maximum in Eq.1 using the above ideas is applying the ALGORITHM shown below starting from the points with highest values of importance and adding in order only one point on each iteration. But this procedure is not good enough since points with very close importance values should be considered as having the same perceptual contribution. Therefore, it is necessary we must do a clustering of the importance values before deciding which of them to choose. For doing this, we apply the method of step 1.

ALGORITHM

```
{
/* Add classes until q ≥ 2 */
  q = number of points in the first class(es)
  p = number of points in the next class
  While (q · g(C^k, q) − (q + p) · g(C^k, q + p) ≤ 0 )
   {
   q = q + p
   p= number of points of next class
  }
  n = q
}
```

3.2. Interpolation Procedure

Once the landmark point set is determined, the idea is to find an interpolation procedure fitting the outline. A very direct interpolation procedure is to link the landmark points with segments, but except with polygonal shapes this procedure would be inappropriate loosing a lot of the information on the behaviour between points. (see Figure 6.C).

In order to get a better interpolation procedure we have to look for a parametric class of functions fitting the behaviour of the curve between landmarks and verifying conditions of regularity on the landmark points. To do this, a new point having zero curvature variation, between each two successive landmark points is taken. We assume that the regularity of the curve of curvatures allows us to suppose the existence of at least one point verifying this condition. In those situations where there are more than one of these points we choose that one more equidistant from the extremities. Now the maximum number of landmark points is $2n$.

The interpolation procedure we use is based on the belief that in between two successive landmark points the behaviour of the curve of curvatures has to verify the following equation for some $\rho > 0$ value

$$C''(s) - \rho^2 C(s) = 0 \quad with \quad C''(s) = d^2 C(s)/ds^2 \tag{3}$$

$C(s)$ and ds being the curve of curvatures and arc element respectively. The curve solution of this equation is called local spline curve in tension ([10, 11] for details). Eq.3 for $\rho \to 0$, (in practice $\rho < 0.001$) we get geometric splines and for $\rho \to \infty$ (in practice $\rho > 10$), expresses a linear behaviour. Consequently, this model allows us to fit a wide range of curves having the same landmark points. The regularity condition to be verified by the whole interpolated curve is that the tangent line and curvature are continuous at the joints.

Estimation of ρ

The fitting procedure works calculating a set of initial values for the $\{\rho_i\}$. Next, it calculates the interpolation values on each segment and calculates the fitting error between observed and interpolated outlines using a distance measure. Using an iterative optimization procedure it calculates a new set of values for the $\{\rho_i\}$ and goes on to another iteration. Eventually it stops when the fitting improvements between two consecutive iterations is not significant. Euclidean and linear distances have been tried obtaining better fits with the linear one.

In order to achieve good time performance in the fitting procedure we calculate the initial values of the $\{\rho_i\}$ from

$$\rho_i = k \exp\{-\frac{1}{a} d_i\} \tag{4}$$

with k being a proportionality constant, a being a constant defining how fast the ρ_i values tend to zero, and d_i being an estimation of the absolute value of the signed area defined by the observed outline and the segment linking the i-th and (i+1)-th landmark points respectively. Values of $k \in [\frac{1}{3}, 1.0]$ and $a \in [5.0, 10.0]$ result adequated.

4. EXAMPLES

Some of the outlines used for testing the algorithm are shown in Figs. 1-6(A) estimated from original images 256x256. 1,5,6 correspond to biological images, 4 to an astronomical image, 3 to a geographical image and 2 is a synthetic image. Window size for smoothing the graph of curvatures was N=20. Points obtained by algorithm are shown in Figs. 1-6(B). We use the clustering criterion in Figs. 1-3 and the optimality criterion in Figs. 4-6. Corresponding linear interpolation to images 4,5,6 are shown in Figs. 4-6(C) and splines in tension in Figs. 4-6(D). Fig. 7(A-B) show a comparative study of errors produced in approximating the contour using linear (-.-) and spline in tension (-+-) interpolation for outlines in Figs. 6 and 3.

Following table resumes the results on number of total points (N_t), the number of points obtained by the algorithm (N_d), and the ratio between them (compress ratio).

Table 1

Image	Fig. 1	Fig. 2	Fig. 3	Fig. 4	Fig. 5	Fig. 6
N_t	729	654	846	471	899	1218
N_d	31	10	30	21	46	15
$(N_d/N_t) \cdot 100$	4.25	1.52	3.54	4.45	5.11	1.23

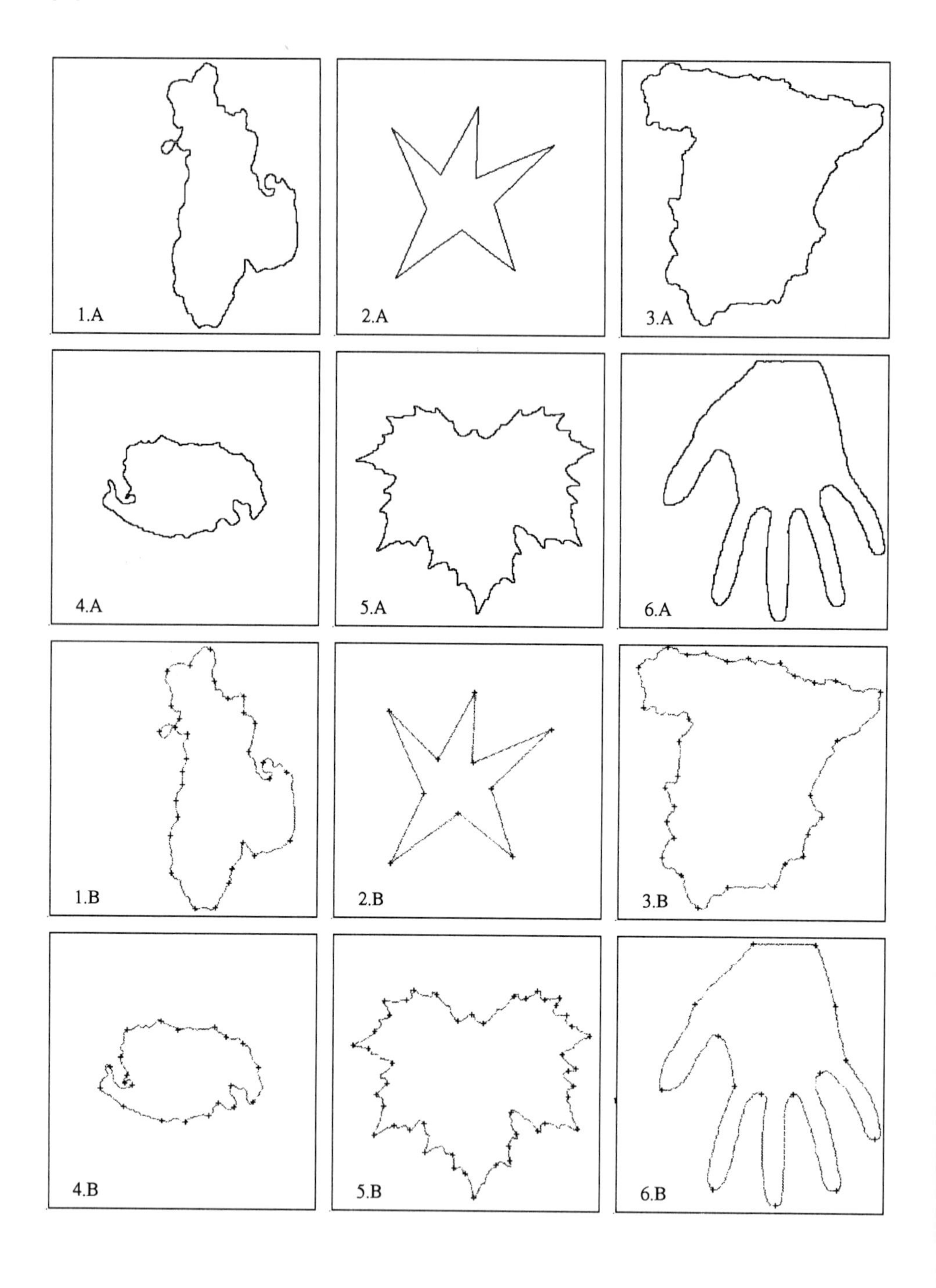
1.A
2.A
3.A
4.A
5.A
6.A
1.B
2.B
3.B
4.B
5.B
6.B

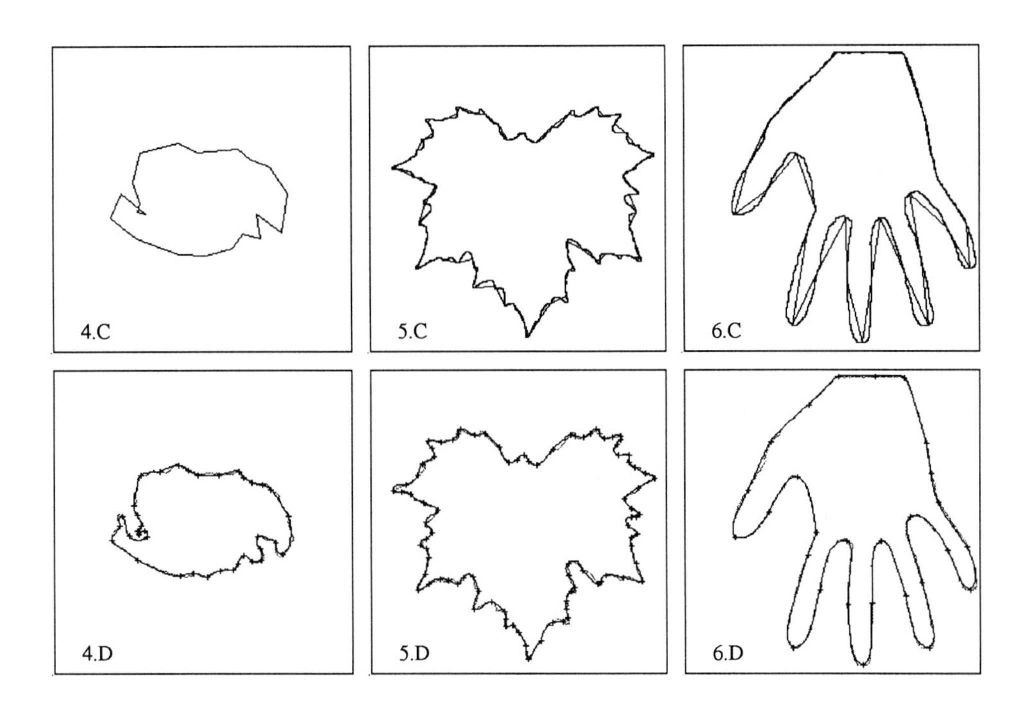
4.C
5.C
6.C
4.D
5.D
6.D

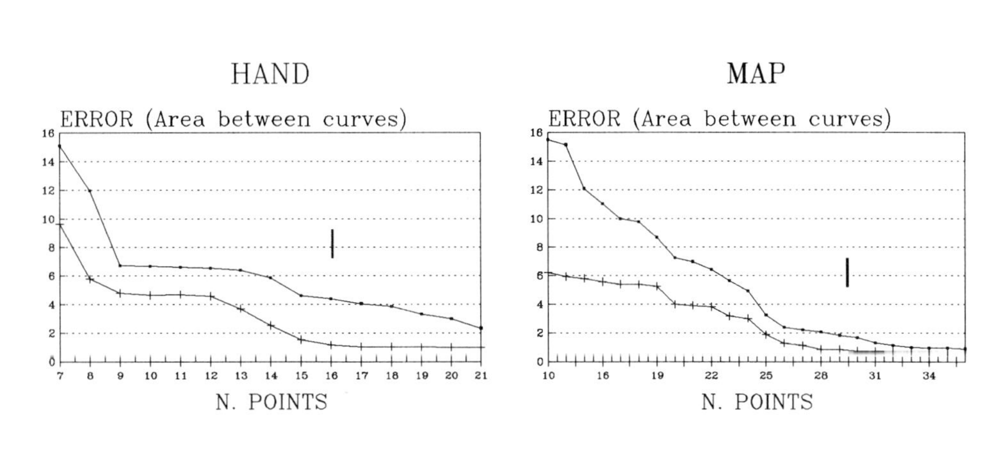
HAND
ERROR (Area between curves)
N. POINTS
MAP
ERROR (Area between curves)
N. POINTS

REFERENCES

1. C. H. Teh and R. T. Chin, *On the detection of dominant points on digital curves.* IEEE Trans. on PAMI-11, pp. 859-872 (1989).
2. H. Asada and M. Brady, *The curvature primal sketch.* IEEE Trans. on PAMI-8, pp. 2-14 (1986).
3. F. Mokhtarian and A. Mackworth , *Scaled-based description and recognition of planar curves and two-dimensional shapes.* IEEE Trans. on PAMI-8, pp. 34-43.(1986)
4. F. L. Bookstein, *The Measurement of Biological Shape and Shape Change.* Lectures Notes in Biomathematics n.24. Springer-Verlag. (1978)
5. F. Attneave and M. D. Arnoult, *The quantitative study of shape and pattern perception.* Psych. Bull. vol. 53, pp. 453-481 (1956).
6. N. Pérez de la Blanca, J. Fdez-Valdivia and R. Molina, *Synthesis and classification of 2-D shapes from their landmark points.* En PRAI series. Perez de la Blanca et al. eds. World Scientific Publish. pp. 35-48 (1992).
7. A. Knoerr, *Globals Models of Natural Boundaries: Theory and Applications.* Report in Pattern Theory 148. Brown University. (1988).
8. J. Fdez-Valdivia and N. Pérez de la Blanca, *Characterization of shapes in microscopical digital images,* 7th SCIA Procc., pp. 701-708 (1991).
9. N. Pérez de la Blanca, J. Fdez-Valdivia and J. A. García, *Characterizing planar outlines.* To be published in Pattern Recognition Letters.
10. D. G. Schweikert, *An interpolation curve using a spline in tension.* J. Math. and Physic, 45. pp. 312-317 (1966).
11. Su Bu-Qing, Liu Ding-Yuan. *Computational Geometry.* Academic Press (1989).

Uncertainty in Intelligent Systems
B. Bouchon-Meunier et al. (Editors)

SEARCHING FOR THE BEST PARTITION BY CLUSTERING

Xianyi ZENG and Christian VASSEUR

Centre d'Automatique Université de Lille I
Cité Scientifique 59650 Villeneuve d'Ascq France
ENSAIT
2, place des Martyrs de la Résistance 59070 Roubaix Cedex 1 France

ABSTRACT: This paper reports a method of searching for the best partition of multivariate data by clustering. One internal index is defined to estimate the true number of classes of observed statistic samples in a multidimensional euclidean space. These samples are clustered to create sequences of partitions. The estimation of the best partition is based on this indice as function of the number of encountered classes.

1. INTRODUCTION

Clustering of multidimensional data is an important subject in pattern analysis and recognition. Its purpose is to group a number of samples, which are usually random vectors in high dimensional space into clusters or classes. Samples in the same class have similar characteristics in some sense and samples in different classes are dissimilar in some sense.

It is important for a clustering to define criteria to mesure separability or resemblance between the classes to cluster. The best partition of the classes corresponds to a situation where the criterion of clustering is the most favourable. The used techniques for the clustering can be classified into 4 following categories:

(a) hierarchical method: This method enables to establish a hierarchy of partitions under the form of classification tree [1]-[3]. In this procedure, we start from the leaves (individual samples) for attaining the root (set of all the observed samples) by carrying out clustering between two similar classes.

(b) typological method: This method enables to define subsets of objects (complete symetric sub graphes) so that in the interior of these subsets, the objects are similar [4].

(c) information method: This method enables to apply the notion of information between variables, called transinformation [5]. The best partition of the classes is obtained by minimizing an entropy function defined a priori.

(d) method based on the minimization of an objective metrical function: This method is more frequently used than the others. Among the application of this method, the K-MEANS or ISODATA algorithm [6] is the best known. We can repetitiously execute the K-MEANS algorithm from different initial classes to select a partition who corresponds to the smallest value of the function associated to the criterion. To obtain an effective solution of this problem, numerious techniques are used to modify the K-MEANS strategy [7], [8].

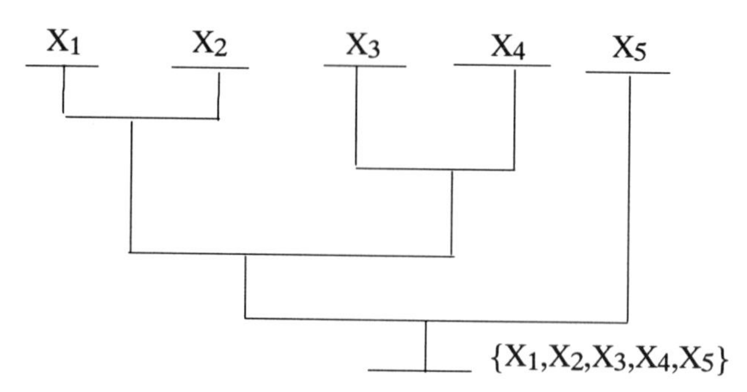

Figure 1: Hierarchical approach

In most of the above presented methods, the number of classes of the best partition is known. If this number is unknown, it is necessary to estimate it. [9] and [10] define internal indexes to find the best number of classes. This number corresponds to a partition where the indices are minimal.

Our paper provides a new clustering method, based on the evolution of an internal index, to test the best partition. This method utilizes a sequence of clusterings of classes from an initial partition. During this sequence, an internal index mesuring separability between classes and compactness inside every class enables to determine the best partition in sense of this index.

Data (observed samples) we discuss here respect the normal distribution laws. Therefore, the form of every class can be represented by an hyper-ellipse [11].

To simplify the clustering procedure, we do the following hypothesis:

- Each of the initial classes contains a sufficient number of samples to estimate the statistical parameters (the mean vector and the covariance matrix). Under this hypothesis, the case of isolated samples is not considered.

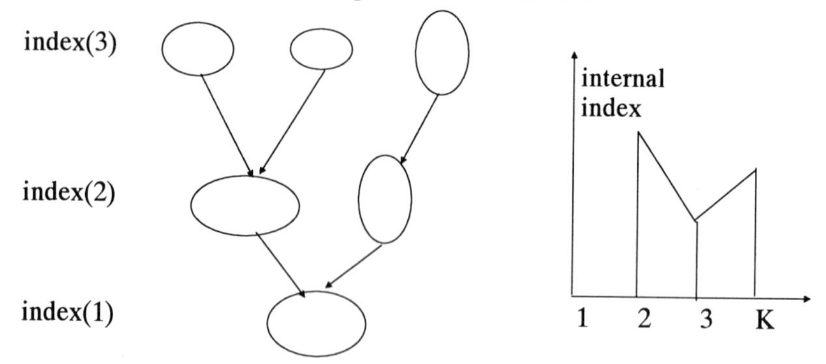

Figure 2: Clustering procedure (K: number of classes)

- After clustering of two classes in the multidimensional euclidean space, the samples of the new class respect also the normal distribution laws.

- Before the clustering procedure, we define a constant k_{max}, as number of classes in the initial partition. The best partition for all the samples is situated in one of the encountered state during the clustering procedure. This state corresponds to the minimal value of the internal index.

In general, a good partition of classes of the samples should satisfy the following conditions:

- Good separability between different classes: In such a partition, each class should show different characteristics from the other classes. In the multidimensional space, these characteristics represent the isolation of every class.

- Good compactness in the interior of every class: The observed samples in the same class should be concentrated on their mean vectors. If there is too much empty space in the hyper-ellipse of this class or there are several aggregations of samples in the interior of this class, the compactness becomes bad.

In Sections 2 and 3, we define respectively the separability degree and the compactness degree for our specified problem. We present an algorithm to find the best partition by minimizing the clustering index in Section 4. This index is defined as a linear combinaison of the separability degree and the compactness degree. Finally, three simulation examples are given in Section 5 to show the performances of the algorithm.

2. DEFINITION OF THE GENERAL SEPARABILITY

Assuming there exists K classes to cluster: C_1, C_2, ..., C_K ($1 \leq K \leq k_{max}$). We define the internal index I(K) according to the compactness degree in the interior of every class and the separability degree between these classes.

We start by defining the resemblance degree R_{ik} between any two classes C_i and C_k ($i,k \in \{1,2,......,K\}$).

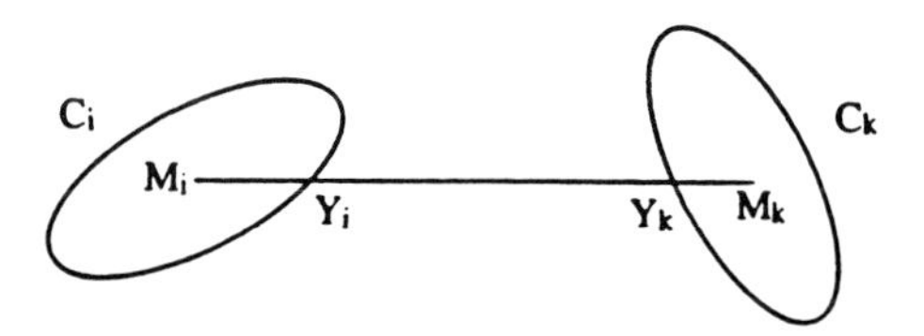

Figure 3: Resemblance between two classes C_i and C_k

The C_i and C_k are respectively characterized by the statistical parameters M_i, S_i and M_k, S_k where M_i, M_k are mean vectors of d dimensions and S_i, S_k covariance matrices of dxd dimensions. The contours of C_i and C_k are represented by the following equations:

$$(Y-M_i)^T S_i^{-1}(Y-M_i)=1 \quad \text{for the contour of } C_i \tag{1}$$

$$(Y-M_k)^T S_k^{-1}(Y-M_k)=1 \quad \text{for the contour of } C_k \tag{2}$$

By drawing a line M_iM_k (Figure 3), we obtains 2 intersection points with the contours of C_i and C_k, noted by Y_i and Y_k.

The resemblance degree R_{ik} is then defined as follows:

$$R_{ik} = \frac{\|M_i-Y_i\|+\|M_k-Y_k\|}{\|M_i-M_k\|} \tag{3}$$

where $\|M_i-M_k\|$ is the euclidean distance between M_i and M_k,

That is: $\|M_i-M_k\| = [(M_i-M_k)^T(M_i-M_k)]^{1/2}$ (4)

Y_i and Y_k being two points on the line M_iM_k, we can write:

$$Y_i-M_i = f_1 \cdot (M_k-M_i) \tag{5}$$

$$Y_k-M_k = f_2 \cdot (M_k-M_i) \tag{6}$$

where f_1 and f_2 are two real numbers

Evidently, Y_i, Y_k satisfy respectively the equations (1) and (2). By carrying (5) into (1), we obtain:

$$f_1^2(M_k-M_i)^T S_i^{-1}(M_k-M_i) = 1 \tag{7}$$

In the same way, by carrying (6) into (2), we obtain:

$$f_2^2(M_k-M_i)^T S_k^{-1}(M_k-M_i) = 1 \tag{8}$$

From where:

$$\begin{cases} |f_1| = \dfrac{1}{[(M_k-M_i)^T S_i^{-1}(M_k-M_i)]^{1/2}} \\ |f_2| = \dfrac{1}{[(M_k-M_i)^T S_k^{-1}(M_k-M_i)]^{1/2}} \end{cases} \tag{9}$$

The resemblance degree can be rewritten under the following form:

$$R_{ik} = |f_1|+|f_2| \tag{10}$$

The covariance matrices can be decomposed under the following form:

$$\begin{cases} S_i = U_i \Lambda_i U_i^T \\ S_k = U_k \Lambda_k U_k^T \end{cases} \tag{11}$$

where $\Lambda_i = \text{diag}(\lambda_{i1}, \ldots\ldots, \lambda_{id})$, $\Lambda_k = \text{diag}(\lambda_{k1}, \ldots\ldots, \lambda_{kd})$ and $U_i^T U_i = I$, $U_k^T U_k = I$.

By noting $\begin{cases} U_i^T (M_k - M_i) = Z_{ik} \\ U_k^T (M_k - M_i) = Z_{ki} \end{cases}$ (12)

It comes:

$$\begin{cases} |f_1| = \dfrac{1}{[Z_{ik}^T \Lambda_i^{-1} Z_{ik}]^{1/2}} \\ |f_2| = \dfrac{1}{[Z_{ki}^T \Lambda_k^{-1} Z_{ki}]^{1/2}} \end{cases} \tag{13}$$

We get:

$$R_{ik} = \frac{1}{[Z_{ik}^T \Lambda_i^{-1} Z_{ik}]^{1/2}} + \frac{1}{[Z_{ki}^T \Lambda_k^{-1} Z_{ki}]^{1/2}} \tag{14}$$

Evidently, the smaller is R_{ik}, the better is separability between the classes C_i et C_k. In other words, C_i and C_k are inseparable if R_{ik} is too big.

We define below the resemblance degree of the class C_k with all the other classes.

$$R_k = \text{maximum}\{R_{ik}\} \quad \text{with } i \neq k \tag{15}$$

R_k is the sparability degree of C_k with the other classes. The smaller is R_k, the better is the separability of C_k with its environment.

The general separability degree for the current K classes S(K) can then be defined as the average of all the R_k (k=1, 2, ..., K):

$$S(K) = \frac{1}{K} \sum_{k=1}^{K} R_k \qquad \text{for } K>1 \tag{16}$$

When K=1, there exists only one class and we define S(1)=0.

3. DEFINITION OF THE GENERAL COMPACTNESS

In a partition of K classes, the compactness degree of one class C_i, noted by P_i ($\forall i \in \{1, 2, \ldots\ldots, K\}$), is defined by the proportion of the number of samples outside the contour of C_i to the total number of samples. A good compactness is interpreted as a small value of P_i.

Let Y be any one sample of C_i. The MAHANALOBIS distance between Y and M_i is calculated by $d_i = (Y - M_i)^T S_i^{-1} (Y - M_i)$

According to the equation (1), we get the following conclusion:

If $d_i(Y)>1$, Y is situated outside the contour of C_i
If $d_i(Y)<1$, Y is situated inside the contour of C_i

Therefore, the definition of P_i can be written as follows:

$$P_i = \frac{Card\{Y | Y \in C_i \text{ and } d_i(Y)>1\}}{Card\{Y | Y \in C_i\}} \tag{17}$$

The general compactness degree is then defined as

$$C(K) = \frac{1}{K}\sum_{i=1}^{K} P_i \qquad (1 \le K \le k_{max}) \tag{18}$$

Assuming that we obtain a new class after the clustering of the classes C_j and C_k at the step K-1. Evidently, nearer are the classes C_j and C_k, more samples the contour of the new clustered class contains and smaller is the value of C(K). This means a good compactness of the samples inside the new class (see Figure 4).

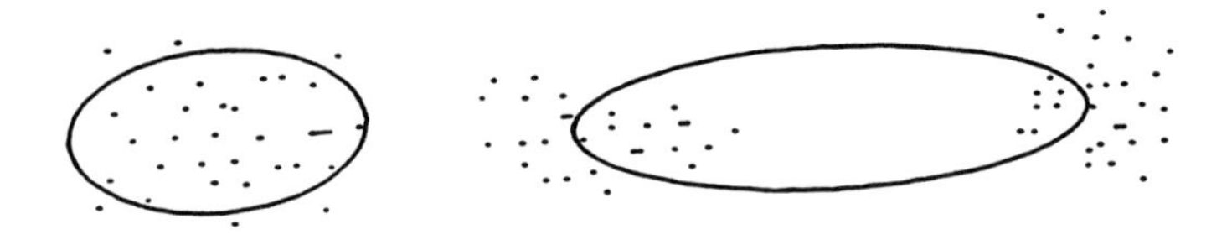

(a) good compactness (b) bad compactness

Figure 4: Different cases of compactness

Generally, C(K) increases during the clustering procedure two by two while the number of classes K decreases. In other words, more clusterings we carry out for remote classes, less compact the samples inside the clustered classes.

4. ALGORITHM OF SEARCHING FOR THE BEST PARTITION BY CLUSTERING

We define the internal index of the partition as follows:

$$I(K) = S(K) + \beta C(K) \tag{19}$$

where β is a positive real number defined a priori.

The best partition is obtained by finding a minimal I(K) for $K=k_{max}$, ..., 2, 1. According to the previous analysis, such a partition has a good separability between the classes and a good compactness inside every class. The clustering algorithm can be stated as follows:

Step 1:
  1.1 Selecting the parameters k_{max} and β.
  1.2 Estimating the parameters M_a, S_a and P_a for a=1, 2, ..., k_{max}.
Step 2: Calculating I(K) for $K=k_{max}$, ..., 1.
For $K=k_{max}$ until 1
  Do
    2.1 Searching for the two classes C_i and C_k so that R_{ik} satisfies the condition below:
    $R_{ik}=\text{maximum}\{R_{ab}\}\ \forall a,b\in\{1, 2, ..., K\}$ and $a\neq b$ (20)
    where the R_{ab} are calculated from M_a, M_b and S_a, S_b.
    2.2 Clustering C_i and C_k and calculating the internal index I(K). Estimating the mean vector, the covariance matrix and the compactness of the new class.
  End_Do
End_For
Step 3: Examining the values of I(K) for K=1, 2, ..., k_{max}. The best partition corresponds to that where I(K) is minimal.

5 SIMULATION RESULTS OF THE CLUSTERING ALGORITHM

The clustering algorithm is executed from $K=k_{max}$ to K=1 according to the recursive procedure presented previously. In each partition K, the two nearest classes are clustered to constitute a new class in the partition K-1. We carry out the simulation on different data distributions with d=2 and d=3. Next, we give the simulation results on 3 examples.

In Example 1, according to Figure 6, it appears that the best partition corresponds to K=1. This is perfectly interpreted in Figure 5 where the classes of the partition are superposed for the states K=2, 3 and 4.

Example 1 (d=2, k_{max}=4, β=20):

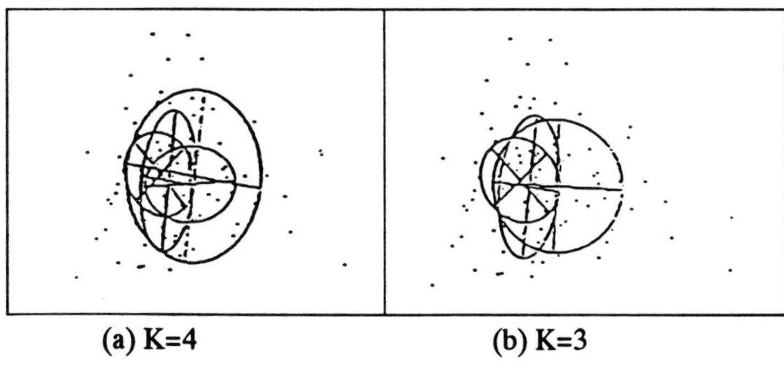

(a) K=4 (b) K=3

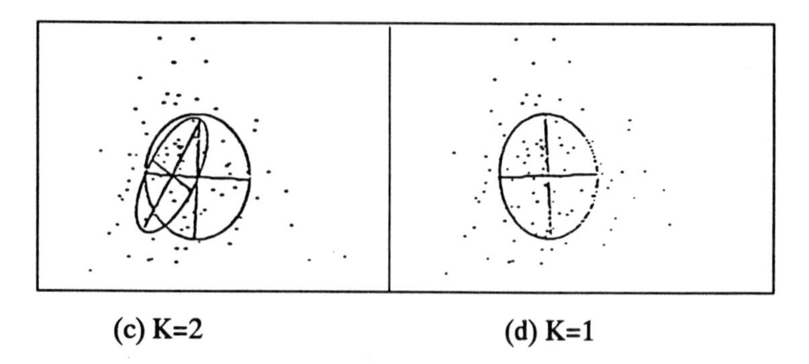

(c) K=2 (d) K=1

Figure 5: Clustering procedure for Example 1

K	4	3	2	1
S(	8.14	4.84	3.01	0
C(K)	0.589	0.550	0.581	0.592
I(K)	19.93	15.84	14.63	11.84

Table 1 Numerical results of clustering for Example 1

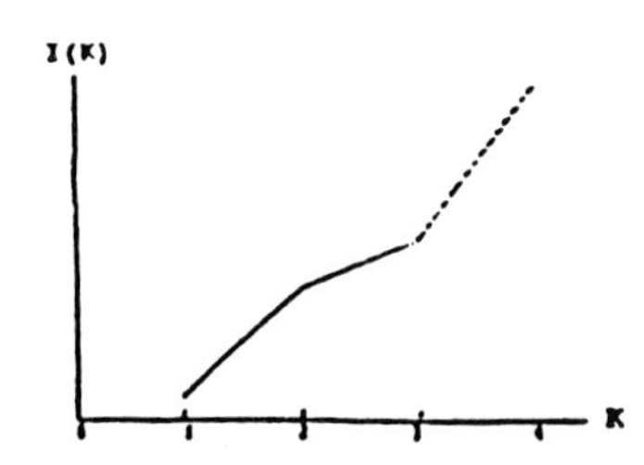

Figure 6: Evolution of the index during the clustering (Ex1)

Example 2 (d=2, k_{max}=5, β=20):

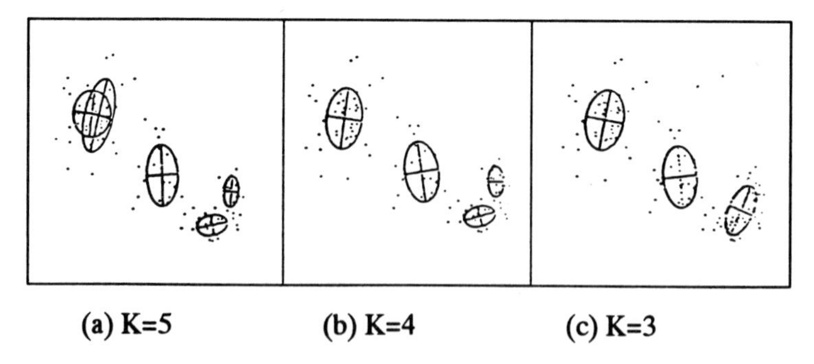

(a) K=5 (b) K=4 (c) K=3

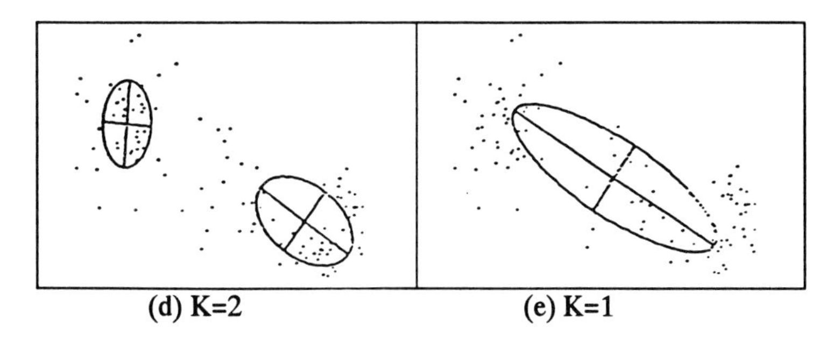

(d) K=2 (e) K=1

Figure 7: Clustering procedure for Example 2

K	5	4	3	2	1
S(K)	2.02	0.478	0.425	0.375	0
C(K)	0.613	0.601	0.623	0.598	0.805
I(K)	14.28	12.49	12.89	12.34	16.10

Table 2 Numerical results of clustering for Ex2

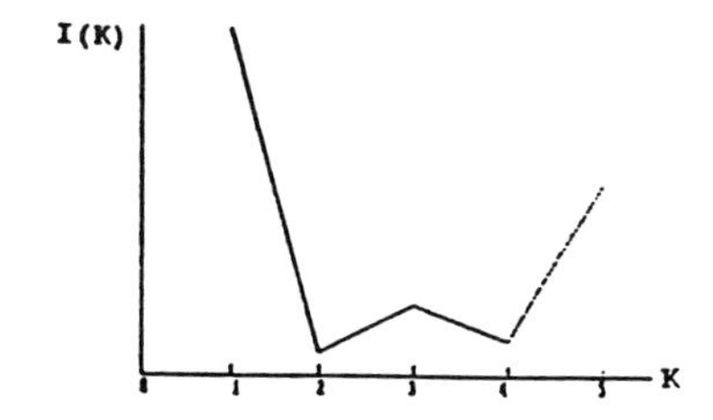

Figure 8: Evolution of the index during the clustering (Ex2)

According to Figure 8, Figure 7(d) represents the best partition corresponding to K=2.

Example 3 ($d=2$, $k_{max}=6$, $\beta=20$):

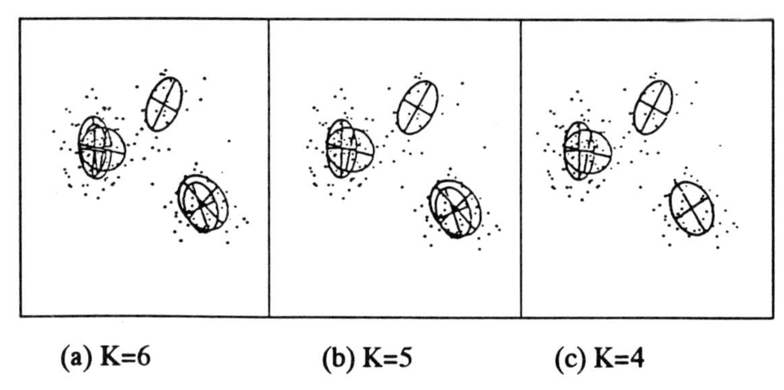

(a) K=6 (b) K=5 (c) K=4

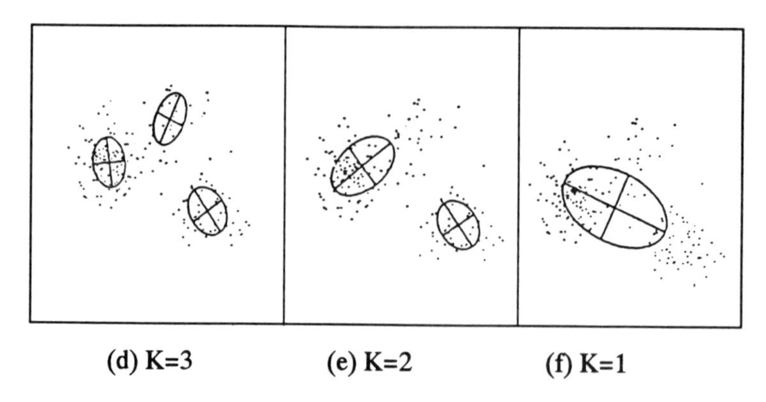

(d) K=3 (e) K=2 (f) K=1

Figure 9: Clustering procedure for Example 3

K	6	5	4	3	2	1
S(K)	9.346	5.259	2.434	0.447	0.381	0
C(K)	0.629	0.634	0.625	0.614	0.606	0.687
I(K)	21.93	17.94	14.92	12.72	12.49	13.74

Table 3 Numerical results of clustering for Ex3

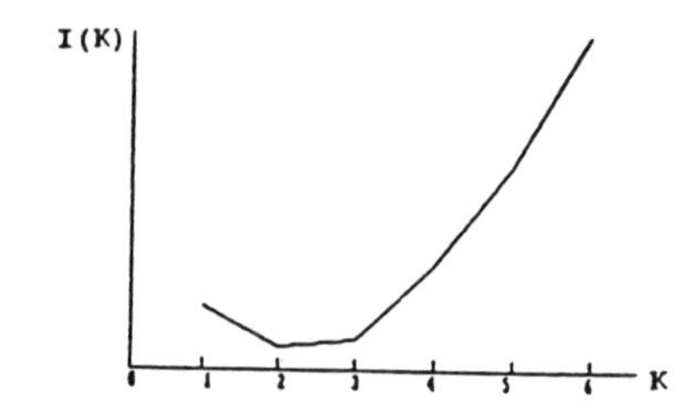

Figure 10: Evolution of the index during the clustering (Ex3)

Remarks: In Example 2, the values of the index I(2) and I(4) are very near (see Figure 8). In this case, the choice of the best partition is multiple (K=2 or K=4).

In the proposed algorithm, the parameter β is selected by user. This parameter is interpreted as the weight between the separability degree and the compactness degree. It risks to drive to an incorrect best partition if it is not adapted to the current situation of data distribution in the multidimensional space. We present below an algorithm for selecting automatically β through a learning procedure.

To determine a reasonable value of the parameter β, a set of examples, constituting a learning base, is provided a priori. In each of the examples, the best partition is known. In these conditions, the value of β is gradually learned by executing the clustering procedure for all the examples of the learning base.

The principle of the algorithm for selecting β by learning of an example is the following:

Let m be the number of classes of the best partition for the presented example and $]b_0, b_1[$ an initial interval of β.

We calculate below the separability degree and the compactness degree for one set of partitions (including the best partition K=m) generated by the clustering procedure. Two lists of S(K) and C(K) are obtained (K=1, 2, ..., m, ...).

As the state K=m represents the best partition, we have: $I(K)>I(m)\ \forall K\neq m$. This can be interpreted as the following inegalities:

$$\begin{cases} \beta < \dfrac{S(K)-S(m)}{C(m)-C(K)} & \text{if } C(m)>C(K) \\ \beta > \dfrac{S(m)-S(K)}{C(K)-C(m)} & \text{if } C(m)<C(K) \end{cases} \tag{21}$$

Therefore, the values of b_0 and b_1 can be modified as follows:

$$\begin{cases} b_0 := \max\{b_0, q_1, \ldots\ldots, q_r\} \\ b_1 := \min\{b_1, q_{r+1}, \ldots\ldots, q_{K-1}\} \end{cases} \tag{22}$$

where $\{q_1, \ldots\ldots, q_r\} = \left\{\dfrac{S(m)-S(K)}{C(K)-C(m)} \mid C(m)<C(K)\}\right\}$

and $\{q_{r+1}, \ldots\ldots, q_{K-1}\} = \left\{\dfrac{S(K)-S(m)}{C(m)-C(K)} \mid C(m)>C(K)\right\}$

If $b_0>b_1$ or if the best partition defined a priori is not encountered during the clustering procedure, then K=m can't be considered as the best partition. In this case, the algorithm fails and the best partition of the example should be redefined.

Otherwise, we get a new interval of β: $]b_0, b_1[$.

VI CONCLUSION

The examples in Section V show that the proposed algorithm can effectively cluster initialy different classes. The internal index defined a priori can reflect a correct situation of separability and of compactness for all the classes of each partition.

In general, this clustering method has the following advantages:

- Simplifying effected claculation for every clustering: In fact, the other methods drive often to very heavy calculations on each sample. This method, based on the normal distribution hypothesis, does not effect directly operations on individual samples (except for the calculation of compactness). It clusters the two classes according to the statistical parameters estimated from the observation samples.

- Obtaining the best partition by accounting for all the samples: In certain clustering methods, the criteria defined a priori drive often to local maxima or minima, corresponding to best partitions from any initial state. The internal index, defined previously, can account for all the samples characterized by the statistical parameters. The minimal value of this index corresponds to the best partition for all the states which evolves from any initial state.

The learning procedure of β based on the clustering algorithm enables to define automatically the weight between the general separability and the general compactness according to the encountered situations. From this procedure, we obtain a correct clustering index. This index can be applied to similar situations to determine the best partition of observation samples.

REFERENCES

[1] F.CAILLIEZ et J.G.PAGE
Introduction à l'analyse des données
SMASH, 1976
[2] P.DUDES et A.K.JAIN
"Validity studies in clustering methodologies"
Pattern Recognition, vol.11, pp.235-254, 1979
[3] M.GONDRAN
"Valeurs propres et vecteurs propres en classification hiérarchique"
RAIRO Informatique Théorique, vol.10, n°3, mars 1976
[4] V.DEGOT et J.M.HUALDE
"De l'utilisation de la notion de clique (sous-graphe complet symétrique) en matière de typologie de population"
RAIRO, janvier 1975, V-1, p.5 à 18
[5] S.WATANABE
"Pattern Recognition as a quest for minimum entropy"
Pattern Recognition, vol.13, n°5, 1981
[6] G.H.BALL et D.J.HALL
"A clustering technique for summarizing multivariate data"
Behav. Sci. 12, 153-155, 1967
[7] J.KITTER et D.PAIRMAIN
"Optimality of reassignment rules in dynamic clustering"
Pattern Recognition, vol.21, n°2, pp.169-174, 1988
[8] M.A.ISMAIL et M.S.KAMEL
"Multidimensional data clustering utilizing Hybrid search strategies"
Pattern Recognition, vol.22, n°1, pp.75-89, 1989
[9] R.C.DUBES
"How many clusters are best? - an experiment"
Pattern Recognition, vol.20, n°6, pp.645-663, 1987
[10] D.L.DAVIES et D.W.BOULDIN
"A cluster separation measure"
IEEE Trans. PAMI, vol.PAMI-1, n°2, april, 1979
[11] J.G.POSTAIRE
De l'image à la décision
Paris, Dunod, 1987

Uncertainty in Intelligent Systems
B. Bouchon-Meunier et al. (Editors)

A posteriori Ambiguity Reject Solving in Fuzzy Pattern Classification using a Multi-step Predictor of Membership Vectors

C. FRELICOT and B. DUBUISSON

URA CNRS 817 Heuristique et Diagnostic des Systèmes Complexes,
Université de Technologie de Compiègne,
B.P. 649, F-60206 Compiègne Cedex, France
Phone +33 44 23 44 23 Ext. 4243, Fax +33 44 23 44 77

Concepts issued from the fuzzy sets theory have improved the ability of pattern recognition methods in dealing with uncertainty. Furthermore, the introduction of reject options and particularly the ambiguity reject option has led to the design of decision systems which are not exclusive. However, if it is important to build up classifiers based on non exclusive discrimination rules, systematic ambiguous decisions are clearly inefficient from an user point of view. In order to solve this dilemma, a new rule is proposed. It is based on the multi-step prediction of the membership vector associated with a pattern to be classified. This rule allows the decision system to solve the ambiguity reject a posteriori, i.e. using predicted membership vectors, but not at any cost. Results obtained on simulated data are presented.

1. PATTERN CLASSIFICATION AND REJECT OPTIONS

1.1. Statistical Pattern Recognition

The aim of pattern recognition [1-4] is to classify objects (or patterns) by comparing them to prototypes. A statistical pattern recognition method is generally designed into two different steps :

• *a learning step* (Figure 1.)

It consists in determining the feature space and the decision space from past recorded data and expert knowledge, that is :

- Select a set of n parameters (or features) which are pertinent enough for the classification problem ; then a pattern is a n-dimensional vector $X = (X^1, X^2, \ldots, X^n)^T$.

- Build up a discrimination rule between c classes $(\omega_1, \omega_2, \ldots, \omega_c)$ assumed to be representative of prototypical patterns and composed of recorded supervised patterns (learning set) ; then such rules lead to decision boundaries between the classes.

• *a decision step* (Figure 2.)

It deals with the class-assignment of a new observed pattern X using the previous discrimination rule, that is identifying which class ω_i among the c possible ones X can be associated with.

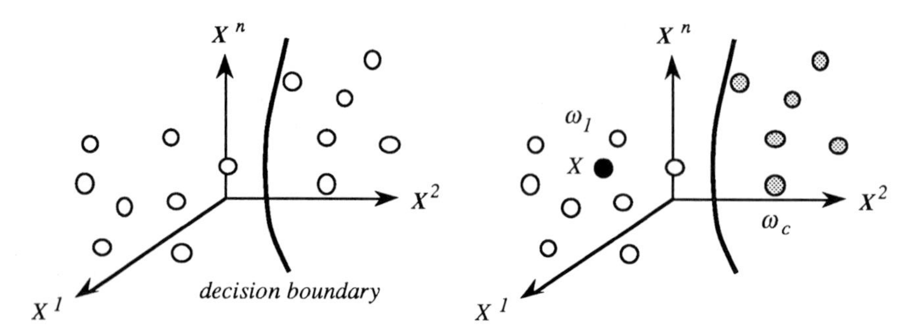

Figure 1. Learning step leading to a decision boundary between c=2 classes

Figure 2. Decision step, a new observed pattern X being assigned to class ω_i

Depending on available a priori knowledge about the learning class probability laws, either the parametric case or the non parametric one has to be considered. Many pattern recognition methods have been developed, e.g. the Bayes rule which minimizes the classification error probability, or the k-Nearest Neighbour rule [4].

1.2. Classification with Reject Options

Such classification methods are exclusive, i.e. a new observation X (or pattern) is assigned to a unique class ω_i among the c learnt ones ($1 \le i \le c$). But in many real applications, it often occurs that a pattern could be associated to several classes equally. Then, an exclusive class-assignment of a vector X can lead to misclassification. In order to decrease the misclassification risk due to this approach, a concept has been introduced : the reject option [5]. This option allows the decision system not to classify a new pattern on a certain cost condition. A distinction between two kinds of rejects has been proposed [6]:

• *the ambiguity reject* (Figure 3. (a))

This type of reject, which corresponds to the previous concept, is applied to patterns that can be assigned to more than one class, that is patterns lying near the decision boundaries.

• *the distance reject* (Figure3. (b))

This type of reject concerns patterns that are far from all the classes.

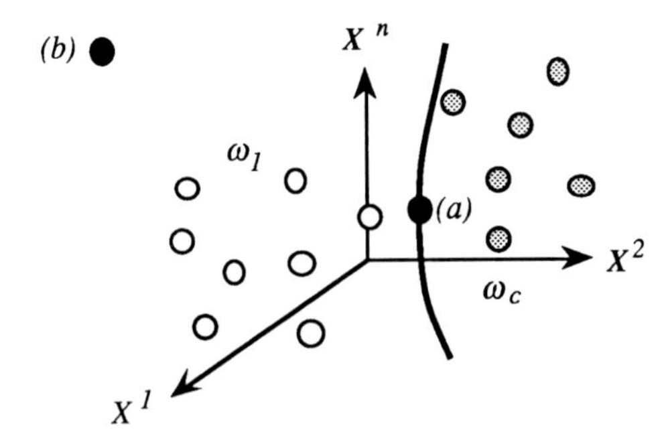

Figure 3. Ambiguity (a) and distance (b) rejected patterns

This latter concept appears as essential in many real applications. Pattern recognition methods are based on the knowledge of all the possible classes a new pattern X may be associated with. In most real applications (e.g. in diagnosis of technological processes), this assumption is not checked because a new pattern may be representative of no learnt class, i.e. may belong to an unknown class. Using this concept of distance reject, an adaptive pattern recognition process increasing the knowledge must be considered. Both parametric and non parametric classification dealing with both ambiguity and distance reject have been considered [6].

2. FUZZY CLASSIFICATION AND A PRIORI REJECT OPTIONS

2.1. Fuzzy Pattern Recognition

The fuzzy sets theory [7] has been introduced in pattern recognition to overcome the source of uncertainty (in both learning and decision tasks) due to classical exclusive assignment [8-10]. The concept of membership function μ^i is used to quantify the degree of belongingness of a pattern X to a class ω_i in order to take a decision. μ^i is an application from the feature space (generally $\Re^n$) to the [0,1] interval which indicates the degree of belongingness of X to ω_i, while this application takes values in the two-valued set {0,1} in the case of classical pattern recognition :

Classical Pattern Recognition

$$\mu^i : \Re^n \to \{0,1\}$$

$$X \mapsto \mu^i(X) = \begin{cases} 1 \text{ if } X \in \omega_i \\ 0 \text{ if } X \notin \omega_i \end{cases}$$

Fuzzy Pattern Recognition

$$\mu^i : \Re^n \to [0,1]$$

$$X \mapsto \mu^i(X)$$

Since the membership function must quantify the degree of belongingness of a pattern X to a class ω_i, it is generally set as a function of similarity or dissimilarity between the pattern and the class. In pattern recognition, two objects are similar if they are close, then a distance function $\mathrm{d}(X,\omega_i)$ may be chosen as the dissimilarity measure, and $\mu^i(X)$ is defined as a monotonic decreasing function of $\mathrm{d}(X,\omega_i)$:

$$\mu^i : \Re^n \to \Re^+ \to [0,1]$$
$$X \mapsto \mathrm{d}(X,\omega_i) \mapsto \mu^i(X)$$

The closer (respectively the farther) to the class ω_i the pattern X is, the closer to the value 1 (respectively 0) its membership function $\mu^i(X)$ is. The distance of the pattern to a class can be computed from the distance of the pattern to prototypes of the class, e.g. k-Nearest Neighbours of X in ω_i or the estimated mean G_i of ω_i.

The basic difference between fuzzy pattern recognition methods and classical pattern recognition ones lies in the fact that each learning or new pattern to be classified is associated with all the possible classes by the way of a c-dimensionnal membership vector $\mu = \mu(X) \in [0,1]^c$:

$$\mu = \mu(X) = (\mu^1(X), \mu^2(X), \ldots, \mu^c(X))^T$$

2.2. Fuzzy Classification Rule dealing with a priori Reject Options

In order to design a decision system under uncertainty, a fuzzy classification rule with reject options can be built up using the membership vector $\mu(X)$ associated with a new pattern X. We describe here a two-stage fuzzy classifier dealing sequentialy with two reject options.

• *membership reject :*

We assume that a fuzzy partition of N recorded patterns $\{X_1, X_2, \ldots, X_N\}$ in c classes exists, issued from a learning step by the use of a clustering procedure. If the clustering method used is a fuzzy one, e.g. Fuzzy C-Means or else [11], a (cxN)-dimensional partition matrix U which elements taking values in [0,1] is available :

$$U = \begin{pmatrix} \ddots & \vdots & \cdot^{\cdot^{\cdot}} \\ \cdots & \mu^{ik} & \cdots \\ \cdot^{\cdot^{\cdot}} & \vdots & \ddots \end{pmatrix} \qquad i=1,c \quad k=1,N$$

where $\mu^{ik} = \mu^i(X_k)$ represents the membership of the kth recorded pattern to the ith class.

If the clustering algorithm involved is a "hard" one, e.g. Dynamical Clustering Method [12] or else [13], the partition matrix U is a hard one (i.e. $\mu^{ik} \in \{0,1\}; \forall i,k$), but a "fuzzified" version can also be computed using a pre-defined membership function.

The membership reject rule we use is based on the diagonal components of a (cxc)-dimensional matrix *min* called the Minimum Membership Discrimination matrix :

$$min = \begin{pmatrix} \ddots & \vdots & \iddots \\ \cdots & min^{ij} & \cdots \\ \iddots & \vdots & \ddots \end{pmatrix} \quad \text{i,j=1,c}$$

with $min^{ij} = \min_{X_k \in \omega_i} \mu^j(X_k)$

The membership reject or the class-assignment of a new pattern X can be expressed in the following fuzzy classification rule :

i) compute $\mu(X)$

ii) **if** $\mu^i(X) < min^{ii} \ \forall i=1,c$

then X is membership rejected

else X is assigned to the set of classes $J = \{\omega_j : \mu^j(X) \geq min^{jj}\}$

endif

The concept of membership reject is clearly similar to the distance reject concept because of the relation between the membership function and the distance of a pattern to a class. In this case, the set of assignment classes is empty ($J = \varnothing$).

• *ambiguity reject :*

We assume now that the new pattern X has not been membership rejected. Then the class assignment of X is either an exclusive one (i.e. X is assigned to a unique class) or a multi-class one (i.e. X is assigned to more than one class). This latter case corresponds to an ambiguity reject.

We qualify these reject options and the related classification rules as ***a priori*** ones because of their systematic nature. It is due to the instantaneous nature of the classification : either $J = \varnothing$ or $\text{card}(J) > 1$. In order to solve the ambiguity reject problem, a systematic rule we could call the *membership maximum rule*, is often used :

X is assigned to the class ω_i for which its membership $\mu^i(X)$ is maximum

When the new pattern X can be associated to several classes equally, this rule is non satisfactory because of possible misclassification. Furthermore, it tends to a non fuzzy decision rule. Our objective is to define a new decision rule aiming at solving the ambiguity reject problem, but not at any cost.

3. A POSTERIORI AMBIGUITY REJECT SOLVING RULE

3.1. Use of a Multi-step Membership Vector Prediction Procedure

A sequence of observed patterns $\{X_0, X_1, \ldots, X_t\}$ can be considered as a discrete-time series assumed to be a realization of a dicrete-time stochastic process. According to the relation between X_t and $\mu(X_t)$, we consider the following discrete-time stochastic non-linear system which state-space representation is given by :

$$\begin{cases} X_{t+1} = F_t X_t \\ \mu_t = \mu(X_t) + w_t \end{cases}$$

with t observation time,
X_t n-dimensional state vector,
F_t state transition matrix,
μ_t c-dimensional measurement vector,
μ known measurement non-linear function : $\Re^n \rightarrow [0,1]^c$
(used in the basic fuzzy decision procedure),
w_t c-dimensional zero-mean white gaussian noise which non-negative symetric covariance matrix satisfies $\mathrm{E}[w_t w_{t'}^T] = R_t \delta_{tt'}$ ($\delta_{tt'}$ is the Kronecker symbol)

We have recently developed an adaptive multi-step membership prediction procedure in $[0,1]^c$ using a Modified Extended Kalman Filter [14]. Such a procedure is able to perform the following conditional expected value :

$$\hat{\mu}_{t+i/t} = \mathrm{E}[\mu_{t+i} / \mu_0, \mu_1, \ldots, \mu_t]$$

After linearization of the measurement equation, the dynamic system is modified in order to estimate the unknown transition matrix F_t during the up-dating step of the Kalman filtering algorithm. Thus, at each observation time t, a sequence of predicted patterns $\{\hat{X}_{t+i/t} : i = 1, k\}$ can be produced using a step-recursive formulation of the Kalman filtering equations (where the non available new observation μ_{t+i} is replaced by its previous estimate). The sequence of predicted membership vectors and associated estimation error's covariance matrices $\{\hat{\mu}_{t+i/t}, \hat{\Sigma}_{t+i/t} : i = 1, k\}$ are also estimated according to :

$$\hat{\mu}_{t+i/t} = \mu(\hat{X}_{t+i/t})$$

$$\hat{\Sigma}_{t+i/t} = \mathrm{E}[(\mu_{t+i} - \hat{\mu}_{t+i/t})(\mu_{t+i} - \hat{\mu}_{t+i/t})^T]$$

At step i, a confidence criterion associated to the predicted membership vector $\hat{\mu}_{t+i/t}$ can be defined considering the volume of the estimation error's hyperellipsoïd given by :

$$V_{\hat{\Sigma}_{t+i/t}} = q(\pi)\prod_{j=1}^{c}\lambda_j$$

where $q(\pi)$ is a scalar and λ_j denotes the *jth* eigenvalue of $\hat{\Sigma}_{t+i/t}$

Since the value of $q(\pi)$ is constant for a fixed dimension c, we propose to use the determinant of matrix $\hat{\Sigma}_{t+i/t}$ as a confidence criterion for the prediction at step i :

$$D_{\hat{\Sigma}_{t+i/t}} = \det\left(\hat{\Sigma}_{t+i/t}\right) = \prod_{j=1}^{c}\lambda_j$$

So, at each observation or decision time t, the sequence of predicted membership vectors $\{\hat{\mu}_{t+i/t}: i = 1, k\}$ can be used to decide about ambiguity rejected points, when either a settled prediction step k is given, or the prediction is confident enough, that is while the determinant $D_{\hat{\Sigma}_{t+i/t}}$ is less than a fixed threshold D_s.

3.2. A Posteriori Membership Ratio Classification Rule

We aim at building up a rule for the ambiguity reject solving problem using the membership prediction procedure when a new pattern X_t has been a priori ambiguity rejected. The first idea is to apply the fuzzy classification rule (section 2.2.) to the predicted membership vectors and look for an exclusive class-assignment. This rule called the *single vote rule* is very easy to implement [15]. In spite of being systematic, the *membership maximum rule* is based on pertinent information because it looks like a membership comparator. So we propose to include the comparison between the components of the predicted membership vectors. Having in mind that the ambiguity can be solved only in favor of the class ω_i for which the membership is maximum, let us define the membership ratio $R_{t+i/t}$ as follows :

$$R_{t+i/t} = \frac{\hat{\mu}^s_{t+i/t}}{\hat{\mu}^f_{t+i/t}}$$

where $\hat{\mu}^f_{t+i/t}$ and $\hat{\mu}^s_{t+i/t}$ denote the first and the second maximum components of $\hat{\mu}_{t+i/t}$:

$$\hat{\mu}^f_{t+i/t} = \max_{j:\omega_j \in J_{t+i/t}} \hat{\mu}^j_{t+i/t}$$

$$\hat{\mu}^s_{t+i/t} = \max_{j:\omega_j \in J_{t+i/t}-\{\omega_f\}} \hat{\mu}^j_{t+i/t}$$

where $J_{t+i/t}$ is the set of expected assignment classes with respect to $\hat{\mu}_{t+i/t}$

$R_{t+i/t}$ has the following interesting properties :

- $0 \leq R_{t+i/t} \leq 1$
- $R_{t+i/t} \to 0 \Leftrightarrow \hat{\mu}^f_{t+i/t} >> \hat{\mu}^s_{t+i/t}$

  $\Rightarrow \hat{\mu}^f_{t+i/t} >> \hat{\mu}^j_{t+i/t} \quad \forall j{:}\, \omega_j \in J_{t+i/t} - \{\omega_f\}$

  So, it seems logic to assign X_t to the dominant class ω_f
- $R_{t+i/t} \to 1 \Leftrightarrow \hat{\mu}^f_{t+i/t} \approx \hat{\mu}^s_{t+i/t}$

  Then, the ambiguity between at least two classes (ω_f and ω_s) is too large to be solved

Thus, the maximum membership can be taken into account by comparing $R_{t+i/t}$ to a fixed threshold R_s ($0 \leq R_s \leq 1$). If $R_{t+i/t}$ is less than R_s, then the class $\omega_f \in J_{t+i/t}$ for which the predicted membership is maximum appears to be a good candidate for the ambiguity solving problem. The new rule we propose just consists in proceeding to a vote for the different candidate classes and a fictive class $\omega_\cap$ indicating the ambiguity reject, counting it at each prediction step i with respect to $J_{t+i/t}$ and $R_{t+i/t}$, and eventually solve the ambiguity in favor of the class for which the vote is maximum. The algorithm can be expressed as follows :

i) initialize D_s and R_s

ii) apply the basic fuzzy decision procedure to the new pattern X_t

iii) **if** X_t is ambiguity rejected **then**

- **repeat**

  $i \leftarrow i+1$

  apply the fuzzy classification rule to $\hat{\mu}_{t+i/t}$ leading to a new set $J_{t+i/t}$

  compute the membership ratio $R_{t+i/t}$

  if $R_{t+i/t} < R_s$ **then** count the vote for the class ω_f

  else count the votes for each class $\omega_j \in J_{t+i/t}$

  and for $\omega_\cap$ if and only if $\text{card}(J_{t+i/t}) > 1$

  endif

  until $(\exists! \omega_j \in J_{t+i/t}{:}\, vote(\omega_j) > vote(\omega_\cap))$ or $(D_{\hat{\Sigma}_{t+i/t}} \geq D_s)$

- **if** $(\exists! \omega_j \in J_{t+i/t}{:}\, vote(\omega_j) > vote(\omega_\cap))$ **then** assign X_t to the class ω_j

  else X_t is still ambiguity rejected

  endif

endif

Note that if the membership ratio threshold R_s is fixed to 0, then the described rule reduces the already mentionned *single vote rule* [15].

3.3. Simulation Results

In order to test the proposed ambiguity solving rule on simulated data, a learning set of $c=3$ overlapping classes composed of 50 2-dimensional patterns each, and a tested sequence composed of $T=60$ vectors in $\Re^2$ (feature space) have been generated (Figure 4.). These patterns $\{X_t : t=0,T-1\}$ are assumed to represent observations issued from a technological process evolving from a particular state (class ω_1) to other states (classes ω_2 and ω_3).

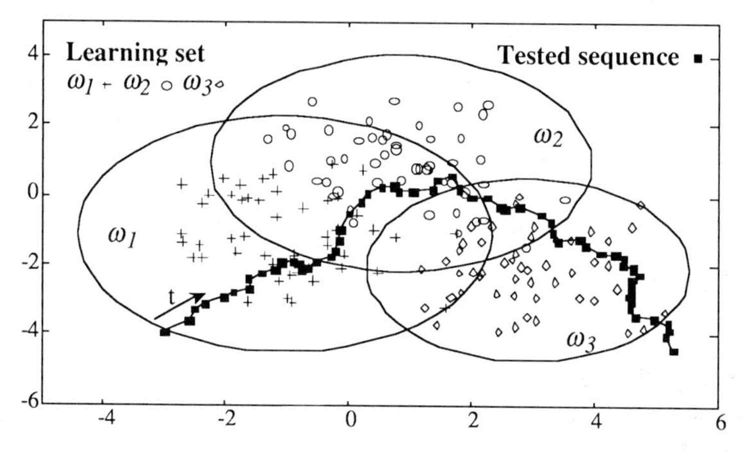

Figure 4. Learning set and tested sequence

Different membership functions have been used :

- $\mu^i(X_t) = \dfrac{1}{1+\mathrm{d}(X_t,\omega_i)} \quad \forall i=1,c$ (inverse function)
- $\mu^i(X_t) = \exp(-\mathrm{d}(X_t,\omega_i)) \quad \forall i=1,c$ (exponential function)

Also different distances have been used :

- $\mathrm{d}^2(X_t,\omega_i) = (X_t - G_i)^T (X_t - G_i) \quad \forall i=1,c$ (euclidian distance)
  where G_i denotes the estimated mean of class ω_i
- $\mathrm{d}^2(X_t,\omega_i) = (X_t - G_i)^T \Sigma_i^{-1} (X_t - G_i) \quad \forall i=1,c$ (Mahalanobis distance)
  where Σ_i denotes the estimated covariance matrix of class ω_i

Every combination of these membership functions and distances has been tested, leading to similar classification results. We just present here the results obtained using the inverse function and the euclidian distance. The classification results performed by the a priori fuzzy classification rule described in section 2.2. have led to an ambiguity reject rate of 36.67% (Figure 5.).

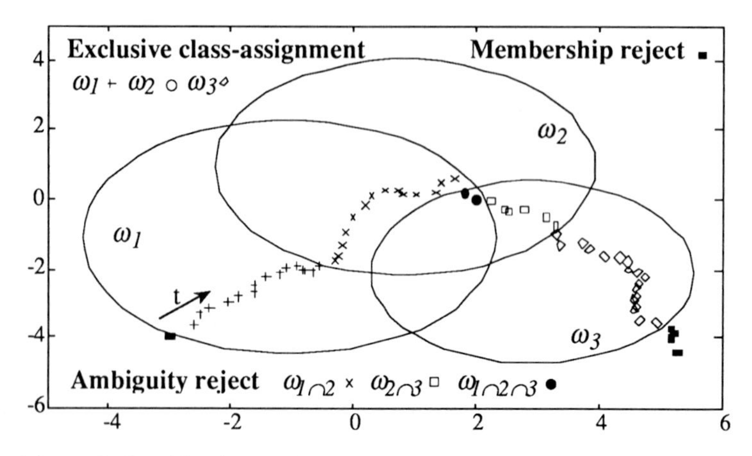

Figure 5. Classification results using the a priori fuzzy classification rule

In order to qualify the ability of the membership ratio rule in solving the ambiguity reject problem, we have tested the proposed algorithm with different values of the ration threshold R_s and the prediction confidence threshold D_s :

- $R_s = 1/2, 2/3, 3/4$
- $D_s = 0.5, 1, 3$

As expected, for a given value of R_s, the greater D_s is, the lower the ambiguity reject rate is. It means that the ambiguity reject may be more solved a posteriori when more predicted membership vectors are taken into account considering that the confidence in the prediction is high. For a given value of D_s, the greater R_s is, the lower the ambiguity reject rate is. It means that the ambiguity reject may be more solved when the superiority of the greatest component of the membership vector needed to assign the pattern to the corresponding class decreases.

Figure 6. shows the evolution of the ambiguity reject rate with D_s for different values of R_s, and classification results obtained using the fuzzy membership ratio rule with $D_s = 1$ and $R_s = 3/4$ are shown on Figure 7 (the patterns for which the ambiguity reject has been solved a posteriori are plotted in bold symbols). In that case the ambiguity reject rate decreased to 5% with no misclassification.

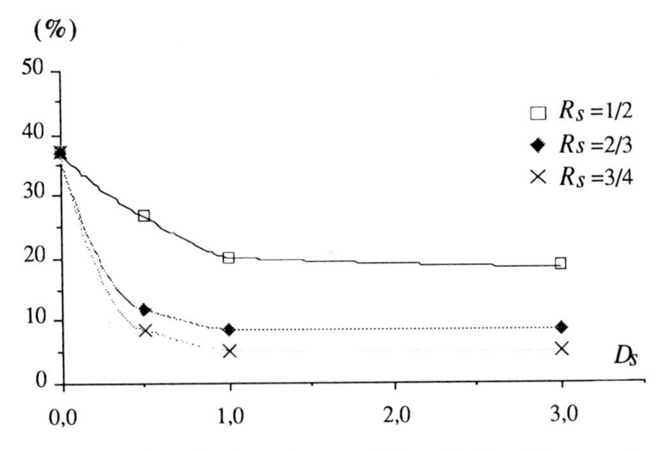

Figure 6. A posteriori ambiguity reject rate (%) vs D_S for different values of R_S

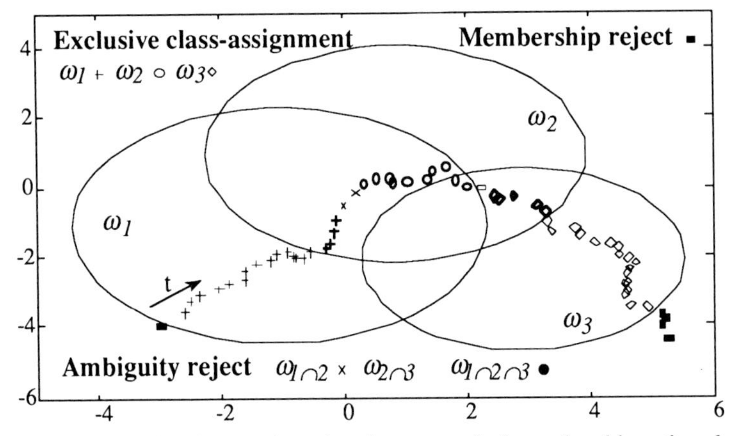

Figure 7. Classification results using the a posteriori membership ratio rule ($D_S = 1; R_S = 3/4$)

4. CONCLUSION

We have presented a new rule dealing with the ambiguity reject problem in fuzzy pattern classification. Its main characteristics are the following ones. It solves the ambiguity reject but not at any cost. In this way, the misclassification risk is reduced. This rule is an a posteriori ambiguity solving rule. It can be easily implemented in decision systems a membership vector multi-step prediction procedure has been added to. In that way, it seems to be well-adapted to applications such as diagnosis problems where a trend has to be detected rapidly. The results we have obtained on simulated data show a significant decrease of the ambiguity reject rate, without leading to misclassification.

REFERENCES :

1. K. Fukunaga, Introduction to statistical pattern recognition, Academic Press, 2nd edition (1990)
2. J.T. Tou and R.C. Gonzalez, Pattern recognition principles, Addison-Wesley (1974)
3. R.O. Duda and P.E. Hart, Pattern classification and scene analysis, Wiley Interscience (1973)
4. P.A. Devijver and J. Kittler, Pattern recognition : a statistical approach, Prentice-Hall (1982)
5. C.K. Chow, An optimum recognition error and reject tradeoff, *IEEE Transactions on Information Theory*, IT-16, 41-46 (1970)
6. B. Dubuisson, Decision with reject option, *Proc. 5th European Signal Processing Conference EUSIPCO'90*, Barcelone (1990)
7. L.A. Zadeh, Fuzzy sets, Information and Control, 8, 338-353 (1965)
8. W. Pedrycz, Fuzzy sets in pattern recognition : methodology and methods, *Pattern Recognition Letters*, 23, 121-146 (1990)
9. S.K. Pal, Fuzzy tools in the management of uncertainty in pattern recognition, image analysis, vision and expert systems, *International Journal of Systems Sciences*, 22, 511-549 (1991)
10. J.C. Bezdek and S.K. Pal (editors), Fuzzy models for pattern recognition, I.E.E.E. Press (1992)
11. J.C. Bezdek, Pattern recognition with fuzzy objective function algorithms, Plenum Press (1981)
12. E. Diday and al., Optimisation en classification automatique, INRIA (1979)
13. A.K. Jain and R.C. Dubes, Algorithms for clustering data, Prentice-Hall (1988)
14. C. Frélicot and B. Dubuisson, K-step ahead prediction in fuzzy decision space - Application to prognosis, *Proc. 1st IEEE International Conference on Fuzzy Systems FUZZ-IEEE'92*, San Diego, CA (1992)
15. C. Frélicot and B. Dubuisson, Solving the ambiguity reject in pattern recognition by multi-step prediction of the membership function, in Signal Processing VI : Theories and Applications, edited by J. Vandewalle and al., Elsevier, 2, 543-546 (1992)

CHAPTER 5:

MEASURES OF INFORMATION

Uncertainty in Intelligent Systems
B. Bouchon-Meunier et al. (Editors)

A unitary treatment of several known measures of uncertainty induced by probability, possibility, fuzziness, plausibility, and belief

Silviu Guiasu

Department of Mathematics and Statistics, York University,
4700 Keele Street, North York, Ontario, M3J 1P3, Canada

The objective of this paper is to give a unitary treatment of some known measures of uncertainty proposed in the recent years when dealing with problems involving possibility, fuzziness, plausibility and belief, by presenting them as combinations of absolute, conditional, and relative classical entropies from information theory. Some new measures of uncertainty are also proposed. A weighted way of combining evidence is also given that contains both Bayes' and Jeffrey's rules as special cases.

1. INTRODUCTION

Many measures of uncertainty induced by probability, possibility, fuzziness, plausibility, or belief have been proposed. Several authors (Klir (1987), Sander (1989), Jumarie (1990)) have underlined the necessity of investigating possible connections and relationships between so many and apparently different measures of uncertainty. On the other hand, the well-known Shannon absolute entropy and its natural extensions, namely the conditional entropy and the relative entropy (known also as the Kullback-Leibler indicator), are the only measures possessing the largest number of properties we can normally expect from a measure of uncertainty. The objective of this paper is to give a unitary treatment of some known measures of uncertainty proposed in the recent years when dealing with problems involving possibility, fuzziness, plausibility and belief, by presenting them as combinations of absolute, conditional, and relative classical entropies from information theory. It is true that the absolute, conditional, and relative entropies are normally assigned to probability distributions while the measures of possibility, fuzziness, plausibility, or belief go beyond probability theory. It is also true, however, that possibility, fuzziness, plausibility, and belief are intimately related to, or induce themselves, different kinds of probability distributions, which makes the above attempt possible.

2. ABSOLUTE, CONDITIONAL, AND RELATIVE ENTROPIES

Let X and Y be finite crisp sets, p_X a probability distribution on X, $p_{Y/X}$ a conditional probability distribution on Y given $x \in X$, and q_X a positive (not necessarily probability) distribution on X. The absolute entropy of p_X, the conditional entropy of $p_{Y/X}$ given $x \in X$, the conditional entropy of $p_{Y/X} = \{p_{Y/X};\ x \in X\}$ given p_X, and the relative entropy of p_X versus q_X are

$$H(p_X) = -\sum_{x \in X} p_X(x) \ln p_X(x), \quad H(p_{Y/X}) = -\sum_{y \in Y} p_{Y/X}(y) \ln p_{Y/X}(y),$$

$$H(p_{Y/X}, p_X) = \sum_{x \in X} H(p_{Y/X})\, p_X(x), \quad H(p_X : q_X) = \sum_{x \in X} p_X(x) \ln [p_X(x)/q_X(x)],$$

respectively. Also, if $p_{X,Y} = p_X\, p_{Y/X}$ is the joint probability distribution, on the Cartesian product $X \times Y$, defined by

$$p_{X,Y}(x,y) = p_X(x)\, p_{Y/X}(y)$$

then the joint (or product) absolute entropy of $p_{X,Y}$ is

$$H(p_{X,Y}) = -\sum_{x \in X} \sum_{y \in Y} p_{X,Y}(x,y) \ln p_{X,Y}(x,y).$$

3. UNCERTAINTY INDUCED BY FUZZINESS

Let $X = \{x_1,\ldots,x_n\}$ be a finite crisp set, A be a fuzzy set with the membership function $\mu\colon X \longrightarrow [0,1]$, and $\bar{A}$ be the fuzzy set with the membership function $\bar{\mu} = 1 - \mu$, where $\mu(x_i)$ is the degree of membership of x_i to A and $\bar{\mu}(x_i)$ is the degree of membership of x_i to $\bar{A}$. According to Yager (1979), Higashi and Klir (1982), the measure of fuzziness should measure the lack of distinction between a given set and its complement.

(a) Let $Y = \{A, \bar{A}\}$, $p_X = (p_1,\ldots,p_n)$, where p_i is the probability that x_i occurs, and $p_{Y/X} = (\mu(x), 1 - \mu(x))$ is interpreted as the conditional probability that $x \in X$ belongs to A or $\bar{A}$, respectively, if x does occur. Then,

$$H(p_{X,Y}) = H(p_X) + H(p_{Y/X}, p_X) =$$

$$= -\sum_i p_i \ln{}_i + \sum_i p_i[- \mu(x_i) \ln \mu(x_i) - (1 - \mu(x_i)) \ln (1 - \mu(x_i))]$$

is Hirota's (1982) entropy of the fuzzy set A. It is the absolute joint entropy, measuring the amount of uncertainty on the occurrence of the elements of X and the membership of these elements to the fuzzy set A or its complement $\overline{A}$, when occurrence and membership are dependent.

(b) If the occurrence of the elements of X and their membership to A or $\overline{A}$ are independent, on one hand, and if the memberships of the elements of X to A or $\overline{A}$ are independent altogether, then taking $Y = \{A,\overline{A}\}^n$ and p_Y to be the joint (product) probability distribution of the marginal probability distributions $(\mu(x_i),1-\mu(x_i))$, $(i=1,\ldots,n)$, we have

$$H(p_{X,Y}) = H(p_X) + H(p_Y) =$$

$$= -\sum_i p_i \ln p_i + \sum_i [- \mu(x_i) \ln \mu(x_i) - (1 - \mu(x_i)) \ln (1 - \mu(x_i))]$$

which is the entropy of the fuzzy set A introduced by De Luca and Termini (1979). It proves to be the absolute joint entropy measuring the amount of uncertainty on the occurrence of the elements of the crisp set X and the independent membership of the elements of X to the fuzzy set A or its complement $\overline{A}$. Also, in the same context,

$$H(p_Y) = -\sum_i [\ \mu(x_i) \ln \mu(x_i) + (1 - \mu(x_i)) \ln (1 - \mu(x_i))] \qquad (1)$$

is just the entropy of the fuzzy set A introduced by De Luca and Termini (1972), which proves to be the absolute entropy measuring the amount of uncertainty on the membership of the independent elements of X to the fuzzy set A or its complement $\overline{A}$.

(c) Let $X = \{x_1,\ldots,x_n\}$ be a finite crisp set. If the elements of X are interdependent, the membership function assigned to a fuzzy set A is a probability distribution χ on the set of all binary sequences with n components $\{A,\overline{A}\}^n$ where, for instance, $\chi(\overline{A},A,A,\overline{A},\ldots,\overline{A},A)$ means the joint probability of the event "$x_1 \notin A$ and $x_2 \in A$ and $x_3 \in A$ and $x_4 \notin A$ and ... and $x_{n-1} \notin A$ and $x_n \in A$." If $c = (c_1,\ldots,c_n) \in \{A,\overline{A}\}^n$ is an arbitrary sequence with n binary (A or $\overline{A}$) components, then the entropy of the fuzzy set A with interdependent elements is

$$H(\chi) = -\sum_{c\in\{A,\bar{A}\}^n} \chi(c) \ln \chi(c). \tag{2}$$

In the particular case when the elements of X are independent and μ: $X \longrightarrow [0,1]$ is the membership function of the elements $x \in X$ to the fuzzy set A, the joint probability distribution χ is completely determined by the membership function μ. Thus, for instance,

$$\chi(\bar{A},A,A,\bar{A},\ldots,\bar{A},A) = [1-\mu(x_1)]\,\mu(x_2)\,\mu(x_3)\,[1-\mu(x_4)]\ldots[1-\mu(x_{n-1})]\,\mu(x_n),$$

and entropy (2) becomes De Luca and Termini's (1972) entropy of the fuzzy set A given by the expression (1).

(d) Let X be a finite crisp set, p_X a probability distribution on X and w a nonnegative (not necessarily probability) distribution on X. The weighted entropy of p_X given the weight w is (Belis and Guiasu (1968), Guiasu (1977)):

$$H_W(p_X) = -\sum_{x\in X} w(x)\, p_X(x) \ln p_X(x).$$

It reduces to the absolute entropy when $w \equiv 1$, which means $w(x) = 1$ for all $x \in X$. Thus, $H_1(p_X) = H(p_X)$. Now, if $A \subset X$ is a fuzzy set with the membership function μ, then Zadeh's (1968) entropy of the fuzzy event A, namely

$$\sum_{x\in X} \mu(x)\, p_X(x) \ln p_X(x)$$

is just $-H_\mu(p_X)$ and should be called the *fuzzy entropy* induced by the fuzzy set A.

4. UNCERTAINTY INDUCED BY BELIEF, AMBIGUITY, PLAUSIBILITY, AND POSSIBILITY

Let X be a finite crisp set and m: $\mathcal{P}(X) \longrightarrow [0,1]$ a basic probability assignment, i.e. a set function satisfying

$$m(\emptyset) = 0, \quad \sum_{A\subseteq X} m(A) = 1,$$

where $\mathcal{P}(X)$ is the crisp set of all the subsets of X. Here $m(A)$ expresses the degree of belief, induced by evidence, that a specific element of X should belong to the set A but not to any special subset of A. Let *Bel* be the associated belief measure, *Amb* the ambiguity measure (relative to the distinction between a set and its complement), and Pl the plausibility measure, all three defined on $\mathcal{P}(X)$ and taking on values in $[0,1]$, that are obtained from m according to the equalities:

$$Bel(A) = \sum_{B \subseteq A} m(B), \quad Amb(A,\bar{A}) = \sum_{\substack{B \cap A \neq \emptyset \\ B \cap \bar{A} \neq \emptyset}} m(B), \quad Pl(A) = \sum_{B \cap A \neq \emptyset} m(B).$$

(a) Let $Y = \mathcal{P}(X)$, $p_Y = m$, and $p_{X/A}$ be a conditional probability distribution on X given the subset $A \in \mathcal{P}(X)$. Let $p_{X,Y}(x,A) = p_{X/A}(x)\ p_Y(A) = p_{X/A}(x)\ m(A)$. Denote by $p_{X/Y} = \{p_{X/A};\ A \subseteq X\}$. Then, the uncertainty induced by the available evidence about the elements of X *and* the subsets of X is measured by the absolute joint entropy

$$H(p_{X,Y}) = H(p_Y) + H(p_{X/Y}, p_Y).$$

With these notations, the expression

$$H(p_Y) = - \sum_{A \subseteq X} m(A) \ln m(A)$$

is the absolute entropy measuring the uncertainty induced by evidence on the subsets of X, used by Nguyen (1985).

(b) Let us take the particular case when

$$p_{X/A}(x) = \begin{cases} 0, & \text{if } x \notin A; \\ 1/|A|, & \text{if } x \in A; \end{cases}$$

where $|A|$ is the number of elements of the crisp subset $A \subseteq X$. Then,

$$H(p_{X/Y}, p_Y) = \sum_{A \subseteq X} m(A) \ln |A|$$

which is the so called U-uncertainty, Dubois and Prade's (1985) generalization of Higashi and Klir's (1982) measure of uncertainty in possibility theory, inspired by Hartley's (1928) logarithmic measure. It appears to be the conditional entropy measuring the maximum uncertainty in discriminating among the elements of the subsets of X once the subsets are selected according to the basic probability assignment m induced by the evidence. In this case, the uncertainty about selecting subsets of X *and* discriminating among the elements of the selected subsets is measured by

$$H(p_{X,Y}) = H(p_Y) + H(p_{X/Y}, p_Y) = - \sum_{A \subseteq X} m(A) \ln [m(A)/|A|] = -H(m:q_Y),$$

where the reference measure is $q_Y(A) = |A|$.

(c) Let us take $Y = \mathcal{P}(X)$, $p_Y = m$, and $q_Y = Bel$. As $Bel(A) = 0$ implies that $m(A) = 0$, this means that p_Y is absolutely continuous with respect to q_Y. Then

$$H(p_Y:q_Y) = \sum_{A \subseteq X} m(A) \ln [m(A)/Bel(A)].$$

Thus, Höhle's (1982) measure of confusion

$$C(m) = -\sum_{A \subseteq X} m(A) \ln Bel(A)$$

proves to be the sum of two entropies, $C(m) = H(p_Y:q_Y) + H(p_Y)$, where the absolute entropy measures the amount of uncertainty contained by the basic probability assignment induced by evidence and the relative entropy measures how much the basic probability assignment m differes on average from the belief measure Bel.

(d) Let us take $Y = \mathcal{P}(X)$, $p_Y = m$, and $q_Y = Pl$. As $Pl(A) = 0$ implies $m(A) = 0$, this means that p_Y is absolutely continuous with respect to q_Y. Then, we have

$$H(p_Y:q_Y) = \sum_{A \subseteq X} m(A) \ln [m(A)/Pl(A)].$$

Thus, Yager's (1983) measure of dissonance

$$E(m) = -\sum_{A \subseteq X} m(A) \ln Pl(A)$$

proves to be the sum of two entropies, $E(m) = H(p_Y:q_Y) + H(p_Y)$, where the absolute entropy measures the amount of uncertainty contained by the basic probability assignment induced by the evidence at hand and the relative entropy measures how much the basic probability assignment m is different on average from the plausibility measure Pl.

(e) Let us take $Y = \mathcal{P}(X)$, $p_Y = m$, $q_Y = Pl$, and

$$p_{X/A}(x) = \begin{cases} 0, & \text{if } x \notin A; \\ 1/|A|, & \text{if } x \in A; \end{cases}$$

Then, Lamata and Moral's (1988) total uncertainty may be written as

$$T(m) = \sum_{A \subseteq X} m(A) \ln [|A|/Pl(A)] =$$

$$= H(p_{X/Y}, p_Y) + H(p_Y : q_Y) + H(p_Y) = H(p_{X,Y}) + H(p_Y : q_Y),$$

measuring the uncertainty on selecting subsets of X according to the basic probability assignment m induced by the available evidence, without discriminating between the inner elements of each subset, plus the average divergence of the basic probability assignment m from the plausibility measure Pl.

(f) Let Z be a finite (crisp) set and take now $X = \mathcal{P}(Z)$, $Y = \mathcal{P}(\mathcal{P}(Z))$, $p_X = m$, and

$$p_{Y/A}(\mathcal{B}) = \begin{cases} Bel(A), & \text{if } \mathcal{B}=\{B; B\subseteq A\}; \\ Amb(A,\bar{A}), & \text{if } \mathcal{B}=\{B; B\cap A\neq\emptyset \text{ and } B\cap\bar{A}\neq\emptyset\}; \\ Bel(\bar{A}), & \text{if } \mathcal{B}=\{B; B\subseteq\bar{A}\}; \\ 0, & \text{for other } \mathcal{B}. \end{cases}$$

Given the subsets of Z supported by the evidence m, the uncertainty about distinguishing between each of them and its respective complement is

$$H(p_{Y/X}, p_X) = \sum_{A \subseteq Z} p_X(A)\, H(p_{Y/A}) =$$

$$= \sum_{A \subseteq Z} m(A)[- Bel(A) \ln Bel(A) - Amb(A,\bar{A}) \ln Amb(A,\bar{A}) - Bel(\bar{A}) \ln Bel(\bar{A})],$$

where $H(p_{Y/A})$ gives the amount of uncertainty in distinguishing between A and $\bar{A}$. As $Pl(\bar{A}) = Amb(A,\bar{A}) + Bel(\bar{A})$, we get

$$H(p_{Y/A}) = - Bel(A) \ln Bel(A) - Amb(A,\bar{A}) \ln Amb(A,\bar{A}) - Bel(\bar{A}) \ln Bel(\bar{A}) =$$

$$= [- Bel(A) \ln Bel(A) - Pl(\bar{A}) \ln Pl(\bar{A})] +$$

$$+ [- Amb(A,\bar{A}) \ln \{Amb(A,\bar{A})/Pl(\bar{A})\} - Bel(\bar{A}) \ln \{Bel(\bar{A})/Pl(\bar{A})\}].$$

In the particular case when the basic probability assignment m has only the elements of the crisp set Z as focal sets, i.e. $m(A) \neq 0$ only if $A = \{z\}$, $z \in Z$, denoting $m(\{z\})$ by $p_Z(z)$, we have

$$Pl(A) = Bel(A) = \sum_{z \in A} p_Z(z) = P(A), \qquad Amb(A,\bar{A}) = 0,$$

where P is the probability on $\mathcal{P}(Z)$ induced by the probability distribution p_Z, and

$$H(p_{Y/Z}) = -p_Z(z) \ln p_Z(z) - [1 - p_Z(z)] \ln [1 - p_Z(z)],$$

representing the uncertainty about distinguishing between the element $z \in Z$ and its complement $Z - \{z\}$, while

$$H(p_{Y/Z}, p_Z) = \sum_{z \in Z} p_Z(z)[-p_Z(z) \ln p_Z(z) - (1 - p_Z(z)) \ln (1 - p_Z(z))]$$

measures the amount of uncertainty about distinguishing between all the elements of the crisp set Z and their complements in Z.

5. COMBINING TWO INDEPENDENT BODIES OF EVIDENCE

Let X be a finite crisp set and m_1 and m_2 two independent basic probability assignments on $\mathcal{P}(X)$. The product basic probability assignment is

$$m_1 \times m_2 : \mathcal{P}(X) \times \mathcal{P}(X) \longrightarrow [0,1],$$

defined for $A \subseteq X$ and $B \subseteq X$ by $(m_1 \times m_2)(A,B) = m_1(A)\, m_2(B)$. Obviously,

$$H(m_1 \times m_2) = H(m_1) + H(m_2).$$

A major problem is how to combine m_1 and m_2, induced by two independent bodies of evidence, in order to get a new combined basic probability assignment m on $\mathcal{P}(X)$? Given a pair (A,B) of focal sets of m_1 and m_2, respectively, define by $c(1,1|A,B)$, $c(1,0|A,B)$, $c(0,1|A,B)$, and $c(0,0|A,B)$ the credibilities of $A \cap B$, $A \cap \overline{B}$, $\overline{A} \cap B$, and $\overline{A} \cap \overline{B}$, respectively. Assume

$$c(i,j|A,B) \geq 0, \quad \sum_{A,B} \left(\sum_{i,j \in \{0,1\}} c(i,j|A,B) \right) m_1(A)\, m_2(B) = 1.$$

The combined basic probability assignment is defined as

$$m(C) = \sum_{A \cap B = C} c(1,1|A,B)\, m_1(A)\, m_2(B) + \sum_{A \cap \overline{B} = C} c(1,0|A,B)\, m_1(A)\, m_2(B) +$$

$$+ \sum_{\overline{A} \cap B = C} c(0,1|A,B)\, m_1(A)\, m_2(B) + \sum_{\overline{A} \cap \overline{B} = C} c(0,0|A,B)\, m_1(A)\, m_2(B).$$

The change of uncertainty induced by combining the basic probability assignments m_1 and m_2 into the new basic probability assignment m is

$$\mathfrak{C}(m_1,m_2;m) = H(m_1) + H(m_2) - H(m).$$

Special cases: (a) Let us notice that by taking

$$c(i,j\,|\,A,B) = \begin{cases} c, & \text{if } i=j=1 \text{ and } A\cap B\neq\varnothing; \\ 0, & \text{elsewhere}; \end{cases}$$

where

$$c = \Big[1 - \sum_{A\cap B=\varnothing} m_1(A)\, m_2(B)\Big]^{-1},$$

we obtain Dempster's rule of combining two independent bodies of evidence.

(b) Another possible case of interest, that takes robustness of the focal sets into account, is

$$c(i,j\,|\,A,B) = \begin{cases} c\,|A\cap B|/|A\cup B|, & \text{if } i=j=1 \text{ and } A\cap B\neq\varnothing; \\ 0, & \text{elsewhere}; \end{cases}$$

where

$$c = \Big[1 - \sum_{A\cap B=\varnothing} (|A\cap B|/|A\cup B|)\, m_1(A)\, m_2(B)\Big]^{-1}.$$

(c) Assume that m_1 and m_2 have the same focal sets A and $\overline{A}$, and the only credibilities different from zero are

$$c(1,1\,|\,A,A) = c(1,0\,|\,A,\overline{A}) = c(0,1\,|\,\overline{A},A) = c(0,0\,|\,\overline{A},\overline{A}) = 1,$$

which means to give credit to A every time when at least one evidence points at A. The corresponding combined basic probability assignment is

$$m(A) = 1 - [1 - m_1(A)][1 - m_2(A)], \quad m(\overline{A}) = m_1(\overline{A})\, m_2(\overline{A}).$$

This is just the rule used for calculating the reliability of a parallel system with independent components in operations research. According to Lindley (1987), this rule for combining probabilities was used by George Hooper in 1685.

(d) Let X and Y be two finite crisp sets. Let m_1 and m_2 be two basic probability assignments on $\mathcal{P}(X\times Y)$, having the focal elements $\{x\}\times Y$, $(x\in X)$ and $X\times\{y\}$, $(y\in Y)$, respectively. Let $\{p_{Y/X};x \in X\}$ be a family of conditional probability distributions on Y given the elements of X. Denote by

$$m_1(\{x\}\times Y) = p_X(x), \quad m_2(X\times\{y\}) = \tilde{p}_Y(y)$$

and take the only credibilities different from zero to be

$$c(1,1|\{x\}\times Y, X\times\{y\}) = p_{Y/X}(y)/p_Y(y), \text{ where } p_Y(y) = \sum_{x\in X} p_{Y/X}(y)\, p_X(x).$$

The combined basic probability assignment is

$$m(\{x\}\times\{y\}) = c(1,1|\{x\}\times Y, X\times\{y\})\, m_1(\{x\}\times Y)\, m_2(X\times\{y\}) =$$

$$= \frac{p_{Y/X}(y)\, p_X(x)}{p_Y(y)}\, \tilde{p}_Y(y). \tag{3}$$

The belief measure induced by m is

$$Bel(\{x\}\times Y) = \sum_{y\in Y} m(\{x\}\times\{y\}) = \sum_{y\in Y} \frac{p_{Y/X}(y)\, p_X(x)}{p_Y(y)}\, \tilde{p}_Y(y)$$

which is Jeffrey's generalization of Bayes' rule for calculating the posterior probabilities. A slightly different way of obtaining Jeffrey's rule can be found in Ichihashi and Tanaka (1989). Particularly, if

$$\tilde{p}_Y(y) = \begin{cases} 1, & \text{if } y = y_0; \\ 0, & \text{if } y \neq y_0; \end{cases}$$

denoting $m(\{x\}\times\{y_0\})$ by p_{X/y_0}, the equality (3) becomes Bayes' classical rule for calculating the posterior probabilities

$$p_{X/y_0}(x) = \frac{p_{Y/X}(y_0)\, p_X(x)}{p_Y(y_0)}.$$

In the extreme case when $\tilde{p}_Y(y_0) = 1$ and $p_{X/y_0}(x) = p_X(x)$, for every $x \in X$, we have $\mathfrak{C}(m_1,m_2;m) = 0$ and there is no change in the initial uncertainty on X. In another extreme case, when $\tilde{p}_Y(y_0) = 1$ and $p_{X/y_0}(x_0) = 1$, we have $\mathfrak{C}(m_1,m_2;m) = H(m_1)$ and the entire initial uncertainty on X is removed.

REFERENCES

Belis M. and Guiasu S. (1968). A quantitative-qualitative measure of information in cybernetic systems. *IEEE Trans.Inform.Theory*, **IT-14**, 593-594.

DeLuca A. and Termini S. (1972). A definition of a nonprobabilistic entropy in the setting of fuzzy sets. *Information and Control*, **20**, 301-312.

De Luca A. and Termini S. (1979). Entropy and energy measures of a fuzzy set. In Gupta M.M., Ragade R.K., and Yager R.R. (eds.), *Advances in Fuzzy Set Theory and Applications*. North-Holland Publishing Company, New York, pp.321-338.

Dubois D. and Prade H. (1985). A note on measures of specificity for fuzzy sets. *Int. J. General Systems*, **10**, 279-283.

Guiasu S. (1977). *Information Theory with Applications*. McGraw-Hill, New York.

Hartley R.V.L. (1928). Transmission of information. *Bell System Tech. J.*, **7**, 535-563.

Higashi M. and Klir G.J. (1982). Measures of uncertainty and information based on possibility distributions. *Int. J. General Systems*, **9**, 43-58.

Hirota H. (1982). Ambiguity based on the concept of subjective entropy. In Gupta M.M. and Sanchez E. (eds.), *Fuzzy Information and Decision Processes*, North-Holland Publishing Company, New York, pp.29-40.

Höhle U. (1982). Entropy with respect to plausibility measures. *Proc.12th IEEE Symp.on Multiple-Valued Logic*, pp.167-169.

Ichihashi H. and Tanaka H. (1989). Jeffrey-like rules of conditioning for the Dempster-Shafer theory of evidence. *Int. J. of Approximate Reasoning*, **3**, 143-156.

Jumarie G. (1990). A theory of information for vague concepts. Outline of application to approximate reasoning. *Kybernetes*, **19**, 15-34.

Klir G.J. (1987). Where do we stand on measures of uncertainty, ambiguity, fuzziness and the like? *Fuzzy Sets and Systems*, **24**, 141-160.

Lamata M.T. and Moral S. (1988). Measures of entropy in the theory of evidence. *Int. J. General Systems*, **14**, 297-305.

Lindley D.V. (1987). Comment: A tale of two wells. *Statistical Science*, **2**, 38-40, 1987.

Nguyen H.T. (1985). On entropy of random sets and possibility distributions. In J.C. Bezdek (ed.), *The Analysis of Fuzzy Information*, CRC Press, Boca Raton, Florida.

Sander W. (1989). On measures of fuzziness. *Fuzzy Sets and Systems*, **29**, 49-55.

Shannon C.E. (1948). A mathematical theory of communication. *Bell System Tech. J.*, **27**, 379-423.

Yager R.R. (1979). On the measure of fuzziness and negation. Part I: Membership in the unit interval. *Int.J. General Systems*, **5**, 221-229.

Yager R.R. (1983). Entropy and specificity in a mathematical theory of evidence. *Int. J. General Systems*, **9**, 249-260.

Zadeh L.A. (1968). Probability measure of fuzzy events. *J. Math. Analysis and Appl.*, **23**, 421-427.

Uncertainty in Intelligent Systems
B. Bouchon-Meunier et al. (Editors)

THE ϕ-ENTROPY IN THE SEQUENTIAL RANDOM SAMPLING

J.A. Pardo[a] and M.L. Vicente[b]

[a]Departamento de de Estadística e I. O., Facultad de Matemáticas, Universidad Complutense de Madrid, 28040-Madrid, Spain.

[b]Departamento de de Estadística e I. O., Escuela Universitaria de Estadística, Universidad Complutense de Madrid, 28040-Madrid, Spain.

ABSTRACT

In this paper we consider the problem of sequential random sampling. We study some applications of ϕ-Entropy measure on the sequential observation by definning a rule based on this measure. Finally, we stablish this process when the parametric space has two elements and we study its relation with the Sequential Probability Ratio Test.

1. INTRODUCTION

Let A be an experiment whose possible results θ belong to a parametric space Θ. Before making any decision, let us suppose that the consequences of the statistic depends on the result θ of the experiment A. We can observe the realization of the experiment X with statistical space $(\mathcal{X}, \beta_{\mathcal{X}}, P_\theta)_{\theta\in\Theta}$ .The observation of X will proportionate information about Θ, and this information will help to make a good decision. Let us suppose that P_θ is absolutely continuous with respect to a countable measure, or with respect to the Lebesgue's measure and we denote, $f(x/\theta) = (dP_\theta/d\mu)(x)$ its density function or probability function. Let us associate to the space Θ a σ-field β_Θ and over this σ-field let us consider an absolutely continuous probability measure with respect to the Lebesgue's measure, or with respect to the countable measure, $p(\theta) = (d\tau/d\lambda)(\theta)$ denotes its density function or probability function. The predictive distribution of X, i.e., the inconditional distribution of X is given by f(x).

Let ϕ be an strictly concave function defined from $\mathbb{R}$ to $\mathbb{R}$, we define the ϕ-entropy functional contained in the prior distribution as

$$H_\phi(p(\cdot)) = \int_\Theta \phi(p(\theta)\ d\lambda(\theta). \tag{1}$$

After performing the X experiment, and observing the values $x_1,\ldots,x_n$ of this experiment, the contained ϕ-Entropy on the posterior distribution is

$$H_\phi(p(\cdot/x_1,\ldots,x_n)) = \int_\Theta \phi(p(\theta/x_1,\ldots,x_n))\ d\lambda(\theta). \tag{2}$$

The expression (2) is called ϕ-Entropy about Θ, with prior knowledge $p(\theta)$, after observing the values $x_1,...,x_n$. The difference between (1) and (2) is known in the literature like ϕ-measure of Jensen Difference after observing the values $x_1,...,x_n$.

The idea of Jensen difference (or information radius) was originally stated by Sibson (1969) for the Shannon's entropy and applied to biological sciences. Burbea and Rao (1982a) extended this idea to the ϕ-entropy as given in (1) and connected it to Fisher's information matrix using differential geometric approach. Some interesting examples of (2) for studying convexity property have been undertaken by Burbea and Rao (1982b). Burbea (1984) extended this idea to Bose-Einstein entropies of degree α. Some applications to Statistics of Jensen difference can be seen in Rao (1982).

In this paper our aim is to study methods of sequential sampling, in order to obtain a prescribed accuracy in the determination of a unknown parameter, based on the expresion (2).

The measure defined in (2), admits many interesting theoretic examples:
Shannon's entropy, ($\phi_1(x) = - x \log_2 x$)
Quadratic entropy, ($\phi_2(x) = x - x^2$)
Cubic entropy, ($\phi_3(x) = x - x^3$)
Genetic entropy, ($\phi_4(x) = x - 2x^2 + 2x^3 - x^4$)
Gamma entropy, ($\phi_5(x) = - \ln \Gamma(1+x)$)
Sine entropy, ($\phi_6(x) = \frac{\sin \pi x}{2} \log_2(\frac{\sin \pi x}{2})$ or $\phi_7(x) = \frac{\sin \pi x}{2}$)
Entropy of degree β, ($\phi_8(x) = (x^\beta - x)/(2^{1-\beta} - 1)$)
Hypoentropy, ($\phi_9(x) = (1+(1/\lambda)) \log(1+\lambda) - (1/\lambda)(1+\lambda x)\log(1+\lambda x)$, $\lambda>0$)
α-log entropy, ($\phi_{10}(x) = - x^\alpha \log_2 x$, $(1/2)\leq\alpha\leq 1$)
Entropy of degree (α,β), ($\phi_{11}(x) = (x^\alpha - x^\beta)/(2^{1-\alpha} - 2^{1-\beta})$, $\alpha\neq\beta$, $0<\alpha\leq 1-\beta$ or $0<\beta\leq 1-\alpha$)
Paired measures, ($\phi_{12}(x) = \phi_k(x) + \phi_k(1-x)$, $k=1, ..., 11$).

The obtained results in this work, are applicable in all these particular cases.

2. A SEQUENTIAL OBSERVATION SCHEME BASED ON THE ϕ-ENTROPY

If the statistician can sequentially take observations and, at each stage, he must decide, at the sight of the amount of the obtained information about Θ, if to stop sampling or to continue sampling and take the next observation, then, the following rule based on the ϕ-Entropy is defined:

Definition 1

The sequential observation rule which states:
To stop observing after the values $x_1,...,x_n$ have been observed if

$$H_\phi(p(\cdot/x_1,...,x_n)) = \int_\Theta \phi(p(\theta/x_1,...,x_n))\, d\lambda(\theta) \leq c$$

where c is a constant which depends on the amount of information required in each particular problem by the statistician, according to subjective criteria, and to continue observing if

$$H_\phi(p(\cdot/x_1,...,x_n)) = \int_\Theta \phi(p(\theta/x_1,...,x_n))\, d\lambda(\theta) > c$$

is called Sequential Observation Scheme based on the ϕ-Entropy,(S.O.S.ϕ-E).

Lindley (1956, 1957) proposed in this context a stopping rule based on the Shannon's entropy which take into account the precision of the posterior distribution. El Sayyad (1969) studied the rule proposed by Lindley with the exponential distribution. Bernardo (1977) proposed a stopping rule based in the logarithmic divergence. Pardo (1986) proposed a stopping rule based on the f-divergence. Finally, Vicente (1990) proposed a stopping rule based on the entropy of degree β (Havrda and Charvat (1967)). Some of the last situations will be particular cases of the stopping rule presented in this paper.

Remark

If we consider that $\Theta \in \mathbb{R}$ and the prior knowledge about θ can be expressed by a member, $g_{a,b}(\theta)$, of the conjugate family, $\mathfrak{F}_C$, that depends on the "a" and "b" parameters, the contained ϕ-Entropy on the $g_{a,b}(\theta)$ distribution, is given by

$$H_\phi(a,b) = \int_\Theta \phi(g_{a,b}(\theta))\, d\lambda(\theta).$$

Since this expression depends on the "a" and "b" values, the set

$$A(c) = \left\{ (a,b) \in \mathbb{R}^2 / \ H_\phi(a,b) = c \right\}$$

represents the set of all distributions of $\mathfrak{F}_C$ family that produces a constant ϕ-Entropy, where c is a value preassigned by the statistician. Therefore, in the plane of "a" and "b", this set represents a curve called Iso-ϕ-Entropy-Curve. In consequence, the S.O.S. ϕ-E, separates in two regions, R_1 and R_2, the parametric space $\mathbb{R}^2$, where

$$R_1 = \left\{ (a,b) \in \mathbb{R}^2 \ / \ H_\phi(a,b) > c \right\}$$

is the region "To continue sampling", and

$$R_2 = \left\{ (a,b) \in \mathbb{R}^2 \ / \ H_\phi(a,b) \leq c \right\}$$

is the region "To stop sampling".

Then, after observing the $x_1,...,x_n$ values, the posterior distribution will have new values of the parameters, "a*" and "b*". So, the sequential sampling will finish when the new values meet in the region R_1.

The obtained results in this paper, will be illustrated with the particular case for ϕ, $\phi(x)= (x^{\beta} - x) / (2^{1-\beta} - 1)$. In this situation, if we consider a exponential population X, with θ parameter and if the prior distribution of θ is a $\Gamma(a,p)$ distribution it is shown in Vicente (1990), that for large values of the p parameter the Iso-ϕ-Entropy-Curve is

$$a^2 = \frac{2\pi(p - 1 + \frac{1}{\beta})}{\beta^{\frac{1}{1-\beta}} \left[c(2^{1-\beta} - 1) + 1\right]^{\frac{2}{1-\beta}}}$$

and, the stopping rule is given by

$$\sum_{i=1}^{n_0} x_i > \frac{(2\pi)^{1/2} (p + n_0 - 1 + \frac{1}{\beta})^{1/2}}{\beta^{\frac{1}{2(1-\beta)}} (c(2^{1-\beta} - 1) + 1)^{1/(1-\beta)}} - a$$

where x_i are the observed values of the X experiment and n_0 is the first stage in wich last inequality is satisfied.

Another important obtained result in this situation is that the sampling procedure is finite.

Next we shall study the behaviour of the S.O.S.ϕ-E when $\Theta = \{\theta_0, \theta_1\}$. In this situation, after observing the values $x_1,..,x_n$, the ϕ-Entropy takes the expression $H_\phi(p(\cdot/x_1,...,x_n)) = \phi(p(\theta_0/x_1,...,x_n)) + \phi(p(\theta_1/x_1,...,x_n))$ and the obtained result is given by the following

Theorem 1

After observing the values $x_1,...,x_n$ the Sequential Observation Scheme based on the ϕ-Entropy when $\Theta = \{\theta_0, \theta_1\}$, indicates to stop observing if

$p(\theta_0/x_1,...,x_n) \in [0, b] \cup [1-b, 1]$

where b, satisfies the equation $c = \phi(b) + \phi(1-b)$, and c, is the prefixed constant by the statistician, belonging to the interval $(0, H_\phi(p(\theta_0/x_1,...,x_n)=1/2))$.

Proof

It is inmediate to prove that $H_\phi(p(\cdot/x_1,...,x_n))$ is a concave function of $p(\theta_0/x_1,...,x_n)$ for a fixed prior distribution $p(\theta) = \{ p(\theta_0), p(\theta_1)\}$. Furthermore, it is a symetric function with respect to 1/2, and $H_\phi(1) = H_\phi(0)$.

The general form of this curve is shown in Fig. 1

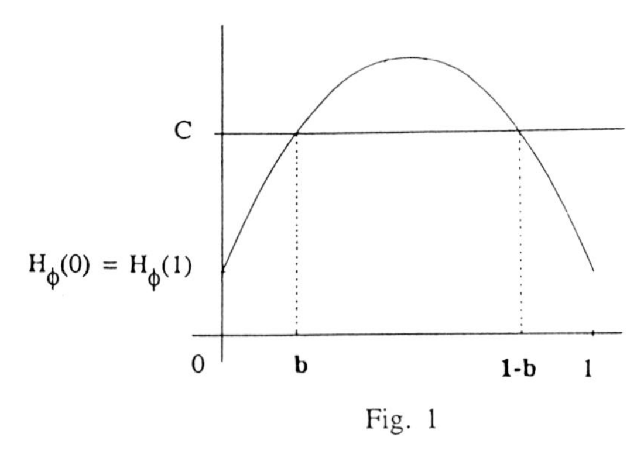

Fig. 1

Then, since $H_\phi(p(\theta_0/x_1,...,x_n))$ is a concave function of $p(\theta_0/x_1,...,x_n)$ we will stop observing when

$H_\phi(p(\theta_0/x_1,...,x_n)) \leq c$

that is, if $p(\theta_0/x_1,...,x_n) \in [0,b]$, or $p(\theta_0/x_1,...,x_n) \in [1-b, 1]$, where b verify that

$c = \phi(b) + \phi(1-b)$

and, we will continue sampling if, and only if

$b < (p(\theta_0/x_1,...,x_n)) < 1-b$.

If we consider $\phi(x) = (x^\beta - x) / (2^{1-\beta} - 1)$ the satisfied equation by b is $(b^\beta + (1-b)^\beta - 1) / (2^{1-\beta} - 1) = c$, and for the particular case $\beta = 2$, we will continue sampling iff

$p(\theta_0/x_1,...,x_n) \in [0, (1-\sqrt{1-c})/2] \cup [(1+\sqrt{1-c})/2, 1]$.

Finally, we study the relation between the stopping rule based on the ϕ-Entropy and the Wald Sequential Probability ratio test. We obtain the following result:

Theorem 2

The region "To continue sampling" given by the sampling scheme based on the ϕ-Entropy, coincides with the region "To continue sampling" given by the Wald Sequential Probability Ratio Test, for testing $H_0: \theta=\theta_0$, against $H_1: \theta = \theta_1$, with stopping bounds:

$$A = \frac{b}{(1-b)} \frac{p(\theta_0)}{p(\theta_1)} \qquad B = \frac{(1-b)}{b} \frac{p(\theta_0)}{p(\theta_1)}$$

where b is the given value in Theorem 1.

Proof

The expression $b < p(\theta_0/x_1,...,x_n) < 1-b$, may be written in terms of the ratio of posterior probabilities for θ_0 and θ_1 in the following way

$$\frac{b}{1-b} < \frac{p(\theta_1/x_1,...,x_n)}{p(\theta_0/x_1,...,x_n)} < \frac{1-b}{b} \tag{3}$$

By Bayes' Theorem, previous expression can be written in the way

$$\frac{b}{1-b}\frac{p(\theta_0)}{p(\theta_1)} < \frac{f(x_1,...,x_n/\theta_1)}{f(x_1,...,x_n/\theta_0)} < \frac{1-b}{b}\frac{p(\theta_0)}{p(\theta_1)}$$

and, in consequence, the sampling scheme is equivalent to the scheme used in the Wald Sequential Probability Ratio Test for testing a simple Hypothesis H_0: $\theta=\theta_0$, against a simple alternative H_1: $\theta=\theta_1$ with stopping bounds

$$A = \frac{b}{1-b}\frac{p(\theta_0)}{p(\theta_1)} \quad \text{and} \quad B = \frac{1-b}{b}\frac{p(\theta_0)}{p(\theta_1)}$$

and stength (α_1, α_2), where

$$\alpha_1 = P_{H_0}(\text{Rejecting } H_0) = \sum_{n=1}^{\infty} \int_{C_n} f(x_1,...,x_n/\theta_0)\, dx_1,...,dx_n ,$$

$$\alpha_2 = P_{H_1}(\text{Accepting } H_0) = \sum_{n=1}^{\infty} \int_{D_n} f(x_1,...,x_n/\theta_1)\, dx_1,...,dx_n$$

and, if we note by N the random variable that represents the number of observed values before stopping sampling, C_n and D_n, are given by:

$$C_n = \left\{(x_1,..,x_n,..)/\ N = n \quad \text{and} \quad \frac{f(x_1,..,x_n/\theta_1)}{f(x_1,..,x_n/\theta_0)} \geq \frac{1-b}{b}\frac{p(\theta_0)}{p(\theta_1)}\right\},$$

$$D_n = \left\{(x_1,..,x_n,..)/\ N = n \quad \text{and} \quad \frac{f(x_1,..,x_n/\theta_1)}{f(x_1,..,x_n/\theta_0)} \leq \frac{b}{1-b}\frac{p(\theta_0)}{p(\theta_1)}\right\}.$$

For $\phi(x) = (x^{\beta} - x) / (2^{1-\beta} - 1)$ and $\beta = 2$, b satisfies $(-2)\,(b^2 + (1-b)^2 - 1) = c$, and consequentely the stopping bounds are given by

$$A = \frac{(1 - \sqrt{1-c})\, p(\theta_0)}{(1 + \sqrt{1-c})\, p(\theta_1)} \qquad B = \frac{(1 + \sqrt{1-c})\, p(\theta_0)}{(1 - \sqrt{1-c})\, p(\theta_1)}.$$

3. CONCLUSION

In this paper, we suggest a general sequential observation scheme based on the ϕ-Entropy. This measure generalizes the studied measures by Shannon (1948), Vajda (1968), Chen (1976), Latter (1973), Kapur (1972, 1988), Sant'anna and Taneja (1985),

Havrda and Charvat (1967), Ferreri (1980), Sharma and Taneja (1975) and De Luca and Termini (1972). This rule is analyzed when the parametric space has two elements and, its relation with the Sequential Probability Ratio Test is stablished.

REFERENCES

Bernardo, J.M. (1.977): "Binomial sampling schemes to obtain a prescribed amount of information". *Pres. to 10th European Meeting of Statisticians. Leuven. Belgium.*

Burbea, J. (1.984): "The Bose-Einstein Entropy of degree α and its Jensen difference". *Utilitas Mathematica, Vol. 25, 225-240.*

Burbea, J. and Rao, R. (1.982a): "Entropy differential Metric, Distance and Divergence Measures in Probability Spaces:A Unified Approach". *J.Mul.Anal. 12, 575-596.*

Burbea, J. and Rao R. (1.982b): "On the convexity of some Divergence Measures based on Entropy functions". *IEEE Trans. on Infor. Theory. Vol. IT-28, No 3, 489-495.*

Chen, C.J. (1976): "On Information and Distance Measures, Error Bounds and Feature Selection". *Information Sciences 10, 159-171.*

De Luca, A. and Termini, S. (1972): "A definition of Nonprobabilistic Entropy in the Setting of Fuzzy Sets Theory". *Information and Control 20, 301-312.*

El-Sayyad, G.N. (1.969): "Information and sampling from the exponential distribution". *Tecnometrics, 11, 40-42.*

Ferreri, C. (1980): "Hypoentropy and related heterogeneity, Divergence and Information Measures". *Statistica, anno XL, n. 2, 155-167.*

Havrda, J. and Charvat, F. (1967): "Quantification Method of Classification Processes: Concept of Styructural a-Entropy". *Kybernetika 3, 30-35.*

Kapur, J.N. (1972): "Measures of Uncertainty,Mathematical Programming and Phisics". *J. Ind. Soc. Agr. Res. Statist. 24, 47-66.*

Kapur, J.N. (1988): "Some new Nonaditive Measures of entropy". *Bull. U.M.I.(7) 2-13, 253-266.*

Latter, B.D.H. (1973): "Measures of Genetic Distance between Individuals and Populations". *Genetic Structure of Populations, 27-39. Univ. Hawai, Honolulu.*

Lindley, D.V. (1.956): "On a measure of information provided by an experiment". *Ann. Math. Statis., 27,986-1005.*

Lindley, D.V. (1.957): "Binomial sampling schemes and concept of information". *Biométrica, 44, 179-186.*

Morales,D., Taneja,I.J. and Pardo,L. (1991): "Comparison of experiments based on ϕ-measures of Jensen difference". *Communicated.*

Pardo, L. (1.986): "The measure of f-Divergence as a stopping rule in the sequential random sampling in a bayesian context". *Statistica, 2, 243-251.*

Rao, C.R. (1.982): " Diversity and Dissimilarity Coefficients: A Unified Approach". *J. Theoret. Pop. Biology, 21, 24-43.*

Sant'anna, A.P. and Taneja, I.J. (1985):"Trigonometric Entropies, Jensen Difference Divergence Measures and Error Bounds". *Information Sciences 35, 145-156.*

Shannon, C.E. (1948): "A Mathematical Theory of Communications". *Bell. Syst. Tech. J. 27, 379-423.*

Sharma, B.D. and Taneja, I.J. (1975): "Entropy of type (α,β) and other Generalized Measures in Information Theory". *Metrika 22, 205-215.*

Sibson, R. (1.969): "Information Radius". *Z. Wahrs. und verw. Geb. 14, 149-160.*

Vajda, I. (1968): "Bounds on the Minimal Error Probability and checking a finite or countable number of Hypothesis". *Inform. Trans. Problems 4, 9-17.*

Vicente, M. L., (1.990): "Aplicaciones Estadísticas de la Entropía no Aditiva de Grado β de Havrda y Charvat". *Tesis Doctoral. Universidad Complutense de Madrid.*

Uncertainty in Intelligent Systems
B. Bouchon-Meunier et al. (Editors)

INFORMATION MEASURES ASSOCIATED TO K-DIVERGENCES*

M. Salicrú[a], M.L. Menéndez[b], D. Morales[c] and L. Pardo[c]

[a]Departamento de Estadística, Universidad de Barcelona, Av. Diagonal 645, 08028 Barcelona, Spain.

[b]Departamento de Matemática Aplicada, E.T.S. de Arquitectura, Universidad Politécnica de Madrid, 28040 Madrid, Spain.

[c]Departamento de Estadística e I.O., Facultad de Matemáticas, Universidad Complutense de Madrid, 28040 Madrid, Spain.

ABSTRACT

In this work, the K-divergence defined by Burbea and Rao (1982b) is applied to construct the parametric measure resulting from parameter pertur-bance in all its coordinates. Based on this parametric measure, a statistic is defined replacing the parameter by its maximum likelihood estimator. Asymp-totic distributions for this statistic are obtained. These results can also be applied to test statistical hipotheses.

1. INTRODUCTION

Several functionals have been proposed in the literature as measures of information. Ferentinos and Papaioannou (1981) gave a convenient way to differentiate among the various measures of information. They proposed to classify them in the following three categories: Parametric, non-parametric and entropy measures of information.

Measures of entropy give the amount of uncertainty concerning the outcome of an experiment. The classical measures of this type are: Shannon's, Renyi's, Havrda and Charvat's, Arimoto's, Sharma and Mittal's. In this context, Burbea and Rao (1982a) introduced the ϕ-entropies and Salicrú, Menéndez, Morales and Pardo (1992) the (h,ϕ)-entropies.

The nonparametric measures of information express the amount of information supplied by the data for discriminating in favour of a probability densi-ty function (p.d.f.) f_1 against another f_2 or measure the distance or affinity between f_1 and f_2. Consider a family $\{P_\theta, \theta\in\Theta\}$ of probability measures on a measurable space $(\mathcal{X},\beta_{\mathcal{X}})$ dominated by a σ-finite measure μ. The parameter space Θ can either be an open subset of the real line or an open subset of an M-dimensional Euclidean space $\mathbb{R}^M$.

For any two p.d.f., f_1 and f_2 the K-Divergence between them, Burbea and Rao (1982b), is defined by

* The research in this paper was supported in part by DGICYT Grants N. PS91-0387 and PB91-0155. Their financial support is gratefully acknowledged.

$$K_\phi(f_1,f_2) = \int_\Gamma (f_1-f_2)\left(\frac{\phi(f_1)}{f_1} - \frac{\phi(f_2)}{f_2}\right) d\mu,$$

where ϕ is a real-valued C^3 function and $\Gamma=\{x\in\mathcal{X}/\ f_1(x)>0,\ f_2(x)>0\}$. When $\phi(x)=x\log x$, we obtain the Jeffreys-Kullback-Leibler's divergence.

Parametric measures of information measure the amount of information supplied by the data about an unknow parameter θ and are functions of θ. In this case the best known measure is Fisher's measure of information.

A method of constructing parametric measures of information from nonpara-metric measures, when $\theta\in\Theta\subset\mathbb{R}$, is given by

$$I(\theta) = \liminf_{t\to 0} \frac{1}{t^2}\, I\left[f(x,\theta),f(x,\theta+t)\right],$$

where $I\left[f(x,\theta),f(x,\theta+t)\right]$ is any nonparametric mesure of information for the densities $f(x,\theta)$ and $f(x,\theta+t)$. This method has been sucessfully employed by many authors such as Kagan (1963), Papaioannou and Kempthorne (1971), Vajda (1973), Aggarwal (1974), Boekee (1979), Ferentinos and Papaionnou (1981), Taneja (1987), Salicrú (1990) and Pardo et al. (1992).

If θ is M-variate, then it is also possible to obtain parametric measures from non-parametric measures of information. In this paper we analyze two methods on the basis of the K-divergence. First we consider a generalization of the standard method for $\theta\in\mathbb{R}$; that is to say ,we consider the measure obtai-ned after perturbing the parameter in all its coordinates. Then we consider the measure of information obtained after perturbing the parameter in the directions to the coordinate axes.

2. PARAMETRIC MEASURES OF INFORMATION ASSOCIATED TO K-DIVERGENCES

Now we shall suppose that the family of the p.d.f. satisfies the following regula-rity properties:

(i) The set $A=\{x\in\mathcal{X}/\ f(x,\theta)>0\}$ does not depend on θ and for all $x\in A$, $\theta\in\Theta$

$$\frac{\partial f_\theta(x)}{\partial\theta_i},\quad \frac{\partial^2 f_\theta(x)}{\partial\theta_i\partial\theta_j} \quad\text{and}\quad \frac{\partial^3 f_\theta(x)}{\partial\theta_i\partial\theta_j\partial\theta_l}\ ;\ i,j,l=1,\dots,M.$$

exist and are finite.

(ii) There exist real valued functions $F(x)$ and $H(x)$ such that

$$\left|\frac{\partial f_\theta(x)}{\partial\theta_i}\right|<F(x),\quad \left|\frac{\partial^2 f_\theta(x)}{\partial\theta_i\partial\theta_j}\right|<F(x),\quad \left|\frac{\partial^3 f_\theta(x)}{\partial\theta_i\partial\theta_j\partial\theta_l}\right|<H(x),$$

where F is finitely integrable and $E[H(X)]<B$, with $B>0$ independent of θ.

(iii) $\left[E\left\{\frac{\partial \log f_\theta(X)}{\partial\theta_i}\,\frac{\partial \log f_\theta(X)}{\partial\theta_j}\right\}\right]_{i,j=1,\dots,M}$ is finite and positive definite.

The measure resulting from parameter perturbance in all its coordinates is defined by

$$I_{\phi}(X,\theta) = \liminf_{t\to 0} \frac{K_{\phi}\left[f(x,\theta),\ f(x,(\theta_1+t,\dots,\theta_M+t))\right]}{t^2}.$$

In the following theorem we obtain an expression of $I_{\phi}(X,\theta)$.

Theorem 1

$$I_{\phi}(X,\theta) = \sum_{i,j=1}^{M} \int_{\mathcal{X}} \psi(f(x,\theta)) \frac{\partial f(x,\theta)}{\partial \theta_i} \frac{\partial f(x,\theta)}{\partial \theta_j}\, d\mu(x),$$

where $\psi(x) = \left[\frac{\phi(x)}{x}\right]'$.

Proof

Let us define the following function

$$g(t) = K_{\phi}\left[f(x,\theta),\ f(x,(\theta_1+t,\dots,\theta_M+t))\right].$$

It can be easily checked that $g(0)=g'(0)=0$ and

$$g''(0) = 2 \sum_{i,j=1}^{M} \int_{\mathcal{X}} \psi(f(x,\theta)) \frac{\partial f(x,\theta)}{\partial \theta_i} \frac{\partial f(x,\theta)}{\partial \theta_j}\, d\mu(x).$$

Thus, applying L'Hopital's rule, the result follows.

Remark 1

(i) If we consider the vectors $e_1=(1,0,\dots,0),\dots,e_M=(0,\dots,1)$ and the probability function $f(x,\theta)$, the measure resulting from parameter perturbance in the directions to the coordinate axes is defined by

$$I_{\phi}^{*}(X,\theta) = \liminf_{t\to 0} \frac{K_{\phi}\left[\ f(x,\theta),\ \frac{1}{M}\sum_{i=1}^{M} f(x,\theta+te_i)\ \right]}{t^2}.$$

It is possible to establish that $M^2 I_{\phi}(X,\theta) = I_{\phi}^{*}(X,\theta)$.

(ii) If we consider $\phi(x)=x\log x+cx$, the amount of information $I_{\phi}(X,\theta)$, veri-fies, $I_{\phi}(X,\theta) = \mathrm{Var}(Y_1+\dots+Y_M)$, with $Y_j = \frac{\partial}{\partial\theta_j}\log f(X,\theta)$. If also $\theta\in\mathbb{R}$, then $I_{\phi}(X,\theta)$ coincides with Fisher's information, $I^F(\theta)$.

(iii) Theorem 1 can be applied to multinomial populations in the following way. Consider the parameter $\theta = P = (p_1,..,p_{M-1})$, where $p_i \geq 0$ (i=1,..,M) and $\sum_{i=1}^{M} p_i=1$. Note that (M-1) is now the parameter dimension and not M. Let X be a random variable taking on the values x_i (i=1,..,M) with probabilities p_i (i=1,..,M); i. e., $f_\theta(x_i) = p_i$. Where we have supposed that μ is a counting measure giving mass one to each of the values x_i of X. In this case we obtain

$$I_\phi(X,\theta) = (M-1)^2 \psi(p_M) + \sum_{s=1}^{M-1} \psi(p_s).$$

3. COMPARISON BETWEEN PARAMETRIC MEASURES ASSOCIATED TO K-DIVERGENCES

In this section we obtain the asymptotic distribution of $I_\phi(X,\hat{\theta})$, where $\hat{\theta}=(\hat{\theta}_1,...,\hat{\theta}_M)$ is the maximum likelihood estimator of θ based on a random sam-ple of size n. This result can be used in various settings to construct confi-dence intervals and to test statistical hipotheses.

Theorem 2

Let $\hat{\theta}$ be the maximum likelihood estimator of θ. If we suppose the usual regularity properties, then

$$n^{1/2}\left[I_\phi(X,\hat{\theta}) - I_\phi(X,\theta) \right] \xrightarrow{L} N(0, T^t I^F(X,\theta)^{-1} T),$$

where $I^F(X,\theta)$ is the Fisher information matrix, $T=(t_1,..,t_M)$, and

$$t_s = \sum_{i,j=1}^{M} \int_{\mathcal{X}} \psi'(f(x,\theta)) \frac{\partial f(x,\theta)}{\partial \theta_s} \frac{\partial f(x,\theta)}{\partial \theta_j} \frac{\partial f(x,\theta)}{\partial \theta_i} d\mu(x) +$$

$$+ 2 \sum_{i,j=1}^{M} \int_{\mathcal{X}} \psi(f(x,\theta)) \frac{\partial^2 f(x,\theta)}{\partial \theta_i \partial \theta_s} \frac{\partial f(x,\theta)}{\partial \theta_j} d\mu(x).$$

Proof

A Taylor's expansion of $I_\phi(X,\hat{\theta})$ around the point θ is given by

$$I_\phi(X,\hat{\theta}) = I_\phi(X,\theta) + \sum_{s=1}^{M} t_s (\hat{\theta}_s - \theta_s) + R_n,$$

where

$$\frac{\partial I_\phi(X,\theta)}{\partial \theta_s} = t_s = \sum_{i,j=1}^{M} \int_{\mathcal{X}} \psi'(f(x,\theta)) \frac{\partial f(x,\theta)}{\partial \theta_s} \frac{\partial f(x,\theta)}{\partial \theta_j} \frac{\partial f(x,\theta)}{\partial \theta_i} d\mu(x) +$$

$$+ 2 \sum_{i,j=1}^{M} \int_{\mathcal{X}} \psi\,(f(x,\theta)) \; \frac{\partial^2 f(x,\theta)}{\partial\theta_i \partial\theta_s} \frac{\partial f(x,\theta)}{\partial\theta_j} \, d\mu(x) \, .$$

As $n^{1/2} R_n \xrightarrow[n\to\infty]{P} 0$ when $\hat{\theta} \longrightarrow \theta$, then the random variables

$$n^{1/2}\left[I_\phi(X,\hat{\theta}) - I_\phi(X,\theta)\right] \qquad \text{and} \qquad n^{1/2}\left[\sum_{s=1}^{M} t_s(\hat{\theta}_s - \theta_s)\right]$$

converge in law to the same distribution.

Let us define $T^t=(t_1,\ldots,t_M)$. As $n^{1/2}\left[\hat{\theta}_1-\theta_1,\ldots,\hat{\theta}_M-\theta_M\right] \xrightarrow[n\to\infty]{L} N(0,\Sigma)$, where $\Sigma = I^F(\theta,X)^{-1}$, then we get $n^{1/2}\left[I_\phi(X,\hat{\theta}) - I_\phi(X,\theta)\right] \xrightarrow[n\to\infty]{L} N(0,T^t\Sigma T)$, provided $T^t\Sigma T>0$.

Remark 2

(i) If we consider $\phi(x) = x\log x + cx$ and $\theta\in\mathbb{R}$, we obtain

$$n^{1/2}\left[I^F(\hat{\theta}) - I^F(\theta)\right] \xrightarrow[n\to\infty]{L} N(0, (I^F(\theta))'(I^F(\theta))^{-1}).$$

(ii) According to remark 1 (iii), theorem 2 can be applied to multinomial populations. It is easy to check that

$$t_s = \psi'(p_s) - (M-1)^2\psi'(p_M).$$

Theorem 2 cannot be applied when $T^tI^F(X,\theta)^{-1}T = 0$. For this last case, we have obtained the following result:

Theorem 3

If $T^tI^F(X,\theta)^{-1}T = 0$, then

$$2n\left[I_\phi(X,\hat{\theta}) - I_\phi(X,\theta)\right] \xrightarrow[n\to\infty]{L} \sum_{i=1}^{M} \beta_i \chi_1^2,$$

where the β_i's are the eigenvalues of the matrix $C\Sigma$, being $\Sigma = I^F(X,\theta)^{-1}$ and

$$C = \left[\frac{\partial^2 I_\phi(X,\theta)}{\partial\theta_r \, \partial\theta_s}\right]_{r,s=1,\ldots,M}.$$

Proof

If $T^tI^F(X,\theta)^{-1}T = 0$, then we must use Taylor's expansion of $I_\phi(X,\hat{\theta})$ around the point θ including the term of order 2, since the first partial derivatives term is equal to zero. In this case we have:

$$I_{\phi}(X,\hat{\theta}) - I_{\phi}(X,\theta) = \frac{1}{2}\sum_{r=1}^{M}\sum_{s=1}^{M}\frac{\partial^2 I_{\phi}(X,\theta)}{\partial\theta_r\,\partial\theta_s}(\hat{\theta}_r - \theta_r)(\hat{\theta}_s - \theta_s) + R_n,$$

with $n\,R_n \xrightarrow[n\to\infty]{P} 0$ when $\hat{\theta} \longrightarrow \theta$.

It is immediate to obtain that $c_{rs} = \dfrac{\partial^2 I_{\phi}(X,\theta)}{\partial\theta_r\,\partial\theta_s} =$

$$= \sum_{i,j=1}^{M}\int_{\mathcal{X}}\psi''(f(x,\theta))\,\frac{\partial f(x,\theta)}{\partial\theta_r}\frac{\partial f(x,\theta)}{\partial\theta_s}\frac{\partial f(x,\theta)}{\partial\theta_i}\frac{\partial f(x,\theta)}{\partial\theta_j}d\mu(x) +$$

$$+ \sum_{i,j=1}^{M}\int_{\mathcal{X}}\psi'(f(x,\theta))\,\frac{\partial^2 f(x,\theta)}{\partial\theta_r\,\partial\theta_s}\frac{\partial f(x,\theta)}{\partial\theta_i}\frac{\partial f(x,\theta)}{\partial\theta_j}\,d\mu(x) +$$

$$+ 2\sum_{i,j=1}^{M}\int_{\mathcal{X}}\psi'(f(x,\theta))\,\frac{\partial f(x,\theta)}{\partial\theta_s}\frac{\partial^2 f(x,\theta)}{\partial\theta_r\,\partial\theta_i}\frac{\partial f(x,\theta)}{\partial\theta_j}\,d\mu(x) +$$

$$+ 2\sum_{i,j=1}^{M}\int_{\mathcal{X}}\psi'(f(x,\theta))\,\frac{\partial f(x,\theta)}{\partial\theta_r}\frac{\partial^2 f(x,\theta)}{\partial\theta_s\,\partial\theta_i}\frac{\partial f(x,\theta)}{\partial\theta_j}\,d\mu(x) +$$

$$+ 2\sum_{i,j=1}^{M}\int_{\mathcal{X}}\psi(f(x,\theta))\,\frac{\partial^3 f(x,\theta)}{\partial\theta_i\partial\theta_s\partial\theta_r}\frac{\partial f(x,\theta)}{\partial\theta_j}\,d\mu(x) +$$

$$+ 2\sum_{i,j=1}^{M}\int_{\mathcal{X}}\psi(f(x,\theta))\,\frac{\partial^2 f(x,\theta)}{\partial\theta_s\,\partial\theta_i}\frac{\partial^2 f(x,\theta)}{\partial\theta_r\,\partial\theta_j}\,d\mu(x)\,.$$

On the other hand, as $W = n^{1/2}(\hat{\theta}-\theta) \xrightarrow[n\to\infty]{L} N(0,\Sigma)$, with $\Sigma = I^F(X,\theta)^{-1}$, then $WCW' \xrightarrow[n\to\infty]{L} \sum_{i=1}^{M}\beta_i\chi_1^2$, where the β_i's are the eigenvalues of the matrix $C\Sigma$, being $\Sigma = I^F(X,\theta)^{-1}$ and $C = (c_{rs})_{r,s=1,\ldots K}$.

Therefore, the result follows.

Remark 3

(i) If we use theorem 3, then we will have to calculate the following probabilities $P\left[\sum_{i=1}^{M}\beta_i\chi_1^2>t\right]$. These probabilities can be computed using the methods given by Kotz et al. (1967). Rao and Scott (1981) suggest to consider the approximate distribution of $\sum\beta_i\chi_1^2$; which is given by $\beta\chi_M^2$, where $\beta = \frac{1}{M}\sum_{i=1}^{M}\beta_i$. In this case we can easily calculate the value of β, since $\sum_{i=1}^{M}\beta_i = tr(C\Sigma)$.

(ii) With the statistics found in theorems 2 and 3, it is possible to test, among others, the following hipotheses:

(a) $I_\phi(X,\theta) = I_0$.

(b) $I_\phi(X,\theta_1) = ... = I_\phi(X,\theta_m) = I_0$.

(c) $\theta = \theta_0$.

(d) $(p_1,...,p_M) = (\frac{1}{M},...,\frac{1}{M})$, in multinomial populations.

(e) $\theta_1 = ... = \theta_m = \theta_0$.

(iii) According to remark 1 (iii), theorem 2 can be applied to multinomial populations. It is easy to check that

$c_{rs} = \psi''(p_s) + (M-1)^2\psi''(p_M)$ if $r=s$
and
$c_{rs} = (M-1)^2\psi''(p_M)$ if $r\neq s$.

REFERENCES

Aggarwal, J. (1974): " Sur l'information de Fisher", In: J. Kampe de Feriet, Ed., *Theories de l'Information,* Springer, Berlin, 111-117.

Boekee, D. (1979): "The D_f-information of order s". In: *Trans 8th Prague Conf. on Information Theory, Statistical Decision Functions and Random Processes*, Prague, 55-66.

Burbea, J. and Rao C.R. (1982a): "On the convexity of some divergence measures based on entropy functions", IEEE *Trans. on Information Theory,* IT-28, 489-495.

Burbea J. and Rao C.R. (1982b): "Entropy differential metric, distance and divergence measures in probability spaces: a unified approach", *J. Multi. Analy.*, 12, 575-596.

Ferentinos, K. and Papaioannou, P.C. (1981): "New parametric measures of information", *Inform. and Control* 51 , 193-208.

Kagan, M. (1963): " On the theory of Fisher's amount of information", *Soviet Math. Dokl.*, 4, 991-993.

Kullback, S. and A. Leibler (1951): "On the information and sufficiency" .*Ann. Math. Statist.*,22,79-86.

Papaioannou, P.C. and Kempthorne O. (1971): "On statistical information theory and related measures of information". Aerospace theory and related measures of Information". *Aerospace Research Laborat. Report ARL* 71-0059. Wright-Patterson A.F.B. Ohio.

Pardo, L., Morales, D. and Taneja I.J. (1992): "Generalized Jensen difference measures and Fisher measure of information". To appear en Kybernetes.

Salicrú, M. (1990): "Measures of information associated with Csiszar's Divergences. Communicated.

Salicrú, M., Menéndez, M.L., Morales, D. and Pardo, L. (1992): "Asymptotic distribution of (h,ϕ)-entropies". To appear in Communictions in Statis-tics.

Shannon, C.E. (1948): "A mathematical theory of communication", *Bell System Tech. J.*,27, 379-423.

Taneja, I.J. (1987): "Statistical aspects of divergence mesures", J. Statist. Planning & Inference, 16, 136-145.

Vadja, I.(1973): "χ^2-divergence and generalized Fisher's information". In: *Trans. 6th Prague Conf. on Inform. Theory, Statistical Decision Functions and Random processes*, Prague, 873-886.

Uncertainty in Intelligent Systems
B. Bouchon-Meunier et al. (Editors)

Local non-probabilistic Information and Questionaries

Carlo Bertoluzza [a] and Carla Poggi[b]

[a]Dipartimento di Informatica e Sistemistica, Universitá, Via Abbiategrasso 209, 27100 PAVIA - ITALY

[b]Dipartimento di Matematica, Università di Pavia, Corso Strada Nuova 65, 27100 PAVIA - ITALY

Summary. We deal in this paper with the questionaries as defined by C.F.Picard [1]. Some modifications have been made in the definition in order to allow us to utilize the axiomatic measures of information and uncertainty instead of the probabilistic ones. In this work we generalize some results due to Picard, Schneider and Bertoluzza-Ronco, concerning the variation of information during an inquiring process.
AMS (MOS) Classification : 94A15 62B10

1. DEFINITIONS AND BASIC NOTATIONS

1. An *information space* [2] is a structure $I = (\Omega, S, \mathcal{E}, \mathcal{K}, H)$, where
Ω is the set of the elementary events,
S is the field of the observable events,
$\mathcal{E}$ is the family of the experiments (partitions of elements of S),
$\mathcal{K}$ is related to the independency notion (without interest here),
$H : \mathcal{E} \to \mathbb{R}^+$ is a function (*uncertainty measure* or *entropy*) such that

$H(\{\Omega\}) = 0 \quad , \quad H(\emptyset) = +\infty$
$\Pi_A \prec \Pi_B \Longrightarrow H(\Pi_A) \leq H(\Pi_B)$
Axiom of independency (without interest here)

where Π_X indicates a partition of the event X, and $\prec$ is the refinement relation.

Among all the possible uncertainties, we consider in this work only those which satisfy the local property, which we give here both in the classical form (1), and in the generalized one (2).

$$H(\Pi_A \vee \Pi_B) = H(\Pi_A \vee \{B\}) + H(\{A\} \vee \Pi_B) - H(\{A\} \vee \{B\}) \tag{1}$$

$$H(\Pi_A \vee \Pi_B) = \Phi\Big[H(\Pi_A \vee \{B\}), H(\{A\} \vee \Pi_B), H(\{A\} \vee \{B\})\Big] \tag{2}$$

where "$\vee$" represents the union of partitions (see e.g.[2]).

2. A *deterministic questionary* [3] is a structure $\mathbf{Q} = (\Omega, G, \sigma)$, where:

Ω is a set representing the population to which we submit the questionary,
$G = (X, \Gamma)$ is a finite connected lattice with an unique root α, without circuits, and such that all the internal vertices have external degree greater than 1,
$\sigma : \Gamma \to \mathcal{P}(\Omega)$ is a map such that

$y, z \in \Gamma(x) \Longrightarrow \sigma(x,y) \cap \sigma(x,z) = \emptyset$,
$\cup_{y\in\Gamma(x)}\sigma(x,y) = \cup_{z\in\Gamma^{-1}(x)}\sigma(z,x)$ for all internal vertices .
$\cup_{y\in\Gamma(\alpha)}\sigma(\alpha,y) = \Omega$.

In the previous formulas $\Gamma(x) = \{y \in X | (x,y) \in \Gamma\}$ and $\Gamma^{-1}(x) = \{z \in X | (z,x) \in \Gamma\}$. Moreover we will use the following notations $E = \{x \in X | \Gamma(x) = \emptyset$, $F = X - E$, $A_y^x = \sigma(x,y)$, $A^x = \bigcup_{y\in\Gamma(x)} A_y^x \ \forall x \in F$, $A^e = \bigcup_{x\in\Gamma^{-1}(e)} A_e^x \ \forall e \in E$.

Let $\mathcal{S}$ be the field generated by $\sigma(\Gamma)$, and suppose that a probability measure P is defined on $(\Omega, \mathcal{S})$. Then we can define a map $p : \Gamma \to [0,1]$ by posing $p(x,y) = P[\sigma(x,y)] \ \forall (x,y) \in \Gamma$. The structure (G,p) is a questionary as defined originally by C.F.Picard in [1].

2. TRANSMISSION OF INFORMATION

There are many kinds of information in a questionary : the global information associated to the final results, the informations locally produced by the questions, and there are also lack of information whenever more than one edge enter in a vertex, because in this case a partial classification will be destroyed. The object of this paper regards the relations between these quantities.

The first result was established by C.F.Picard. He defined the global information (H_T), and the informations produced, $h(x)$, and lost, $a(x)$, in the vertices of the questionary, using the Shannon entropy. More precisely

$$H_T = -\textstyle\sum_{e\in E} p(e) \log p(e)$$
$$h(x) = -\textstyle\sum_{y\in\Gamma(x)} p(y|x) \log p(y|x) \quad \forall x \in F \ , \quad h(e) = 0 \ \ \forall e \in E$$
$$a(x) = -\textstyle\sum_{z\in\Gamma^{-1}(x)} p(z|x) \log p(z|x) \quad \forall x \in X \backslash \{\alpha\} \ , \quad h(\alpha) = 0.$$

He proved also that

$$H_T = \textstyle\sum_{x\in X} p(x) \cdot h(x) - \sum_{x\in X} p(x) \cdot a(x) = H_P - H_A \ , \qquad (3)$$

where the terms H_P and H_A are the informations globally produced and lost by the questionary.

Now the question is : what happens when we have to use informations which are not derived from the Shannon entropy?

A first attempt to answer this question was proposed by M.Schneider [4]. He assumed that a classical additive measure μ was defined on $(\Omega, \mathcal{S})$, so that it was possible to associate to each edge $(x,y) \in \Gamma$ and to each vertex $x \in X$ a measure ν by posing $\nu(x,y) = \mu(A_y^x)$ and $\nu(x) = \mu(A^x)$. Then he defined the quantities $H_T, h(x), a(x), H_P, H_A$, by means of the following relations

$$H_T = \sum_{e_i \in E} G[\nu(\alpha), \nu(e_i), \sum_{r=i+1}^{n} \nu(e_r)] ,$$
$$h(x) = \sum_{y_i \in \Gamma(x)} G[\nu(\alpha), \nu(x,y_i), \sum_{r=i+1}^{b_x} \nu(x,y_r)] ,$$
$$a(x) = \sum_{z_j \in \Gamma^{-1}(x)} G[\nu(\alpha), \nu(z_j,x), \sum_{s=j+1}^{c_x} \nu(z_s,x)] ,$$
$$H_P = \sum_{x \in X} h(x) \quad , \quad H_A = \sum_{x \in X} a(x) ,$$

and he proved that

$$H_T = H_P - H_A . \tag{4}$$

In the previous relations b_x and c_x are respectively the external and the internal degree of x, and obviously $a(\alpha) = 0$, $h(e) = 0 \ \forall e \in E$.

We will remark that the given definitions are coherent with the notion of branching entropy as introduced by B.Forte and N.Pintacuda.

In [3] we gave another generalization of the result of Picard, which generalizes also the Schneider case. It regards cases where the informations associated to the questionary are expressed in terms of axiomatic uncertainty measures as defined by Forte and Kampé de Fériet. The first problem we encountered has been to define the global and local (produced and lost) informations. In order to solve this problem, we associate to the internal vertices $X - E - \{\alpha\}$ of the graph G the three following partitions of the set Ω :

$$\Pi_x^1 = \{A_x^z, \Omega \backslash A^x | z \in \Gamma^{-1}(x)\} , \tag{5}$$
$$\Pi_x = \{A_x, \Omega \backslash A_x\} , \tag{6}$$
$$\Pi_x^2 = \{A_y^x, \Omega \backslash A_x | y \in \Gamma(x)\} , \tag{7}$$

whereas we can associete to the root α and to each of the terminal vertices $e \in E$ only the partitions Π_α, Π_α^2 and Π_e^1, Π_e. To the set of the terminal vertices of G corresponds the partition of Ω defined by

$$\Pi = \{A^e | e \in E\} . \tag{8}$$

We may observe that at each point $x \in X$ we lose some informations by destroying the classification Π_x^1 (passage from Π_x^1 to Π_x), and we produce information, by realyzing a new classification in the transition from Π_x to Π_x^2. So that the natural way to define $a(x)$ and $h(x)$ is:

$$a(x) = H(\Pi_x^1 | \Pi_x) \quad \forall x \notin E \ , \quad a(\alpha) = 0 \tag{9}$$
$$h(x) = H(\Pi_x^2 | \Pi_x) \quad \forall x \neq \alpha \ , \quad h(e) = 0, \ \forall e \in E \tag{10}$$

Then, posing

$$H_T = H(\Pi) \quad , \quad H_P = \sum_{x \in X} h(x) \quad , \quad H_A = \sum_{x \in X} a(x) ,$$

we proved the following result :

Theorem 1. The relation (4) $(H_T = H_P - H_A)$ holds if and only if the uncertainty measure H used to define these quantities, fulfils the classical local condition (1).

This result contains, as particular cases, the ones exposed above: nevertheless its main interest consists in the fact that this scheme allows us to attach the informations directly to the classifications determined by the questions (or by the whole questionary) without passing trough a (some time fictitious) probability or measure assignment.

3. THE GENERAL BALANCE OF INFORMATION

The last result of the previous paragraph establishes that a balance between the global information (H_T), the produced information (H_P) and the lost information (H_A) is strongly related to the branching property.

Nevertheless there exists a relevant family of uncertainty spaces which does not have this property: among these the entropies of order α (introduced by Rényi) and those of order α and type β which are the best known besides the Shannon one.

We will examine here whether it is possible to find a more general form of information balance, of the type

$$H_T = F(H_P, H_A) \quad \text{with} \quad F(x,0) = x \tag{11}$$

(with suitable choice of the function F and of the informations H_T, H_P, H_A), which could be used also in the cases where the classical relation (4) does not hold.

In order to answer to this question, it is necessary to give a suitable definition of the informations involved in the problem, that is of the quantities H_T , $h(x)$, $a(x)$, H_P , H_A .

It is evident that the global information associated to the whole questionary must be the uncertainty of the partition determined by the final vertices $e \in E$; then we pose

$$H_T = H(\{A^e | e \in E\}) \ . \tag{12}$$

In order to define $h(x)$ and $a(x)$ we suppose that they depend respectively on the partitions Π_x^2, Π_x and Π_x^1, Π_x, only through their uncertainties H. The condition that $a(x) = 0$ when in x enters only one edge, implies that $a(x)$ has to be the relative information of Π_x^1 given Π_x, and the same holds for $h(x)$. Thus, using the characterization theorem proved in [5], we have the following expressions :

$$a(x) = H(\Pi_x^1 | \Pi_x) = g^{-1}[g\{H(\Pi_x^1)\} - g\{H(\Pi_x)\}] \tag{13}$$

$$h(x) = H(\Pi_x^2 | \Pi_x) = g^{-1}[g\{H(\Pi_x^2)\} - g\{H(\Pi_x)\}] \ , \tag{14}$$

$$a(\alpha) = 0 \quad , \quad h(e) = 0 \ \ \forall e \in E$$

where g is a strictly increasing function with $g(0) = 0$, $g(+\infty) = +\infty$. The informations H_A and H_P depend on the local informations $a(x)$ and $h(x)$, that is $H_A = F_a[a(x)|x \in X]$, $H_P = F_p[h(x)|x \in X]$. The form of the functions F_a , F_p will be determined later.

In order to search the general form of the balance (11) we require that the function F reduces to the classical (4) in the cases where the hypotesis of theorem 1 hold. This forces us to consider only uncertainty spaces with the generalized local property (2), but this is really a weak limitation because it has been proved in [6] that this class contains all the uncertainties of interest. Unfortunately not all the permitted forms of the function Φ are compatible with the expressions (13)(14) of the relative information. In fact, by comparing the results established in [5] and [6], we can recognize that the function Φ which we can use, must have the form

$$\Phi(u,v,z) = g^{-1}[g(u) + g(v) - g(z)] \,. \tag{15}$$

Proposition 1. By posing $\xi_i = h(x_i)$, $i = 1 \dots n = |X|$ we have

$$H_P = F_p(\xi_1 \dots \xi_n) = g^{-1}\left[\sum_{i=1}^{n} g(\xi_i)\right] \,. \tag{16}$$

To prove this we begin by considering a questionary whose graph is a tree with two questions, α and x (see figure). Then we pose $\Pi_B = \{A^e | e \in \Gamma(x)\}$ and $\Pi_A = \{A^e | e \in E - \Gamma(x)\}$ so that $H_T = H(\Pi_A \vee \Pi_B)$. Clearly in this questionary $H_A = 0$. Then we can apply the condition $F(x, 0) = x$, and remembering that H has the local property in the form (15), we obtain

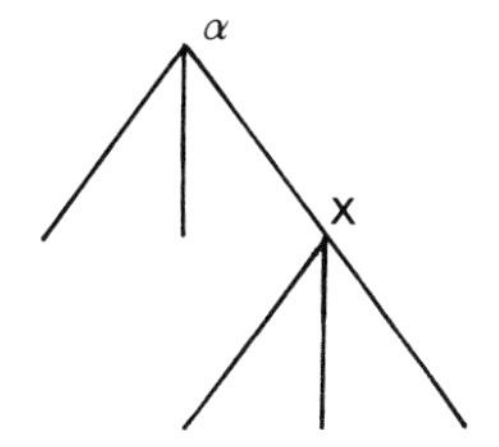

$$g^{-1}[g(a) + g(b) - g(c)] \;=\; F_p[h(\alpha), h(x)]$$

where $a = H(\Pi_A \vee \{B\})$, $b = H(\{A\} \vee \Pi_B$, $c = H(\{A\} \vee \{B\}$. This proves (16) in the case $n = 2$ because it is easy to recognize that $h(\alpha) = a$ and $h(x) = g^{-1}[g(b) - g(c)]$. We obtain the proof of (16) in the general case without difficulty by recursion over questionaries with trees as support.

Proposition 2. If the questionary **Q** has height $l = 2$, then $F_p(\xi_1 \dots \xi_n) = g^{-1}[\sum_{i=1}^{n} g(\xi_i)]$ and $F(\xi, \eta) = g^{-1}[g(\xi) - g(\eta)]$, so that

$$H_A = g^{-1}\left[\sum_{i=1}^{n} g\{a(x_i)\}\right] \quad , \quad H_T = g^{-1}[g(H_P) - g(H_A)] \,. \tag{17}$$

We will prove this result for the particular questionary showed in figure, where $a(\beta) \neq 0$ and $a(x) = 0 \; \forall x \neq \beta$, but the proof still holds, with few modifications, in the general case.

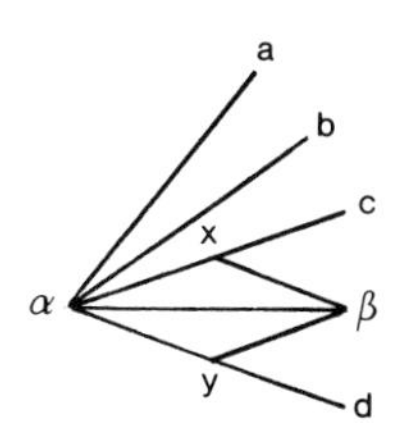

The partition $\Pi = \{A_a^\alpha, A_b^\alpha, A_c^x, A_\beta^x, A_\beta^\alpha, A_\beta^y, A_d^y\}$ is the union of the two disjoint partitions $\Pi_A = \{A_a^\alpha, A_b^\alpha, A_c^x, A_d^y\}$ and $\Pi_B = \{A_\beta^x, A_\beta^\alpha, A_\beta^y\}$, where $A = A^a \cup A^b \cup A^c \cup A^d$ and $B = A^\beta$. As H is a local uncertainty of type (15) we have

$$H(\Pi) = g^{-1}\{g[H(\{A^a, a^b, A^c, a^d, A^\beta\})] + g[H(\{A^x_\beta, A^\alpha_\beta, A^y_\beta, A\})] - \\ -g[H(\{A, A^\beta\})]\} . \tag{18}$$

On the other hand, it is easy to verify, by straightforward computations, that

$$H(\Pi) = g^{-1}g\{h(\alpha)\} + g\{h(x)\} + g\{h(y)\}] . \tag{19}$$

But $H(\{A^a, A^b, A^c, A^d, A^\beta\}) = H_T$, and $a(\beta) = g^{-1}\{g[H(\{A^x_\beta, A^\alpha_\beta, A^y_\beta, A\})]$- $g[H(\{A^\beta, A\})]\}$. Then, by comparing (18) and (19) we obtain (17).

Proposition 3. The relations (17) hold for all questionaries.

This result may be proved by recursion over the height of the questionaries. We suppose that it holds for the questionaries of height n, and we reason as in proposition 2 to demonstrate that (17) holds also for the questionaries of height $n+1$. In doing this we use the following well known result, holding for the local uncertainties of the type (15) :

$$H(\vee_{i=1}^n \Pi_i) = g^{-1}\left[g[H(\Pi_i \vee \{\cup_{j\neq i} A_j\})] - g[H(\{A_i \vee \{\cup_{j\neq i} A_j\})]\right] .$$

We have just proved that the local property (15) is a sufficient condition for the existence of a balance of information. It is easy to verify that if we search a balance of the form (17), then the local property is also a necessary condition. The proof is quite similar to those developed in [3] for the particular case (4).

4. FINAL REMARKs

Remark 1. A natural question arises : "is it possible to have a balance of information in a form which is different from (17)?". We give here some partial answers to this question.

1. If the uncertainty H is local, but the "locality law Φ" does not have the form (15), then a balance is in general impossible, except in some particular cases in which H assumes a bounded set of values. The characterization of this cases depends on the form of the law Φ and does not present any difficulty.

2. If the uncertainty measure is totally composable, then the balance exists, provided that the composition law has the form characterized in [7]. The form of the balance is quite similar to the one presented here, and may be obtained either directly or by a reduction, via a suitable transformation, to the local case. What happens when the composition law is not of this form, is still an open problem.

Remark 2. It is easy to recognize that the balance of information which we established in the previous paragraph holds for a large class of uncertainty measures. In particular we proved that it holds, besides of the Shannon's entropies, also for the following ones : Renyi, Varma, Hvarda-Charvat*, Belis-Guiasu*, Arimoto, Sharma-Mittal, Taneja*, Picard*, Ferreri* Picard-van der Pyl, Bertoluzza-Schneider et al. The entropies marked with *, have the balance in the classical form.

REFERENCES

1. C.F.Picard, Théorie des questionnaires, Gautier-Villars, Paris (1965)
2. B.Forte, Measure of information: the general axiomatic theory, R.I.R.O, 3, 2 (1969) 63-90
3. C.Bertoluzza and C.Ronco, Trasmissione dell'informazione in un questionario deterministico, Boll.U.M.I (5) 14-B (1977) 267-284
4. M.Schneider, Information généralisée et questionnaires, Thése de Docteur de Specialité, Lyon (1970)
5. C.Poggi, Informazione relativa in uno spazio con legge d'indipendenza qualsiasi, Stochastica, vol VI, n.1 (1982)
6. C.Poggi, Entropie G-diramative ed entropie G-locali in spazi d'informazione con legge d'indipendenza di tipo generale, Statistica, XLIII, n.2 (1983) 289-299
7. C.Bertoluzza and M.Schneider, Informations totalement composables, in Lecture Notes in Mathematics, n.398, Springer (1974) 90-97

Uncertainty in Intelligent Systems
B. Bouchon-Meunier et al. (Editors)

Questionaries and Decision Processes

Franco Barbaini[a] and Carlo Bertoluzza[b]

[a]Dipartimento di Matematica, Università di Pavia, Corso Strada Nuova 65, 27100 Pavia - Italy

[b]Dipartimento di Informatica e Sistemistica, Università, Via Abbiategrasso 209, 27100 Pavia - Italy

Summary. We present here some results regarding the possibility of optimizing questionaries with respect to some criteria based on generalized entropies. After recalling the notions of questionary and uncertainty, we present the problem in terms of standard questionary notations. Then we translate it in probabilistic terms in order to apply the results previously obtained in studying the optimal stopping times.

1. DEFINITIONS

1. An *uncertainty space* (see e.g.[1]) is a structure $(\Omega, \mathcal{S}, \mathcal{E}, H)$, where

$(\Omega, \mathcal{S})$ is a measurable space,
$\mathcal{E}$ is a family of partitions of elements of $\mathcal{S}$,
$H : \mathcal{E} \to \mathbb{R}^+$ is a function (*Entropy*) with suitables properties.

In this paper we consider only entropies which are of the type

$$H(A_1, \ldots A_n) = \sum_{i=1}^{n} v(A_i)\varphi[p(A_i)] - \varphi(1) \quad (1)$$

where p and v are respectively a probability measure and a finite measure on $(\Omega, \mathcal{S})$, and $\varphi : [0,1] \to \mathbb{R}^+$ is a strictly decreasing function. Later we will use mainly the relative entropy $H(\Pi_1|\Pi_2)$, defined as

$$H(\Pi_1|\Pi_2) = H(\Pi_1 \wedge \Pi_2) - H(\Pi_2). \quad (2)$$

2. A *questionary* (see e.g.[2]) is a structure (Ω, G, σ), where

Ω is the set of the individuals which answer to the questionary, $G = (X, \Gamma)$ is a finite d-graph with a root α, n terminal points and whitout circuits, whose vertices and edges represent questions and answers, $\sigma : \Gamma \to \mathcal{P}(\Omega)$ is a function which associates to each edge the elements of Ω which gave the corresponding answer. Properties of σ are given in order to ensure coherence.

2. THE PROBLEM

Let Ω be a set of elements ω. We want to construct a questionary Q such that the sets associated to its terminal vertices correspond, as better as possible, to a given partition $\Pi_O = \{\Omega_1 \ldots \Omega_n\}$ of Ω. To reach this purpose we have a finite family of questions $\mathcal{Q} = \{q_i\}$. Each question q_i determines a partition of Ω

$$\Pi^i = \{\Omega_1^i \ldots \Omega_{a(i)}^i\}.$$

The construction of the questionary $\mathbf{Q}$ proceeds by recursion. Firstly we construct the questionary $\mathbf{Q_1}$ composed by a single questiono $q_1 \in \mathcal{Q}$. Then for realize $\mathbf{Q_{i+1}}$ we choose a question $q_{i+1} \in \mathcal{Q}_i = \mathcal{Q} - \{q_1, \ldots q_i\}$ and we apply this question to one or more subsets $A^{(i)}$ associated to one or more of the terminal vertices of $\mathbf{Q}_i$. This means that in $\mathbf{Q}_{i+1}$ each $A^{(i)}$ to which we apply the question q_{i+1} would be subdivided in the subsets $\{A_j^{(i+1)} = A^{(i)} \cap \Omega_j^{(i)}\}$. If all the $q_i \in \mathcal{Q}$ are applied to all the final subsets of $\mathbf{Q}_i$, then the partition Π_M associated to the "complete" questionary $\mathbf{Q}_M$ is the product of the partitions $\Pi^{(i)}$, whereas on the contrary the partition Π_i associated to the terminal vertices is "contained" in Π_M. Clearly the questionary $\mathbf{Q}_M$ furnish the maximum of information, but at the same time it has the largest possible lenght.

The aim of this work is to describe an algorithm which realize a "good" compromise between two opposite exigences: maximum information and minimum lenght.

Suppose that $(\Omega, \mathcal{S}, \mathcal{E}, H)$ is an information space such that $H(\Pi)$ is defined for all the partitions associated to the vertices of the $\mathbf{Q}_i$, and establish to construct $\mathbf{Q}_{i+1}$ by applying the question q_{i+1} to a single termina subset $A^{(i)}$ of $\mathbf{Q}_i$.

The construction algorithm is realized by selecting at each step the question $q_i \in \mathcal{Q}_i$ in such a way that the quantity

$$H(\Pi_O|\Pi_i) - H(\Pi_O|\Pi_M) \tag{3}$$

attains its minimum value, and the construction will be stopped when

$$H(\Pi_O|\Pi_i) - H(\Pi_O|\Pi_M) < \epsilon. \tag{4}$$

This algorythm has a natural meaning, because $H(\Pi|\Pi^*)$ represents the uncertainty on the classification Π given the classification Π^* and then if Π_M is a refinement of Π_i it will be natural to request that $H(\Pi_O|\Pi_i) \geq H(\Pi_O|\Pi_M)$. If the uncertainty space fulfills this condition, then the algorythm we just described is satisfactory. On the contrary an alternative criterion may be represented by the following : choose the question such that would be maximized the quantity

$$H(\Pi_i|\Pi_{\mathcal{O}}) \tag{5}$$

and stop the construction process when

$$H(\Pi_M|\Pi_{\mathcal{O}}) - H(\Pi_i|\Pi_{\mathcal{O}}) < \epsilon \, . \tag{6}$$

It must be remarked that this procedure can be used to costruct a questionary in the classical (Picard) sense as well as to contruct a pseudo-questionary as defined by M.Terrenoire. In the first case we must minimize (3) [or maximize (5)] with respect to the choice both of the question q_i and of the subset to which q_i is applied, whereas in the second case the subset is completely determined and the optimization regards only the choice of the question. In this latter case all the entities involved in the construction process are random variables on the space $(\Omega, \mathcal{S})$. In particular the number of questions posed before stopping is a random variable and we can search, among the questionary which realize the condition (6) the one which corresponds to the minimum number of posed questions.

Alternatively we can fix the maximum number of allowed questions and search the questionary which minimize the difference on the left side of (4). In this case our task consists in determining a questionary $\mathbf{Q}$ such that

$$h(\mathbf{Q}_*) \leq \alpha \text{ and } H(\Pi_{\mathcal{O}}|\Pi_M) - H(\Pi_{\mathcal{O}}|\Pi_*) = \min[H(\Pi_{\mathcal{O}}|\Pi_M) - H(\Pi_{\mathcal{O}}|\widetilde{\Pi}), \tag{7}$$

where the minimum has to be searched between the questionaries with mean height (maximum number of posed questions) not greater than α.

3. INFORMATION PROCESSES

Let Y , $X_1 \ldots X_k$ be a family of finite random variables on a probability space $(\Omega, \mathcal{S}, P)$. We will indicate by $\mathcal{F}_i$, $\mathcal{O}$ and $\mathcal{F}(i[1]\ldots i[j])$ the fields generated respectively by the random variables X_i , Y and $(X_{i[1]} \ldots X_{i[j]})$. To each succession $X_{i[1]} \ldots X_{i[k]}$ we can associate the succession of fields $\mathcal{F}(i[1])$, $\mathcal{F}(i[1], i[2])$, $\mathcal{F}(i[1]\ldots i[k])$ and consider stopping times T over this sequence.

To each stopping time T' we can associate a conditional entropy $H(\mathcal{O}|\mathcal{F}_{T'})$ following a way similar to those developped in [3], [4]. As in [3] we will determine the strategy for finding the α-optimal stopping time T, that is a stopping time T such that

$$E(T) = \alpha \quad , \quad H_T(\mathcal{O}) = H(\mathcal{O}|\mathcal{F}_T) = \min_{\sigma, T'} H(\mathcal{O}|\mathcal{F}_{T'})$$

where σ are all the possible orders in which the random variables X_i can be rearranged. In order to introduce the information process associated to the questionary, we consider a sequence of σ-fields $\mathcal{F}_n \subset \mathcal{F}_{n+1}$, $n = 0, ..., \infty$ and for each n a triplet

$\mathcal{I}_n = (I_n, H_n, \mathcal{F}_n)$

where I_n and H_n are respectivly

$I_n(A) = \mu(A)\varphi[P(A)] \quad , \quad \mu(A) = v(A)/P(A) \qquad \forall A \in \mathcal{F}_n$ and

$P(A) \neq 0$, $\lim_{P(A)\to 0} \mu(A) < +\infty$

$H_n = \sum_{A\in\mathcal{F}_n} P(A) I_n(A)$

In [4] we proved that the sequence of random variables

$X_n = \sum_{A\in\mathcal{F}_n} \chi_A I_n(A)$

is a sub-martingale, and it satisfies the relation $E(X_n) = H_n$ and consequently

$H_n \leq H_{n+1} \qquad \forall n \in \mathbf{N}$

We also proved that, under suitable conditions, $\mathcal{I}_{\inf(T,n)}$ converges to

$\mathcal{I}_T \; = \; (I_T \; , \; H_T(\mathcal{O}) \; , \; \mathcal{F}_T)$

where I_T is defined by

$I_T(A) = \mu(A)\varphi[P(A)] \quad , \quad \mu(A) = v(A)/P(A), \qquad \forall A \in \mathcal{F}_T$

and $H_T = H_T(\mathcal{O})$. It is easy to recognize that *the questionary described in the paragraph 2 is equivalent to the information process* $\mathcal{I}_T$. In fact the questions q_i correspond to the random variables X_i and the paritions Π^i and Π_k have as corresponent respectively the fields $\mathcal{F}_i$ and $\mathcal{F}(i[1]\ldots i[k])$. Finally the questionaries $\mathbf{Q}_j$ correspond to the processes constructed by means of the random variables (defined by recursion)

$\widehat{X}_1 = X_1, \quad \widehat{X}_j \; = \; \sum_{(x_1\ldots x_i)} X_{j(\widehat{X}_1=x_1\ldots\widehat{X}_i=x_i)} I_{(\widehat{X}_1=x_1\ldots\widehat{X}_i=x_i)} \; , \;$ with $j(.) > i$,

if the variables used in the process, till to instant j are $X_1 \ldots X_i$. The stopping time is constructed recursively by posing

$T_0 = 0 \; , \; T_1 = T_0 + I_{A_1} \; , \ldots \; T_{n+1} = T_n + I_{A_n}$

where $A_n \in \mathcal{F}_{T_n}$ must be chosen so that

$H(\mathcal{O}|\mathcal{F}_{T_{n+1}}) \; = \; \min H(\mathcal{O}|\mathcal{F}_{T_n+I_A})$

where, in computing the minimum A varies in $\mathcal{F}_{T_n}$ and X_j in the subfamily of the variables which are not jet used till the time T_n. The procedure stops when

$$E(T_{k+1}) > \alpha \quad \text{and} \quad E(T_k) \leq \alpha \; .$$

Then, by posing $T = T_k$, we have

$$H(\mathcal{O}|\mathcal{F}_T) \;=\; \min_{t \in \mathcal{C}} H(\mathcal{O}|\mathcal{F}_t)$$

where $\mathcal{C}$ is the class of the stopping times with expectation less than or equal to α. Then we say that T is α–optimal.

When the construction of T presents too many difficulties, it is reasonable to consider a easier problem, that is to search stopping times which are sub-optimal in the sense that

$$E(T) \leq \alpha \quad \text{and} \quad H(\mathcal{O}|\mathcal{F}_T) \leq \min_{t \in \mathcal{C}} H(\mathcal{O}|\mathcal{F}_t) + \epsilon(\alpha) \; ,$$

where $\epsilon(\alpha)$ is a suitable positive real number depending on α.

It is easy to verify that the random variable T is really a stopping time associated with the sequence of fields which define the random process just described (this because the finiteness of the variables X_i; otherwise some cautions are needed). In fact T_0 is trivially a stopping time, and the same holds for T_1 because

$$(T_1 = 1) = \Omega \in \mathcal{F}_0 \subset \mathcal{F}_1 .$$

Moreover if T_{k-1} is a stopping time, then the set $(T_k = T_{k-1}) \in \mathcal{F}_{k-1}$ and

$$(T_k = T_{k-1} + 1) = A_{k-1} \in \mathcal{F}_{T_k} \subset \mathcal{F}_{T_k+1} .$$

The fact that T is measurable is not surprising because we proved in [4] that, for every bounded stopping time, we have $H_T = E[A_T]$, where A_T is the increasing process associated to the martingale which characterizes the entropy H.

Remark 1. The stopping time T corresponds to a process which constructs a questionary to solve the second problem presented in the second paragraph. It is interesting to remark that T is almost certainly finite and also the number of questions to be posed to complete the interrogation is almost certainly finite. This enable us to construct the optimal questionary by selecting the questions in an a. s. finite set $\mathcal{Q}$.

Remark 2. The questionary corresponding to the stopping time T is obtained step by step by optimizing locally the entropy H. As we proved in [3] the final questionary optimizes globally the problem.

Remark 3. It is possible to construct an absolute upper bound for the sub-optimal stopping time T when the entropies associated to the questions q_i have some natural properties. In this case we have

$T \leq (\alpha + 1)\frac{h^*}{h_*}$

where h^* and h_* are the maximum and the minimum of the relative entropies associated to the questions. Infact, where T is sub-optimal, we can stop in the limit where $T = 1$ or $T = n$. (In this procedure we use the bancing technique). Consequentily we have

$\alpha h_* \leq nh_* - \beta \leq \alpha h^*$

where $\beta \geq 0$ and n is the greater integer such that $n\alpha h_* = \alpha h^*$, obviously

$n\alpha h_* \leq \alpha h^* + \beta \leq (\alpha + 1)h^*$

that is

$n \leq (\alpha + 1)\frac{h^*}{h_*}$

In the same way we can establish an absolute upper bound for the optimal stopping time which solves the first problem.

REFERENCES

1. B. Forte, Measures of information: The general axiomatic theory, R.I.R.O. 3e année, R-2, pp. 63-90 (1969)
2. C. Bertoluzza and C. Ronco, Trasmissione dell'informazione in un questionario applicato, Boll. U.M.I. (5), 14B, pp. 267-284 (1977)
3. F. Barbaini and C. Bertoluzza, Strategie aleatorie basate sull'entropia nel disegno d'esperimenti, Rend.Mat. VII, vol.II pp. 529-540 (1991)
4. F. Barbaini, Martingale e processi d'informazione, Rend.Mat. VII, vol. IX, pp. 131-143 (1989)
5. J. Neveu, Martingales á temps discret, Masson, Paris (1972)

Uncertainty in Intelligent Systems
B. Bouchon-Meunier et al. (Editors)
1993 Elsevier Science Publishers B.V.

Decision tree pruning using an additive information quality measure

L. Wehenkel†
Department of Electrical Engineering - Institut Montefiore
University of Liège - Sart Tilman B 28, B 4000 Liège - Belgium

Abstract. An additive quality measure based on information theory is introduced for the inductive inference of decision trees. It takes into account both the information content and the complexity of a tree, combined so as to evaluate the tree on the basis of its learning sample. The additivity of the quality measure with respect to the decomposition of a tree into subtrees, allows to formulate an efficient recursive backward pruning algorithm to maximize the quality. Simulation results are provided on the ground of a real life problem related to electric power system operation and a synthetic digit recognition problem.

Keywords. Decision trees, inductive inference, information quantity, pruning.

1. INTRODUCTION

A very popular approach to inductive inference is based on the automatic construction of decision - or class probability - trees on the basis of a learning set of objects of known classification [1]. Generally, tree induction methods have an irrevocable top down approach, growing a tree by successively developing its test nodes, starting from the top node and ending with its terminal nodes. The latter may be used in order to predict the class probabilities of any unseen object in terms of its observed attributes and hence to classify it. To induce a tree, the decision tree (DT) building algorithms choose intermediate splits in order to increase as much as possible the class purity of successor nodes, by using heuristic "purity measures" evaluated in the learning set [2, 3, 4]. A well known difficulty within this framework is that DTs fitting too perfectly the data contained in their learning set are generally overly complex and less reliable than simpler ones. A good tree should express a compromise between the amount of detail provided about the attribute vs. class relationship observed in the learning set, and the possibility to extrapolate its information correctly to unseen objects. Too large trees will overfit the data, whereas too small ones will underexploit the information contained in the learning set.

To avoid building unnecessarily huge trees, a stop splitting rule may be used during the tree growing [4]. Another approach consists of growing initially a maximal tree from the learning data, then generating therefrom a sequence of simpler trees of decreasing complexity, by using an appropriate pruning method, and selecting one of the pruned trees on the basis of its ability to correctly classify unseen states. Many different stop splitting and pruning methods have been proposed, some of them being more or less equivalent [5]. The most efficient ones reduce the trees size dramatically, by factors of 2 to 10, and at the same time improve their generalization

†Senior Research Assistant F.N.R.S.

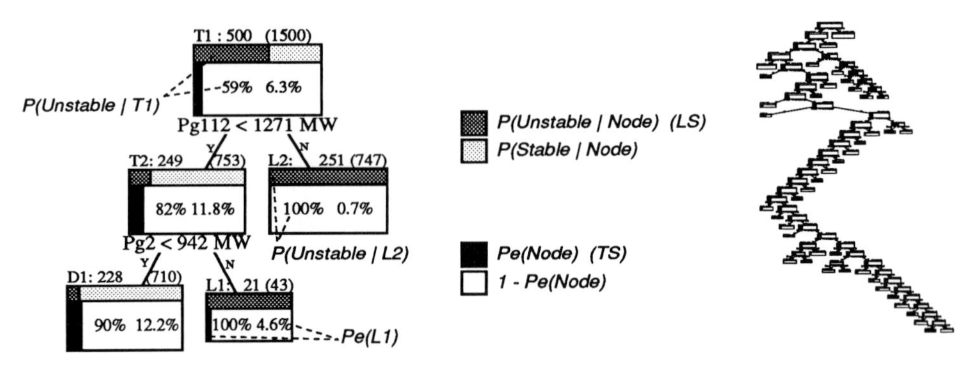

Figure 1. Example of pruned and unpruned DTs

capability to unseen cases [5, 6]. In the context of data analysis applications, the practical benefit is principally to simplify the interpretation of trees, by reducing their complexity and their statistical variability.

The aim of this paper is to present a tree *quality* measure derived from information theory [7, 8] and to propose an optimal pruning method maximizing this quality. Its performances in terms of complexity and reliability of the resulting DTs is investigated on three test problems. The two first are related to the (real life) problem of electric power systems security assessment and control, studied in [6, 9]. The last one concerns the classical LED digit recognition problem of Breiman et al. [2].

2. DECISION TREE METHODOLOGY [7, 8]

2.1. Objects and decision trees

Let us denote by

- o an object; $\boldsymbol{a}(o) = (a_1(o) \ldots a_n(o))$ its attribute values; $c(o) \in \{c_1, \ldots, c_m\}$ its class;
- DT a decision tree, which is considered as a hierarchical model of the attribute conditional class probabilities $P(c(o)|\boldsymbol{a}(o))$;
- $\boldsymbol{LS}$ a learning sample composed of N independently observed objects;
- $\boldsymbol{TS}$ a test set of M other independently observed objects;
- $\mathcal{N}$ a node of the DT, $\mathcal{R}$ its root, $\mathcal{N}_l$ one of its terminal nodes, $\mathcal{N}_t$ one of its test nodes;
- $DT(\mathcal{N})$ the subtree of DT of root $\mathcal{N}$, containing $\mathcal{N}$ and all its successor nodes;
- $\boldsymbol{\mathcal{N}}(DT)$ (or simply $\boldsymbol{\mathcal{N}}$) the set of all nodes of a tree, $\boldsymbol{\mathcal{N}}_l(DT)$ (or $\boldsymbol{\mathcal{N}}_l$) the set of its terminal nodes, and $\boldsymbol{\mathcal{N}}_t(DT)$ (or $\boldsymbol{\mathcal{N}}_t$) the set of its test nodes.

Anticipating on the results of Section 4.1, we illustrate in Fig. 1 a simple DT, built in the context of power systems transient stability assessment [6]. A learning set composed of 500 randomly generated operating states of a test power system (composed of 31 generators, 128 nodes and 253 lines) was built up. Its states were preanalyzed with respect to the occurrence of a three-phase short-circuit nearby generator 2, by simulating numerically the system behavior for a fault duration of 320ms and classifying the states into the stable or unstable class by observing their transient behavior, during approximately 2 seconds after the fault clearance. The candidate attributes used to build the tree were chosen as the active powers of generators 2 and 112.

The right hand part of Fig. 1 sketches the full tree, composed of 85 nodes, grown so as to classify correctly its learning set. The optimal splitting rule used for this purpose consists of choosing at each step the test that most reduces the mean misclassification cost, estimated in the learning set [2], and assuming a non-detection cost five times larger for stable states than for unstable ones. The left hand tree is its optimally pruned version (see §§3 and 4), composed of only 5 nodes. Each node is represented by a box, the upper part of which corresponds to the proportions of stable and unstable *learning* states relative to this node.

Such a tree may be used to classify unseen objects (i.e. to analyze the stability of new states of the power system with respect to the occurrence of the considered fault) by directing them towards its terminal nodes, starting at the root (T1), applying the tests encountered along the path, and directing the states to the appropriate successor, according to the value taken by their test attributes, "Pg112" at node T1 and "Pg2" at node T2. Upon reaching a terminal node, the conditional class probabilities may be estimated by the proportions of states of each class contained in the local learning set (see §2.2). Accordingly, assuming that the states are classified into the most likely class, those reaching nodes L1 or L2 are classified unstable, whereas those arriving at node D1 (the root of a pruned subtree) are classified stable.

Comparing the tree classification with the true classification on a set of 1500 independently drawn *test* states yields an error rate of $P_e = 6.3\%$. This is shown in Fig. 1, where the black area in the lower part of the node box indicates the *local* test set error rate, defined by the proportion of test states arriving at a node, and which are misclassified by the subtree of this node. The value of the error probability is also indicated inside the node box, together with the estimated probability of the majority class at the node. On the top of each nodes' box is also indicated its name along with its number of learning and test states.

Concerning reliability, let us observe that the majority (> 90%) of misclassifications are concentrated at node D1, where 12.2% of states are unstable and erroneously classified stable. As one may guess from the right hand part of Fig. 1, in the unpruned version of the tree this node is the root of a subtree composed of 81 nodes. Moreover, the error rate of the latter subtree is found to be of 18.3%, yielding a mean error probability of 9.1% for the overall tree. Here, pruning has improved significantly the generalization capability of the tree and has reduced at the same time its complexity by a factor of 17.

2.2. Information qualities

We consider a tree as a model for conditional class probabilities, and denote by $P(c_i|\mathcal{N})$ $(\forall i \leq m\ ,\ \mathcal{N} \in \boldsymbol{\mathcal{N}})$ the estimated probabilities attached to its nodes. For a given tree structure these parameters are estimated by the corresponding relative frequencies in the learning set [8]. Thus, we assume that for every node $\mathcal{N}$:

$$P(c_i|\mathcal{N}) \triangleq \frac{N_i(\mathcal{N})}{N(\mathcal{N})}\ ,\quad \forall i \leq m\ ,\ \mathcal{N} \in \boldsymbol{\mathcal{N}}, \tag{1}$$

where $N_i(\mathcal{N})$ denotes the number of learning states of class c_i at node $\mathcal{N}$, and $N(\mathcal{N}) = \sum_{i=1}^{m} N_i(\mathcal{N})$, the total number of learning states at node $\mathcal{N}$.

Let us further define by

- $H_C(\mathcal{N}) = -[\sum_{i=1}^{m} P(c_i|\mathcal{N}) * \log_2 P(c_i|\mathcal{N})]$ the entropy of the learning set of node $\mathcal{N}$, expressed in $bits$;
- $H_C(\boldsymbol{LS}) = H_C(\mathcal{R})$, the prior entropy of the learning set;
- $H_C(\boldsymbol{LS}|DT) = \sum_{\mathcal{N}_l} \frac{N(\mathcal{N}_l)}{N} * H_C(\mathcal{N}_l)$, the mean residual entropy of the $\boldsymbol{LS}$, given the DT;
- $I_C^{DT}(\boldsymbol{LS}) = H_C(\mathcal{R}) - H_C(\boldsymbol{LS}|DT)$ the mean information provided by the DT on the classification of the objects contained in the $\boldsymbol{LS}$;
- $C(DT) = \#\boldsymbol{\mathcal{N}}_l - 1$ the complexity of the tree, one less than its number of terminal nodes (for a binary tree, $C(DT) = \#\boldsymbol{\mathcal{N}}_t$, its number of test nodes);
- $q \geq 0$ a weighting factor;
- $Q(DT; \boldsymbol{LS}) = N * I_C^{DT}(\boldsymbol{LS}) - q * C(DT)$, the quality of the DT, evaluated in the $\boldsymbol{LS}$.

Given subjective prior probabilities $P(DT)$ for the DTs, which are (i) independent of the values of its nodewise class probabilities (eqn (1)), (ii) decreasing exponentially when the complexity $C(DT)$ increases, and (iii) under the hypothesis of independent learning states, one shows that the *quality* $Q(DT; \boldsymbol{LS})$ is proportional to the logarithm of the posterior probability $P(DT|\boldsymbol{LS})$ of the DT given the classification of the learning states [8].

In other words, under these conditions *maximizing the quality of a DT* is equivalent to *maximizing its posterior probability.* Within this framework high q values correspond to low a priori complexity expectations, and vice versa.

2.3. Inductive inference

We restate the objective of inductive inference as the search for a decision tree of maximal quality. Consequently, we reformulate the general tree building methodology as follows.

1. **Grow a tree,** by successively splitting its nodes and choosing at each step a test maximizing the local quality improvement.
2. **Stop splitting,** as soon as a maximum of quality is reached, namely at nodes where the best split cannot provide a large enough amount of information to compensate for the incumbent increase in complexity. Full growing is achieved by using $q = 0$, otherwise q is chosen on the basis of the a priori expected (or desired) complexity, and the tree is expanded up to the first *local* maximum of quality.
3. **Prune the tree,** by simplifying it in a way allowing to improve its quality as much as possible, ideally by extracting its pruned tree of *absolute* maximal quality.

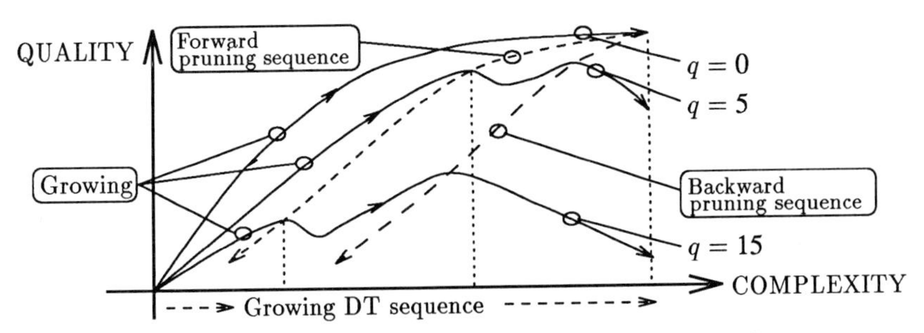

Figure 2. Quality variation during tree growing and pruning

The intuitive behavior of this procedure is sketched in Fig. 2, suggesting also that different q values might be used at the different steps, in order to provide more or less complex trees. The solid line curves correspond to the quality vs. complexity relationship of the sequence of growing trees, generated by the algorithm, for different values of q. Stop splitting consists of keeping the tree corresponding to the first maximum of quality. The curves reflect the fact that the higher the value of q the lower the quality of the trees, and the lower the complexity at the different local and global maxima of the curves. Two different pruning methods are suggested by the dashed line curves, showing the sequence of pruned trees that would be generated from the maximal tree, built by using $q = 0$. The figure suggests that backward pruning allows one to extract the tree corresponding to the *global* maximum of the quality, whereas forward pruning (similarly to stop splitting) extracts the first *local* optimum of minimal complexity.

It should be mentioned that, for a certain value of q, doing both stop splitting and pruning is redundant. Nevertheless, stop splitting may be used with a rather low value of q, to avoid developing nodes corresponding to very small samples, to save time during the tree growing. The pruning approach is used, with appropriate (larger) q values so as to extract an optimal tree. This is further discussed in the next section.

3. PRUNING METHODS

The pruning of a DT relies on the elementary operation of test node *contraction*. Contracting a node consists of replacing its subtree by a terminal node.

Let us define the following nodewise complexities and information quantities :

- $C(\mathcal{N}_t) = \#SUCC(\mathcal{N}_t) - 1$, the complexity of a test node, where $SUCC(\mathcal{N}_t)$ denotes the set of its direct successor nodes (e.g. for binary trees $C(\mathcal{N}_t) = 1, \ \forall \ \mathcal{N}_t$);
- $I_C^{\mathcal{N}_t} = H_C(\mathcal{N}_t) - \sum_{\mathcal{N} \in SUCC(\mathcal{N}_t)} \frac{N(\mathcal{N})}{N(\mathcal{N}_t)} * H_C(\mathcal{N})$, the mean information provided by a test node;
- $Q(\mathcal{N}_t; \boldsymbol{LS}) = N(\mathcal{N}_t) * I_C^{\mathcal{N}_t} - q * C(\mathcal{N}_t)$, the quality of a test node.[1]

A consequence of the definition of the information quantity and the complexity measure is the additivity of the tree quality. Thus, $Q(DT(\mathcal{N}_t); \boldsymbol{LS})$, *the total quality of a subtree of root*

[1]The information, complexity, and quality of a terminal node are naturally equal to zero.

$\mathcal{N}_t$, *is equal to the sum of the qualities of its test nodes*. Similarly, for any decomposition of a tree into subtrees, its quality is equal to the sum of the qualities of these subtrees.

The contraction of a test node modifies therefore the tree quality by an amount equal to the quality of its subtree, and pruning intuitively consists of contracting the weakest test nodes (of negative quality) to improve the tree as much as possible. As suggested in Fig. 2, the resulting tree depends of course on the value of q, but also on the order according to which the nodes are considered for contraction. Two strategies are considered below.

3.1. Backward (optimal) pruning

By definition, optimal pruning consists of pruning a DT in a *globally* optimal fashion, i.e. of generating a pruned DT of *maximal* quality. It is remarkable that due to the additivity of the quality measure, optimal pruning of a DT can be achieved in a single bottom up pass as described by the following recursive algorithm.

Backward pruning algorithm.

- *If the* DT *is trivial (i.e. reduces to a single node) then no pruning can be achieved; the pruned tree is the original* DT *itself.*

- *Otherwise, the subtrees of each of the successors of its top node are first pruned optimally in a recursive fashion, and the quality improvement of the tree is computed. This simply consists of adding the (necessarily non-negative) quality improvements resulting from pruning the subtrees.*

- *If, after this step the quality of the new* DT *is strictly positive then the so pruned* DT *is the optimally pruned overall tree.*

- *Otherwise, the* DT *is pruned by contracting its root and the final tree will reduce to a single terminal node.*

Hence, for a given value of q the result of pruning a DT is either a trivial DT, of quality equal to zero, or a non-trivial DT of (strictly) positive quality and whose non-trivial subtrees are also of strictly positive quality. For example, if the DT contains only one test node, then the quality of this node must be positive. If however the tree contains several levels, then only its deepest test nodes must be of positive quality, whereas the intermediate test nodes may be of negative quality, provided this is compensated by the total quality of their successors.

It is interesting to observe that a necessary condition for a test node to be contracted in the backward pruning approach is that both its own and its overall subtree's initial qualities are non-positive. This condition is not sufficient, however, since the successor subtrees may have a sufficiently positive overall quality, due to their own pruning, to compensate the initially negative quality of the test node, and prevent its contraction.

3.2. Forward (sub-optimal) pruning

The preceding recursive algorithm is globally optimal and thus produces the best pruned tree. It is moreover efficient since its computational complexity is linear in the number of nodes of a tree. For comparison purposes, we will describe the so-called "forward pruning method" which allows to simulate the stop splitting rule used during the tree growing, formulated as follows.

Stop splitting rule.

Don't split a node for which the incremental variation of quality corresponding to its best test t^*, $\Delta Q^* = N(\mathcal{N}_{t^*}) * I_C^{\mathcal{N}_{t^*}} - q * C(\mathcal{N}_t^*)$, *is not strictly positive.*

In the case of two-class problems and fixed branching factors, this rule is equivalent to the χ^2 hypothesis test described in [4].

Rather than fixing the value of q at the tree growing stage, and using this criterion, one may obtain the same result in a more flexible way by first building a full tree, and subsequently simplifying it by contracting all its test nodes which have a non-positive quality. This type of pruning may be achieved in a top down, *forward* fashion, since a test node is contracted on the basis of its own quality, without taking into account its successors.[2]

To obtain a set of trees for different values of q, in the stop splitting approach several trees would have to be grown, whereas in the pruning approach a single maximal tree may be successively pruned for growing values of q. This is clearly advantageous, since pruning is a very efficient procedure, of linear complexity in the overall tree size, almost negligible with respect to the computing burden involved in building a tree, especially if large numbers of learning states and attributes are considered.

It is interesting to note that, as for backward pruning, all subtrees of a DT produced by forward pruning have a positive quality. In addition, all its test nodes are of positive quality, which is not necessarily true for backward pruned trees. Hence, for fixed q, forward pruning produces simpler - and of course of lower quality - trees than optimal pruning. Both approaches are easily implemented as a post-processing tool, for the generation of all the relevant pruned subtrees of a tree, for an increasing sequence of critical q values. These pruning sequences, considered in the next section, may furnish an appropriate pruned tree, selected on the basis of its complexity and its ability to correctly classify unseen states.

3.3. Pruning sequences

Whatever the method used to prune, since there exist only a finite number of test nodes to contract in a DT, there are only a finite number of possible pruning combinations. Actually, we will see that both methods can produce at most as many different pruned trees (for different q values) as there are test nodes in a tree.

Let us denote by

- $Pr(DT, q)$, the tree obtained by pruning the tree DT, for a given value of q, using either the forward or the backward approach as indicated by the context;
- $Q(DT, q)$ the quality of a DT, explicitly depending on q and implicitly on the $\boldsymbol{LS}$;
- $DT' \prec DT$ a tree which is included in - but not necessarily a fully pruned version of - DT (i.e. DT' results from contracting at least one test node of DT). This relation is a partial order.

---

[2]Therefore, forward pruning does not guarantee to improve the tree's overall quality.

Let us prove that for both forward and backward pruning

$$\forall\, DT, q_1, q_2 \;:\; q_1 \leq q_2 \;\Rightarrow Pr(DT, q_2) \preceq Pr\,(DT, q_1)\,. \tag{2}$$

Equation (2) states the intuitively obvious fact that nodes pruned away for a given value of q, remain pruned away for all higher values. In the case of forward pruning, this is of course trivial, since increasing the value of q will decrease each test node's quality, and thus can only lead to additional test nodes with a negative quality.

Denoting by DT_i the pruned tree $Pr(DT, q_i)$,let us first notice that eqn. (2) implies that

$$Q(DT_1, q_1) \geq Q(DT_2, q_2). \tag{3}$$

Indeed, eqn. (2) implies that DT_1 may be decomposed into a part identical to DT_2 and a (possibly empty) collection of subtrees which are not yet pruned away for $q = q_1$, but are so for $q = q_2$. Denoting by $Q(DT_1 - DT_2, q_1)$ the total quality of these subtrees, we may decompose the left hand side of eqn. (3) in the following way :

$$Q(DT_1, q_1) = Q(DT_2, q_1) + Q(DT_1 - DT_2, q_1).$$

$Q(DT_1 - DT_2, q_1) \geq 0$, since the quality of any subtree of DT_1 is necessarily positive, when $q = q_1$, the latter being optimally pruned for $q = q_1$. This implies that

$$Q(DT_1, q_1) \geq Q(DT_2, q_1). \tag{4}$$

Moreover, according to its definition the quality measure decreases monotonically when q increases, and since $q_1 \leq q_2$, we have

$$Q(DT_2, q_1) \geq Q(DT_2, q_2), \tag{5}$$

and property (3) follows from eqns. (4) and (5).
□

Let us now prove property (2) by using an induction argument with respect to the complexity of the trees. For the base case tree composed only of its root ($C(DT) = 0$), the property holds trivially, since for such a tree

$$Pr(DT, q_1) = Pr(DT, q_2) = DT.$$

For a non-trivial tree of complexity $C(DT) = p > 0$, we derive the property from the assumption that it holds for all trees DT' such that $C(DT') < p$. Indeed, consider the root node of DT and of its two pruned versions DT_1 and DT_2. If neither in DT_1, nor in DT_2 the root has been contracted then the property derives directly by induction. If the root is contracted both in DT_1 and DT_2 then $DT_1 = DT_2$ and the property holds trivially. If only the root of DT_2 is contracted, it is of course true that $DT_2 \preceq DT_1$.

So, it is sufficient to show that if the root is contracted for $q = q_1$ then it must also be contracted for $q = q_2$. Let us denote by $DT^1, DT^2 \ldots DT^{r+1}$ the subtrees corresponding to the direct successors of the root of the initial DT, and by $DT_1^1, DT_1^2 \ldots$ (resp. $DT_2^1, DT_2^2 \ldots$) their corresponding pruned version for $q = q_1$ (resp. $q = q_2$).

The root of DT is contracted for $q = q_1$, if and only if

$$N(\mathcal{R}) * I_C^{\mathcal{R}} - q_1 * r + \sum_{i=1}^{r+1} Q(DT_1^i, q_1) \leq 0, \tag{6}$$

which implies that

$$N(\mathcal{R}) * I_C^{\mathcal{R}} - q_1 * r + \sum_{i=1}^{r+1} Q(DT_2^i, q_2) \leq 0, \tag{7}$$

according to inequality (3), implied by the induction hypothesis for each subtree DT^i. Inequality (7) implies also that

$$N(\mathcal{R}) * I_C^{\mathcal{R}} - q_2 * r + \sum_{i=1}^{r+1} Q(DT_2^i, q_2) \leq 0, \tag{8}$$

since $q_2 \geq q_1$. Consequently, the root node must be contracted for $q = q_2$.
□

A straightforward, but important consequence of property (2) is that there exists a *strictly* increasing and finite sequence, $q_0^c < q_1^c < \ldots < q_{i_{max}}^c$ of *critical* q values, corresponding to a *strictly* decreasing sequence of pruned trees, $DT_0 \succ DT_1 \succ \ldots \succ DT_{i_{max}}$ (where $DT_0 = DT$ and $DT_{i_{max}}$ is the the trivial DT), which are defined by

$$q_0^c = 0 \text{ and } q_{i+1}^c \triangleq \inf\{q \geq q_i^c \mid Pr(DT, q) \overset{\neq}{\prec} Pr(DT, q_i^c)\} \text{ ; } DT_i \triangleq Pr(DT, q_i^c). \tag{9}$$

Moreover, since two successive pruned trees must at least differ for one of the test nodes of the initial DT, $i_{max} \leq \#\mathcal{N}_t$. It is worthwhile to mention that in general the number of critical q values is much lower than this upper bound.

Remark. A simple recursive algorithm exists for the direct computation of the complete (forward or backward) pruning sequence of a tree, in a single bottom-up pass. This algorithm (not reported for reasons of space limitation) was used in the context of the simulations of §4.

4. SIMULATION RESULTS

4.1. Power systems security applications

We consider first two real life applications, relating to power systems security assessment. They are typical examples of classification problems characterized by continuous numerical attributes and where the classification is a *causal* consequence of the attribute values, i.e. prediction problems.

4.1.1. Voltage security

Figure 3 illustrates a typical tree built for preventive voltage security assessment of a real power system, with respect to the loss of one of its important generating units, providing normally in addition to its active power generation, an important local reactive power support, within the Brittany region of North-Western France. (This region has experienced voltage security problems in the past [10].) The alert (resp. non-vulnerable) states correspond to precontingency situations unable (resp. able) to withstand the loss of the generating unit, without voltage

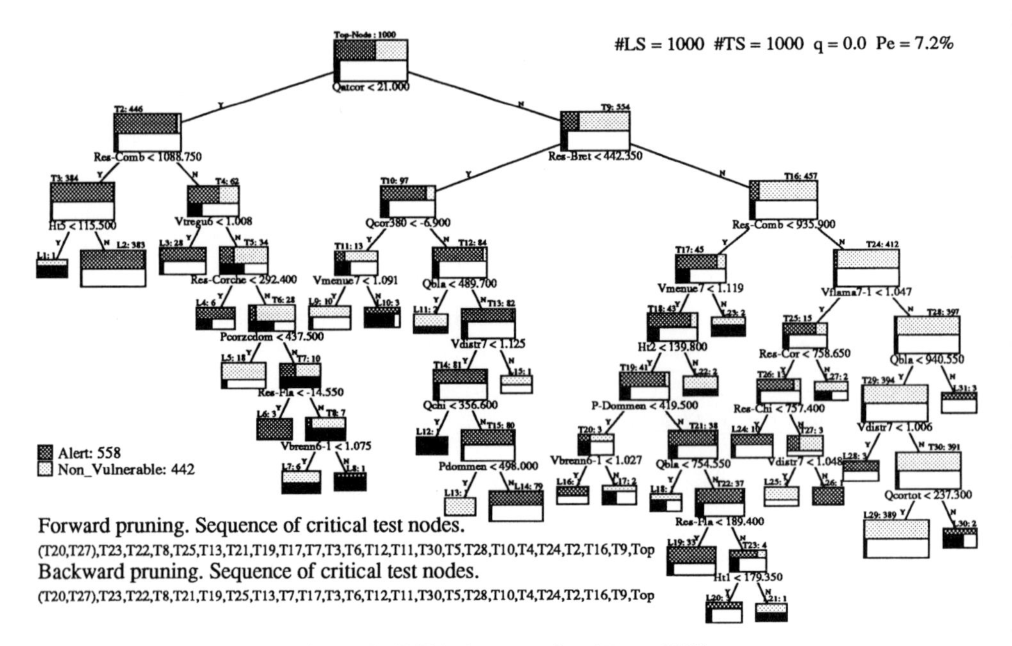

Figure 3. *DT* before pruning (From [10]).

collapse. The states are described by attributes characterizing the precontingency operating state of the system, such as voltage magnitudes, power flows, and reactive generation reserves. They were preclassified by using a loadflow program, to simulate the effect of the contingency.

The tree was grown on the basis of 1000 such states, by using the optimal splitting criterion based on the normalized information gain described in [4], and continuing to split nodes until the learning states of different classes were completely separated. This yields a complexity of 30, i.e. a total of 61 nodes. The tree was subsequently tested on the basis of 1000 independent states, yielding an error rate of 7.2%. On the lower part of Fig. 3 are indicated the two sequences of critical test nodes, successively contracted for increasing values of q, for either forward or backward pruning.

Figure 4 shows the plots of the complexity, test set error rate, quality, and total information of the decreasing sequences of pruned trees. These curves are typically decomposed into two parts : (i) when $q \in [0 \ldots q^*[$, one observes first a fast reduction in tree complexity, and a slow decrease in error probability; (ii) closely above q^*, generally in the range of $[5 \ldots 100]\frac{bits}{node}$ the error probability starts quickly to increase, expressing the fact that crucial nodes are unduly pruned away. One may also observe that the two pruning methods give very similar results. In particular, the same *minimum error tree* is obtained, for $q^* \approx 13$; its complexity is of 10 (21 nodes) and its error rate of 5.2%.

4.1.2. Transient stability

Coming back to the *DT* of Fig. 1, let us first recall that pruning allows to decrease its complexity from 42 to 2 test nodes, while increasing its reliability from 91.9% to 93.7% of correct classifications. Table 1 describes its pruning sequences. Each line specifies, for

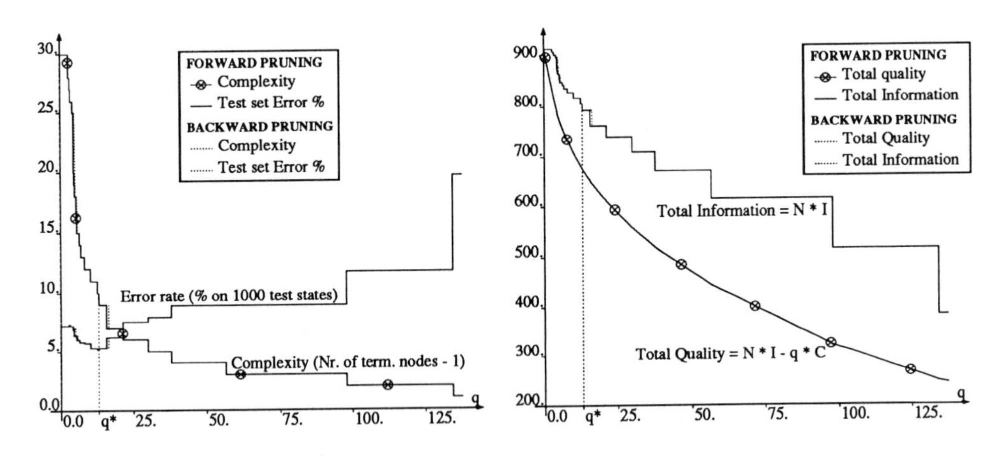

Figure 4. Forward and backward pruning curves of the DT of Fig. 3 (From [9])

the successively pruned tree, its range $[q^c_{i-1} \ldots q^c_i[$, its complexity C, its information quantity $I_Q = N * I_C^{DT}(\boldsymbol{LS})$, and its test set error rate P_e. The last column indicates the critical test node $\mathcal{N}_{tc}$, which will be contracted at the following step, i.e. for $q = q^c_i$. One can make the following observations :

- for low q values ($q \in [0.0 \ldots 2.77[$) the two pruning sequences differ significantly. For example, for $q \in [1.08 \ldots 1.76[$, and forward pruning, the nodes T19, T10 and T39 are contracted, and the tree reduces to 21 nodes ($C(DT) = 10$). Its quality is $(415 - 10 * q) \in]397 - 404]bits$. For the same q values, the backward pruned tree is still complete and its quality is $(488 - 42 * q) \in]414 - 442]bits$, which is naturally larger than the previous value.
- for higher q values ($q \in [2.77 \ldots \infty[$), the two pruning sequences become identical : the same nodes are successively pruned, for the same critical q values. The optimal q value lies

Table 1. Pruning sequences of the DT of Fig. 1

Forward pruning					Backward pruning				
$[q^c_{i-1} \cdots q^c_i[$ bit/nds	C nds	I_Q bit	P_e %	$\mathcal{N}_{tc}$	$[q^c_{i-1} \cdots q^c_i[$ bit/nds	C nds	I_Q bit	P_e %	$\mathcal{N}_{tc}$
					0.00...2.14	42	488	9.1	T31
0.00...0.57	42	488	9.1	T19	2.14...2.17	36	475	8.6	T17
0.57...0.73	24	448	7.8	T10	2.17...2.27	22	444	7.8	T15
0.73...1.08	14	425	7.1	T39	2.27...2.35	20	440	7.5	T38
1.08...1.76	10	415	7.5	T38	2.35...2.45	15	428	7.9	T10
1.76...2.38	9	413	7.5	T6	2.45...2.77	9	413	7.5	T6
2.38...3.20	5	402	7.4	T5	2.77...3.20	5	402	7.4	T5
3.20...9.05	4	399	6.9	T4	3.20...9.05	4	399	6.9	T4
9.05...13.05	3	390	6.9	T3	9.05...13.05	3	390	6.9	T3
13.05...59.05	2	377	6.3	T2	13.05...59.05	2	377	6.3	T2
59.05...317.99	1	318	8.9	T1	59.05...317.99	1	318	8.9	T1
317.99...∞	0	0	42.0		317.99...∞	0	0	42.0	

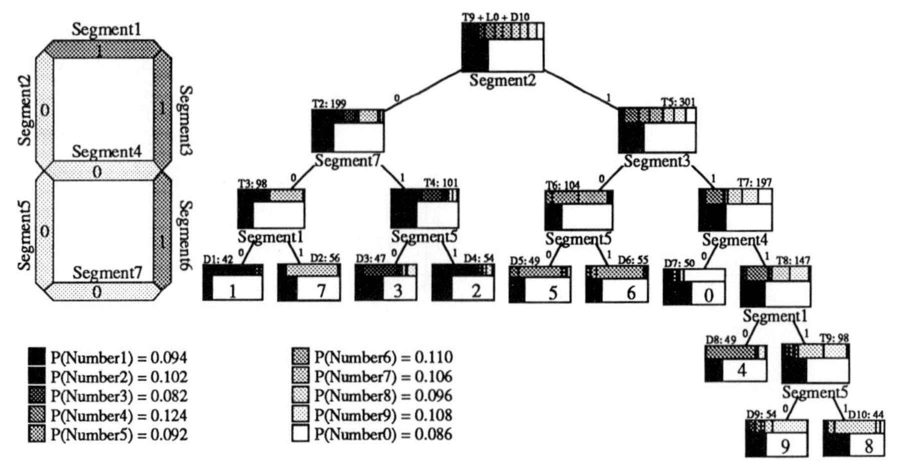

Figure 5. LED digit recognition. $P_{mis} = 10\%$, $N = 500$, $M = 1000$, $P_e = 31.3\%$

in this range, where the most random nodes have already been contracted.

4.2. Digit recognition

This is an artificial application, initially proposed in [2] and used by many other authors [5, 11]. A seven segment representation of a decimal digit, as commonly used by LED displays, is considered. The attributes are the states ("on" or "off") of the seven segments of the display, and the class is the represented digit ($c \in \{0, \ldots, 9\}$). This is representative of diagnostic type problems together with discrete attributes, the latter being here a causal consequence of the class. In order to depart from deterministic behavior, noise is introduced by simulating a misoperation of the display segments, each one being subject to random misoperation, with a probability P_{mis} of being in the wrong state.

Figure 5 represents a LED display, specifying the meaning of the attributes, and showing a decision tree corresponding to a noisy case, for $P_{mis} = 10\%$. To build this tree, a data base of 5000 states was first randomly drawn. A subset of 2000 states was reserved for testing and decomposed into two subsets of equal size. Out of the remaining 3000, a random learning set of 500 states was used to grow the tree and to generate its backward pruning sequence. For the tree growing, the same normalized information gain as in the example of §4.1.1 was used. For the tree selection, the reliability of the pruned trees was estimated on the basis of the first test sample, and the final tree was selected by using the so called *1 Standard Error rule* [2]. This consists of choosing the smallest tree such that $P_e \leq P_e^* + \sqrt{P_e^*(1 - P_e^*)M^{-1}}$, where P_e^* is the minimal test set error rate of the pruned trees. Finally, in order to avoid optimistic bias, the reliability of the selected tree was reestimated on the basis of the second independent test sample. The resulting tree represented in Fig. 5, is similar to the tree obtained in the noiseless case; it has exactly one terminal node corresponding to each one of the 10 digits.

Similarly, a systematic simulation was conducted, for variable noise levels and numbers of learning states. For each of the 16 combinations of $P_{mis} \in \{0\%, 5\%, 10\%, 20\%\}$ and $N \in \{125, 250, 500, 1000\}$, Table 2 indicates the complexity and error probability of the initial

Table 2. Digit recognition - Effect of pruning

			Learning set size							
			$N = 125$		$N = 250$		$N = 500$		$N = 1000$	
Noise level	Tree characteristics		μ	σ	μ	σ	μ	σ	μ	σ
$P_{mis} = 0\%$	Initial	$C(DT)$	9.0	0.0	9.0	0.0	9.0	0.0	9.0	0.0
		$P_e(DT)$	0.0	0.0	0.0	0.0	0.0	0.0	0.0	0.0
	Pruned	$C(DT)$	9.0	0.0	9.0	0.0	9.0	0.0	9.0	0.0
$P_e^{Bayes} = 0.0\%$		$P_e(DT)$	0.0	0.0	0.0	0.0	0.0	0.0	0.0	0.0
$(\sigma = 0.0\%)$		$\frac{1}{2}(q_i^c + q_{i-1}^c)$	8.8	2.4	20.6	1.8	43.9	3.8	83.2	5.1
$P_{mis} = 5\%$	Initial	$C(DT)$	22.8	2.7	33.6	3.0	47.0	3.5	59.3	3.1
		$P_e(DT)$	17.9	0.4	16.5	0.8	15.9	1.0	13.7	0.6
	Pruned	$C(DT)$	10.7	3.2	11.1	4.0	12.3	5.2	29.3	4.9
$P_e^{Bayes} = 14.0\%$		$P_e(DT)$	17.6	0.6	17.2	0.6	17.0	1.0	14.1	0.7
$(\sigma = 0.5\%)$		$\frac{1}{2}(q_i^c + q_{i-1}^c)$	16.8	2.9	16.3	7.1	26.2	13.1	6.5	3.1
$P_{mis} = 10\%$	Initial	$C(DT)$	35.5	4.0	54.7	4.1	69.4	2.8	86.7	2.8
		$P_e(DT)$	33.7	2.0	31.1	1.6	29.4	1.4	27.7	0.9
	Pruned	$C(DT)$	12.9	2.9	11.3	2.3	16.2	7.8	29.5	6.5
$P_e^{Bayes} = 26.8\%$		$P_e(DT)$	31.8	1.1	31.7	1.0	30.4	1.1	27.8	0.9
$(\sigma = 0.6\%)$		$\frac{1}{2}(q_i^c + q_{i-1}^c)$	6.8	2.7	14.1	4.0	20.9	13.4	11.1	4.8
$P_{mis} = 20\%$	Initial	$C(DT)$	54.2	1.6	78.1	2.4	99.2	3.5	116.1	3.2
		$P_e(DT)$	61.1	1.7	59.1	2.8	55.7	2.0	52.1	1.5
	Pruned	$C(DT)$	16.4	10.1	14.5	5.2	18.7	7.2	22.9	8.1
$P_e^{Bayes} = 48.6\%$		$P_e(DT)$	60.5	1.9	56.8	2.6	54.4	2.5	52.4	1.8
$(\sigma = 0.7\%)$		$\frac{1}{2}(q_i^c + q_{i-1}^c)$	6.5	3.1	11.1	3.2	14.5	5.3	18.0	9.3

and of the selected pruned tree, as well as its mean q^c value. The figures are average values (and their standard deviation) of 10 trees built for 10 random learning sets.

In order to assess the effectiveness of the decision tree approach, the error probability of the Bayes rule was estimated on the basis of the 5000 states of the data base, and is indicated in the table.[3] One can see that for the noisy cases with $P_{mis} \leq 10\%$, the reliability of the decision trees is very close to this lower bound, as soon as $N = 1000$. For the noisier situations, only little room is left for improvement, e.g. by using slightly larger learning sets.

The Table reflects the ability of the pruning method to simplify very systematically the decision trees, and preserve their generalization ability. For $P_{mis} = 0\%$, all the trees are of course identical, similar to the tree of Fig. 5. They classify all the test and learning states correctly. For increasing noise levels, one observes an important growing of the complexity of the initial trees, much less important however for the pruned ones. At the same time the error probability of the pruned trees remains always very close to the optimal value.

Another observation is that the optimal q^c values are only marginally influenced by the noise level and the number of learning states. This is in good agreement with the theoretical conjecture that the optimal q^c is in direct relation with the prior tree complexity expectation, which is here

[3]The Bayes rule here simply consists of choosing the digit having the largest number of segment states identical to those of the observed state.

always in the range of 10 to 30 terminal nodes. Moreover, extensive experimental investigations across different domains, not reported here, suggest that the q values in the range of $[10 \dots 30]$ generally provide a very good compromise, allowing to reduce enough the complexity and to obtain nearly optimal error probabilities [7]. This is an interesting practical outcome, meaning that no additional test set is actually required to select the pruned tree, since an appropriate q value may be chosen a priori. Thus all available objects may be used in the learning set, during tree growing *and* subsequent pruning.

Incidentally, for the 120 pruned trees built for $P_{mis} \in \{5\%, 10\%, 20\%\}$, we have compared the two error probabilities, estimated either on the first test sample, also used for the tree selection, or on the basis of the second fully independent test set, corresponding to the values given in Table 2. In principle, the first estimate could be optimistically biased since it uses a test set also used in order to select the tree. It was however found that the mean bias was only of 0.1%, which is negligible with respect to the standard deviation of test set error probability estimates ($\sigma(P_e) \in [1.0\% \dots 1.6\%]$, for $M = 1000$).

5. CONCLUSION

A novel additive tree quality measure based on information theory has been proposed and used to devise forward and backward pruning algorithms. Their comparison on some real life and synthetic problems of different types shows that both approaches are effective in controlling the size of decision trees and in improving their generalization and interpretability features.

Although the resulting trees could in principle differ in the two methods, they turn out to be often quite close to each other, especially in the interesting range of values of the parameter q, where the same critical test nodes are generally identified as the weak points of a tree. All in all this is quite natural, since the test nodes of a tree are always developed in order to maximize the quality gain. This consists of searching preferably for the most salient *first-order* correlations between attributes and classes. It is however intuitively clear that the backward pruning method could outrank significantly forward pruning only if *second* (or higher) order correlations could predominate in a tree.

From the performance point of view, especially concerning the reduction in tree complexity, the simulation results given in this paper suggest that the method compares favorably with the most effective other methods described in the literature [5]. Additional systematic explorations carried out on standard and real life data sets have confirmed this impression on a broader range of problems.

A quite interesting and original feature of the quality measure, stemming from our investigations, is the very good stability of the "optimal" value of the weighting parameter q. This suggests that for a rather wide range of applications, this value may be fixed a priori, and the pruned tree need not be selected indirectly on the basis of a test set of independent objects. This could be advantageously exploited by using more efficiently the available data as learning states, hence enhancing the reliability of the resulting trees.

More generally, the results given in this paper confirm the well foundedness of the proposed quality measure. Its ability to appropriately prune trees is a good indication of its aptitude to correctly express the complexity vs. informativity compromise. In addition, many practical applications may be foreseen, such as honest node split evaluation in the context of variable branching factors, as well as comparison of multi-level tests, in the context of generalized search strategies, such as lookahead techniques. Among other appealing possibilities let us mention the generalization of the method to "variable complexity splits" (e.g. involving functional combinations of several attributes) and to more general graph structures than trees [12].

REFERENCES

[1] J. N. Morgan and J. A. Sonquist. Problems in the analysis of survey data, and a proposal. *J. of the Amer. Stat. Ass.* **58**, pp. 415–434, 1963.

[2] L. Breiman, J. H. Friedman, R. A. Olshen, and C. J. Stone. *Classification and Regression Trees*. Wadsworth International (California), 1984.

[3] J. R. Quinlan. Induction of decision trees. *Machine Learning* **1**, pp. 81–106, 1986.

[4] L. Wehenkel, T. Van Cutsem, and M. Ribbens-Pavella. An artificial intelligence framework for on-line transient stability assessment of power systems. *IEEE Trans. on Power Syst.* **PWRS-4**, pp. 789–800, 1989.

[5] J. Mingers. An empirical comparison of pruning methods for decision tree induction. *Machine Learning* **4**, pp. 227–243, 1989.

[6] L. Wehenkel and M. Pavella. Decision trees and transient stability of electric power systems. *Automatica* **27**, no. 1, pp. 115–134, 1992.

[7] L. Wehenkel. *Une approche de l'intelligence artificielle appliquée à l'évaluation de la stabilité transitoire des réseaux électriques*. PhD thesis, University of Liège - Belgium, May 1990. In French.

[8] L. Wehenkel. A probabilistic framework for the induction of decision trees. Submitted for publication, 1992.

[9] L. Wehenkel and M. Pavella. Decision tree approach to power system security assessment. *Int. J. of Elec. Power and Energy Syst.* **15**, no. 1, 1993.

[10] Y. Harmand, M. Trotignon, J. F. Lesigne, J. M. Tesseron, C. Lemaître, and F. Bourgin. Analyse d'un cas d'écroulement en tension et proposition d'une philosophie de parades fondées sur des horizons temporels différents. In *CIGRE Report 38/39-02 Paris*, Aug. 1990.

[11] J. R. Quinlan. Simplifying decision trees. *Int. J. of Man-Mach. Studies* **27**, pp. 221–234, 1987.

[12] A. Zighed, J.P. Auray, and G. Duru. *SIPINA. Méthode et Logiciel*. Alexandre Lacassagne - Lyon, 1992.

Uncertainty in Intelligent Systems
B. Bouchon-Meunier et al. (Editors)

An explicit formula for fractional entropy

Antonella Delmestri, Anna Fioretto and Andrea Sgarro

Dipartimento di Scienze Matematiche
Università di Trieste - 34100 Trieste (Italy)

The fractional entropy $H_M(X)$ is an information measure derived from rate-distortion theory which measures the uncertainty contained in random experiment X when the pragmatic experimenter is not interested in the exact value of X, but is contented with a set of M possible outcomes where the actual outcome belongs (for M=1 one finds Shannon entropy). So far only an iterative algorithm (the general algorithm for rate-distortion functions) was available to compute $H_M(X)$ to the desired degree of accuracy. We give here an explicit (and perspicuous) formula for $H_M(X)$ which returns its exact value without computational trammels by expressing it as the difference of two usual Shannon entropies.

1. INTRODUCTION

Shannon entropy H(X)=H(P) is a deservedly famous information measure derived from information theory (more precisely: from source coding theory); it is meant to measure the uncertainty contained in the random variable, or random experiment, X which has probability distribution P; ($X \in \{x_1, x_2, ..., x_K\}$, $P=(p_1, p_2, ..., p_K)$, $p_i = P(x_i) = \text{Prob}\{X=x_i\}$; as a general reference to source coding we suggest e.g. [1]).

However, as it has often been observed, *Shannon entropy*, in spite of its great merits, has also serious drawbacks when used as an information measure, in particular because it *ignores the pragmatic facets of information*. Now, it is well-known that rate-distortion theory is a deep and well-developed pragmatic generalization of source coding theory, the place of Shannon entropy being taken by a more general functional called the *rate-distortion function*. By a self-imposing analogy, rate distortion theory appears then to be a natural source of inspiration for providing *pragmatic information measures based on the rate-distortion function.*

In [2] the last author introduced such a pragmatic measure calling it the *fractional entropy* $H_M(X)=H_M(P)$: this is meant to represent the uncertainty contained in X when the pragmatic experimenter is not interested in the exact value of X, but is contented with a set of M possible outcomes where the actual outcome belongs ($1 \le M \le K$; for $M=1$ one re-finds Shannon entropy). Even if fractional entropy was introduced in the specific context of cryptology, it is thought to have a larger impact in uncertainty management and representation (recently, possible application to statistical methodology has been kindly pointed out to us by G. Klir). We defer a formal definition of $H_M(X)=H_M(P)$ to Section 3; we soon mention, however, that a sizeable list of properties is available for the fractional entropy, which support its use as an information measure; as a specimen we just retain:

$$0 \le H_M(X) \le \log \frac{K}{M}$$

Equality holds on the left iff X takes on at most M values with positive probability; on the right iff X is uniform; for $M=1$ one re-finds well-known properties of Shannon entropy. (Cf [2-4] for more properties of $H_M(X)$.)

In Section 3 an *implicit* formula expressing fractional entropy through a minimum will be given, which is directly derived from rate-distortion theory; this formula, however, is rather cumbersome from the computational point of view and hampers the use of fractional entropy in practical applications. The contribution of this paper consists in giving an *explicit* formula for $H_M(P)$ for all values of M. The proof is lengthy, but not at all obscure. This paper is rather technical than philosophical: for a philosophical discussion we refer to [2-5]. Our theorem, however, perspicuously expresses $H_M(P)$ as a difference of two usual Shannon entropies, and so does have a philosophical feedback.

2. A REMINDER ON INFORMATION-THEORETIC MEASURES

Below we provide a short reminder on entropy, mutual information and the rate distortion function; we shall be very concise, or even coarse: the interested reader is referred e.g. to [1]. We stress that the quantities below are all well-established in the context of coding.

Let X and Y be two finite (arbitrarily correlated) random variables. Let δ be a binary "distortion" matrix row-indexed in the outcome-space of X and column-indexed in the outcome-space of Y ($\delta(x,y) \in \{0,1\}$; $\delta(x,y)=1$ means that y is a "distorted", i.e. unacceptable, reproduction for x; we require that in each row of δ there is at least a zero, so that each x is reproducible. A list of definitions and comments follows.

Shannon entropy: $H(X) = - \sum_x \text{Prob}\{X=x\} \log \text{Prob}\{X=x\} = - \sum_i p_i \log p_i$

Mutual information: $I(X;Y) = H(X) + H(Y) - H(XY)$

$I(X;Y)$ measures the amount of stochastic dependence between X and Y and is at its lowest ($I(X;Y)=0$) iff X and Y are independent. The joint distribution of XY can be described by giving the probability vector P (the marginal probability distribution of X) and the stochastic matrix W row-indexed in the outcome space of X and column-indexed in the outcome space of Y whose (x,y)-entry is the conditional probability $Prob\{Y=y|X=x\}$. This given, the notation I(X;Y) can be replaced by I(P,W).

Rate-distortion function: $R(X,\delta) = \min I(X;Y) = \min I(P,W)$, the minimum being taken with respect to all random couples XY s.t. $Prob\{\delta(X,Y)=0\}=1$, i.e. with respect to all stochastic matrices W which are constrained to have zeroes in positions (x,y) where $\delta(x,y)=1$. $R(X,\delta)$ is a natural and important generalization of Shannon entropy, this being re-found when the outcome spaces of X and Y coincide and the distortion measure is the binary complement of Kronecker's δ ($\delta(x,y)=0$ iff $x=y$), called also Hamming distortion. $R(X,\delta)$ can be considered as a pragmatic measure of uncertainty, the experimenter being now contented to describe the output x of X through any y which has zero-distortion from x. (Readers conversant with Shannon theory will have noticed that we are here confining ourselves to describe the rate-distortion function in the case when the allowed distortion level is zero.)

3. THE THEOREM

By definition, (cf [2]), the fractional entropy $H_M(X)=H_M(P)$ equals the rate-distortion function $R(X,\delta)$ when the secondary outcome-space (the one for Y) is made up of all the $\binom{K}{M}$ primary subsets of M primary outcomes and $\delta(x,y)=0$ iff $x \in y$. Shannon entropy is re-found for M=1. An expression for $H_M(X)$ is soon obtained from that for $R(X,\delta)$:

$$H_M(X) = H_M(P) = \min I(P,W) \tag{1}$$

the minimum being taken with respect to all stochastic matrices W which are constrained to have zeroes in positions (x,y) where $x \notin y$; (W has K rows and $\binom{K}{M}$ columns). An iterative algorithm is available to approximate the value of the rate-distortion function to the desired degree of accuracy (cf [1]); in some special "symmetric" cases, however, one can solve for the minimum and obtain an explicit formula which returns the exact value of $R(X,\delta)$ without computational trammels. As we show below, this is fortunately the case for $H_M(X)$. So far an implicit formula was known only for M=1 (Shannon entropy) and M=2 (*semientropy*, cf [2]).

Without real loss, we restrict our attention to probability vectors (p.v.'s) P such that $p_1 \geq p_2 \geq \ldots \geq p_M \geq \ldots \geq p_K$. Let R_i, $0 \leq i \leq M-1$, be the region of p.v.'s defined as follows:

$$R_i = \{P : p_i \geq \frac{1-p_1-p_2-\ldots-p_{i-1}}{M-(i-1)}\} = \{P : p_i \geq \frac{p_i+p_{i+1}+\ldots+p_K}{M-(i-1)}\}$$

(The "void" condition defining R_0 holds unconditionally; we interpret a void summation as zero, so that $R_1=\{P: p_1 \geq \frac{1}{M}\}$). One has, as an easy check shows:

$$R_0 \supset R_1 \supset ... \supset R_{M-1} \tag{2}$$

Informally, R_i-R_{i+1} *is the region of p.v.'s for which the first* i *components are "comparatively large", while the remaining components are "comparatively small"*. Set now:

$$P_i = (p_1, p_2, ..., p_i, \frac{1-p_1-p_2-...-p_i}{M-i}, ..., \frac{1-p_1-p_2-...-p_i}{M-i}) \tag{3}$$

Thus P_i is a p.v. with M components, the last (M-i) being constant; in particular P_0 is uniform (we recall that P instead has K components, K≥M). For the corresponding Shannon entropies, one has:

$$H(P_0) = \log M \geq H(P_1) \geq ... \geq H(P_{M-1})$$

These inequalities derive from elementary properties of Shannon entropy, P_i being "more uniform" than P_{i+1}; we do not insist on this since they are also derivable as a straightforward corollary to (2) and lemma 1 below.

We are now ready to state our theorem, which expresses fractional entropy as a difference of two usual Shannon entropies:

Theorem: If P belongs to R_i - R_{i+1}, then $H_M(P) = H(P) - H(P_i)$

(Actually, if the condition defining R_{i+1} holds with equality, $P_i=P_{i+1}$, just as the continuity of $H_M(P)$ requires.) The proof of the theorem will be preceded by a few lemmas. From now on we assume knowledge of basic facts concerning the manipulation of information-theoretic measures (entropies, conditional entropies, informational divergences), as can be derived, say, from chapter 1.3 in [1].

Lemma 1. If P belongs to R_i, then $H_M(P) \geq H(P) - H(P_i)$

Proof. Instead than as a minimum, the general rate-distortion function $R(X,\delta)$ can also be expressed as a maximum (cf [1]; actually, this is the starting point for the iterative approximation algorithm we have hinted at). In the case of fractional entropy, this alternative expression reads as:

$$H_M(P) = - \min_Q \{D(P;Q) + \log S(Q)\}$$

where Q is a probability vector with K components, S(Q) is the sum of the M largest components of Q, and $D(P;Q) = \sum_i p_i \log(p_i/q_i)$ is the *informational divergence* (called also *cross-entropy*, or *Kullback's discrimination*: this is yet another information measure which is non negative and equals zero iff P=Q; cf [1,2]). So, any "test" probability vector Q gives automatically a lower bound to $H_M(P)$. We suggest taking Q with components proportional to

$$(p_1,p_2,\ldots,p_i,\frac{1-p_1-p_2-\ldots-p_i}{M-i},\ldots,\frac{1-p_1-p_2-\ldots-p_i}{M-i})$$

The last (K-i) components of Q are equal (we stress that Q is *not* the same as P_i, since it has K components, rather than M components: this is why we need a normalization coefficient, α, say). The fact that P belongs to R_i soon implies that the components of Q are in non-increasing order of magnitude, from largest to smallest, which allows to compute S(Q). A trivial computation shows now that $D(P;Q) = -H(P) + H(P_i) - \log S(Q)$, which gives the lemma (to save time, do not compute explicitly α, since it cancels anyway). QED

We shall need the following important theorem of liner algebra (cf e.g. [6]):

Farkas lemma. Let C be an (H,K)-matrix, and $\underline{b}$ a K-vector; then one and only one of the systems below admits of solutions ($\underline{x}$ is un unknown K-vector, and $\underline{y}$ is an unknown H-vector):

$$\begin{cases} C\,\underline{x} \leq \underline{0} \\ \underline{b}\,\underline{x} > \underline{0} \end{cases} \qquad \begin{cases} C^T\,\underline{y} = \underline{b} \\ \underline{y} \geq \underline{0} \end{cases}$$

Lemma 2. If $p_1 \leq \frac{1}{M}$, then $H_M(P) = H(P) - \log M$

Proof. This amounts to proving the theorem for i=0. In view of lemma 1, one has $H_M(P) \geq H(P) - \log M$. Actually, this bound can be proven directly by using expression (1), writing mutual information as $I(X;Y) = H(X) - H(X|Y) = H(P) - H(X|Y)$, and recalling that, given Y=y, X can take at most M values with positive probability, these being the M primary outcomes making up subset y, so that $H(X|Y) \leq \log M$ (H(X|Y) denotes *conditional entropy*). To have $H(X|Y) = \log M$, we need to find in the minimization set a stochastic matrix W such that $H(X|Y) = \log M$, that is such that the conditional distribution of X given Y=y is uniform (when defined, i.e. for Prob{Y=y}>0). We set R(y) = Prob{Y=y}; with this notation, $\text{Prob}\{X=x|Y=y\} = \frac{1}{M}$ holds iff $W(y|x) = \frac{R(y)}{M\,P(x)}$. Suppose that R has been fixed as the marginal distribution of Y. A W yielding that R exists iff, for all x:

$$\sum_{y:x\in y} \frac{R(y)}{M\,P(x)} = 1$$

(the rows of a stochastic matrix must sum to 1; non-negativity for W is ensured by non-negativity for P). Therefore, the bound $H(X|Y) \leq \log M$ is attained with equality iff one can solve the system:

$$\begin{cases} \forall\, x \quad \Sigma_{y:x\in y} \;\; R(y) = M\, P(x) \\ \quad\;\; \forall\, y \qquad\quad R(y) \geq 0 \end{cases}$$

(the R(y)'s automatically sum to 1 because $\Sigma p_i=1$). For $p_1 \leq \frac{1}{M}$, the system does admit of solutions as shown by use of Farkas lemma. All we need to prove is that the "dual" system in the unknown T(x)

$$\begin{cases} \forall\, y \quad \Sigma_{x:x\in y} \;\; T(x) \leq 0 \\ \Sigma_x \; M\, P(x)\, T(x) \;>\; 0 \end{cases}$$

is instead impossible. Actually, $\Sigma_x \; M\, P(x)\, T(x) \leq M\, p_1\, \Sigma_x \; T(x) \leq \Sigma_x \; T(x)$, and so the last inequality would imply $\Sigma_x \; T(x) > 0$. Summing the remaining inequalities over y one obtains instead $\Sigma_x \; T(x) \leq 0$. QED

We are now ready to prove the theorem.

Proof of the theorem. In view of the two lemmas it is enough to prove $H_M(P) \leq H(P) - H(P_i)$ for $i \geq 1$. Recalling (1) and writing again mutual information as $I(X;Y) = H(P) - H(X|Y)$ it will be enough to provide an admissible stochastic matrix W such that $H(X|Y) = H(P_i)$. Now we do this. For convenience, we introduce some slang. We recall that, informally, $P \in R_i - R_{i+1}$ means that the first i components of P are "large", while the last K-i components are "small". From now on we shall refer to outcomes (elements) $a_1, a_2, \ldots, a_i$ as to the "likely" elements; the M-subsets which comprise *all* the i likely elements will be called themselves the "likely" subsets; observe that to identify a likely subset it is enough to specify its M-i *unlikely* elements. The stochastic matrix that we are going to build has the form

$$W \quad = \quad \begin{bmatrix} A & \\ & C \\ B & \end{bmatrix}$$

where:

i) the i rows corresponding to A are headed to the likely elements, while the K-i rows corresponding to B are headed to the unlikely elements; C has K rows;

ii) the columns corresponding to A and B are headed to the likely M-subsets, while those corresponding to C are headed to the unlikely subsets.

We describe C, B and A, in this order. The entries of C are all zero. To construct B we make use of lemma 2. Let us introduce a binary random variable E, which takes the value 1 iff X is an unlikely element. Let P* be the conditional distribution of X given E=1: $P^*(a_{i+j}) = \text{Prob}\{X^*=a_{i+j}\} = \text{Prob}\{X=a_{i+j}|E=1\}$, $1 \leq j \leq K-i$. Now we use lemma 2, with K-i rather than K (delete the i likely elements), with P* rather than P, and M-i rather than M (using the lemma is legitimate since $P \notin R_{i+1}$ implies $P^*(a_{i+1}) \leq \frac{1}{M-i}$). We obtain $H_{M-i}(P^*) = I(P^*,W^*) = H(P^*)$ -

$\log(M-i)$ for a suitable W* which achieves the minimum in the definition of $H_{M-i}(P^*)$. It is precisely this W* which will be inserted into W at position B (the rows of W* are already headed to the unlikely elements; as for columns, identify each likely subset with the corresponding (M-i)-subset comprising its (M-i) unlikely elements). We still have to construct A. The rows of A will be all equal to one another; more precisely we do as follows. Take the rows of W*=B and form a linear (convex) combination of these by weighting each row with the corresponding probability $P^*(a_{i+j})$: the resulting row will be repeated i times to form A. This completes the construction of W.

We prove that, with this W, $H(X|Y) = H(P_i)$. Preliminarily, we observe that our construction of A ensures that E and Y are independent random variables (the trivial check is omitted), and so $H(E|Y)=H(E)$. Since E is a deterministic function of X, the following chain is now obvious

$$H(X|Y) = H(XE|Y) = H(E|Y) + H(X|EY) =$$
$$= H(E) + \text{Prob}\{E=0\}\ H(X|Y,E=0\} + \text{Prob}\{E=1\}\ H(X|Y,E=1\}$$

We deal with the two conditional entropies at the end of the chain. Yet another trivial probability manipulation shows that $\text{Prob}\{X=x|Y=y,E=0\}$ is independent of y (use the fact that the first i rows of W are equal to one another): so $H(X|Y,E=0) = H(X|E=0)$ and the chain simplifies to

$$H(X|Y) = H(E) + \text{Prob}\{E=0\}\ H(X|E=0\} + \text{Prob}\{E=1\}\ H(X|Y,E=1\} \qquad (4)$$

As for $H(X|Y,E=1\}$ we go back to the construction of B, and write $H_{M-i}(P^*)$ as $H_{M-i}(X^*) = I(X^*;Y^*) = H(X^*) - H(X^*|Y^*) = H(P^*) - \log(M-i)$, where Y* is a random subset of M-i unlikely elements: however, $H(X^*|Y^*) = H(X|Y,E=1)$, and so, by comparison, $H(X|Y,E=1) = \log(M-i)$ (to see that $H(X^*|Y^*) = H(X|Y,E=1)$ observe first that the fact that the columns of submatrix C are all zero constrains Y to be a likely set, and so we can identify the ranges of Y and Y* in the by now obviuos way; this observed, the equality $\text{Prob}\{X^*=x|Y^*=y\} = \text{Prob}\{X=x|Y=y,E=1\}$ becomes again a matter of trivial probability manipulation; in a way, for E=1, Y and Y* are the "same" random variable). Chain (4) now simplifies to

$$H(X|Y) = H(E) + \text{Prob}\{E=0\}\ H(X|E=0\} + \text{Prob}\{E=1\}\ \log(M-i) \qquad (5)$$

However, $H(P_i)$ is equal precisely to the second side of (5) (insert the components of P_i as given by (3) into the formula for Shannon entropy to see this). So $H(X|Y) = H(P_i)$, which proves the theorem. QED

REFERENCES

1. I. Csiszár and J.Körner, *Information theory*, Academic Press, New York, 1982
2. A. Sgarro, A measure of semiequivocation, in *Advances in cryptology*, ed. by Ch. G.

Günther, Springer Verlag, Lectures Notes in Computer Science 330 (1988), pp. 375-387

3. G. Longo and A. Sgarro, A pragmatic way-out of the maze of uncertainty measures, in *Uncertainty in knowledge bases*, ed by B. Bouchon-Meunier, R.R. Yager, L.A. Zadeh, Springer Verlag, Lectures Notes in Computer Science 521 (1991), pp.370-376
4. A. Fioretto and A. Sgarro, Joint fractional entropy, in *Eurocode '90*, ed. by G. Cohen, P. Charpin, Lectures Notes in Computer Science 514, Springer Verlag (1991), pp. 292-297
5. A. Sgarro, A rate-distortion approach to information measures, Proc. of IFSA 91, Brussels, Computer, management & systems science, pp. 248-251
6. O.L. Mangasarian, *Nonlinear programming*, Mc Grow Hill, 1969

Uncertainty in Intelligent Systems
B. Bouchon-Meunier et al. (Editors)

Finalized Entropy for Fuzzy Questionnaires

P. Benvenuti, D. Vivona and M. Divari

Dept. Metodi e Modelli Matematici per le Scienze Applicate

Università di Roma "La Sapienza"

INTRODUCTION

The research about the axioms of fuzzy partitions leads to the notion of questionnaire and its generalizations by means of fuzzy questions. With every fuzzy questionnaire we can associate an algebra of crisp sets, which represents the reachable properties of the questionnaire.

By making use of the corresponding algebras, questionnaires are compared, and it's possible to give an axiomatization of their entropy.

Then we study the properties of a kind of entropy which, in the usual probabilistic setting, suitably faces the problem of classifying the events of the algebra associated to the questionnaire, when conditioned to an other algebra, whose events represent the classification which is in the purposes of the questionnaire.

1. FUZZY QUESTIONNAIRES

Let $(\Omega,\mathcal{F})$ be a measurable space.

DEFINITION 1.1. A *fuzzy question* q is a measurable map from Ω to $[0,1]$ and can be viewed as the fuzzy set whose membership function is q.

DEFINITION 1.2. A *fuzzy questionnaire* Q is a measurable map from Ω to $[0,1]^n$.

In fact, any questionnaire Q consists of a finite number of questions q_i $(i = 1,\dots,n)$ and then it can be considered as a finite collection of fuzzy sets whose membership functions are q_i.

With every fuzzy questionnaire Q there is an associated algebra $\mathcal{C}$ consisting of those crisp sets which are inverse images of Borel sets of $[0,1]^n$ through the map Q.

In general, this associated algebra is not finite; it is the minimal algebra containing all the algebras associated to the single questions.

In particular, the questions q_i can correspond to crisp sets, and then they produce a usual questionnaire, i.e. a finite partition of Ω. In this case, the algebras associated to the single questions are $\mathcal{C}_{q_i} \equiv \{\emptyset, A_i, A_i^c, \Omega\}$, and $\mathcal{C}$ is an atomic algebra, whose atoms are all the non-empty intersections of some sets A_i and the complements of the others.

The questionnaires can be classified by means of their associated algebras. This leads to an equivalence relation among questionnaires and to an order relation too.

DEFINITION 1.3. Two questionnaires $Q\colon \Omega \to [0,1]^n$, and $Q'\colon \Omega \to [0,1]^m$ are *equivalent* ($Q \simeq Q'$) if their associated algebras coincide, i.e.

$$Q \simeq Q' \iff \mathcal{C}_Q \equiv \mathcal{C}_{Q'} .$$

DEFINITION 1.4. The fuzzy questionnaire Q is *finer* ($\succeq$) than the fuzzy questionnaire Q' when $\mathcal{C}_Q \supset \mathcal{C}_{Q'}$, i.e.

$$Q \succeq Q' \iff \mathcal{C}_Q \supset \mathcal{C}_{Q'} .$$

For the sub-algebras of $\mathcal{F}$, the partial order has a minimum in the trivial algebra 2, consisting of the whole set Ω and the empty set $\emptyset$. This algebra is associated with the constant questionnaire to which everybody gives the same answer, it can be seen as the minimal questionnaire.

The algebra $\mathcal{F}$, on the contrary, is the maximum in the given order, but it is not always true that it is associated with any questionnaire.

The partial order relation, when restricted to *crisp* questionnaires, gives back the well-known notion of refinement for partitions [2].

2. ENTROPY OF A FUZZY QUESTIONNAIRE

With every questionnaire two different algebras can be associated. The first is a finite algebra $\mathcal{A}$ linked to the classification we would like to obtain from the questionnaire. It is the algebra generated by a finite partition Π_A of Ω (of course, the elements of Π_A are the atoms of A).

The second algebra $\mathcal{G}$ consists of all the events that are reachable through the questionnaire. If the questionnaire is a crisp one, then it consists of binary questions, to which crisp sets correspond, and $\mathcal{G}$ is finite.

On the other hand, the questions from a fuzzy questionnaire are characterized by an algebra (a σ-algebra, actually), which is not finite in general, i.e. by the family of all counter-images of the Borel sets through the membership function of the fuzzy question.

Then the questionnaire is represented by these two algebras: the first is called "*algebra of purposes*" ($\mathcal{A}$) and the second "*algebra of experiments*" ($\mathcal{G}$).

It may happen that $\mathcal{A}$ is contained in $\mathcal{G}$: then the questionnaire distinguishes the atoms of $\mathcal{A}$. If this is the case, the questionnaire gives an amount of information which coincides with the entropy of Π_A.

Otherwise, if $\mathcal{A}$ is not contained in $\mathcal{G}$, the questionnaire doesn't recognize Π_A and on this partition it provides an amount of information that can be evaluated by the difference between the entropy of Π_A and the remaining entropy of Π_A when it is conditioned to $\mathcal{G}$. In fact, through the questionnaire one cannot in general decide if ω belongs to one or another event of Π_A; but if we know that ω belongs to a set of $\mathcal{G}$, we can modify our uncertainty about Π_A, and the conditional entropy measures a kind of average residual uncertainty, [3].

In general, the questionnaire ties the algebra $\mathcal{A}$ with the algebra $\mathcal{G}$: through it, an entropy can be associated, depending on the questionnaire through the two algebras. This entropy must be viewed finalized to study the algebra $\mathcal{A}$.

A first example of entropy, associated with a fuzzy questionnaire, is given in the probability space $(\Omega, \mathcal{F}, P)$ by the following formula

$$H_{\mathcal{A}}(\mathcal{G}) = H(\mathcal{A}) - H(\mathcal{A}|\mathcal{G}) \tag{2.1}$$

where $H(\mathcal{A})$ is the Shannon entropy for the finite algebra $\mathcal{A}$ and $H(\mathcal{A}|\mathcal{G})$ is the conditional entropy of $\mathcal{A}$ given any σ-algebra $\mathcal{G} \subset \mathcal{F}$: here

$$H(\mathcal{A}) = E\left\{\sum_{A} \eta[P(A)]\right\} \tag{2.2}$$

$$H(\mathcal{A}|\mathcal{G}) = E\left\{\sum_{A} \eta[P(A|\mathcal{G})]\right\}, \tag{2.3}$$

where $\eta(t) = -t \log t$, $\eta(0) = \eta(1) = 0$.

The entropy $H_{\mathcal{A}}(\mathcal{G})$ was introduced by Yaglom, [3], when both $\mathcal{A}$ and $\mathcal{G}$ are finite. In this case, (2.1) defines a symmetric function, in fact from (A_1') of [1] it holds

$$H_{\mathcal{A}}(\mathcal{G}) = H_{\mathcal{G}}(\mathcal{A}) = H(\mathcal{A}) + H(\mathcal{G}) - H(\mathcal{A} \vee \mathcal{G}). \tag{2.4}$$

The extension of $H(\mathcal{A}|\mathcal{G})$ to the case when $\mathcal{G}$ is any σ-algebra, $\mathcal{A}$ being still finite, (see [1]), allows to extend definition (2.1) to more general case too. In this way, for every fixed finite algebra $\mathcal{A} \subset \mathcal{F}$, generated by a finite partition Π_A, the expression (2.1) gives a function defined on the family of all σ-algebras of $\mathcal{F}$, finite or not.

We can deduce, in general, the following properties:

(I) $\quad H_{\mathcal{A}}(\mathcal{G}) \in \mathbb{R}^+$

(II) $\quad H_{\mathcal{A}}(2) = 0$

(III) $\quad \mathcal{G}_1 \subset \mathcal{G}_2 \implies H_{\mathcal{A}}(\mathcal{G}_1) \leq H_{\mathcal{A}}(\mathcal{G}_2)$.

These properties are immediate, and show that, for every fixed $\mathcal{A}$, $H_{\mathcal{A}}(\mathcal{G})$ is an entropy according with Forte, [2]. Moreover, the entropy is bounded, because of

(IV) $\quad H_{\mathcal{A}}(\mathcal{G}) \leq H_{\mathcal{A}}(\mathcal{F}) = H(\mathcal{A})$.

From theorem 12.1 of [1], immediately it follows

(V) $$\mathcal{G}_n \nearrow \mathcal{G} \implies H_{\mathcal{A}}(\mathcal{G}_n) \nearrow H_{\mathcal{A}}(\mathcal{G})\,.$$

When $\mathcal{G}$ is kept fixed, it holds the following:

PROPOSITION 2.1.

(VI) $$\mathcal{A}_1 \subset \mathcal{A}_2 \implies H_{\mathcal{A}_1}(\mathcal{G}) \leq H_{\mathcal{A}_2}(\mathcal{G})\,,$$

(VII) $$H_2(\mathcal{G}) = 0\,.$$

PROOF. From: $H(\mathcal{A}_1 \vee \mathcal{A}_2|\mathcal{G}) = H(\mathcal{A}_1|\mathcal{G}) + H(\mathcal{A}_2|\mathcal{A}_1 \vee \mathcal{G})$, (see[1], eq. C_1), as $\mathcal{A}_1 \vee \mathcal{A}_2 = \mathcal{A}_2$, one has $H(\mathcal{A}_2|\mathcal{G}) = H(\mathcal{A}_1|\mathcal{G}) + H(\mathcal{A}_2|\mathcal{A}_1 \vee \mathcal{G})$; for $\mathcal{G} = 2$ it becomes $H(\mathcal{A}_2) = H(\mathcal{A}_1) + H(\mathcal{A}_2|\mathcal{A}_1)$.

From the two last equalities, one gets:
$H(\mathcal{A}_2) - H(\mathcal{A}_2|\mathcal{G}) = H(\mathcal{A}_1) - H(\mathcal{A}_1|\mathcal{G}) + H(\mathcal{A}_2|\mathcal{A}_1) - H(\mathcal{A}_2|\mathcal{A}_1 \vee \mathcal{G})$, as: $H(\mathcal{A}_2|\mathcal{A}_1 \vee \mathcal{G}) \leq H(\mathcal{A}_2|\mathcal{G})$, (see [1], eq. C_3), one gets the inequality (VI).

Condition (VII) espresses that the minimum value of $H_{\mathcal{A}}(\mathcal{G})$, as $\mathcal{A}$ varies, is null. The proof is immediate as $H(2) = 0$ and $H(2|\mathcal{G}) = 0$.

Moreover, properties on links between $\mathcal{A}$ and $\mathcal{G}$ are valid:

PROPOSITION 2.2.

(VIII) $$\mathcal{A} \subset \mathcal{G} \implies H_{\mathcal{A}}(\mathcal{G}) = H(\mathcal{A})\,,$$

(IX) $$\mathcal{G} \subset \mathcal{A} \implies H_{\mathcal{A}}(\mathcal{G}) = H(\mathcal{G})\,.$$

PROOF. For $\mathcal{A} \subset \mathcal{G}$, one has $H(\mathcal{A}|\mathcal{G}) \leq H(\mathcal{G}/\mathcal{G}) = 0$, whence the thesis follows. For $\mathcal{G} \subset \mathcal{A}$, one obtains $H(\mathcal{A}/\mathcal{G}) = H(\mathcal{G}) - H(\mathcal{A})$, and so the assertion follows.

For independent algebras, the following theorem holds:

THEOREM 2.3. *The following are equivalent:*

a) $\mathcal{A}$ *and* $\mathcal{G}$ *are independent,*

b) $H_{\mathcal{A}}(\mathcal{G}) = 0\,.$

PROOF. If $\mathcal{A}$ and $\mathcal{G}$ are independent, $H(\mathcal{A}|\mathcal{G}) = H(\mathcal{A})$, and then $H_{\mathcal{A}}(\mathcal{G}) = 0$.

Viceversa, $\forall A \in \mathcal{A}$, $G \in \mathcal{G}$, set $A_1 = A$, $A_2 = A^c$ and $G_1 = G$, $G_2 = G^c$, $\mathcal{A}' = \{\emptyset, A_1, A_2, \Omega\}$, and $\mathcal{G}' = \{\emptyset, G_1, G_2, \Omega\}$.

Assuming that $H_{\mathcal{A}}(\mathcal{G}) = 0$, by monotonicity, we obtain $H_{\mathcal{A}'}(\mathcal{G}') = 0$, i.e. $H(\mathcal{A}') = H(\mathcal{A}'|\mathcal{G}')$. Therefore, set $p_i = P(A_i)$, $q_j = P(G_j)$, and $r_{ij} = P(A_i \cap G_j)$ it follows:

$$\sum_i p_i \ln p_i = \sum_{i,j} r_{ij} \ln \frac{r_{ij}}{q_j}\,,$$

and then

$$\sum_{i,j} r_{ij} \ln r_{ij} = \sum_{i,j} r_{ij} \ln(p_i q_j)\,,$$

so, we obtain $r_{ij} = p_i q_j$, because the function $\ln x$ is concave. Then, $\forall A \in \mathcal{A}$ and $\forall G \in \mathcal{G}$ it holds $P(A \cap G) = P(A)P(G)$.

The independence condition can be expressed by means of finalized entropy also for not necessarily finite algebras.

THEOREM 2.4. *The following are equivalent:*

a) $\mathcal{G}_1$ *and* $\mathcal{G}_2$ *are independent,*

b) $\forall \mathcal{A}_1$ *finite,* $\mathcal{A}_1 \subset \mathcal{G}_1$ *and* $\forall \mathcal{G}_2$ *finite,* $\mathcal{A}_2 \subset \mathcal{G}_2$ *it is* $H_{\mathcal{A}_1}(\mathcal{A}_2) = 0$.

In this theorem the perfect symmetry of the independence conditions becomes evident. As for finite algebras it holds:

$$H_{\mathcal{A}_1}(\mathcal{A}_2) = H_{\mathcal{A}_2}(\mathcal{A}_1) = H(\mathcal{A}_1) + H(\mathcal{A}_2) - H(\mathcal{A}_1 \vee \mathcal{A}_2),$$

it is immediate that the two conditions:

$$H_{\mathcal{A}_1}(\mathcal{A}_2) = 0, \; H_{\mathcal{A}_2}(\mathcal{A}_1) = 0 \quad \text{and} \quad H(\mathcal{A}_1 \vee \mathcal{A}_2) = H(\mathcal{A}_1) + H(\mathcal{A}_2)$$

are equivalent.

3. FURTHER PROPERTIES OF $H_{\mathcal{A}}(\mathcal{G})$

The finalized entropy $H_{\mathcal{A}}(\mathcal{G})$ is defined for every algebra $\mathcal{G}$. But interesting and meaningful properties hold when $\mathcal{G}$ is restricted to a finite questionnaire. Besides the already seen symmetry property, the following hold true:

PROPOSITION 3.1. *For every finite algebra* $\mathcal{G}_1$ *and* $\mathcal{G}_2$, *one has:*

$$H_{\mathcal{A}}(\mathcal{G}_1 \vee \mathcal{G}_2) \leq H_{\mathcal{A}}(\mathcal{G}_1) + H(\mathcal{G}_2)\,, \tag{3.1}$$

$$H_{\mathcal{A}}(\mathcal{G}_1 \vee \mathcal{G}_2) \leq H_{\mathcal{A}}(\mathcal{G}_2) + H(\mathcal{G}_1)\,. \tag{3.2}$$

PROOF. For (2.4):

$$H_{\mathcal{A}}(\mathcal{G}_1 \vee \mathcal{G}_2) = H_{\mathcal{G}_1 \vee \mathcal{G}_2}(\mathcal{A}) = H(\mathcal{G}_1 \vee \mathcal{G}_2) - H(\mathcal{G}_1 \vee \mathcal{G}_2 | \mathcal{A}),$$

being $H(\mathcal{G}_1 \vee \mathcal{G}_2 | \mathcal{A}) \geq H(\mathcal{G}_1 | \mathcal{A})$ and $H(\mathcal{G}_1 \vee \mathcal{G}_2) \leq H(\mathcal{G}_1) + H(\mathcal{G}_2)$,

one gets $H_{\mathcal{A}}(\mathcal{G}_1 \vee \mathcal{G}_2) \leq H(\mathcal{G}_1) + H(\mathcal{G}_2) - H(\mathcal{G}_1 | \mathcal{A})$,

whence the assertion follows.

In analogous way one proves (3.2) by the monotonicity of the conditional entropy:

PROPOSITION 3.2. *If* $\mathcal{A}$ *and* $\mathcal{G}_1$ *are independent, it holds*

$$H_{\mathcal{A}}(\mathcal{G}_1 \vee \mathcal{G}_2) \leq H(\mathcal{G}_2)\,, \tag{3.3}$$

likewise, if $\mathcal{A}$ *and* $\mathcal{G}_2$ *are independent, it is*

$$H_{\mathcal{A}}(\mathcal{G}_1 \vee \mathcal{G}_2) \leq H(\mathcal{G}_1)\,. \tag{3.4}$$

PROPOSITION 3.3. *If $\mathcal{G} = \mathcal{G}_1 \vee \mathcal{G}_2$ is finite and $\mathcal{G}_1 \subset \mathcal{A}$, or $\mathcal{G}_2 \subset \mathcal{A}$ one gets*

$$H_{\mathcal{A}}(\mathcal{G}_1 \vee \mathcal{G}_1) \leq H_{\mathcal{A}}(\mathcal{G}_1) + H_{\mathcal{A}}(\mathcal{G}_2). \tag{3.5}$$

PROOF. If $\mathcal{G}_1 \subset \mathcal{A}$, one has $H_{\mathcal{A}}(\mathcal{G}_1) = H_{\mathcal{G}_1}(\mathcal{A}) = H(\mathcal{G}_1)$, being $H(\mathcal{G}_1|\mathcal{A}) = 0$. Then, from (3.2) one obtains the thesis. In analogous way one gets (3.5) if $\mathcal{G}_2 \subset \mathcal{A}$.

PROPOSITION 3.4. *If $\mathcal{G}_1$ and $\mathcal{G}_2$ are finite and independent algebras, it holds*

$$H_{\mathcal{A}}(\mathcal{G}_1 \vee \mathcal{G}_2) \geq H_{\mathcal{A}}(\mathcal{G}_1) + H_{\mathcal{A}}(\mathcal{G}_2). \tag{3.6}$$

PROOF. Being $\mathcal{G}_1$ and $\mathcal{G}_2$ independent,

$$H_{\mathcal{A}}(\mathcal{G}_1 \vee \mathcal{G}_2) = H(\mathcal{G}_1) + H(\mathcal{G}_2) - H(\mathcal{G}_1 \vee \mathcal{G}_2|\mathcal{A})$$

and for [1, cap. 3], one has $H(\mathcal{G}_1 \vee \mathcal{G}_2|\mathcal{A}) \leq H(\mathcal{A}_1|\mathcal{G}) + H(\mathcal{A}_2|\mathcal{G})$. From this relation, the thesis is true.

COROLLARY 3.5. *If $\mathcal{G}_1$ and $\mathcal{G}_2$ are finite and independent algebras and $\mathcal{G}_1 \subset \mathcal{A}$ (or $\mathcal{G}_2 \subset \mathcal{A}$) the following equality holds*

$$H_{\mathcal{A}}(\mathcal{G}_1 \vee \mathcal{G}_2) = H_{\mathcal{A}}(\mathcal{G}_1) + H_{\mathcal{A}}(\mathcal{G}_2). \tag{3.7}$$

Moreover, as the algebra of purposes varies, some properties are valid.

PROPOSITION 3.6. *If $\mathcal{A} = \mathcal{A}_1 \vee \mathcal{A}_2$ and $\mathcal{G}$ is any algebra, the following inequalities hold:*

$$H_{\mathcal{A}_1 \vee \mathcal{A}_2}(\mathcal{G}) \leq H_{\mathcal{A}_1}(\mathcal{G}) + H_{\mathcal{A}_2}(\mathcal{A}_1 \vee \mathcal{G}), \tag{3.8}$$

$$H_{\mathcal{A}_1 \vee \mathcal{A}_2}(\mathcal{G}) \leq H_{\mathcal{A}_2}(\mathcal{G}) + H_{\mathcal{A}_1}(\mathcal{A}_2 \vee \mathcal{G}). \tag{3.9}$$

PROOF. Being $H(\mathcal{A}_1 \vee \mathcal{A}_2) \leq H(\mathcal{A}_1) + H(\mathcal{A}_2)$ and $H(\mathcal{A}_1 \vee \mathcal{A}_2|\mathcal{G}) = H(\mathcal{A}_1|\mathcal{G}) + H(\mathcal{A}_2|\mathcal{A}_1 \vee \mathcal{G})$ [1, cap. 4], one has $H_{\mathcal{A}_1 \vee \mathcal{A}_2}(\mathcal{G}) = H(\mathcal{A}_1 \vee \mathcal{A}_2) - H(\mathcal{A}_1 \vee \mathcal{A}_2|\mathcal{G}) \leq H(\mathcal{A}_1) + H(\mathcal{A}_2) - H(\mathcal{A}_1|\mathcal{G}) - H(\mathcal{A}_2|\mathcal{A}_1 \vee \mathcal{G}) = H_{\mathcal{A}_1}(\mathcal{G}) + H_{\mathcal{A}_2}(\mathcal{A}_1 \vee \mathcal{G})$.

PROPOSITION 3.7. *If $\mathcal{A} = \mathcal{A}_1 \vee \mathcal{A}_2$, $\mathcal{A}_1$, $\mathcal{A}_2$ independent and $\mathcal{G}$ is any algebra, it holds:*

$$H_{\mathcal{A}_1 \vee \mathcal{A}_2}(\mathcal{G}) = H_{\mathcal{A}_1}(\mathcal{G}) + H_{\mathcal{A}_2}(\mathcal{A}_1 \vee \mathcal{G}).$$

These properties enlighten some formal difference between Shannon and finalized entropies. This becomes particularly evident when comparing entropies and considering independent algebras.

4. A PROBABILISTIC EXAMPLE

One can have a first example of entropy, associated with a fuzzy questionnaire, in a probabilistic context. Let $(\Omega, \mathcal{F}, P)$ be a probability space; the simpler questionnaire one can have is given by one question $q\colon \Omega \longrightarrow [0,1]$. The reachable properties are represented by the sets of counter-images of Borel σ-algebra of $[0,1]$:

$$\mathcal{G} = q^1(\mathcal{B}) \subset \mathcal{F}\,.$$

For every $B \subset \mathcal{B}$, let $p(B) = P(q^{-1}B)$ and $p_i(B) = P(A_i \cap q^{-1}B)$. As $p_i \leq p\ \forall i$, there exists a measurable function $\varphi_i\colon [0,1] \longrightarrow [0,1]$ such that

$$p_i(B) = \int_B \varphi_i(x)dp \qquad \forall B\,.$$

By definition one has

$$P(A_i|\mathcal{G}) = \varphi_i[q(\omega)]\,.$$

In fact, $\varphi_i[q(\omega)]$ is $\mathcal{B}$-measurable and integration by substitution gives:

$$\int_{q^{-1}B} P(A_i|\mathcal{G})dP = \int_B \varphi_i(x)dp = p_i(B) = P(A_i \cap q^{-1}B)\,.$$

For the conditional entropy one obtains:

$$H(\mathcal{A}|\mathcal{G}) = \int_\Omega \sum_i \eta[P(A_i|\mathcal{G})]dP = \sum_i \int_{[0,1]} \eta[\varphi_i(x)]dp$$

and then,

$$H_{\mathcal{A}}(\mathcal{G}) = \textstyle\sum_i \eta[p_i([0,1])] - \sum_i \displaystyle\int_{[0,1]} \eta[\varphi_i(x)]dp\,,$$

with $H_{\mathcal{A}}(\mathcal{G}) \geq 0$, because η is a convex function and $p_i(A_i) = \int_{[0,1]} \varphi_i(x)dp$.

One gets that $H_{\mathcal{A}}(\mathcal{G}) = 0$ iff $\forall i$, $\vee x$, $\varphi_i(x) \equiv P(A_i)$, i.e.

$$P(A_i \cap q^{-1}B) = P(A_i) \cdot P(q^{-1}B) \quad \forall B \in \mathcal{B}.$$

REFERENCES

[1] P. BILLINGSLEY: *Ergodic theory and information*, Wiley, New York, 1965.

[2] B. FORTE: *Measures of information: the general axiomatic theory*, R.A.I.R.O. 3° annèe, N R-2, (1969), 63-90.

[3] A.M. YAGLOM, I.M. YAGLOM: *Probability and information*, Riedel Publishing company, 1983.

CHAPTER 6:

UNCERTAINTY IN SOCIAL AND BEHAVIOURAL SCIENCES

Uncertainty in Intelligent Systems
B. Bouchon-Meunier et al. (Editors)

A Stochastic Theory for System Failure Assessment

Jean-Claude Falmagne* and Jean-Paul Doignon†

Summary. This paper presents an overview of some work by Doignon, Falmagne and their colleagues (Doignon & Falmagne, 1985; Falmagne et al., 1990) aimed at the development of a comprehensive theory for the efficient assessment of the state of a system. The paradigm motivating this research comes from computer-assisted instruction: the system is a student, whose knowledge state consists of all notions mastered at a given time. A computer program sequentially chooses questions to ask, and checks whether the student's responses are correct or false, gradually narrowing down the possible knowledge states. In general, the state of a system can be described by a collection of features, indicating the parts of the system that properly work. The problem is to efficiently uncover all these parts: in case of system failure, a number of tests on individual features must be performed that is as small as possible. Both combinatorial and probabilistic aspects of our theory will be illustrated here.

The basic concept is a system, the state of which can be described by a collection of features. In turn, these features indicate the parts of the system that correctly operate. The assessment problem consists in quickly identifying the state of the system, among the family of all feasible states. Before describing some possible assessment procedures, we shall consider in some details the collection of system states. A paradigmatic example comes from knowledge evaluation in computer-assisted instruction.

1. KNOWLEDGE SPACES

Let Q be a finite set of *questions* or *problems* in a particular area of information, let $\mathcal{K}$ be a distinguished family of subsets of Q, and suppose that both $\varnothing$ and Q are in $\mathcal{K}$. Then, we say that the pair $(Q, \mathcal{K})$ is a *knowledge structure* or, equivalently, that $\mathcal{K}$ is a *knowledge structure (on Q)*. (Since we always have $Q = \cup\mathcal{K} \in \mathcal{K}$, explicit reference to the set Q can be omitted without ambiguity.) The set Q is the *domain* of the knowledge structure, and the elements of $\mathcal{K}$ are refered to as *(knowledge) states*. An important axiom is that $\mathcal{K}$ is closed under union. (Motivation will be given below.) A knowledge structure satisfying this axiom is a *knowledge space*. A knowledge state is

*School of Social Sciences, University of California, Irvine, CA 92717. This work is supported by Grant BNS 8919-068 to the first author at the University of California, Irvine.
†Université Libre de Bruxelles, C.P. 216 Bd. du Triomphe 1050 Bruxelles Belgium.

thus a set containing all the questions that some subject is capable of solving correctly. The goal of the assessment procedure is to uncover a subject's knowledge state by asking appropriate questions. The terminology associated with 'knowledge' and 'questions' will be used throughout. The reader should have no difficulty in translating our concepts and results in the context of more general systems: Q is the set of all observable features of the system, and any K from $\mathcal{K}$ is a possible state of the system.

Two other concepts have been investigated, each of which provide exactly the same information as a knowledge space. Let Q be as above, and let σ be a mapping associating to each element q of Q, a nonempty family $\sigma(q)$ of subsets of Q. The function σ is used to formalize some inferences that can be drawn from observing that a subject has mastered some question $q \in Q$. Specifically, the subject must also have mastered all the questions in at least one of the elements of $\sigma(q)$. For any q in Q, each C in $\sigma(q)$ is a *clause for q*. The pair (Q, σ) is called a *a surmise system* if the following three conditions are satisfied for any $q \in Q$:

(1) $q \in C$ for any $C \in \sigma(q)$.
(2) If $q' \in C \in \sigma(q)$, then $C' \subseteq C$ for some $C' \in \sigma(q')$.
(3) If $C' \subseteq C$ for some $C, C' \in \sigma(q)$, then $C = C'$.

The function σ is then said to be a *surmise function* on Q. The third concept is that of an *entail* relation, that is, a transitive extension $\mathcal{P}$ of the inclusion relation $\subseteq$ on 2^Q, satisfying:

$$A\mathcal{P}B \quad \Longleftrightarrow \quad A\mathcal{P}\{x\} \text{ for all } x \in B\,.$$

The interpretation of $A\mathcal{P}B$ is as follows: if a student has not mastered any of the questions in A, this student will also fail all the questions in B. (Chance factors, such as lucky guesses or careless errors, are ignored at this stage of the work, and will be considered later in this paper.) We have the following result.

Theorem 1. *Let Q be a finite set. There are three one-to-one correspondences linking, pairwise, the following three sets:*

(i) *the set of all knowledges spaces on Q;*
(ii) *the set of all surmise functions on Q;*
(iii) *the set of all entail relations on 2^Q.*

This theorem is a straightforward consequence of more general results by Doignon and Falmagne (1985) and Koppen and Doignon (1990) expressed in terms of Galois connections (see also Müller, 1989; Dowling, 1991).

The linkage between knowledge spaces and entail relations provides the theoretical basis for a computer routine, called QUERY, designed to build a knowledge space on a given set of questions by querying a human expert, e.g. an experienced teacher (Koppen, in press). The assumption is that an expert's conception of the educational material is organized in terms of a personal underlying knowledge space which can be rendered explicit by an application of the QUERY routine.

In practice, the expert is asked to respond to queries such as: "*Suppose that a student has failed to solve questions* $q_1, \ldots, q_n$. *Would this student also fail to solve question* q_{n+1}*?*" (A positive response to this query correspond the statement $\{q_1, \ldots, q_n\}\mathcal{P}\{q_{n+1}\}$ of the entail relation $\mathcal{P}$.) The unique knowledge space consistent with the responses to **all** these questions is constructed by progressive elimination of subsets of Q which cannot be states. Various inference rules make use of the considerable redundancy between queries, so that, in fact, only a minute fraction of all the possible queries is actually asked. This routine has been used by Kambouri et al. (1991) to construct the knowledge spaces of five expert-teachers, for 50 questions in highschool mathematics. These experts were capable of completing the task in 10-15 hours of questioning by the computer. The total number of queries, for any expert, was of the order of only a few thousand. Even more remarkable is the number of resulting states, which ranged from 881 to 7,932 (thus never exceeding $10^{-9}\%$ of 2^{50}, the number of subsets in the set of 50 questions).

2. PROBABILISTIC ASSESSMENT PROCEDURES

Applying these concepts in knowledge assessment requires a probabilistic framework. For example, we must suppose that the knowledge states occur with different frequencies in the population of reference. Moreover, the state of the subject does not necessarily determine the responses: a subject may, with some probability, commit a careless error in responding to a question that has in fact been mastered (i.e., a question contained in his or her state). Similarly, a subject could, in some cases, guess the correct response to a question not mastered. Accordingly, probabilities must enter in several ways into the formulation of knowledge assesssment procedures: an efficient assessment routine must make use of the a priori probabilities of the states in the population, and could also take into account the fact that a response to a question may be a noisy reflexion of the subject's state. Two examples of such probabilistic assessment procedures, taken from Falmagne and Doignon (1988a, b), are briefly described here.

Both examples involve Markov processes in discrete time, with $0, 1, \ldots, n, \ldots$ denoting the successive time points. It is supposed that the student's state may vary over time. Accordingly, this state is symbolized at time n by a random variable $\mathbf{K}_n$ taking its values in the family $\mathcal{K}$ of knowledge states. We assume, however, that there exists a probability distribution π on $\mathcal{K}$, such that $\mathbb{P}(\mathbf{K}_n = K) = \pi(K)$, independent of n. The only observable effect of the distribution π concerns the validity of the response provided to the questions asked. The notations $\mathbf{Q}_n$ and $\mathbf{R}_n$ stand for random variables formalizing the question asked and the response provided at time n. Thus, $\mathbf{Q}_n \in Q$ and $\mathbf{R}_n \in \{0,1\}$, with $\mathbf{R}_n = 1$ symbolizing a correct response. All the random variables $\mathbf{K}_n, \mathbf{Q}_n, \mathbf{R}_n$, $n = 0, 1, \ldots$ are jointly distributed. At time n, the conditional probability of a correct response, given the question asked and the subject's state, is specified by parameters β_q, δ_q, $q \in Q$, according to the relation:

$$\mathbb{P}(\mathbf{R}_n = 1 \,|\, \mathbf{Q}_n = q, \mathbf{K}_n = K) = \begin{cases} 1 - \beta_q & \text{if } q \in K, \\ \delta_q & \text{if } q \notin K. \end{cases} \tag{1}$$

Thus, β_q and δ_q are 'careless error' and 'lucky guess' parameters attached to question q. Next we specify how the value $\mathbf{Q}_n = q$ is selected by the assessement procedure. Here, our two probabilistic assessment procedures part company because they rely on different encodings of the information accumulated about the student's knowledge state.

The first procedure to be described (Falmagne & Doignon , 1988b) will lead to a Markov chain. At time n, we have a random variable $\mathbf{M}_n$ taking as values the subsets of $\mathcal{K}$. The meaning of $\mathbf{M}_n = m_n$ is that $m_n \subseteq \mathcal{K}$ is the collection of all the knowledge states which are plausible at time n. This subset m_n is called the *marker*. We then choose the question to be asked so as to split the marker into two parts as equal as possible; specifically, we select a question q that minimizes the absolute difference of the number of states that contain q and the number of states that do not contain q. (If there are several such questions, we uniformally draw among them.) This rule is applied as long as the marker contains more than one state. When $\mathbf{M}_n$ contains a single state K, we first add to the marker the collection $N(K)$ all the states of $\mathcal{K}$ that differ from K by exactly one question. The family $N(K)$ is called the *neighborhood* of K. We then apply the same selection principle to the resulting family of states $\{K\} \cup N(K)$.

The final stage of this first procedure specifies how the marker $\mathbf{M}_{n+1}$ is obtained from the marker $\mathbf{M}_n$ at the previous step. There are two cases.

(1) The marker contains more than one state. In this case, we just keep the states that are confirmed by the subject's response at step n: if $\mathbf{R}_n = 1$, we remove from the marker all the states not containing the question asked; otherwise, we remove all the states containing that question.

(2) The marker contains a single state K. If that state is confirmed by the observed response, we leave it unchanged. Thus, if the question asked is in K and the response is correct, or if the question asked is not in K and the response is incorrect, we set $\mathbf{M}_{n+1} = \{K\}$. If the state K is refuted, we consider its neighborhood $N(K)$ and add to the marker the elements of $N(K)$ that are confirmed.

To summarize, we have a discrete parameter stochastic process $(\mathbf{X}_n)$, $n = 0, 1, \ldots,$

$$\mathbf{X}_n = (\mathbf{R}_n, \mathbf{Q}_n, \mathbf{K}_n, \mathbf{M}_n),$$

with state space $\{0,1\} \times Q \times \mathcal{K} \times 2^{\mathcal{K}}$. ('State' is used here with its standard meaning in stochastic processes. This usage will not reoccur in this paper.) The process must be capable of uncovering (or at least approximating in some way) the knowledge state producing the responses, whatever that state may be. For that reason, we require that the process be (theoretically) able to visit, starting from any state, any other state by taking unit steps, consisting of replacing a state K with a state in $N(K)$. Under the assumption that $(Q, \mathcal{K})$ is a space, an easily formulated requirement is as follows: for any two states $K, K' \in \mathcal{K}$, there is at least one state K'' in $N(K)$ with either

$$K'' = K \cup \{q\}, \text{ and } q \in K',$$

or

$$K'' = K \setminus \{q\}, \text{ and } q \notin K'.$$

(Thus there is a neighbor state K'' of K, with K'' 'closer' to K' than K is). A knowledge space satisfying the last requirement for all distinct states K and K' is *well graded*. Notice in passing that a well graded knowledge space $(Q, \mathcal{K})$ satisfies the following property: any maximal chain of states is a maximal chain of subsets of Q.

The assessment procedures $\mathbf{X}_n = (\mathbf{R}_n, \mathbf{Q}_n, \mathbf{K}_n, \mathbf{M}_n)$ just defined on a well graded knowledge space are said to be *unitary* in Falmagne and Doignon (1988b). We quote a few results from this paper.

Theorem 2. *The following processes are homogeneous Markov chains:*

$$(\mathbf{X}_n) = (\mathbf{R}_n, \mathbf{Q}_n, \mathbf{K}_n, \mathbf{M}_n), \qquad (\mathbf{M}_n), \qquad (\mathbf{Q}_n, \mathbf{M}_n).$$

Theorem 3. *Suppose that $\beta_q = \delta_q = 0$ for each question q, and that the distribution π on $\mathcal{K}$ representing the student is concentrated on one state K_0. Then*

$$\lim_{n \to \infty} \mathbb{P}(\mathbf{M}_n = \{K_0\}) = 1\,.$$

This situation is almost deterministic. Nevertheless it is worth considering it: we have thus checked that our assessment procedures are robust against 'sporadic' perturbations.

Theorem 4. *Suppose $\delta_q = 0$ for any question q, and that the distribution π is concentrated on one state K_0. Then, the Markov chain $(\mathbf{M}_n)$ has a unique ergodic set $\mathcal{E}$, consisting only of $\{K_0\}$ and possibly some Markov states $\{K\}$ with $K \subseteq K_0$. Moreover, if $\beta_q > 0$ for each q in Q, then $\mathcal{E}$ is in fact the family of all those Markov states such that $K \subseteq K_0$.*

The uniqueness of the ergodic set can be extended to other situations.

Theorem 5. *Suppose $\beta_q = \delta_q = 0$ for each question q, and let π denote the distribution on the set of states, representing the student. Then, the Markov chain $(\mathbf{M}_n)$ has the unique ergodic set consisting of Markov states $\{K\}$ with*

$$\bigcap\{L \in \mathcal{K} \,|\, \pi(L) > 0\} \subseteq K \subseteq \bigcup\{L \in \mathcal{K} \,|\, \pi(L) > 0\}.$$

Other results by Falmagne and Doignon (1988b) show that it is possible, in some cases, to estimate the distribution π just from the observed frequencies of Markov states of the chain $(\mathbf{M}_n)$. However, it is also shown that this is not always true. Here is a brief explanation: we observe only $|Q|$ parameters (validities of responses to questions), while

π has $|\mathcal{K}| - 1$ parameters. This result can be used when, as may sometimes be the case, π is concentrated on a small number of states.

In a different type of probabilistic assessment procedure (Falmagne and Doignon, 1988a), the role of the marker is replaced by a function $K \mapsto \mathbf{L}_n(K)$, assigning, at each time n, a numerical likelihood to each of the states in $\mathcal{K}$. For each n, $\mathbf{L}_n$ is a random vector taking its values in the $(|\mathcal{K}| - 1)$-dimensional simplex.

The principle governing the choice of the question at each step n is similar in spirit to that used in the preceding procedure. We set $\mathbf{Q}_n = q$ so as to split the mass of $\mathbf{L}_n$ into two parts as equal as possible. In other words a question q is picked that minimizes the quantity:

$$|\sum_{K\in\mathcal{K},\text{ with } K\ni q} \mathbf{L}_n(K) \quad - \quad \sum_{K\in\mathcal{K},\text{ with } K\not\ni q} \mathbf{L}_n(K)|\,.$$

(If several questions achieve the minimum, a random choice is performed, with uniform probabilities.) Another rule for question selection, also described in Falmagne and Doignon (1988a), is based on the minimization of the expected entropy of the distribution $\mathbf{L}_{n+1}$.

Regarding the updating of the likelihood function, two formulas are investigated in the paper, that compute $\mathbf{L}_{n+1}$ in termes of $\mathbf{L}_n$ and the information collected from the subject's response. We will not go into details here. The main result asserts that, when the distribution π is concentrated on one state K_0, then $\mathbf{L}_n(K_0) \to 1$ as $n \to \infty$.

3. STOCHASTIC LEARNING PATHS

We now turn to a rather different aspect of the theory. One of the conditions of a successful educational system is a close monitoring of student's progress, providing responses to various questions such as: Exactly what has been learned by a typical student entering the kth grade? What are the most travelled learning paths through the material? What are the most difficult concepts? What is the average time between this and that stage of the learning process, etc. The basic concepts of knowledge space theory provide an appropriate framework to that effect. The rest of this paper is devoted to a stochastic process describing the progress of a sample of subjects (for example, students learning a particular field) over a period of time (Falmagne, in press). The succession of observable response patterns will be explained by the transitions of the subjects through the states of a knowledge structure.

The basic paradigm is one in which the subjects in a sample have been tested repeatedly, using the same set Q of problems, at times $t_1 < t_2 < \ldots < t_n$. We write $\mathbf{A}_t$ for the pattern of responses (i.e. the subset of problems correctly solved), observed at time t. Thus, for any real number $t \geq 0$, $\mathbf{A}_t$ is a random variable taking its values in 2^Q. We have a continuous parameter stochastic process $(\mathbf{A}_t)_{t\geq 0}$, and the main task for a theory is to provide a prediction, fully specified except for the values of some parameters, for the joint probabilities

$$\mathbb{P}(\mathbf{A}_{t_1} = A_1, \mathbf{A}_{t_2} = A_2, \ldots, \mathbf{A}_{t_n} = A_n) \tag{2}$$

that a (randomly selected) subject tested at times $t_1, t_2, \ldots, t_n$ would provide successive response patterns $A_1, A_2, \ldots, A_n$, the times of the test being chosen arbitrarily. Among the potentially relevant factors are: the skills or learning rates of the subjects, the difficulty of the problems, the possibility or careless errors or lucky guesses, the knowledge structure itself, the probabilities of taking one or the other of the possible learning paths. The story behind the joint probabilities in (2) is obviously intricate. The brief outline of the theory given here—which is all that can be given in the allotted space—will be made in the form of comments on a concrete example, involving the knowledge space

$$\mathcal{F} = \{\varnothing, \{1\}, \{2\}, \{1,2\}, \{1,2,3\}, \{1,2,4\}, \{1,2,3,4\}, \{1,2,3,5\}, Q\} \tag{3}$$

on $Q = \{1,2,3,4,5\}$. This knowledge space is pictured in Fig. 1. To simplify matters, we shall only consider the case of a single test: $n = 1$, $t = t_1$ in (2).

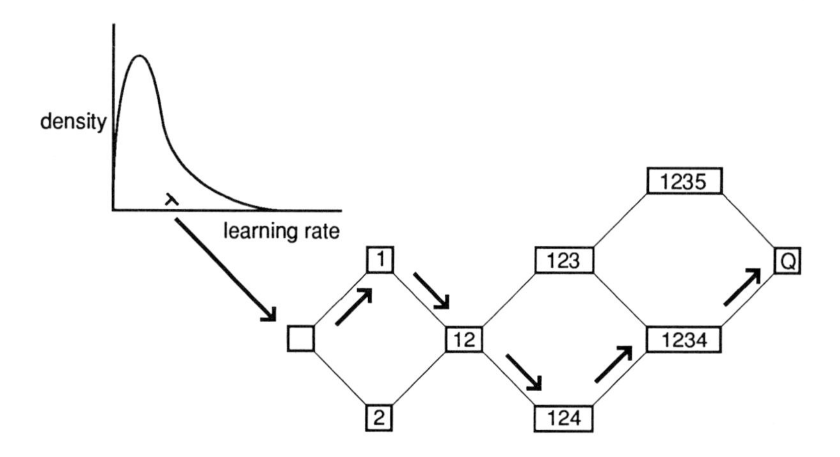

Fig. 1. The knowledge space $\mathcal{F}$ of Eq. 3

Scanning Fig. 1 from left to right, we discuss the learning history of a subject, sampled from a specified population. We suppose that this subject is equipped with a learning rate $\lambda > 0$, this number being a sample value of some random variable $\mathbf{L}$ associated with that particular population of subjects. (The density of $\mathbf{L}$ is pictured at the upper left corner of the figure.) We also suppose that this subject adopts the learning path

$$\varnothing \subseteq \{1\} \subseteq \{1,2\} \subseteq \{1,2,4\} \subseteq \{1,2,3,4\} \subseteq Q\,, \tag{4}$$

indicated by the arrows linking these states on the figure. There are six possible learning paths. These are, with obvious abbreviations: 12354, 12345, 12435, 21354, 21345, 21435. We assume that the choice between them is governed by a probability distribution, specified by parameters to be estimated. Thus, the subject starts the learning process in the knowledge state $\varnothing$. For a subject moving along path 12435, the first question to be mastered is 1, and we suppose that the time required to master that question is a random variable $\mathbf{T}_{\lambda,1}$ depending on the learning rate λ of the subject, and of the difficulty of Question 1. The next question encountered by this subject is 2, which is mastered in a random time $\mathbf{T}_{\lambda,2}$. Then successively come 4, 3 and finally 5, with their associated learning time random variables $\mathbf{T}_{\lambda,i}$. The probability that this particular subject is in state $\{1,2\}$ at some time t of a test is thus

$$\mathbb{P}(\mathbf{T}_{\lambda,1} + \mathbf{T}_{\lambda,2} \leq t < \mathbf{T}_{\lambda,1} + \mathbf{T}_{\lambda,2} + \mathbf{T}_{\lambda,4}) \\ = \mathbb{P}(\mathbf{T}_{\lambda,1} + \mathbf{T}_{\lambda,2} \leq t) - \mathbb{P}(\mathbf{T}_{\lambda,1} + \mathbf{T}_{\lambda,2} + \mathbf{T}_{\lambda,4} \leq t). \quad (5)$$

Distributional assumptions are required to derive explicit predictions for the joint probabilities in (2). We suppose that the learning rate $\mathbf{L}$ is distributed as a (general) gamma random variable, with parameters $\alpha > 0$ and $\xi > 0$, and that the random variables $\mathbf{T}_{\lambda,i}$ are also distributed gamma, with parameters $\lambda > 0$, $\eta_i > 0$, $i = 1, \ldots, 5$, all these random variables being independent. These particular assumptions make for an easy development of the model, but do not appear to be critical. For example, applications suggest that the predictions of the model may be robust with respect to such distributional assumptions (see Falmagne, 1989). In view of the additivity property of the gamma random variables, the right member of (5) becomes

$$\int_0^t \frac{\lambda^{\eta_1+\eta_2} e^{-\lambda\tau} \tau^{\eta_1+\eta_2-1}}{\Gamma(\eta_1+\eta_2)} d\tau - \int_0^t \frac{\lambda^{\eta_1+\eta_2+\eta_4} e^{-\lambda\tau} \tau^{\eta_1+\eta_2+\eta_4-1}}{\Gamma(\eta_1+\eta_2+\eta_4)} d\tau . \quad (6)$$

To obtain the probability that **any** subject following the learning path (4) be in the state $\{1,2\}$ at time t, we have to multiply (6) by the density of $\mathbf{L}$, and integrate over λ. Writing, as usual, with $x, y > 0$ and $0 < \theta < 1$,

$$B(x,y) = \frac{\Gamma(x)\Gamma(y)}{\Gamma(x+y)}$$

$$I_\theta(x,y) = \frac{1}{B(x,y)} \int_0^\theta \tau^{x-1}(1-\tau)^{y-1}\, d\tau$$

for the *beta* function, and for the *incomplete beta function ratio*, respectively (Johnson and Kotz, 1970), we obtain, after multiplying (6) by the gamma density, and integrating over λ,

$$I_{\frac{t}{t+\xi}}(\eta_1+\eta_2, \alpha) - I_{\frac{t}{t+\xi}}(\eta_1+\eta_2+\eta_4, \alpha). \quad (7)$$

The probability that any subject is in state $\{1,2\}$ at time t (whatever path is taken) is the weighted sum of six such probabilities (since all six paths go through $\{1,2\}$), with the weights being the probabilities p_ν of the paths ν. The probabilities that the subject is in any other state can be derived by similar computations. Take, for instance, any state $K \neq \varnothing, Q$. Denoting by K^ν the state immediately following K along learning path

ν and writing $\mathcal{C}(K)$ for the set of all learning paths ν going through K, and $\mathbf{B}_t$ for the state of the subject at time t, we have, with $\eta_K = \sum_{i\in K} \eta_i$,

$$\mathbb{P}(\mathbf{B}_t = K) = \sum_{\nu\in\mathcal{C}(K)} p_\nu[I_{\frac{t}{t+\xi}}(\eta_K, \alpha) - I_{\frac{t}{t+\xi}}(\eta_{K^\nu}, \alpha)]\,. \tag{8}$$

The probabilities of a particular response pattern A at time t, is obtained from

$$\mathbb{P}(\mathbf{A}_t = A) = \sum_{K\in\mathcal{F}} \mathbb{P}(\mathbf{A}_t = A \,|\, \mathbf{B}_t = K)\mathbb{P}(\mathbf{B}_t = K)\,. \tag{9}$$

In this equation, the state probabilities are computed as exemplified above in (8), and the conditional probabilities are defined by parameters, in the style of Eq. (1). The developments in the general case of n test, as in Eq. (2), are similar, but more care is required. A number of axioms are given in Falmagne (in press), defining a full-fledge stochastic process. This model is currently being subjected to empirical tests. In one recent experiment (Taagepera, Potter, Miller, & Lakshminarayan, submitted for publication), students were tested before and after a lecture covering a basic physical concept, and the model was used to evaluated various aspects of the learning process, with promising results.

REFERENCES

Doignon, J.-P. & Falmagne, J.-Cl., *Spaces for the Assessment of Knowledge*, International Journal of Man-Machine Studies **23** (1985), 175–196.

Dowling, C.E., *Constructing Knowledge Spaces from Judgements with Different Degrees of Certainty*, in "Mathematical Psychology: Current Developments," J.-P Doignon & J.-Cl. Falmagne (Eds.), Springer-Verlag, New York/Berlin, 1991, pp. 221-231.

Falmagne, J.-Cl., *A Latent Trait Theory via Stochastic Learning Theory for a Knowledge Space*, Psychometrika **53** (1989), 283–303.

Falmagne, J.-Cl., *Stochastic Learning Paths in a Knowledge Structure*, Journal of Mathematical Psychology (in press).

Falmagne, J.-Cl., *Finite Markov Learning Models for Knowledge Structures*, Technical report MBS 92-XX, Irvine Research Unit in Mathematical Behavioral Sciences. Submitted for publication.

Falmagne, J.-Cl. & Doignon, J.-P., *A Class of Stochastic Procedures for the Assessment of Knowledge*, The British Journal of Mathematical and Statistical Psychology (1988a), 1-23.

Falmagne, J.-Cl. & Doignon, J.-P., *A Markovian Procedure for Assessing the State of a System*, Journal of Mathematical Psychology **3** (1988b), 232–258.

Falmagne, J.-Cl., Koppen, M., Villano, M., Doignon, J.-P. & Johanessen, L., *Introduction to Knowledge Spaces: How to Build, Test and Search Them*, Psychological Review **97 (2)** (1990), 201-224.

Johnson, N.L., & Kotz, S., "Distributions in Statistics. Continuous Univariate Distributions, I, II," Houghton Mifflin, New York, 1970.

Kambouri, M., Koppen, M., Villano, M. & Falmagne, J.-Cl., *Knowledge assessment: Tapping human expertise by the QUERY routine*, Technical report MBS 91-20, Irvine Research Unit in Mathematical Behavioral Sciences. Submitted for publication.

Koppen, M. & Doignon, J.-P., *How to Build a Knowledge Space by Querying an Expert*, Journal of Mathematical Psychology **34** (1990), 311-331.

Koppen, M., *Extracting Human Expertise for Constructing Knowledges Spaces: an Algorithm*, Accepted for publication in the *Journal of Mathematical Psychology*.

Müller, C.E., *A Procedure for Facilitating an Expert's Judgement on a Set of Rules*, in "Mathematical Psychology in Progress," Edw. E. Roskam (Ed.), Springer-Verlag, Berlin/New York, 1989, pp. 157-170.

Taagepera, M., Potter, F. , Miller, G. E., & Lakshminarayan, K., *Defining the critical learning path from knowledge state analysis of the density concept.*, Submitted for publication.

Uncertainty in Intelligent Systems
B. Bouchon-Meunier et al. (Editors)

Combining probability densities or random variables underlying individual or social choice

A.A.J. Marley[a]

[a]Department of Psychology, McGill University, 1205 Dr. Penfield Avenue, Montreal, Quebec, H3A 1B1*

There are many situations where an individual or group of individuals has to *select* the *best* option or options from some available set of unidimensional or multidimensional options according to some single criterion or multiple criteria. Such choices or decisions are frequently *probabilistic.* It is also frequently necessary in such situations to *combine or aggregate* the selection data obtained in different contexts or from different individuals. In this paper, it is assumed that a functional relation exists between the selection probabilities over the multidimensional options and the selection probabilities over the associated component unidimensional options. It is shown that if that function satisfies a *simple marginalization property* then it is essentially an arithmetic mean, and if the function satisfies a *likelihood independence property* then it is a weighted geometric mean. It is also shown that aggregation of *random variables* over a common probability space is, except in trivial cases, incompatible with aggregation at the level of selection probabilities.

1. INTRODUCTION

There are many situations where an individual or group of individuals has to *select* the *best* option or options from some available set according to some single criterion or multiple criteria. For instance, a consumer has to select the best used car within a specified budget; a voter has to select the best subset of candidates in an election; a committee has to select the best rank order of a set of job applicants. It is also frequently necessary in such situations to *combine* or *aggregate* the selection data obtained in different contexts or from different individuals. For instance, one might wish to predict (or prescribe) an individual's overall selections on a set of options from the individual's selections on a set of component dimensions of those options; or one might wish to determine a consensus or group selection on a set of options from a set of individual selections on those options. In these and similar situations the choices are frequently *stochastic*, that is repeated decisions in the 'same' situation will not necessarily lead to identical choices. Since *deterministic* data and models can frequently be seen as special cases of *probabilistic* (or stochastic) data and models, I concentrate on the latter.

Turning to notation, we have a set A of options, which I assume finite for simplicity. At different times, I wish to model selection of a *single* element from i) the set A, ii) the power set $\mathcal{P}(A)$ of A, that is the set of all subsets of A, and iii) $\mathcal{R}(A)$, the set of permutations of A, that is the set of all rank orders of A. Since for a given finite set A, each of A, $\mathcal{P}(A)$, $\mathcal{R}(A)$ are themselves finite sets, I use a generic set X and element $x \in X$ to represent all these cases.

*This work was supported by the Natural Sciences and Engineering Research Council of Canada.

DEFINITION 1. A *structure of selection probabilities* is a pair (X,P) where X = {x,y,...} is a finite set and P is a function from X × {X} to [0,1] such that $\sum_{x \in X} P(x,\{X\}) = 1$.

P(x,{X}), which I henceforth simplify to P(x:X), is interpreted as the probability of selecting the element x of X. There is a slight problem with the definition as the empty set ∅ is normally considered an element of *any* set X, and thus one could have P(∅:X)>0 even though the empty set is not listed as a member of X. I avoid this difficulty by implictly assuming that P(∅:X) = 0 unless ∅ is explicitly listed as a member of X - as it would be in the case when X = $\mathcal{P}$(A) for some set A. In the relevant aggregation contexts, we have or wish to construct a structure of *overall* selection probabilities (X,P) satisfying Definition 1 and we have a set of structures of *component* selection probabilities (X,P_i), i=1,...,n, each of which also satisfies Definition 1. The goal is to suggest plausible conditions on the relation of the component structures (X,P_i), i=1,...,n, to the overall structure (X,P). The paper presents two such sets of conditions, each of which leads to a particular *aggregation function* provided X has at least 3 elements; more general solutions exist when X has 2 elements. I show that if the relevant relation, which I call an *aggregation function*, satisfies a *simple marginalization property* (plus various regularity conditions) then it is essentially an arithmetic mean, and if the relation satisfies a *likelihood independence principle* (plus various regularity conditions) then it is a weighted geometric mean.

The above results place no restrictions (beyond the regularity conditions) on the *form* of the component selection probabilities. Going beyond this, one can require that the component selection probabilities satisfy one of the classic stochastic choice models - for instance, the *random function model* or the *constant function model* (Marley, 1991b) - and ask whether the overall selection probabilities generated by each of the above aggregation functions then also satisfy a model in the same class. I call a model (or a class of models) *closed* when all of the component *and* the overall selection probabilities satisfy the 'same' model (or class of models). Later in the paper I present examples of constant function models that are closed under weighted geometric aggregation and prove that random function models on a common probability space are closed only under trivial aggregation functions.

Special cases of the results in this paper on arithmetic mean and weighted geometric mean aggregation appear in Marley (1991a, 1993). Before turning to these results I present some comments on, and extensions of, Definition 1. I frequently need to consider distinct structures of selection probabilities on the same set X. I normally use (X,P), (X,Q), etc, to denote such structures. A class of such structures of selection probabilities (X,P), (X,Q), etc, on a fixed set X (satisfying certain properties) is called a *class of selection probabilities on the set X.* I also need to expand the notation of Definition 1 to explicitly mention the dimensional representation of the objects of choice. So let {1,...,n} be the set of components and let P_i(x:X) be the probability of selecting x as the 'best' option in X on component i. For notational completeness, I should now write $(X,P,P_1,...,P_n)$ for such a structure of component and overall choice probabilities. However, for notational simplicity, I continue to write (X,P) with the understanding that $P_1,...,P_n$ are also defined, and I continue to refer to such (X,P), (X,Q), etc, on a fixed set X as a *class of selection probabilities on the set X.*

2. AGGREGATION FUNCTIONS SATISFYING THE SIMPLE MARGINALIZATION PROPERTY

I now state the various assumptions that lead to arithmetic mean aggregation. The major condition is the first, with the others constraining the solution in appropriate ways for the current situation. Marley (1991a) includes detailed motivation and discussion of these assumptions, especially the 'technical' ones concerning regularity and existence. Remember, X is a fixed finite set.

ASSUMPTION M1 (simple marginalization property). There exists a function F_X such that for all structures of selection probabilities (X,P) and for each $x \in X$,

$$P(x:X) = F_X[x,P_1(x:X),...,P_n(x:X)].$$

I call Assumption M1 simple marginalization because it is a special case of the marginalization property considered in the context of opinion aggregation (McConway, 1981, and Genest, 1984). Clearly, aggregation by simple marginalization is similar to the way univariate or marginal distributions are aggregated to form multivariate distributions. However, it is important to realize that here the P_i are not necessarily the marginals of P.

The first main result of this section is that provided $|X| > 2$, the simple marginalization property, plus some regularity and existence conditions, imply that F_X is essentially an arithmetic mean with weights that can depend on X. However, when $|X| = 2$, the class of solutions is much larger - see Marley (1991a) for examples.

ASSUMPTION M2 (solvability). For any n-dimensional real vectors $(r_1,...,r_n)$, $(s_1,...,s_n)$, with $r_i,s_i,r_i+s_i \in [0,1]$, i=1,...,n, and for X with $|X| \geq 3$, it is possible to select a structure of choice probabilities (X,Q) and $x,y,z \in X$ such that for i=1,...,n,

$$Q_i(x:X) = r_i,\ Q_i(y:X) = s_i,\ Q_i(z:X) = 1 - (r_i+s_i).$$

ASSUMPTION M3 (dominance principle). For all $x \in X$ and structures of selection probabilities (X,P), (X,Q), if $P_i(x:X) \leq Q_i(x:X)$ for i=1,...,n, then $P(x:X) \leq Q(x:X)$.

ASSUMPTION M4 (zero preservation property). For all $x \in X$ and structures of selection probabilities (X,P), if $P_i(x:X) = 0$ for i=1,...,n, then $P(x:X) = 0$.

THEOREM 1. *If a class of selection probabilities on a finite set X with $|X| \geq 3$ satisfies Assumptions M1 and M2, then there exist weights $w_X(i)$, $i = 1,...,n$, and a fixed probability measure $\psi(.:X)$ such that $\sum_{j=1}^{n} w_X(j) \leq 1$ and for each structure of selection probabilities (X,P) and $x \in X$,*

$$P(x:X) = \sum_{i=1}^{n} w_X(i)P_i(x:X) + \left[1 - \sum_{i=1}^{n} w_X(i)\right]\psi(x:X).$$

If Assumptions M2 and M3 hold, then the above representation has $w_X(i) \in [0,1]$, and if Assumptions M1, M2, and M4 hold, then $w_X(i) \in [0,1]$ and $\sum_{i=1}^{n} w_X(i) = 1$.

The proof of Theorem 1 parallels that given in the Appendix of Marley (1991a). The theorem might appear to have no content, in that for a given structure of selection probabilities (X,P) one can always set $\psi(x{:}X) = P(x{:}X)$ for each $x \in X$. However, remember that the representation is to hold for *all* structures of selection probabilities (X,P) satisfying the relevant conditions of Theorem 1, and this particular solution says that the overall selection probabilities are constant, and in particular have no dependence on the component unidimensional selection probabilities. This is, of course, a possible solution but not one of much interest for the relevant empirical domains. Also note that the theorem gives no constraint on the form of the (nonnegative) weights $w_X(i)$ and the probability distribution ψ, only that they exist. The obvious interpretation is that $w_X(i)$ is the probability of 'attending' to dimension i, and basing the choice on that dimension alone, and that $1 - \sum_{i=1}^{n} w_X(i)$ is the probability of choosing 'randomly' according to the distribution ψ. It is important to note that since the weights depend on the context X, such a strategy can lead to quite complex patterns of selections, which will not superficially appear to be determined (at any given selection opportunity) by a single dimension.

3. AGGREGATION FUNCTIONS SATISFYING THE LIKELIHOOD INDEPENDENCE PROPERTY.

To avoid excessive technical detail, I restrict attention in this section to structures of nonzero selection probabilities, that is structures (X,P) such that for each $x \in X$, $P(x{:}X) \neq 0$ and $P_i(x{:}X) \neq 0$, $i = 1,\dots,n$.

As in the previous section, I present various assumptions, for which detailed motivation and discussion are given in Marley (1991a). First, for a structure of nonzero selection probabilities (X,P), and $x, y \in X$, let

$$L_X^P(x,y) = \frac{P(x{:}X)}{P(y{:}X)},$$

that is, $L_X^P(x,y)$ is the likelihood or odds ratio for choosing x versus y in context X according to the measure P. Similar notation is used for the corresponding component odds ratio.

ASSUMPTION L1 (likelihood independence property). There exists a function F_X such that for all structures of nonzero selection probabilities (X,P) and for $x,y \in X$,

$$L_X^P(x,y) = F_X[L_X^{P_1}(x,y),\dots,L_X^{P_n}(x,y)].$$

Clearly, this condition states that it does not matter in calculating likelihood ratios whether one first calculates them on the component selection probabilities, and then combines these ratios over components, or simply calculates likelihood ratios on the overall selection probabilities. There are various, roughly equivalent, forms of this assumption, some of which have been used in the aggregation literature (see Marley, 1991a).

The next main result is that provided $|X|>2$, the likelihood independence property, plus some regularity and existence conditions, imply that F_X is a weighted geometric mean with weights that can depend on X. Again, when $|X|=2$, the class of solutions is much larger (Marley, 1991a).

ASSUMPTION L2 (solvability). For any n-dimensional positive real vectors $(r_1,...,r_n)$, $(s_1,...,s_n)$, it is possible to select a structure of nonzero selection probabilities (X,P) and $x,y,z \in X$, such that for $i=1,...,n$,

$$L_X^{P_i}(x,y) = r_i \ , \ L_X^{P_i}(y,z) = s_i.$$

Note that this condition requires $|X| \geq 3$.

ASSUMPTION L3 (dominance principle). For structures of selection probabilities (X,P), (X,Q), and $x,y \in X$,

$$\text{if} \quad L_X^{P_i}(x,y) \leq L_X^{Q_i}(x,y), \ i=1,...,n$$

$$\text{then} \quad L_X^{P}(x,y) \leq L_X^{Q}(x,y) \ .$$

THEOREM 2. *If a class of selection probabilities on a finite set X with $|X| \geq 3$ satisfies Assumptions L1 and L2 then there exist constants $w_X(i)$, $i=1,...,n$, such that for each structure of selection probabilities (X,P) and for $x \in X$,*

$$P(x:X) = \frac{\prod_{i=1}^{n} P_i(x:X)^{w_X(i)}}{\sum_{y \in X} \prod_{i=1}^{n} P_i(y:X)^{w_X(i)}} \ .$$

If Assumption L3 also holds then the constants $w_X(i)$ are nonnegative.

The proof of Theorem 2 parallels that given in the Appendix of Marley (1991a). Clearly, it does not make sense for P(x:X) to be a decreasing function of $P_i(x:X)$ for any $i=1,...,n$, so examples would have nonnegative $w_X(i)$, $i=1,...,n$.

4. A DETAILED EXAMPLE INVOLVING GEOMETRIC MEAN AGGREGATION OF CHOICE AND RANKING PROBABILITIES

So far I have used a generic notation for the selection of an 'element' x from a finite set X. As mentioned in the introduction, there are at least three important interpretations of the results. Given a finite set A, we might wish to model the selection of a *single* element from i) the set A, ii) the power set $\mathcal{P}(A)$ of A, that is the set of all subsets of A, and iii) $\mathcal{R}(A)$, the set of all permutations of A, that is the set of all rank orders of A. I now present some results and open problems concerning relations between cases i), iii); case ii) is discussed in Marley (1993). I motivate these results using ideas from *probabilistic social choice* (Marley, 1993).

We have a finite m element set X of options, $m \geq 2$, and a finite n element set of voters (dimensions). [The set X here actually corresponds to the base set A of the above paragraph. I now use X rather than A for ease of reference to my previous papers which use X in the way I now use it]. We are interested in two distinct structures of selection probabilities and their relations to one another. First, for a given set X and element $x \in X$, the component choice probability $P_i(x:X)$, $i=1,...,n$, is interpreted as the probability that voter i selects option x as the 'best' option in X.

Then the overall choice probability $P(x:X)$ denotes the probability that the 'society' (consisting of the n voters) selects option x from X. Thus, the ***probabilistic social choice problem*** involves suggesting plausible forms for the $P(x:X)$, $x \in X$, in terms of the $P_i(x:X)$, $x \in X$, $i=1,\ldots,n$. Obvious candidate forms are the previously presented arithmetic mean and weighted geometric mean aggregation rules.

Second, social choice is often discussed in terms of rank orders on the set X. For this case some additional notation is required. $\mathcal{R}(X)$ denotes the set of rank orders of X (no ties allowed). For $\rho \in \mathcal{R}(X)$, ρ_g is the element of X that is ranked in the gth position of ρ, and $\rho(x)$ denotes the rank order position of $x \in X$ under the ranking ρ. $\rho=\rho_1\ldots\rho_m$, $\sigma=\sigma_1\ldots\sigma_m$, and $\tau=\tau_1\ldots\tau_m$, are arbitrary elements of $\mathcal{R}(X)$. For each i, there is a probability distribution P_i over the m! rank orders of X with $P_i(\rho:\mathcal{R}(X))$ denoting the probability that voter i rank orders the elements of X from 'best' to 'worst' according to the rank order ρ. Then the overall ranking probability $P(\rho:\mathcal{R}(x))$ denotes the probability that the 'society' (consisting of the n voters) rank orders the elements of X from 'best' to 'worst' according to the rank order ρ. Similarily to the above probabilistic social choice problem, the ***probabilistic social ranking*** problem involves suggesting plausible forms for the $P(\rho:\mathcal{R}(X))$, $\rho \in \mathcal{R}(X)$, in terms of the $P_i(\rho:\mathcal{R}(X)$, $\rho \in \mathcal{R}(X)$, $i=1,\ldots,n$. Candidate forms are again the previously presented arithmetic mean and weighted geometric mean aggregation rules.

There are numerous further problems of interest in the domain of probabilistic social choice and ranking (Marley, 1993). For instance, one can explore possible relations between probabilistic social choice and ranking on a given set and on its subsets. To discuss such relations, it is necessary to apply the results developed thus far to arbitrary subsets of some fixed m element set X. For instance, a set of ranking probabilities $Q(\rho:\mathcal{R}(X))$, $\rho \in \mathcal{R}(X)$, satisfies *L(uce)-decomposability* (Critchlow et al, 1991) provided there exist choice probabilities $Q(y:Y)$, $y \in Y \subseteq X$, such that for each $\rho=\rho_1\ldots\rho_m \in \mathcal{R}(X)$,

$$Q(\rho:\mathcal{R}(X)) = Q(\rho_1:\{\rho_1,\ldots,\rho_m\})\, Q(\rho_2:\{(\rho_2,\ldots,\rho_m\})\, Q(\rho_{m-1}:\{\rho_{m-1}, \rho_m\}).$$

Note that the term on the left is a ranking probability, whereas each term on the right is a choice probability, called an *induced* choice probability. The motivation of this condition is the assumption that the probability of a particular rank order ρ occuring is the probability of a particular sequence of choices occurring: First, ρ_1 must be selected as the best element in the set $X = \{\rho_1,\ldots,\rho_m\}$; ρ_2 must then be selected as the best element in the set $X - \{\rho_1\} = \{\rho_2,\ldots,\rho_m\}$; etc; until ρ_{m-1} is selected as the best element in the set $\{\rho_{m-1}, \rho_m\}$.

Notice that in the above I did not specify whether the choice probabilities Q were component choice probabilities (for individual voters) or overall choice probabilities (for the society). Clearly they can be either. Thus, bringing together the various ideas presented thus far in the paper (focussing on geometric mean aggregation as it turns out to give the most interesting results), we need to explore: weighted geometric mean combination of ranking probabilities; weighted geometric mean combination of choice probabilities; and L-decomposability of component and overall choice and ranking probabilities. The obvious next topic is which ranking and choice models satisfy some or all of these properties. This leads to a number of fascinating questions and characterizations on which I have numerous results (Marley, 1991a, 1991b, 1993). As illustration, I now present results on a particular class of models where the ranking probabilities are induced from paired comparisons.

For distinct $x,y \in X$, $p(x,y)$ denotes the probability that item x is preferred to item y in a paired comparison of these two items. Assume that every distinct pair of items can be compared, and that if the results are consistent with a rank order then that rank order is selected; otherwise, the entire process is repeated until a ranking is obtained. This process yields, for $\rho = \rho_1 \dots \rho_m \in \mathcal{R}(X)$,

$$P(\rho:\mathcal{R}(X)) = \frac{\prod_{1 \le g < h \le m} p(\rho_g, \rho_h)}{\sum_{\sigma \in \mathcal{R}(X)} \prod_{1 \le j < k \le m} p(\sigma_j, \sigma_k)} .$$

For notational simplicity I write this as

$$P(\rho:\mathcal{R}(X)) \sim \prod_{1 \le g < h \le m} p(\rho_g, \rho_h)$$

(with similar notation for all such 'normalized' representations). Clearly, if the above representation holds for each dimension i, i = 1,...,n, and if weighted geometric mean combination holds with weights that do not depend on X, then a representation of the same form holds for the overall ranking probabilities $P(\rho:\mathcal{R}(X))$, that is

$$P(\rho:\mathcal{R}(X)) \sim \prod_{1 \le g < h \le m} p(\rho_g, \rho_h)$$

where

$$p(\rho_g, \rho_h) \sim \prod_{i=1}^{n} p_i(\rho_g, \rho_h)^{w(i)} .$$

The unidimensional version of this model satisfies L-decomposability (Marley, 1968; Critchlow et al., 1991). Since the multidimensional version has the same form, it also satisfies L-decomposability. However, the choice probabilities of the decomposition (explicitly stated in Marley, 1968) do not combine according to weighted geometric mean combination when $|X| \ge 3$. (I have not proved this result, which seems obviously correct). Therefore, an open problem in this area concerns whether or not there is a ranking model that satisfies L-decomposability on each component and overall, with both the ranking probabilities *and* the induced choice probabilities satisfying weighted geometric mean combination. Finally, Marley (1993) shows that a special case of this model, which places further restrictions on the binary choice probabilities, is compatible with a 'classical' ranking by scores method.

5. AGGREGATION OF RANDOM UTILITIES

The above probabilistic models of choice and ranking assume that the probabilities for the overall choices can be written as a function of the corresponding probabilities on the component dimensions. For *random utility models*, also known as order statistics models (Critchlow et al., 1991), an alternative (in general incompatible - see below) combination process is plausible. Remember, in a random utility model (generalized to multiple components), for each x in the master set T and for each component i, i=1,...,n, there is a random variable $\mathbf{t}_i(x)$ such that for each $x \in X$ and each $\rho=\rho_1 \dots \rho_m \in \mathcal{R}(X)$, $X \subseteq T$, $|X| \ge 2$,

$$P_i(x:\mathcal{R}(x)) = \Pr\left[\mathbf{t}_i(x) = \max_{y \in X} \mathbf{t}_i(y)\right],$$

and

$$P_i(\rho:\mathcal{R}(X)) = \Pr[\mathbf{t}_i(\rho_1) > \dots > \mathbf{t}_i(\rho_m)] ,$$

and also there are overall random variables $\mathbf{t}(x)$, $x \in T$, such that for each $x \in X$ and each $\rho=\rho_1 \dots \rho_m \in \mathcal{R}(X)$, $X \subseteq T$, $|X| \ge 2$,

$$P(x:X) = \Pr[\mathbf{t}(x) = \max_{y \in X} \mathbf{t}(y)],$$

and

$$P(\rho:\mathcal{R}(X)) = \Pr[\mathbf{t}(\rho_1) > \dots > \mathbf{t}(\rho_m)] .$$

Before continuing I note three technical points. First, for models of choice and ranking it is customary to assume that $\Pr[\mathbf{t}_i(x) = \mathbf{t}_i(y)] = \Pr[\mathbf{t}(x) = \mathbf{t}(y)] = 0$, $i=1,\dots,n$. Second, I implicitly assume that $\mathbf{t}$, $\mathbf{t}_i$, $i=1,\dots,n$, are random variables on a *common* probability space; without this assumption the following negative results do not hold. Thirdly, in the remainder of this section I confine attention to a *fixed* set X. For such a fixed set X, any reasonably well-behaved single set of choice or ranking probabilities can be represented by a random utility model (Marley and Colonius, 1992). However, this fact has no direct relevance to the importance of the following negative result.

With the above forms for random utility models of choice and ranking in front of us, there is an obvious plausible combination model relating the component and overall random utilities: assume that there is a function G such that for each $x \in T$,

$$\mathbf{t}(x) = G[\mathbf{t}_1(x), \dots, \mathbf{t}_n(x)] .$$

As stated above, G is constrained in such a way that $\mathbf{t}$ is a random variable on the *same* probability space as the $\mathbf{t}_i$, $i=1,\dots,n$ (Alsina, 1987).

It is now natural to ask whether any such aggregate random utility model for choice or ranking yields choice or ranking probabilities that satisfy either arithmetic mean or weighted geometric mean aggregation. I show next that this is *not* possible for two element sets. In fact, on two element sets there are no nontrivial combinations of (possibly different) aggregation rules that coexist at the level of random utilities and at the level of choice or ranking probabilities. Since on two element sets choice and ranking are the same I do not need to distinguish between them. I have not yet extended the following result to larger sets.

THEOREM 3. *Assume the following.*

i) *We have a two element set $X = \{x,y\}$ and two components i, $i=1,2$.*
ii) *The choice probabilities on each component and overall satisfy random utility models, that is there are random variables $\mathbf{t}$ and $\mathbf{t}_i$, $i = 1,2$ such that*

$$p_i(x,y) = P_i(x{:}\{x,y\}) = Pr[\mathbf{t}_i(x) > \mathbf{t}_i(y)],$$
$$p(x,y) = P(x{:}\{x,y\}) = Pr[\mathbf{t}(x) > \mathbf{t}(y)].$$

iii) *$\mathbf{t}_i$, $i=1,2$ are independent.*

iv) *There is a function G with $G(a,b)=G(c,d)$ only if $a=c$ and $b=d$ such that $\mathbf{t}=G[\mathbf{t}_1,\mathbf{t}_2]$.*

v) *There is a function F such that for $z \in X$, $P(z{:}X) = F[P_1(z{:}X),P_2(z{:}X)]$.*

Then one of the following four cases holds:

1. *For all $a, b \in [0,1]$, $F(a,b) = a$.*
2. *For all $a, b \in [0,1]$, $F(a,b) = b$*
3. *For all $a, b \in [0,1]$, $F(a,b) = 1-a$.*
4. *For all $a, b \in [0,1]$, $F(a,b) = 1-b$.*

PROOF. Consider a random utility model with the following structure, where a, b $\in (0,1)$.

$$p_1(x,y) = \Pr(\mathbf{t}_1(x)=1, \mathbf{t}_1(y)=0) = a,$$

$p_1(y,x) = Pr(t_1(x)=0,\ t_1(y)=1) = 1\text{-}a,$
$p_2(x,y) = Pr(t_2(x)=1,\ t_2(y)=0) = b,$
$p_2(y,x) = Pr(t_2(x)=0,\ t_2(y)=1) = 1\text{-}b.$

Let $\mathbf{t} = (\mathbf{t}_1,\mathbf{t}_2)$. Since we are assuming the $\mathbf{t}_i$, i=1,2 are independent, the possible pairings of random variable values and their probabilities are:

$t_1(x)$	$t_2(x)$	$t_1(y)$	$t_2(y)$	Probability
1	0	0	1	a(1-b)
1	1	0	0	ab
0	0	1	1	(1-a) (1-b)
0	1	1	0	(1-a)b

By the assumptions of the theorem, we also have

$$\begin{aligned} F(a,b) &= F[p_1(x,y),p_2(x,y)] \\ &= P(x,y) \\ &= Pr\Big[G\big(t_1(x),t_2(x)\big) > G\big(t_1(y),t_2(y)\big)\Big] \end{aligned}$$

Because of the assumed constraints on G (assumption iv) we only have to consider four cases.

Case 1. G(1,1) > G(0,0) and G(1,0) > G(0,1).

Then using the above table,

$$\begin{aligned} F(a,b) = P(x,y) &= Pr\Big[G\big(t_1(x),t_2(x)\big) > G\big(t_1(y),t_2(y)\big)\Big] \\ &= Pr\big[t(x) = (1,1), t(y) = (0,0)\big] + Pr\big[t(x) = (1,0), t(y) = (0,1)\big] \\ &= ab+a(1\text{-}b) \\ &= a \end{aligned}$$

Similar arguments for the remaining cases give

Case 2. G(1,1) > G(0,0) and G(0,1) > G(1,0) imply F(a,b) = b

Case 3. G(0,0) > G(1,1) and G(1,0) > G(0,1) imply F(a,b) = 1-b

Case 4. G(0,0) > G(1,1) and G(0,1) > G(1,0) imply F(a,b) = 1-a □

This result parallels those of Alsina and Schweizer (1988) and Alsina, Nelsen, and Schweizer (1993). It is important to reiterate that the result depends strongly on the assumption that the **t**, $\mathbf{t}_i$, i=1,...,n are defined on a *common* probability space. Without that assumption, the arithmetic mean aggregate of selection probabilities each of which satisfies a random utility model can be shown to satisfy a random utility model (Alsina and Schweizer, 1988; Marley, 1993). The latter result and representation builds on the interpretation (given after Theorem 1) of the arithmetic mean weights $w_X(i)$, i=1,...,n, of that theorem as the probability of 'attending' to dimension i.

REFERENCES

Alsina, C., 1987. Synthesizing judgements given by probability distribution functions. *Note di Matematica, 7*, 29-40.

Alsina, C. and B. Schweizer, 1988. Mixtures are not derivable. *Foundations of Physics Letters, 1*, 171-174.

Alsina, C., R.B. Nelsen and B. Schweizer, 1993. On the characterization of a class of binary operations on distributions functions. *Statistics and Probability Letters* (In Press).

Critchlow, D.E., M.A. Fligner and J.S. Verducci, 1991. Probability models on rankings. *Journal of Mathematical Psychology, 35*, 277-293.

Falmagne, J.C., 1981. On a recurrent misuse of a classical functional equation result. *Journal of Mathematical Psychology, 23*, 190-193.

Genest, C., 1984. Pooling operators with the marginalization property. *The Canadian Journal of Statistics, 12*, 153-163.

Marley, A.A.J., 1968. Some probabilistic models of simple choice and ranking. *Journal of Mathematical Psychology, 5*, 311-332.

Marley, A.A.J., 1991a. Aggregation theorems and multidimensional choice models. *Theory and Decision, 30*, 245-272.

Marley, A.A.J., 1991b. Context-dependent probabilistic choice models based on measures of binary advantage. *Mathematical Social Sciences, 21*, 201-231.

Marley, A.A.J., 1993. Aggregation theorems and the combination of probabilistic rank orders. In D.E. Critchlow, M.A. Fligner and J.S. Verducci (Eds.). *Probability Models and Data Analysis for Ranking Data.* New York, NY: Springer Verlag, pp 216-240.

Marley, A.A.J., and H. Colonius, 1992. The "horse race" random utility model for choice probabilities and reaction times, and it competing risks interpretation. *Journal of Mathematical Psychology, 36*, 1-20.

McConway, K.J., 1981. Marginalization and linear opinion pools. *Journal of the American Statistical Association, 76*, 410-414.

Uncertainty in Intelligent Systems
B. Bouchon-Meunier et al. (Editors)

On the Median Procedure

Fred S. Roberts
Department of Mathematics, Rutgers University
New Brunswick, NJ 08903, USA

1. INTRODUCTION

Among the most important problems in the social and behavioral sciences is the problem of group decisionmaking. Two of the most useful group decisionmaking procedures are the median procedure and the closely related mean procedure. These procedures are used in aggregating preferences in social choice problems; in voting; in operations research as solutions to location problems; in statistical analysis as concepts of centrality; in molecular biology in the process of finding consensus patterns in molecular sequences; and in mathematical taxonomy in finding the consensus in classification problems. Recently, there have been many new results about these procedures, motivated by applications. In this paper, we will present a general framework for speaking about these procedures, and use it to organize some of the recent results and to present some of the important open questions in the study of these procedures. Recent surveys on medians and means are those by Barthélemy and Monjardet [1981,1988] and Barthélemy [1989]. Some summaries about these procedures are given in Mirkin [1979] and Roberts [1976,1989].

In a standard framework for group decisionmaking, each member of a group gives an opinion and then we seek a consensus among these opinions. Sometimes the opinion is just a vote for a first choice among a set of alternatives. In other contexts, the opinion is more complicated, involving for example a ranking of all of the possible alternatives. By a ***ranking*** we shall mean either a linear order (simple order) or a ***weak order***, a linear order with ties. There is a long literature on group consensus starting from rankings, going back several hundred years and spurred by the fundamental work of Arrow [1951]. Arrow's famous Impossibility Theorem is a negative result, saying that there is no method of group consensus that satisfies several reasonable axioms.

Among the most important directions of research in the theory of group consensus after Arrow's Impossibility Theorem is the idea of Kemeny [1959] and Kemeny and Snell [1962] that we can obtain a group consensus by first finding a way to measure the distance between any two rankings and then finding a ranking or set of rankings which minimize the sum of the distances to all of the expressed rankings. Kemeny and Snell call such a set of rankings the ***median*** of the group's expressed rankings. The median is of course dependent on the particular distance function used. Among the early results in the literature is a set of axioms of Kemeny and Snell which give rise to a unique distance measure, the symmetric difference distance (to be defined below), and then lead to a very specific concept of median. It should be noted that the axioms characterize the distance function, but do not attempt to characterize the median procedure among possible group consensus procedures. A related concept is that of the ***mean***, which is a ranking or set of rankings which minimize the sum of the squares of the distances to all of the expressed rankings.

The problem of finding a median or mean makes sense in the general setting of a space M in which we can measure distance between two opinions with a nonnegative, symmetric distance function $d(x,y)$ defined for all pairs of elements x, y of M. In this general context, we can define the ***median*** of a vector $(a_1, a_2, \ldots, a_n)$ of points from M as the set of all points x in M which minimize $d(x) = \Sigma_{i=1}^{n} d(a_i, x)$ and the ***mean*** as the set of all points x in M which minimize $D(x) = \Sigma_{i=1}^{n} d(a_i, x)^2$. More generally, suppose we have two sets M and N, and we have a nonnegative distance function d defined on $M \times N$, i.e., so that $d(x,y)$ is defined for all $x \in M$, $y \in N$. Then the ***(M,N)-median*** of a vector $(a_1, a_2, \ldots, a_n)$ from M is the set of all $x \in N$ which minimize $d(x)$ and the ***(M,N)-mean*** is the set of all $x \in N$ which minimize $D(x)$. In what follows, we shall usually use the terms median and mean for the (M,N)-median and (M,N)-mean, respectively.

There are two kinds of problems of great interest in connection with various group consensus procedures, and in particular the median and the mean. The first is to give conditions under which one should use a particular procedure. The traditional approach to this problem has been to give axioms for a group decisionmaking procedure, and accept a procedure if and only if it satisfies the axioms. Hopefully one can add enough axioms so as to completely characterize the procedure. We shall discuss axioms which characterize the median or mean in different contexts. The second problem in connection with group consensus procedures is to find ways to calculate or compute them. There has been considerable interest in this direction of work in recent years. For instance, Bartholdi, Tovey, and Trick [1989] have shown that certain consensus procedures used in some practical situations, in particular the Dodgson winner, are not easy to compute, indeed, are NP-complete. When applied to elections, the conclusion is that it is NP-complete to compute the winner of the election held under certain rules! In the special case of the median when M and N are sets of rankings, it has been shown by Wakabayashi [1986] that the calculation of this group consensus function is an NP-complete problem when we consider the symmetric difference distance, whether or not we allow ties in our rankings. (The same result without ties is also shown by Bartholdi, Tovey, and Trick [1989].) However, medians and means can be calculated efficiently when more structure is introduced into the problem and when we seek a median or mean in a different context than that of rankings. For instance, suppose the space M consists of the points in a partially ordered set P and d(x,y) is the shortest path metric in the diagram of the partial order when M = N = P. If P is a certain kind of meet semilattice called a median semilattice, then Barthélemy and Janowitz [1991] show that a polynomial algorithm allows the computation of a median. Marcotorchino and Michaud [1981] have explored a number of heuristics for computing medians. However, in general, no good approximation algorithms of a general nature are known. In this paper, we concentrate on axiomatic questions.

In the rest of this section, we put the group consensus problem in a general framework. Suppose A is a set of alternatives. *Throughout this paper, A will be assumed to be finite.* We shall distinguish two kinds of sets relative to A, a finite set I(A) of ***input structures*** and a finite set O(A) of ***output structures***. The term median will subsequently be used to mean (M,N)-median, where M = I(A) and N = O(A), and similarly for mean. In many of the traditional applications, the input structures and the output structures are rankings of the elements of A. In these applications, each individual in the group ranks all of the alternatives and then the group consensus function provides a group ranking. In other applications, the input structures are rankings and the output structures are individual elements of A; here, each individual provides a ranking, but the group consensus procedure provides only a winner. Input and output structures can be more complex; for instance, among the structures studied have been partial orders, trees, n-trees, directed graphs, signed directed graphs, lattices, etc. As we have been suggesting, the group consensus will depend on the opinions of all members of the group. We formalize this idea by speaking of a ***profile***, which is a finite sequence of input structures, one for each member of the group. Profiles are sometimes restricted to be sequences of a given length (the size of a particular group) and sometimes they are allowed to be sequences of arbitrary finite length (so that the consensus procedures can be applied to all possible groups). We denote by P(A) the collection of profiles. Then, a ***group consensus function*** is a function $F:P(A) \longrightarrow 2^{O(A)}$, a function assigning to each profile a set of output elements thought of as the group's consensus. In many cases we seek a function F which assigns a single outcome element to each profile. However, this is not the case with the median or mean, necessarily. *In this paper,* $F(a_1, a_2, \ldots, a_n)$ *will always be assumed to be nonempty.* This assumption is called by Hansen and Roberts [1992] the axiom of **EXISTENCE**.

In many group consensus problems, we find a way to measure the distance between two elements of I(A) or between an element of I(A) and an element of O(A). Sometimes distances are part of the natural input of the problem. Other times, they are derived from the input, for instance on the basis of axioms as in the original approach of Kemeny and Snell [1962]. In any case, we can often usefully represent the group decisionmaking problem by taking the elements of I(A)∪O(A) to be vertices of an undirected graph, joining any element in I(A) to any element in O(A) by an edge, and letting the edge carry a nonnegative real number representing the distance between the corresponding vertices. It is very useful to do this, for instance, in computing Kemeny-Snell medians when the number of alternatives being ranked is small. Other graph-theoretical representations of the problem will also be useful. Conversely, many group decisionmaking problems already arise in graph-theoretical form. For instance, in the social and behavioral sciences, it is sometimes useful to consider networks which have as points both individuals and opinions they might have about different things or experiences they might have had. Then we estimate the distance between an individual and an opinion or experience. For instance, this is the case in Guttman scaling in psychology. (See Guttman [1944], Roberts [1979], and Cozzens and Leibowitz [1984,1987].) The individuals are elements of set I(A) and the opinions or experiences are elements of set O(A). Then we might be given a group of individuals (a profile) and ask for a consensus opinion or experience or consensus group of opinions or experiences somehow centrally located for all of them. An analogous problem occurs in location theory in operations research. Here, we have a variety of users and we wish to locate a facility or facilities which is or are convenient to the users. We often have a location problem given as a network or graph, with the users at certain places of the network and the potential facility sites as certain places of the network as well. We seek a consensus location or locations for the facility. Both the median and mean are widely used as consensus locations. It will turn out to be useful to distinguish various cases for where the users and solution locations can be. In some cases, they are restricted to be at vertices of the network. Then I(A) = O(A) = the set of vertices. In other cases, they can be anywhere along an edge of the network. (For a recent survey of location theory, see Hansen, et al. [1987].)

In this paper, we consider the problem of axiomatizing the median and the mean in three different settings, that where I(A) = O(A) = the set of rankings on A; that where I(A) and O(A) appear as vertices or along edges of a network; and one where I(A) and O(A) are words in a certain alphabet.

2. MEDIANS AND MEANS FOR RANKINGS

In this section, we present an axiomatization of the Kemeny-Snell median. To do so, let us first define the ***symmetric difference distance*** $d_k(a,b)$ between two rankings a and b of A. For every pair u,v of elements from A, let c(u,v) be 2 if one of the rankings has u over v and the other v over u, 0 if the rankings both have u over v or both have v over u, and 1 otherwise. The latter applies if rankings allow ties. Then $d_k(a,b)$ is the sum of the c(u,v) over all u,v. As we have remarked above, Kemeny and Snell [1962] give axioms on a distance function on the set of rankings (weak orders) which characterize this distance function. Bogart [1973,1975] gives axioms which characterize this and other distance functions on more general sets of rankings, including partially ordered sets and asymmetric relations.

Suppose I(A) and O(A) consist of all rankings of A. Then a group consensus function F is the ***Kemeny-Snell median*** if

$$F(a_1, a_2, \dots, a_n) = \{x \in O(A): \sum_{i=1}^{n} d_k(a_i, x) \text{ is minimum}\}.$$

The ***Kemeny-Snell mean*** is defined similarly. We now present axioms which are clearly necessary conditions for the Kemeny-Snell median. Since the work of Arrow [1951] and May [1952], the following two axioms have been standard assumptions in group consensus theory. The second assumes, as we do throughout this paper, that A is finite. **ANONYMITY** holds if for all permutations p of {1,2,...,n},

$$F(a_{p(1)}, a_{p(2)}, \dots, a_{p(n)}) = F(a_1, a_2, \dots, a_n)$$

for all profiles $(a_1, a_2, \dots, a_n)$. **NEUTRALITY** holds if for all permutations σ of A,

$$F(\sigma(a_1), \sigma(a_2), \dots, \sigma(a_n)) = \bar{\sigma}[F(a_1, a_2, \dots, a_n)],$$

where $\sigma(a_i)$ is obtained from a_i by replacing x by $\sigma(x)$ and $\bar{\sigma}(X) = \{\sigma(x): x \in X\}$. Another axiom which is widely used is the following: We say that **CONSISTENCY** holds if whenever $F(a_1, a_2, \dots, a_n) \cap F(b_1, b_2, \dots, b_m) \neq \phi$, then

$$F(a_1, a_2, \dots, a_n, b_1, b_2, \dots, b_m) = F(a_1, a_2, \dots, a_n) \cap F(b_1, b_2, \dots, b_m).$$

This axiom has played an important role in social choice theory. Young [1975] proved that if I(A) is the set of linear orders on A and O(A) = A, then F is given by a scoring procedure (F is a so-called *scoring rule*) if and only if F satisfies the conditions of anonymity, neutrality, and consistency. Under the same assumptions about I(A) and O(A), these three axioms also played a role in the characterizations by Richelson [1978] and Roberts [1991] of the group consensus function called the plurality rule. Where I(A) and O(A) are both A, the axioms played a role in the characterization by Roberts [1991] of the so-called plurality function.

In the case where I(A) consists of all linear orders of A, it is useful to use the notation $n_{uv}(\mathbf{a})$ to mean the number of rankings in profile $\mathbf{a} = (a_1, a_2, \dots, a_n)$ in which element u is ranked above element v minus the number in which v is ranked above u. With this definition, we can now present the next axiom, which only makes sense in the setting where both I(A) and O(A) are sets of rankings on A. We say that the **CONDORCET** axiom holds if whenever $n_{uv}(\mathbf{a}) > 0$, then in $F(\mathbf{a})$, v is not ranked directly above u; and if whenever $n_{uv}(\mathbf{a}) = 0$, then in $F(\mathbf{a})$, u is not ranked directly above v and v is not ranked directly above u.

Theorem 1 (Young and Levenglick [1978]). If both I(A) and O(A) are the set of linear orders on A, then a group consensus function is the Kemeny-Snell median if and only if it is neutral, consistent, and Condorcet.

In the situation where I(A) and O(A) are both sets of weak orders, no similar result is known. Also, when I(A) and O(A) are both sets of rankings, linear or weak, no similar result is known for the Kemeny-Snell mean. No similar results are known for median or mean under other distance functions on rankings, in particular the distance functions axiomatized by Bogart [1973,1975]. However, results of a nature similar to Theorem 1 have been obtained by Barthélemy and McMorris [1986] when I(A) and O(A) are the n-trees which arise in classification problems in mathematical taxonomy. Such results have also been obtained by Barthélemy and Janowitz [1991] in the general setting when I(A) and O(A) are the points of a median semilattice. An important special case of a median semilattice arises when the points are hierarchical trees and here the Barthélemy-McMorris result about n-trees comes out as a special case.

3. MEDIANS AND MEANS ON NETWORKS

In this section, we consider the case where we start with a network given in the form of a graph. We consider four special cases: The ***V-V case*** is where both I(A) and O(A) are the set of vertices of the graph. The ***E-E case*** is where both I(A) and O(A) consist of the set of ***points*** in the network, i.e., vertices or points on edges. Here the vertices play no distinguished role. The ***V-E case*** is where I(A) is the set of vertices of the graph and O(A) is the set of points of the network. The ***E-V case*** is the opposite of the V-E case. We shall limit the discussion to where the distance d(a,b) between two points in I(A) or between a point of I(A) and a point of O(A) is given by the length of the shortest path between the points in the network.

Recently, Holzman [1990], motivated by problems of location theory, observed that in solving such problems, we usually assume a priori what the objective function is. He set out to take an axiomatic approach, and concentrated on a special class of networks, the trees. He characterized the mean in this situation in the E-E case.

To state Holzman's Theorem, we present the needed axioms. **UNIQUENESS** holds if for every profile $\mathbf{a}$, $F(\mathbf{a})$ consists of a single element. **UNANIMITY** holds if for all $a \in I(A)$, $F(a,a,..,a) = \{a\}$. The next axiom assumes the axiom of uniqueness. It makes sense for trees, where we can speak about two points being at the same distance and in the same direction from a given point, but not for general networks. **INVARIANCE** is said to hold if the following is true: Suppose $F(a_1, a_2, \ldots, a_n) = x^*$ and $F(b_1, b_2, \ldots, b_n) = y^*$. If $a_j = b_j$, $j \neq i$, and b_i is at the same distance and in the same direction from x^* as a_i, then $x^* = y^*$. The next axiom also assumes the axiom of uniqueness. We say that the **LIPSCHITZ CONDITION** holds if whenever $a_j = b_j$, $j \neq i$, $F(a_1, a_2, \ldots, a_n) = x^*$, $F(b_1, b_2, \ldots, b_n) = y^*$, then

$$d(x^*, y^*) \leq \frac{1}{n} d(a_i, b_i).$$

<u>**Theorem 2 (Holzman [1990])**</u>. In the E-E case where the network is a tree and F satisfies the axiom of uniqueness, F is the mean if and only if it satisfies the axioms of unanimity and invariance and the Lipschitz condition.

An alternative characterization of the mean in trees is the following result, which is identical to Holzman's Theorem except that the invariance axiom is replaced by consistency.

<u>**Theorem 3 (Vohra [1990])**</u>. In the E-E case where the network is a tree and F satisfies the axiom of uniqueness, F is the mean if and only if it satisfies the axioms of unanimity and consistency and the Lipschitz condition.

Hansen and Roberts [1992] set out to see if Holzman's theorem could be generalized to settings other than trees. They discovered that axioms very similar to Holzman's in fact were inconsistent for any connected network with a cycle, i.e., for all connected networks other than those to which Holzman's Theorem applies. This is an impossibility theorem in the spirit of the group consensus literature. We state the Hansen-Roberts result next. First, we need some axioms. We restate these axioms in the notation of this paper. It is routine to verify that the next theorem as stated follows from that in Hansen and Roberts. We say that the **AXIOM OF DISTANCE DETERMINATION** holds if, given $a_1, a_2, \ldots, a_n$, $F(a_1, a_2, \ldots, a_n)$ is dependent only on the vectors of distances

$$(d(a_1, x), d(a_2, x), \ldots, d(a_n, x))$$

for elements x of $O(A)$. We say that **PARETO OPTIMALITY** holds if whenever $d(a_i, x) \leq d(a_i, y)$ for all i and $d(a_i, x) < d(a_i, y)$ for some i, then $y \notin F(a_1, a_2, \ldots, a_n)$. If uniqueness is not assumed the Lipschitz condition can be stated in various forms. We shall need only a much weaker Lipschitz axiom, the following axiom for some value of t, no matter how small, and independent of n. The **WEAK LIPSCHITZ AXIOM $L(t)$** holds for t a fixed positive number if whenever $a_j = b_j$, $j \neq i$, and $x^* \in F(a_1, a_2, \ldots, a_n)$, then there is $y^* \in F(b_1, b_2, \ldots, b_n)$ so that

$$d(x^*, y^*) \leq \frac{1}{t} d(a_i, b_i).$$

In fact, in one version of the result, we shall only need this when $n = 3$. We say that the **WEAK LIPSCHITZ AXIOM** $L_3(t)$ holds for t a fixed positive number if whenever $a_j = b_j$, $j \neq i$, and $x^* \in F(a_1, a_2, a_3)$, then there is $y^* \in F(b_1, b_2, b_3)$ so that

$$d(x^*, y^*) \leq \frac{1}{t} d(a_i, b_i).$$

Theorem 4 (Hansen and Roberts [1992]): In the E-E case where the network has a cycle, there is no group consensus function that satisfies the axioms of distance determination, Pareto optimality, and anonymity, and the weak Lipschitz axiom $L_3(t)$ for some $t > 0$.

Note that in Hansen and Roberts [1992], the axiom of existence is stated explicitly in this theorem. In this paper, existence is assumed throughout. Hansen and Roberts also prove that without the axiom of existence, the same theorem holds if instead of axiom $L_3(t)$, we assume the stronger weak Lipschitz axiom $L(t)$ for some $t > 0$. Note that in Theorems 2 and 3, the existence axiom is not needed, because uniqueness is assumed and this implies existence. However, existence is needed for Theorems 5 and 6 below.

It is not known whether similar impossibility results occur in the E-V, V-E, or V-V cases. It is also not known if such impossibility results occur in generalizations of Vohra's result, in particular, if consistency is assumed and, say, Pareto optimality is dropped. The same question does not make sense with invariance instead of consistency, since invariance only makes sense for trees.

Vohra [1990] has proven a theorem similar to Theorem 3 for the case of the median. To state his result, we need another axiom, which only makes sense for trees, where one can speak of the unique path from point a to point b. We say that **CANCELLATION** holds if whenever profile $\mathbf{a}$ has half of its entries equal to a and the other half equal to b, then $F(\mathbf{a})$ consists of all points of $O(A)$ in the unique path from a to b in the network.

Theorem 5 (Vohra [1990]). In the E-E case where the network is a tree, F is the median if and only if it satisfies the axioms of anonymity, consistency and cancellation.

Vohra does not state the anonymity axiom since his formalization has the function F defined on unordered sets (multisets) of elements of $I(A)$. Also, as McMorris and Roberts [1992] note, the axiom of unanimity, added to the other axioms by Vohra, is not needed, since it follows from consistency and cancellation.

McMorris and Roberts [1992] have also considered the median on trees, and have found a simpler result, which also applies to four different cases. It only makes sense if the distance function $d(x,y)$ is defined for all x,y in $I(A)$, even if $I(A) \neq O(A)$. To state their result, we need another axiom. We say that point x of $O(A)$ is ***between*** points a and b of $I(A)$ if $d(a,b) = d(a,x) + d(x,b)$. **BETWEENNESS** holds if $F(a,b)$ is the set of all x in $O(A)$ which are between a and b.

Theorem 6 (McMorris and Roberts [1992]). In the E-E, V-V, V-E, and E-V cases where the network is a tree, F is the median if and only if it satisfies the axioms of anonymity, betweenness, and consistency.

McMorris and Roberts note that Theorem 5 is a corollary of Theorem 6, and that it holds not only for the E-E case, but also for the V-V, V-E, and E-V cases. They also note that Theorem 6 holds for more general networks than trees, though specifying the most general setting for it is still an open question. The theorem definitely fails for some networks. For instance, McMorris and Roberts show that in the V-V case, the median is not the only group consensus function satisfying the axioms of Theorem 6 if the network is the complete graph of three vertices with all sides having length one. It is an open question to state Theorem 6, and also Theorems 2-5, in their most general abstract setting.

4. MEDIANS AND MEANS FOR COLLECTIONS OF WORDS IN SEQUENCES

In many problems of the social sciences, data is presented as a sequence or "word" from some alphabet Σ. We seek a consensus sequence or set of sequences. The sets $I(A)$ and $O(A)$ are some sets of sequences. We measure $d(x,y)$ for $x \in I(A)$, $y \in O(A)$, in various ways. For instance, if x and y are sequences of the same length, then a common way to define $d(x,y)$ is to take it to be the number of mismatches between the sequences. If y is longer than x, then $d(x,y)$ could be the smallest number of mismatches in all possible alignments of x as a consecutive subsequence of y. We call this the ***mismatch distance***. For example, the mismatch distance between words 0011 and 111010 is 2, which can be achieved if 0011 starts in the third position. Alternatively, $d(x,y)$ could be the smallest number of mismatches between y and a sequence obtained from x by inserting gaps in appropriate places (where a mismatch between a letter of Σ and a gap is counted as an ordinary mismatch). Other ways of treating insertions or deletions from sequences or counting mismatches differently if they involve insertions or deletions lead to other notions of distance.

Motivated by problems of molecular genetics, Waterman [1989] and others study the situation where Σ is a finite alphabet, k is a fixed finite number, $I(A)$ is all words (sequences) of length L from Σ where $L \geq k$ and L is either fixed or variable, $O(A)$ is all words of length k from Σ, and $d(x,y)$ is the mismatch distance. They consider nonnegative parameters λ_d assumed to be decreasing in size with increasing d and choose for $F(a_1, a_2, \ldots, a_n)$ all those words x in $O(A)$ which maximize $\Sigma_{i=1}^{n}\lambda_{d(x,a_i)}$. We call such an F a ***Waterman consensus***. In particular, Waterman and others use the parameters $\lambda_d = (k-d)/k$. Mirkin and Roberts [1992] observe that with this choice of λ_d, F is just the median. Moreover, they prove the following theorem:

Theorem 7 (Mirkin and Roberts [1993]). Suppose that Σ is an alphabet of at least 2 letters, k is fixed, $k \geq 2$, $\lambda_1 < \lambda_0$, and F is the Waterman consensus. Then F is the median if and only if $\lambda_j = aj + c$, $a < 0$; and F is the mean if and only if $\lambda_j = aj^2 + b$, $a < 0$.

Of course, this result does not give a characterization of the median or the mean since it depends on the Waterman consensus method. No results are known in the sequence context which characterize axiomatically those consensus functions which are the median or the mean, either when $L = k$ or $L \geq k$. Also, no results are known which characterize axiomatically the Waterman consensus.

5. CLOSING REMARKS

The median and mean procedures, as well as some of their variants, should continue to have important new practical applications. Two kinds of results are needed for this work to be most useful in the future. One kind is the axiomatic kind, of the type presented here. We have mentioned open axiomatic problems throughout this paper. A second kind is the computational kind. As we have pointed out only briefly, there has been considerable effort in developing ways to calculate medians and means, but much more work is needed in this direction, with practical computational procedures needed in almost all contexts in which medians and means are used.

REFERENCES

Arrow, K., ***Social Choice and Individual Values***, Cowles Commission Monograph 12, Wiley, New York, 1951.

Barthélemy, J.P., "Social Welfare and Aggregation Procedures: Combinatorial and Algorithmic Aspects," in F.S. Roberts (ed.), ***Applications of Combinatorics and Graph Theory in the Biological and Social Sciences***, IMA Vols. in Mathematics & its Applications, Vol. 17, Springer-Verlag, New York, 1989, pp. 39-73.

Barthélemy, J.P., and Janowitz, M.F., "A Formal Theory of Consensus," ***SIAM J. Disc. Math.***, 4 (1991), 305-322.

Barthélemy, J.P., and McMorris, F.R., "The Median Procedure for n-Trees," ***J. Classification***, 3 (1986), 329-334.

Barthélemy, J.P., and Monjardet, B., "The Median Procedure in Cluster Analysis and Social Choice Theory," ***Math. Soc. Sci.***, 1

(1981), 235-268.

Barthélemy, J.P., and Monjardet, B., "The Median Procedure in Data Analysis: New Results and Open Problems," in H.H. Bock (ed.), ***Classification and Related Methods of Data Analysis***, North Holland, Amsterdam, 1988, pp. 309-316.

Bartholdi, J.J. III, Tovey, C.A., and Trick, M.A., "Voting Schemes for Which it can be Difficult to Tell who Won the Election," ***Social Choice and Welfare***, 6 (1989), 157-165.

Bogart, K., "Preference Structures I: Distances between Transitive Preference Relations," ***J. Math. Sociology***, 3 (1973), 49-67.

Bogart, K., "Preference Structures II," ***SIAM J. Appl. Math.***, 29 (1975), 254-262.

Cozzens, M.B., and Leibowitz, R., "Threshold Dimension of Graphs," ***SIAM J. Alg. Disc. Meth.***, 5 (1984), 579-584.

Cozzens, M.B., and Leibowitz, R., "Multidimensional Scaling and Threshold Graphs," ***J. Math. Psychol.***, 31 (1987), 179-191.

Guttman, L., "A Basis for Scaling Qualitative Data," ***Amer. Sociol. Rev.***, 9 (1944), 139-150.

Hansen, P., Labbé, M., Peeters, P., Thisse, J-P, and Henderson, J.V., "Facility Location Analysis," ***Fundamentals of Pure and Applied Economics***, 22 (1987), 1-70.

Hansen, P., and Roberts, F.S., "An Impossibility Result in Axiomatic Location Theory," RUTCOR Research Report RRR 1-92, Rutgers Center for Operations Research, Rutgers University, New Brunswick, NJ, January 1992. (Stronger results in revised version, mimeographed, May 1992.)

Holzman, R., "An Axiomatic Approach to Location on Networks," ***Math. of Oper. Res.***, 15 (1990), 553-563.

Kemeny, J.G., "Mathematics without Numbers," ***Daedalus***, 88 (1959), 575-591.

Kemeny, J.G., and Snell, J.L., ***Mathematical Models in the Social Sciences***, Blaisdell, New York, 1962. (Reprinted by MIT Press, Cambridge, MA, 1972.)

Marcotorchino, P., and Michaud, P., "Heuristic Approach of the Similarity Aggregation Problem," ***Meth. of Oper. Res.***, 43 (1981), 395-404.

May, K.O., "A Set of Independent Necessary and Sufficient Conditions for Simple Majority Decision," ***Econometrica***, 20 (1952), 680-684.

McMorris, F.R., and Roberts, F.S., "Medians in Trees," in preparation, Department of Mathematics, Rutgers University, New Brunswick, NJ, May 1992.

Mirkin, B.G., *Group Choice* (P.C. Fishburn, ed.), Wiley, New York, 1979.

Mirkin, B.G., and Roberts, F.S., "Consensus Functions and Patterns in Molecular Sequences," *Bull. Math. Biol.*, 55 (1993), in press.

Richelson, J.T., "A Characterization Result for the Plurality Rule," *J. Econ. Theory*, 19 (1978), 548-550.

Roberts, F.S., *Discrete Mathematical Models, with Applications to Social, Biological, and Environmental Problems*, Prentice-Hall, Englewood Cliffs, NJ, 1976.

Roberts, F.S., *Measurement Theory, with Applications to Decisionmaking, Utility, and the Social Sciences*, Addison-Wesley, Reading, MA, 1979.

Roberts, F.S., "Applications of Combinatorics and Graph Theory in the Biological and Social Sciences: Seven Fundamental Ideas," in F.S. Roberts (ed.), *Applications of Combinatorics and Graph Theory in the Biological and Social Sciences*, Springer-Verlag, New York, 1989, pp. 1-37.

Roberts, F.S. "Characterizations of the Plurality Function," *Math. Soc. Sci.*, 21 (1991), 101-127.

Vohra, R.V., "An Axiomatic Characterization of Some Locations in Trees," mimeographed, Faculty of Management Sciences, Ohio State University, Columbus, OH, March 1990.

Wakabayashi, Y., *Aggregation of Binary Relations: Algorithmic and Polyhedral Investigations*, Ph.D. thesis, Augsburg, 1986.

Waterman, M.S., "Consensus Patterns in Sequences," in M.S. Waterman (ed.), *Mathematical Methods for DNA Sequences*, CRC Press, Boca Raton, FL, 1989, pp. 93-115.

Young, H.P., "Social Choice Scoring Functions," *SIAM J. Appl. Math.*, 28 (1975), 824-838.

Young, H.P., and Levenglick, A., "A Consistent Extension of the Condorcet Election Principle," *SIAM J. Appl. Math.*, 35 (1978), 285-300.

Acknowledgements: The author thanks the U.S. National Science Foundation and U.S. Air Force Office of Scientific Research for their support under grants IRI-89-02125, AFOSR-89-0512, and AFOSR-90-0008 to Rutgers University, and he thanks Denise Sakai for her helpful comments.

Uncertainty in Intelligent Systems
B. Bouchon-Meunier et al. (Editors)

Diverse applications of a simple functional equation method in the social and behavioural sciences

János Aczél[1]

Faculty of Mathematics, University of Waterloo, Waterloo, Ont. Canada, N2L 3G1

One can deal with the following kinds of problems, among others, with aid of functional equations and (for 3.) of their stability.

1. A fixed amount of resources should be allocated to certain projects. A committee of decision makers makes recommendations. These should be aggregated by project-specific functions into a consensus decision about the final allocations.

2. Equalization payments from a higher level, federal government to each lower level, state government, including transfers from other states, are supposed to depend upon the hypothetical and real tax revenues of that state, both adding up to the same total, and upon a planned federal grant. The resulting budget allocations of the states should add up to the total of state tax incomes plus the total planned federal grant.

3. Inputs contribute to the outputs of producers. Each producer's output depends upon the inputs to that producer through a producer-specific production function. Can both the outputs and each kind of inputs be aggregated so that the aggregated output depends only upon the aggregated inputs?

1. ALLOCATION

A fixed amount C of resources (money, raw materials, ...) should be allocated to n projects. A committee of m decision makers makes recommendations. These should be aggregated by project-specific functions into a *consensus* decision about the final *allocations.*

This problem has been solved when the final allocation to a project depends only upon the recommendations on that project (Aczél-Wagner 1981, Aczél-Ng-Wagner 1984): For $n > 2$ the consensus allocation must be a weighted arithmetic mean of the recommended allocations, where the weights may depend on the individual decision makers but not upon the projects. We spell out our assumptions (for (I) see Table 1):

[1]This research has been supported in part by a Natural Sciences and Engineering Research Council of Canada grant.

Table 1
Allocations of resources to projects, aggregated from recommendations by several decision makers

Decison makers	Projects 1	2	...k...	n	Sums
1	x_{11}	x_{12}	$\dots x_{1k} \dots$	x_{1n}	C
$\vdots$	$\vdots$	$\vdots$	$\vdots\vdots\vdots$	$\vdots$	$\vdots$
j	x_{j1}	x_{j2}	$\dots x_{jk} \dots$	x_{jn}	C
$\vdots$	$\vdots$	$\vdots$	$\vdots\vdots\vdots$	$\vdots$	$\vdots$
m	x_{m1}	x_{m2}	$\dots x_{mk} \dots$	x_{mn}	C
Column vectors	$\mathbf{x}_1$	$\mathbf{x}_2$	$\dots \mathbf{x}_k \dots$	$\mathbf{x}_n$	$\begin{pmatrix} C \\ \vdots \\ C \end{pmatrix} = C\mathbf{1}$
Aggregates	$f_1(\mathbf{x}_1)$	$f_2(\mathbf{x}_2)$	$\dots f_k(\mathbf{x}_k) \dots$	$f_n(\mathbf{x}_n)$	C

(I) If $\sum_{k=1}^{n} x_{jk} = C\ (j = 1, \dots, m)$ then $\sum_{k=1}^{n} f_k(x_{1k}, \dots, x_{mk}) = C\ (C > 0,\ n > 2,$ $m > 1$ fixed; $x_{jk} \in [0, C]$; $j = 1 \dots, m$; $k = 1, \dots, n)$.

(II) $f_k(0, \dots, 0) = 0$ ($k = 1, \dots, n$: if all decision makers reject a project then no resources are allocated to it).

(III) $f_k(x_1, \dots, x_n) \in [0, C]$ for $x_1, \dots, x_n \in [0, C]$ (the consensus allocations are non-negative and do not allocate more than there is).

The mathematical arguments are similar to those which will come in section **2.** – For $n = 2$ there are many more solutions.

The problem is essentially solved also when final allocations may depend on all recommendations for all projects (Aczél, unpublished; again many solutions). It is unsolved when final allocation to a project may depend only upon the recommended allocations for that project and for the two neighbouring projects.

2. EQUALIZATION

Equalization payments from a higher level ("federal") government to each lower level ("state") government, including transfers from other states, are supposed to depend upon hypothetical (t_k) and real (r_k) tax revenues of that state (both adding up, by adjustments if necessary, to the same total C, which we can even keep fixed) and upon a planned amount of federal grant (subvention: s_k). The resulting budget allocations (funds: $f_k(r_k, t_k, s_k)$) of the states should add up to the total of state tax incomes (C) plus the total planned federal grant (S).

This has been solved (Buhl-Pfingsten 1986; Aczél-Pfingsten, in preparation): For more than two states, the budget of each state is the weighted arithmetic mean of its hypothetical and real tax revenue plus the planned federal grant and a state-specific constant; these constants add up to zero; the weight is the same for all states). Here we present the calculations. Our suppositions are, in detail, the following. (Compare Table 2 for (i).)

Table 2
Equalization payments to states of a federation

	States				Sums
	1	2	$\ldots k \ldots$	n	
Real tax revenue	r_1	r_2	$\ldots r_k \ldots$	r_n	C
Hypothetical tax revenue	t_1	t_2	$\ldots t_k \ldots$	t_n	C
Federal grant (planned?)	s_1	s_2	$\ldots s_k \ldots$	s_n	
"Equalized" funds	$f_1(r_1,t_1,s_1)$	$f_2(r_2,t_2,s_2)$	$\ldots f_k(r_k,t_k,s_k) \ldots$	$f_n(r_n,t_n,s_n)$	$C+S$
Equalization payments	$f_1(r_1,t_1,s_1)$ $-r_1$	$f_2(r_2,t_2,s_2)$ $-r_2$	$\ldots f_k(r_k,t_k,s_k)$ $-r_k \ldots$	$f_n(r_n,t_n,s_n)$ $-r_n$	S

(i) $\sum_{k=1}^{n} r_k = \sum_{k=1}^{n} t_k = C$ and $\sum_{k=1}^{n} s_k = S$ imply $\sum_{k=1}^{n} f_k(r_k, t_k, s_k) = C + S$ ($n > 2$ and $C > 0$ fixed, all other quantities variable).

(ii) $\sum_{k=1}^{n} f_k(0,0,0) = 0$ (if there is no tax income - hypothetical or real - and no subvention, then the total of state funds is 0).

(iii) at least one of the f_k, say f_2, is locally increasing in the broader sense in the tax variables r_k and t_k (higher taxes cannot lead to lower funds).

If we write $\beta_k = f_k(0, \ldots, 0)$ then (ii) gives

$$\sum_{k=1}^{n} \beta_k = 0 \tag{1}$$

Now, (1) and (i) with $r_1 = t_1 = C$, $s_1 = S$, $r_\ell = t_\ell = s_\ell = 0\,(\ell > 1)$ give

$$f_1(C, C, S) = C + S - \sum_{\ell=2}^{n} \beta_\ell = C + S + \beta_1;$$

similarly

$$f_k(C, C, S) = C + S + \beta_k \ (k = 1, \ldots, n). \tag{2}$$

Again (i), this time with $r_1 = C - r$, $t_1 = C - t$, $s_1 = S - s$, $r_2 = r$, $t_2 = t$, $s_2 = s$, $r_\ell = t_\ell = s_\ell = 0\ (\ell > 2)$ yields

$$f_1(C - r, C - t, S - s) = C + S - f_2(r, t, s) - \sum_{\ell=3}^{n} \beta_\ell$$

$$= C + S - f_2(r, t, s) + \beta_1 + \beta_2. \tag{3}$$

Finally, we get from (i) with $r_1 = C - r_2 - r_3$, $t_1 = C - t_2 - t_3$, $s_1 = S - s_2 - s_3$, $r_\ell = t_\ell = s_\ell = 0\,(\ell > 3$; here we used $n > 2)$:

$$f_1(C - r_2 - r_3, C - t_2 - t_3, S - s_2 - s_3) + f_2(r_2, t_2, s_2) + f_3(r_3, t_3, s_3) + \sum_{\ell=4}^{n} \beta_\ell = C + S.$$

Using (3), this becomes

$$f_2(r_2 + r_3, t_2 + t_3, s_2 + s_3) + \beta_3 = f_2(r_2, t_2, s_2) + f_3(r_3, t_3, s_s)$$

$$(r_2, r_3, s_2, s_3, r_2 + r_3, s_2 + s_3 \in [0, C]; t_2, t_3 \geq 0) \tag{4}$$

which is a *conditional Pexider equation.* Putting into it $r_2 = t_2 = s_2 = 0$ gives

$$f_3(r_3, t_3, s_3) = f_2(r_3, s_3, t_3) + \beta_3 - \beta_2 \tag{5}$$

and this reduces (4) to

$$f_2(r_2+r_3, t_2+t_3, s_2+s_3) = f_2(r_2, t_2, s_2) + f_2(r_3, t_3, s_3) - \beta_2.$$

This is a (conditional) Cauchy equation for the three-place function $f_2(r,t,s) - \beta_2$. So

$$f_2(r,t,s) = \alpha^1(r) + \alpha^2(t) + A(s) + \beta_2,$$

where the α^1, α^2 and A satisfy *conditional Cauchy equations* for one-place functions:

$$\alpha^i(x+y) = \alpha^i(x) + \alpha^i(y) \quad (x, y, x+y \in [0, C]) \text{ for } i = 1, 2;$$

$$A(u+v) = A(u) + A(v) \quad (u \geq 0, v \geq 0). \tag{6}$$

By (iii), α^1 and α^2 are locally increasing in the broader sense, so (Aczél-Dhombres 1989)

$$f_2(r,t,s) = cr + \gamma t + A(s) + \beta_2, \quad \text{where } c \geq 0, \gamma \geq 0.$$

By (5), $f_3(r,t,s) = cr + \gamma t + A(s) + \beta_3$. Similarly, in general,

$$f_k(r,t,s) = cr + \gamma t + A(s) + \beta_k \quad (k = 1, 2, \ldots, n).$$

Putting this into (2) yields

$$(c+\gamma)C + A(S) = C + S.$$

Since $C \neq 0$ and, from (6), $A(0) = 0$, this is possible only if $A(S) = S, c + \gamma = 1$. So we finally have

$$f_k(r_k, t_k, s_k) = (1-\gamma)r_k + \gamma t_k + s_k + \beta_k \quad (k = 1, 2, \ldots, n; 0 \leq \gamma \leq 1)$$

and, with (1), this is the result stated above. The equalization payments will be

$$f_k(r_k, t_k, s_k) - r_k = \gamma(t_k - r_k) + s_k + \beta_k \quad (k = 1, 2, \ldots, n)$$

The value of γ (and of $\beta_1, \ldots, \beta_n$ as long as $\beta_1 + \cdots + \beta_n = 0$: this is a zero-sum-game) is determined by political considerations. The extreme cases are: $\gamma = 0$ yielding

$$f_k(r_k, t_k, s_k) - r_k = s_k + \beta_k \quad (k = 1, 2, \ldots, n)$$

as equalization payment (independent of taxes, both real and hypothetical) and $\gamma = 1$ resulting in

$$f_k(r_k, t_k, s_k) - r_k = (t_k - r_k) + s_k + \beta_k \quad (k = 1, 2, \ldots, n)$$

as equalization payment. Notice that $t_k - r_k$ is the difference between the hypothetical and real tax income of the k-th state.

3. AGGREGATION

Inputs (goods and services) of n kinds contribute to the outputs of m producers. The j-th producer's output depends upon the inputs $x_{j1}, \ldots, x_{jn}$ to that producer through a possibly producer-specific production function. Do there exist *aggregator functions for the outputs* (F) *and for* each kind of *inputs* (f_k; $k = 1, \ldots, n$) *such that the aggregated output depends only upon the n aggregated inputs?*

Table 3
Aggregation of inputs and outputs

Producers	Inputs(goods and services) 1	…	k	…	n	Row Vectors	Outputs (production functions)
1	x_{11}	…	x_{1k}	…	x_{1n}	$\mathbf{x}_1'$	$y_1 = g_1(\mathbf{x}_1')$ $= g_1(x_{11}, \ldots, x_{1n})$
⋮	⋮	⋮	⋮	⋮	⋮	⋮	⋮
j	x_{j1}	…	x_{jk}	…	x_{jn}	$\mathbf{x}_j'$	$y_j = g_j(\mathbf{x}_j')$ $= g_j(x_{j1}, \ldots, x_{jn})$
⋮	⋮	⋮	⋮	⋮	⋮	⋮	⋮
m	x_{m1}	…	x_{mk}	…	x_{mn}	$\mathbf{x}_m'$	$y_m = g_m(\mathbf{x}_m')$ $= g_m(x_{m1}, \ldots, x_{mn})$
Column Vectors	$\mathbf{x}_1$	…	$\mathbf{x}_k$	…	$\mathbf{x}_n$		$\mathbf{y}$
Aggregates	$z_1 = f_1(\mathbf{x}_1)$		$z_k = f_k(\mathbf{x}_k)$		$z_n = f_n(\mathbf{x}_n)$	$\mathbf{z}'$	$F(\mathbf{y}) = F(y_1, \ldots, y_m) =$ $F(g_1(\mathbf{x}_1'), \ldots, g_m(\mathbf{x}_m')) \stackrel{?}{=}$ $\stackrel{?}{=} G(f_1(\mathbf{x}_1), \ldots, f_n(\mathbf{x}_n))$ $= G(z_1, \ldots, z_n) = G(\mathbf{z}')$

As we see from Table 3, the problem is to solve

$$F(g_1(\mathbf{x}'_1),\ldots,g_m(\mathbf{x}'_m)) \;=\; G(f_1(\mathbf{x}_1),\ldots,f_n(\mathbf{x}_n)) \tag{7}$$

where $\mathbf{x}'_j$ $(j = 1,\ldots,m)$ and $\mathbf{x}_k$ $(k = 1,\ldots,n)$ are the row and column vectors, respectively, of a variable matrix $\mathbf{X}$.

The main problem is that the production functions $g_1,\ldots,g_m$ cannot be conveniently predetermined. If, for instance, the outputs are given by their monetary values then it is tempting to aggregate them by adding: $F(y,\ldots,y_m) = y_1 + \cdots + y_m$. In this case (7) becomes

$$g_1(\mathbf{x}'_1) \;+\; \ldots \;+\; g_m(\mathbf{x}'_m) \;=\; G(f_1(\mathbf{x}_1),\ldots,f_n(\mathbf{x}_n)). \tag{8}$$

If all inputs (all goods and services) are "totally separated", then also

$f_k(\mathbf{x}_k) = f_k(x_{1k},\ldots,x_{mk}) = x_{1k} + \cdots + x_{mk}$ $(k = 1,\ldots,n)$,
resulting in

$$G(x_{11} + \ldots + x_{m1},\ldots,x_{1n} + \ldots + x_{mn}) = G(\mathbf{x}'_1 + \ldots + \mathbf{x}'_m) = g_1(\mathbf{x}'_1) + \ldots + g_m(\mathbf{x}'_m) \tag{9}$$

this is again a *Pexider equation* and under very weak regularity conditions (for instance local boundedness: Aczél-Dhombres 1989) the only solutions are (" $\cdot$ " is the scalar product)

$$g_j(\mathbf{x}') \;=\; \mathbf{a}\cdot\mathbf{x}' \;+\; b_j\,(j = 1,\ldots,m),\;\; G(\mathbf{x}') \;=\; \mathbf{a}\cdot\mathbf{x}' \;+\; \sum_{j=1}^{m} b_j \tag{10}$$

$(\mathbf{a} \geq 0;\, b_j \geq 0;\, j = 1,\ldots,m)$. So *not for all production functions* $g_1,\ldots,g_m$ *does there exist a G for which (8) is satisfied,* for instance not for the popular Cobb-Douglas functions $g(x_1,x_2,\ldots,x_m) = Cx_1^{\alpha_1}x_2^{\alpha_2}\ldots x_m^{\alpha_m}$.

In this case not even *weakening* of (9) to

$$|G(\mathbf{x}'_1 \;+\; \ldots \;+\; \mathbf{x}'_m) - g_1(\mathbf{x}'_1 + \ldots + g_m(\mathbf{x}'_m)| < \epsilon$$

helps. This *"stability"* (closeness) *problem of Pexider equations is solved:* The solutions are still *"close"* to the linear (really: "affine") functions (10). Maybe the "total separation" supposition is too strong. For (8) and even more so for (7) (where also the first sum assumption is dropped) the corresponding stability problems ("small deficit" rather than "zero deficit")

$$|G(f_1(\mathbf{x}_1),\ldots,f_n(\mathbf{x}_n)) - g_1(\mathbf{x}'_1) - \cdots - g_m(\mathbf{x}'_m)| < \epsilon,$$

$$|G(f_1(\mathbf{x}_1),\ldots,f_n(\mathbf{x}_n)) - F(g_1(\mathbf{x}'_1),\ldots,g_m(\mathbf{x}'_m))| < \epsilon$$

may have reasonable solutions for many given $g_1, \ldots, g_m$ ("lability"?). But this problem is not solved yet.

Thus the answer to the question asked at the start of this section is yes (Pokropp 1972, van Daal-Merkies 1987; Aczél-Eichhorn, in preparation) for some production functions but not for the most popular ones (for instance Cobb-Douglas). Maybe the right question is not equality but "closeness". This is solved only in the simplest case of aggregation by summation, which also leaves out certain production functions. Further work is in progress.

REFERENCES

J. Aczél - J. Dhombres, Functional Equations in Several Variabales with applications to mathematics, information theory and to the natural and social sciences, Cambridge University Press, Cambridge-New York, 1989.

J. Aczél - C.T. Ng - C. Wagner, Aggregation theorems for allocation problems, SIAM J. Algebraic Discrete Methods, 5 (1984), 1-8.

J. Aczél - C. Wagner, Rational group decision making revisited: The case of several unknown functions, C.R. Math. Rep. Acad. Sci. Canada, 3 (1981), 138-142.

H.U. Buhl - A. Pfingsten, Eigenschaften und Verfahren für einen angemessenen Länderfinanzausgleich in der Bundesrepublik Deutschland, Finanzarchiv, NF 44 (1986), 98-109.

J. van Daal - A.H.Q. M. Merkies, The problem of aggregation of individual economic relations, consistency and representativity in a historical perspective, Measurement in Economics, Physica, Heidelberg, 1987, pp. 607-637.

F. Pokropp, Aggregation von Produktionsfunktionen; Klein-Nataf-Aggregation ohne Annahmen über Differenzierbarkeit und Stetigkeit, Springer, Berlin-New York, 1972.